Hadron and Nuclear Physics 09

Hadron and Nuclear Physics 09

Osaka University, Japan, 16 – 19 November 2009

editors

Atsushi Hosaka
RCNP, Osaka University, Japan

Takayuki Myo
Osaka Institute of Technology, Japan

Hideko Nagahiro
Nara Women's University, Japan

Kanabu Nawa
RCNP, Osaka University, Japan

World Scientific

NEW JERSEY • LONDON • SINGAPORE • BEIJING • SHANGHAI • HONG KONG • TAIPEI • CHENNAI

Published by

World Scientific Publishing Co. Pte. Ltd.

5 Toh Tuck Link, Singapore 596224

USA office: 27 Warren Street, Suite 401-402, Hackensack, NJ 07601

UK office: 57 Shelton Street, Covent Garden, London WC2H 9HE

Library of Congress Cataloging-in-Publication Data

Hadron and nuclear physics 09 : Osaka University, Japan, 16-19 November 2009 / edited by Atsushi Hosaka ... [et al.].

p. cm.

Includes bibliographical references.

ISBN-13: 978-981-4313-92-6 (hardcover : alk. paper)

ISBN-10: 981-4313-92-0 (hardcover : alk. paper)

1. Hadrons--Congresses. 2. Quantum chromodynamics--Congresses. 3. Particles (Nuclear physics)--Chirality--Congresses. I. Hosaka, Atsushi.

QC793.5.H328H325 2010

539.7'216--dc22

2010024276

British Library Cataloguing-in-Publication Data

A catalogue record for this book is available from the British Library.

Printed in Singapore by B & JO Enterprise

PREFACE

This volume of the proceedings collects papers presented at the international workshop "Hadron-Nuclear Physics 09 (HNP09)" held at Osaka University, November 16-19, 2009. The series of the workshop started from a collaboration-meeting between the hadron physics groups of RCNP (Osaka University) and Pusan National University in 1999. This is the 7th successful meeting held. From 2006's Kizu (Kyoto) meeting at JAEA, Chinese physicists were also invited to have people of wider spectrum and physics interests, covering topical subjects in hadron and nuclear physics. Since hadrons and nuclei are systems governed by the strong interaction and are eventually described by the Quantum Chromodynamics (QCD), there are much interference between the two research fields. In this regards, the workshop have provided opportunities to discuss common interests of hadron and nuclear physics.

Hadrons and nuclei show up in different layers of subatomic phenomena governed by the same dynamics dictated by QCD. The basic building blocks of matter, quarks and gluons, are confined in hadrons, generating their masses dynamically and breaks chiral symmetry spontaneously. The latter is the origin of the pion which determines the essential part of the nuclear interaction. Therefore, the common key words would be chiral symmetry and pions.

The reports here cover current achievements in hadron physics such as exotic multiquark states, various production reactions and non-linear dynamics of hadrons, and those in nuclear physics such as clustering phenomena, exotic neutron rich nuclei and pions in nuclei. As related subjects, applications to astronuclear physics, accelerator physics and laser physics are also discussed. Although objectives of these subjects appear different, the underlying concepts and methods share common aspects. For instance, the clustering phenomenon well known in nuclear physics is expected to occur also in hadrons of exotic multiquark states. Collective phenomena also show up in different manner in field theories for hadron physics and in many-body quantum mechanics in nuclear physics.

We hope that the present proceedings will be useful particularly for young researchers who wish to start their research career in the related fields how similar concepts and methods are applied to the phenomena in different systems. We hope also that this volume will provide opportunities for the discussions among the researchers in hadron and nuclear physics.

This workshop was supported in part by Research Center for Nuclear Physics (RCNP), Osaka University and by a research grant of "Elucidation of New Hadrons with a Variety of Flavors (New hadrons)" of MEXT: http://www.hepl.phys.nagoya-u.ac.jp/public/new_hadron/index-e.html.

May 2010

Editors:
Atsushi Hosaka, *RCNP, Osaka University*
Takayuki Myo, *Osaka Institute of Technology*
Hideko Nagahiro, *Nara Women's University*
Kanabu Nawa, *RCNP, Osaka University*

CONTENTS

HADRON PHYSICS

NUCLEAR PHYSICS

SPECIAL TOPICS

RECENT RESULTS FROM BESII AND BESIII

XIAOYAN SHEN and MING YANG (FOR BES COLLABORATION)

Institute of High Energy Physics, Chinese Academy of Sciences
Beijing 100049, China
**E-mail: shenxy@ihep.ac.cn*
yangming@ihep.ac.cn

We reported the study of $Y(2175)$ in J/ψ decays, the results from a partial wave analysis of $J/\psi \to p\bar{p}\pi^0$ and the observation of charged κ, based on 5.8×10^7 J/ψ events collected at BESII. We also reported the precision R values at the energy points of 2.60, 3.05 and 3.65 GeV/c^2 and the measurement of the line shape for the cross sections of $e^+e^- \to hadrons$ in the energy range of 3.650 - 3.872 GeV/c^2. The status of BESIII running, as well as the preliminary results from BESIII are also presented.

Keywords: decay; ; partial wave analysis; line shape.

1. Introduction

The Standard Model (SM), regarded as the theory of the elementary particle physics, has been very successful in describing all the relevant physics phenomena. However, there are still many unanswered questions remained. The QCD, as a fundamental theory of the strong interaction, has been well tested at high energies, but is not well understood at low energies due to its non-perturbative features. The Beijing Electron Positron Collider (BEPC)/Beijing Electromagnetic Spectrometer (BES) is a facility running at the energy region of 2-4.6 GeV/c^2, in which many of the charmonium states and charm mesons, as well as the τ lepton can be produced. The physics program of BES covers the spectroscopy of light hadrons, charmonium physics, $D(Ds)$ physics, τ physics and precision measurement of R values. It also allows the searches for possible new physics.

The BESI started running in 1989. In 1997, the BESI detector was upgraded to BESII detector. With BESII detector, the R values at 91 energy points in the energy region of 2-5 GeV were measured with a factor of 2-3 improvement of the errors. In addition, the $5.8 \times 10^7 J/\psi$ events, $1.4 \times 10^7 \psi(2S)$ events and 33 pb^{-1} $\psi(3770)$ data were accumulated.

There have been many discussions on the future of BEPC/BES since 1995. The project of BEPCII/BESIII was approved officially in 2003. BEPCII is a major upgrade of the previous collider BEPC. In BEPC, the electrons and positrons shared the same vacuum tube in a single ring of magnets, and this arrangement limited the rate at which interesting particles are produced. The major change of BEPCII is the installation of a second ring of magnets that allows the electron and positron beams to be stored separately, and therefore leads to the designed peak luminosity of 10^{33}, an order of two to that of BEPC. BESIII detector is completely new with a number of major improvements. These include the superconducting magnet, which produces a magnetic field of 1 Tesla throughout the detector. BESIII detector consists of a drift chamber (MDC) which has a small cell structure filled with a helium-based gas, an electromagnetic calorimeter (EMC) made of CsI(Tl) crystals, time-of-flight counters (TOF) for particle identification made of plastic scintillators, a muon system made of Resistive Plate Chambers (RPC) and a super conducting magnet. The designed single wire spatial resolution, the resolutions of dE/dx and momentum for the MDC are 130μm, 6% and 0.5 % $\sqrt{1+p^2}$ (p in GeV/c), respectively. The time resolution for TOF is $\sigma_{TOF} = 100$ ps for Bhabha events and the energy resolution for photons is $\sigma_E/E \simeq 2.5\ \% /\sqrt{E}$ (E in GeV) in EMC. To handle the huge data rates expected in the BESIII detector, a specialized state-of-the-art high speed data communication system has been developed and implemented.

In this talk, the recent results from BESII are presented. The status of BESIII running and some preliminary results at BESIII are reported as well.

2. Study of $Y(2175)$ in J/ψ decays

A new 1^{--} structure, denoted as $Y(2175)$ and with the mass and width of $2.175 \pm 0.010 \pm 0.015$ GeV/c^2 and $58 \pm 16 \pm 20$ MeV/c^2, was observed by the BABAR experiment in the $e^+e^- \to \gamma_{ISR}\phi f_0(980)$ initial-state radiation(ISR) process.[1,2] This observation stimulated some theoretical speculation that this state may be an s-quark version of the $Y(4260)$ since both of them are produced in e^+e^- annihilation and decay to similar final states.[3] The $Y(2175)$ has correspondingly been interpreted as a $s\bar{s}g$,[4] a 2^3D_1 $s\bar{s}$ state or a tetraquark $s\bar{s}s\bar{s}$ state.[5] As of now, none of these interpretations have either been established or ruled out by experiment.

The decays of $J/\psi \to \eta\phi f_0(980)$, with $\eta \to \gamma\gamma$, $\phi \to K^+K^-$, $f_0(980) \to \pi^+\pi^-$ are analyzed at BESII.[6] The $\phi f_0(980)$ invariant mass spectrum for the selected events is shown as points with error bars in Fig. 1, where a clear enhancement is seen around 2.18 GeV/c^2.

We fit the $\phi f_0(980)$ invariant mass spectrum and the total sidebands simultaneously. The statistical significance of the signal is 5.5 σ. The mass and width obtained from the fit (shown as smooth curves in Fig. 1) are $M = 2.186 \pm 0.010(stat) \pm 0.006(syst)$ GeV/c^2 and $\Gamma = 0.065 \pm 0.023(stat) \pm 0.017(syst)$ GeV/c^2, and the product branching ratio is measured to be $Br(J/\psi \to \eta Y(2175)) \cdot Br(Y(2175) \to \phi f_0(980)) \cdot Br(f_0(980) \to \pi^+\pi^-) = (3.23 \pm 0.75(stat) \pm 0.73(syst)) \times 10^{-4}$.

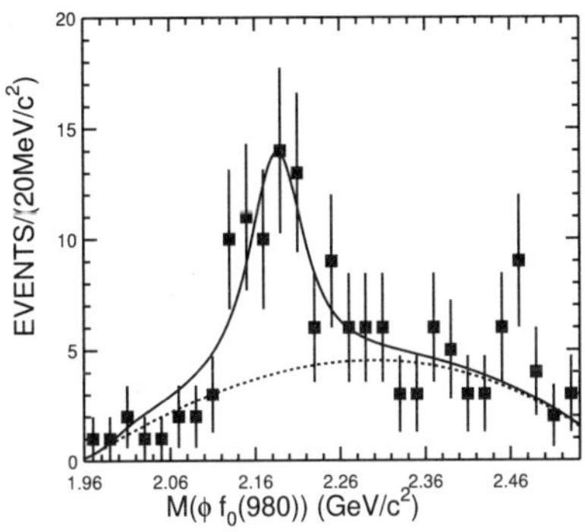

Fig. 1. The solid curve is the fit to the data (points with error bars).

We also studied the decays of $J/\psi \to \eta K^{*0}\bar{K}^{*0}$, with $K^{*0} \to K^{\pm}\pi^{\mp}$ in a J/ψ sample of 58 million events collected with the BESII detector. No evidence of $Y(2175) \to K^{*0}\bar{K}^{*0}$ is seen. Fig. 2 is the invariant mass spectrum of $K^{*0}\bar{K}^{*0}$ recoiling against η. An upper limit of $Br(J/\psi \to \eta Y(2175))Br(Y(2175) \to K^{*0}\bar{K}^{*0}) < 2.52 \times 10^{-4}$ is set at the 90% confidence level.

3. Partial wave analysis of $J/\psi \to p\bar{p}\pi^0$

J/ψ decays provide a good laboratory for studying excited baryon states. For $J/\psi \to N\bar{N}\pi$ and $N\bar{N}\pi\pi$ decays, the $N\pi(\bar{N}\pi)$ and $N\pi\pi(\bar{N}\pi\pi)$ systems are expected to be dominantly isospin 1/2 due to that the isospin conserving three-gluon annihilation of the constituent c-quarks dominates over the isospin violating decays via intermediate photon for the baronic final states. This makes the study of N^* resonances from J/ψ decays less complicated, compared with πN and γN experiments which have states that are a mixture of isospin 1/2 and 3/2. In a recent analysis of $J/\psi \to p\bar{n}\pi^- + c.c.$,[7] a

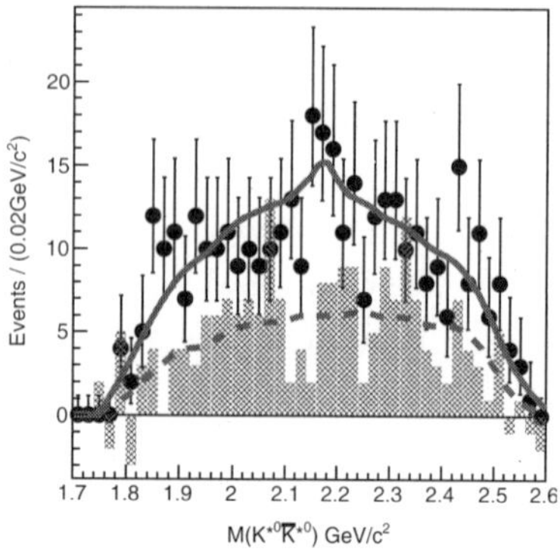

Fig. 2. The $K^{*0}\bar{K}^{*0}$ invariant mass spectrum, where points with error bars are candidate events, the dashed histogram is from MC phase space for $J/\psi \to \eta K^{*0}\bar{K}^{*0}$, the shaded histogram is from side-band events, and the solid curve is the fitting result, where the Y(2175) shape used is from MC simulation.

'missing' N^* at around 2.0 GeV/c^2 named $N_x(2065)$ was observed, based on $5.8 \times 10^7 J/\psi$ events collected with BESII. The mass and width for this state are determined to be $2065 \pm 3^{+15}_{-30}$ MeV/c^2 and $175 \pm 12 \pm 40$ MeV/c^2, respectively, from a simple Breit-Wigner fit.

We analyzed $J/\psi \to p\bar{p}\pi^0$ from BESII J/ψ data.[10] The Dalitz plot of this decay is shown in Fig. 3(left), and some N^* bands are evident. The Dalitz plot exhibits an asymmetry for $m_{p\pi^0}$ and $m_{\bar{p}\pi^0}$, which is mainly caused by different detection efficiencies for the proton and antiproton.

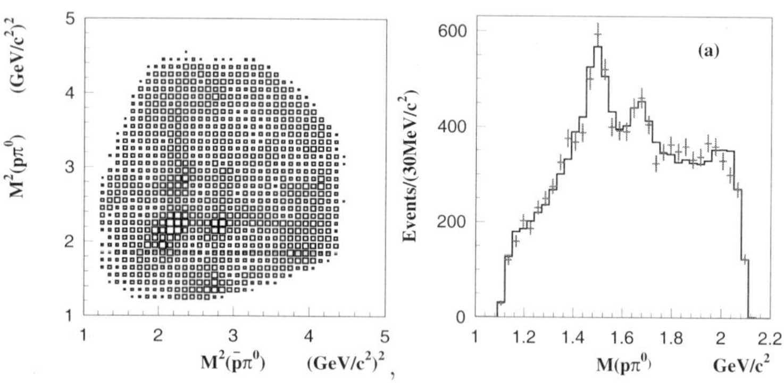

Fig. 3. Left: the Dalitz plot of $J/\psi \to p\bar{p}\pi^0$. Right:the $p\pi^0$ and $\bar{p}\pi^0$ invariant mass spectra after optimization of masses and widths. Where the crosses are data and histograms are fit results.

A partial wave analysis (PWA) is performed to study the N^* states in this decay. The sequential decay process can be described by $J/\psi \to \bar{p}N^*(p\bar{N}^*)$, $N^*(\bar{N}^*) \to p\pi^0(\bar{p}\pi^0)$. The amplitudes are constructed using the relativistic covariant tensor amplitude formalism,[8,9] and the maximum likelihood method is used in the fit. In the analysis, the well established N^* states, such as $N(1440)$, $N(1520)$, $N(1535)$, $N(1650)$, $N(1675)$ and $N(1680)$ are included. In addition, $N_x(2065)$, a long-sought 'missing' N^* which was observed recently by BES,[7] is also included. We find that the $N_x(2065)$ is definitely needed in all fits, and its spin-parity favors $\frac{3}{2}^+$. The mass and width are determined to be $2040^{+3}_{-4} \pm 25$ MeV/c^2 and $230^{+8}_{-8} \pm 52$ MeV/c^2, respectively, which are consistent with those from $J/\psi \to p\bar{n}\pi^- + c.c.$ within errors. We also obtain the masses and widths of other N^* resonances in the low mass region. Fig. 3(right) shows the comparison of data and PWA fit, in which the agreement is reasonable.

4. Observation of charged κ in $J/\psi \to K^\pm K_s \pi^\mp \pi^0$

The existence of σ and κ has been controversial for a long time, due to the facts that the total phase shifts in the lower mass region are much less than 180° and they do not fit into ordinary $q\bar{q}$ meson nonets.

Evidence for κ particles firstly comes from re-analysis of the old $K\pi$ scattering data,[11–18] though some theorists are not convinced, and then the production decay process $D^+ \to K^-\pi^+\pi^-$,[19] $J/\psi \to \bar{K}^*(892)^0 K^+\pi^-$,[20] $D^0 \to K^-\pi^+\mu^+\nu$.[21] A $K\pi$ S-wave component is found in the decay $D^0 \to K^-K^+\pi^0$[22] and $\tau \to K_s\pi^-\nu_\tau$.[23] But no evidence for the κ in $D^0 \to K^-\pi^+\pi^0$[24] or for the charged κ in $D^0 \to K^-K^+\pi^0$[25] is seen. The present status of the κ is summarized by the Particle Data Group (PDG08).

The σ and κ particles have been studied with BES data, where evidence for σ and κ particles is quite clear .[20,26,27] In BES, the neutral κ particle was found in the decay $J/\psi \to \bar{K}^*(892)^0 K^+\pi^-$.[20] Because of the isospin symmetry, a charged κ is expected. Here, we report the charged κ in $J/\psi \to K^\pm K_s \pi^\mp \pi^0$.

The combined $K_s\pi^\mp$ and $K^\mp\pi^0$ invariant mass distribution recoiling against the $K^*(892)^\pm$, after the event selection, is shown in Fig. 4. Where, a clear narrow peak at 892 MeV/c^2 and a wider peak at about 1430 MeV/c^2 are seen. This is the sum of the following four decays.

$$J/\psi \to K^*(892)^+ K_s\pi^- \to K^+\pi^0 K_s\pi^- \quad (1)$$
$$J/\psi \to K^*(892)^- K_s\pi^+ \to K^-\pi^0 K_s\pi^+ \quad (2)$$
$$J/\psi \to K^*(892)^+ K^-\pi^0 \to K_s\pi^+ K^-\pi^0 \quad (3)$$
$$J/\psi \to K^*(892)^- K^+\pi^0 \to K_s\pi^- K^+\pi^0 \quad (4)$$

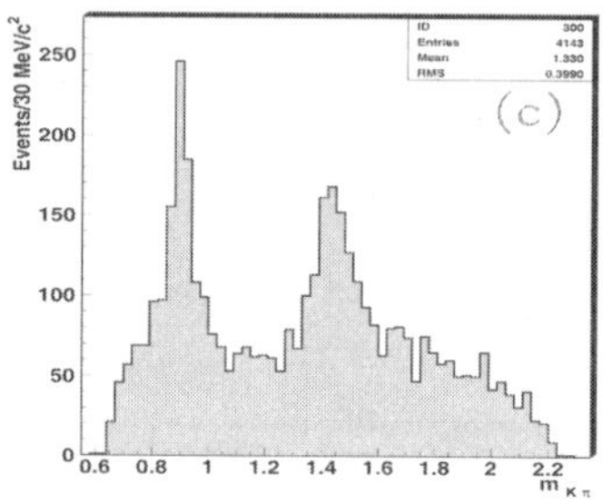

Fig. 4. Invariant mass distribution of $K^{\pm}\pi^0$ and $K_s\pi^{\pm}$ recoiling against $K^*(892)^{\mp}$.

A partial wave analysis, which is based on the covariant helicity amplitude analysis is performed. We add the likelihoods of all four channels together, and find the minimum of the sum of likelihood functions in the fit. If we use a Breit-Wigner function of constant width to parameterize the κ, its pole locates at $(849 \pm 77^{+18}_{-14}) - i(256 \pm 40^{+46}_{-22})$ MeV$/c^2$. Three different parameterizations are tried in the fit, and the obtained pole positions are consistent with those of neutral κ and are also in good agreement with those in the analysis of $K\pi$ scattering phase shifts.

5. R measurement

The R ratio is defined as the lowest level hadronic cross section normalized by the theoretical $\mu^+\mu^-$ production cross section in e^+e^- annihilation

$$R = \frac{\sigma^0_{had}(e^+e^- \to \gamma^* \to \text{hadrons})}{\sigma^0_{\mu\mu}(e^+e^- \to \gamma^* \to \mu^+\mu^-)}, \quad (5)$$

and it is an important input parameter for precision tests of the Standard Model (SM). The errors on R value measurements below 5 GeV have a strong influence on the uncertainties of the calculated QED running electromagnetic coupling constant $\alpha(s)$, the muon anomalous magnetic moment $(g-2)$ and global SM fits for the Higgs mass.[28–30]

In 1998 and 1999, R value measurements were made at 91 energy points between 2 and 5 GeV by the BESII experiment.[31,32] The average statistical errors of the R values are $2-4\%$, and the systematical errors are $5-8\%$ depending on the energy point. In 2003, the R values at 68 energy points in 3.650-3.872 GeV energy region were measured from a dedicated $\psi(3770)$ scan data, with the statistical errors of 3-4% and systimatic errors of about 4%.[33]

In this talk, the R measurements at center-of-mass energies of 2.60, 3.07 and 3.65 GeV are reported. The data were taken in 2004 and have a total integrated luminosity of 10.0 pb^{-1}. An additional 65.2 nb^{-1} was accumulated at 2.2 GeV for the purpose of tuning the parameters of the hadronic event generator. Improvements in the event selection, tuning of generator parameters and luminosity measurement have been made in order to decrease the systematic errors. The results are shown in Fig. 5, together with previous measurements. The errors on the R values reported here are about 3.5%. The R values are consistent within errors with the prediction of perturbative QCD.[35]

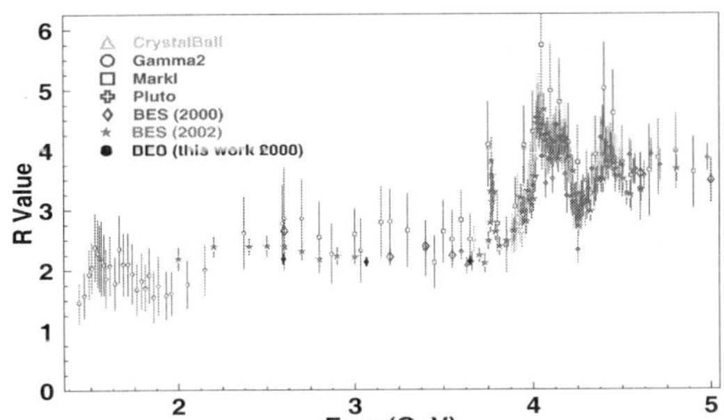

Fig. 5. R values reported here together with prevous measurements below 5 GeV.

6. Anomalous Line-Shape of Cross Sections for $e^+e^- \to$ Hadrons in the Center-of-Mass Energy Region between 3.650 and 3.872 GeV

In the energy range from 3.700 to 3.872 GeV, the well established $\psi(3770)$ resonance is believed to be the only observed structure, and has been identified as a mixture of D-wave and S-wave of angular momentum eigenstates of the $c\bar{c}$ system. In addition, the $\psi(3770)$ resonance is expected to decay into $D\bar{D}$ meson pairs with a branching fraction greater than 98%. However, there is a long-standing puzzle in the existing measurements of $\psi(3770)$ production and decays. BESII experiment measured the branching fraction of $\psi(3770)$ decays to $D\bar{D}$ to be $B[\psi(3770) \to D\bar{D}] = (85 \pm 5)\%$ and directly measured the non-$D\bar{D}$ branching fraction of $\psi(3770)$ decay to be around 15% under the assumption of only one simple $\psi(3770)$ resonance in the energy region between 3.700 and 3.872 GeV .[36–39]

The $e^+e^- \to$ hadrons total cross sections in the energy region between 3.700 and 3.872 GeV are measured from the data samples taken with the BESII detector. The observed inclusive hadronic cross sections obtained from the cross section scan data taken in March 2003 and in December 2003 are shown in Fig. 6. The line shape of the observed inclusive hadronic cross sections shows that the slopes on the two sides of the peak are quite different: the slope of the high energy side of the peak is substantially larger than that of the low energy side. It conflicts with the expectations for only one resonance in this energy region.

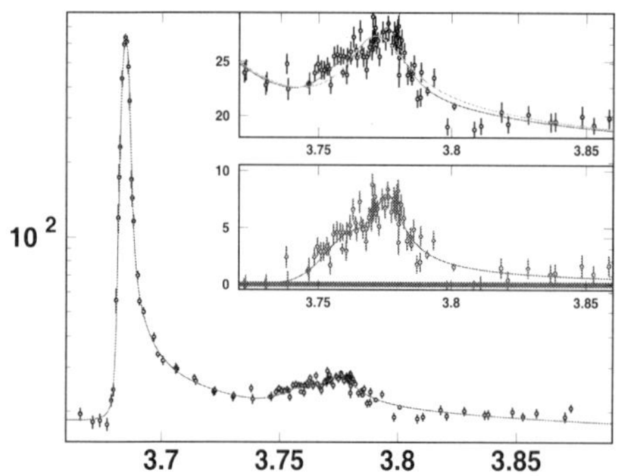

Fig. 6. The measured inclusive hadronic cross sections versus the c.m. energy for the two data sets taken in March and December 2003; the fit is done with two incoherent amplitudes.

we tried to fit the 3.770 GeV energy region with two pure P-wave Breit-Wigner amplitudes with energy-dependent total widths or one amplitude, respectively. The curves in Fig. 6 are the fits with two incoherent amplitudes. The analysis shows that the fit is inconsistent with the explanation for only one simple $\psi(3770)$ resonance at 7σ statistical significance, which indicates either there is likely a new structure in addition to the $\psi(3770)$ resonance around 3.773 GeV, or there are some physics effects reflecting the $D\bar{D}$ production dynamics. A fine scan in this energy region is needed to further clarify the line shape of the inclusive hadronic cross section.

7. Status of BESIII

The construction of BESIII detector started in 2003. By the end of 2007, the construction and installation of BESIII were accomplished. Cosmic ray experiments were then started and continued till March of 2008 to test each parts, including detectors, electronics, trigger, data acquisition and slow control system. In early May of 2008, BESIII was pushed into the collision point, and the precise regulation was finished at the point. The measurement showed that the location accuracy was better than the design requirements. Joint commissioning of BESIII and the accelerator began in June. On July 19, 2008, the first physics event produced by the electron-positron collision was observed. In Nov. of 2008, about 20 pb^{-1} $\psi(2S)$ events were accumulated for understanding the performance of both detector and software. The detectors worked properly and stably, and the performance has reached the design requirements.

On March 14, 2009, BESIII finished accumulating her first large data set of over 100M $\psi(2S)$ events. This is the world's largest $\psi(2S)$ data set. The J/ψ data taking started on June 12 and ended on July 28, with totally about 200M J/ψ events collected. Here are some preliminary results from BESIII 100M $\psi(2S)$ events.

The spin-singlet P wave charmonium state h_c was first observed by CLEOc experiment in $\psi(2S) \to \pi^0 h_c, h_c \to \gamma\eta_c$ decays. Using BESIII 100M $\psi(2S)$ events, the h_c state is confirmed. The width of h_c, the production rate $Br(\psi(2S) \to \pi^0 h_c$ and the decay branching ratio of $Br(h_c \to \gamma\eta_c)$ are measured for the first time. Fig. 7 shows the π^0 recoiling mass spectrum in $\psi(2S) \to \pi^0 h_c$ after the background subtraction, where the h_c signal is evident.

In 2003, an anomalous enhancement, $X(1860)$, near the mass threshold in the $p\bar{p}$ invariant mass spectrum from $J/\psi \to \gamma p\bar{p}$ decays was reported, with 58M J/ψ events at BESII. This resonance is not compatible with any known meson. The experimental observations stimulated a number of theoretical speculations, including the baryonium explanation. The confirmation of this state, the observation of more decay modes and the determination of the J^{PC}'s become crucial in clarifying its nature. With 100M $\psi(2S)$ events collected at BESIII, the $p\bar{p}$ threshold enhancement $X(1860)$ is confirmed in the decays of $\psi(2S) \to \pi^+\pi^- J/\psi, J/\psi \to \gamma p\bar{p}$. The mass and width of $X(1860)$ are consistent with those from BESII data. Fig. 8 shows the distribution of $M - M_{p\bar{p}}$ from $\psi(2S) \to \pi^+\pi^- J/\psi, J/\psi \to \gamma p\bar{p}$. The solid curve

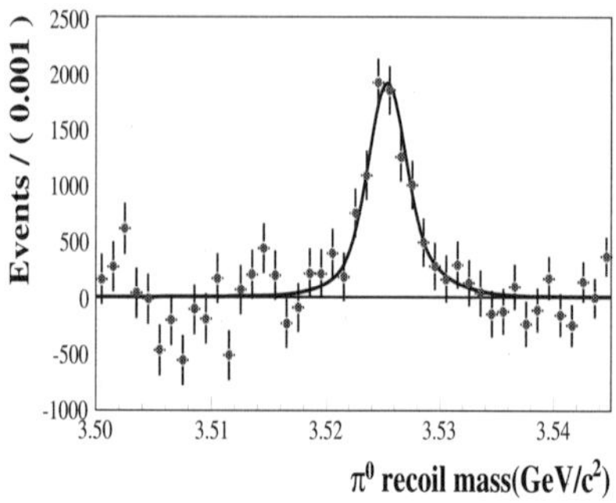

Fig. 7. The π^0 recoiling mass spectrum in $\psi(2S) \to \pi^0 h_c$ after the background subtraction.

is the fit.

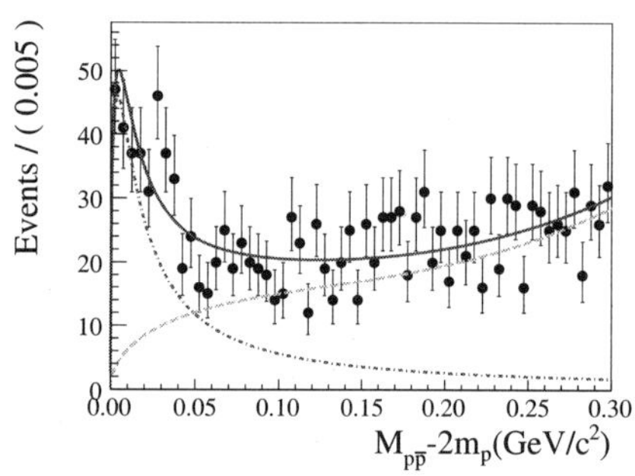

Fig. 8. The confirmation of $p\bar{p}$ threshold enhancement at BESIII.

8. Acknowledgments

The BES collaboration thanks the staff of BEPC and computing center for their hard efforts. This work is supported in part by the National Natural Science Foundation of China under contracts Nos. 10625524, 10821003, the Chinese Academy of Sciences under contract No. KJCX2-YW-N29, the Ministry of Science and Technology (MOST 973 project) under contract No. 2009CB825200.

References

1. BABAR Collaboration, Phys. Rev. D **74**, 091103(R) (2006).
2. BABAR Collaboration, Phys. ReV. D **76**, 031102 (2007).
3. B. Aubert et al. [BABAR collaboration], Phys. Rev. Lett. **95**, 142001 (2005).
4. Gui-Jun Ding, Mu-lin Yan, Phys. Lett. B **650**, 390-400 (2007).
5. Zhi-Gang Wang, Nucl. Phys. A **791**, 106-116 (2007).
6. M. Ablikim *et. al.* (BES Collaboration), Phys. Rev. Lett., **100**, 102003 (2008).
7. Ablikim *et. al.* (BES Collaboration), Phys. Rev. Lett., **97**, 062001 (2006).
8. W. Rarita and J. Schwinger, Phys. Rev. 60, 61 (1941).
9. W. H. Liang, P. N. Shen, J. X. Wang and B. S. Zou, J. Phys. G28, 333 (2002).
10. Ablikim *et. al.* (BES Collaboration), Phys. Rev. D. **80**, 052004 (2009).
11. E. van Beveren, et al., Z. Phys. **C 30** (1986)651.
12. D. Aston, et al., Nucl. Phys. **B 296** (1988)253.
13. S. Ishida, et al., Prog. Theor. Phys. **98** (1997)621.
14. D.Black, et al., Phys. Rev. **D58** (1998)054012.
15. J.A. Oller, E. Oset, Phys. Rev. **D 60** (1999)074023.
16. M.J. Jamin, et al., Nucl. Phys. **B 587** (2000)331.
17. Z.Xiao, H.Q.zheng, Nucl. Phys. A 695 (2001) 273.
18. D. Bugg, Phys. Lett. **B 572** (2003)1.
19. E.M. Aitala, et al., E791 Collaboration, Phys. Rev. Lett. 89 (2002)121801.
20. M. Ablikim, et al., BES collaboration, Phys. Lett. **B 633** (2006) 681.
21. J.M. Link, et al., FOCUS Collaboration, Phys. Lett. **B535** (2002)430.
22. C. Cawlfield, et al., CLEO Collaboration, Phys. Rev. **D 74** (2006) 031108R.
23. D. Epifanov, et al., BELLE Collaboration, Phys. Lett. **B 654** (2007) 65.
24. S. Anderson, et al., Phys. Rev. **D 63** (2001)09001.
25. B. Aubert, et al., BABAR Collaboration, Phys. Rev. **D 76** (2007) 011102R. Phys. Nucl. Phys. 28 (2004) 215.
26. M. Ablikim, et al., BES collaboration, Phys. Lett. **B 598** (2004) 149.
27. M. Ablikim, et al., BES collaboration, Phys. Lett. **B 645** (2007) 19.
28. B. Pietrzyk, Nucl. Phys. B (Proc. Suppl.) **162**, 18 (2006).
29. F. Jegerlehner, Nucl. Phys. B (Proc. Suppl.) **162**, 22 (2006).
30. F. Ambrosino *et al.*, hep-ex/0603056, and references therein.
31. J.Z. Bai et al., BES Collaboration, Phys. Rev. Lett. **84**, 594 (2000)
32. J.Z. Bai et al., BES Collaboration, Phys. Rev. Lett. **88**, 101802 (2002).
33. J.Z. Bai et al., BES Collaboration, Phys. Rev. Lett. **97**, 262001 (2006).
34. J.Z. Bai et al., BES Collaboration, Phys. Lett. **B 677**, 239 (2009).
35. C. Amsler *et al.*, Phys. Lett. B 667 , 122 (2008).
36. M. Ablikim *et al.* (BES Collaboration), Phys. Rev. Lett. 97, 121801 (2006).
37. M. Ablikim *et al.* (BES Collaboration), Phys. Lett. B 641, 145 (2006).
38. M. Ablikim *et al.* (BES Collaboration), Phys. Rev. D. 76, 122002 (2007).
39. M. Ablikim *et al.* (BES Collaboration), Phys. Lett. B 659, 74(2008).

X(3872): A CHARMONIUM-MOLECULE HYBRID WITH ISOSPIN SYMMETRY BREAKING

M. TAKIZAWA

Showa Pharmaceutical University,
Machida, Tokyo 194-8543, Japan
E-mail: takizawa@ac.shoyaku.ac.jp

S. TAKEUCHI

Japan College of Social Work,
Kiyose, Tokyo 204-8555, Japan
E-mail: s.takeuchi@jcsw.ac.jp

In order to understand the structure of the $X(3872)$, we have studied the effects of the $c\bar{c}$ charmonium core state coupling to the $D^0\ \overline{D^{0*}}$ and $D^+\ D^{-*}$ molecule states. We determine the coupling strength so as to reproduce the observed mass of the the $X(3872)$. The isospin symmetry breaking is introduced by the mass differences of the neutral and charged D mesons. The obtained structure of the $X(3872)$ is about 15% of $c\bar{c}$ charmoniumu, 72% of the isoscalar $D\ \overline{D^*}$ molecule and 13% of the isovector $D\ \overline{D^*}$ molecule which explains observed properties of the $X(3872)$ well.

Keywords: Exotic hadrons; Charmonium; X(3872); Isospin symmetry breaking.

1. Introduction

The $X(3872)$ state was first observed in 2003 by Belle[1] and confirmed by CDFII,[2] D0[3] and BaBar[4] collaborations. Since then, it has received much attention because the $X(3872)$ seems to be difficult to explain its structure by a simple $c\bar{c}$ bound state using the quark potential model and is considered as the candidate of the exotic state.[5] Many structures of the $X(3872)$ have been suggested such as a tetraquark structure and a $D^0\bar{D}^{0*}$ molecule.

Another improtant property of the $X(3872)$ is its isospin structure. Belle have measured the branching fractions[6]

$$\frac{Br(X \to \pi^+\pi^-\pi^0 J/\Psi)}{Br(X \to \pi^+\pi^- J/\Psi)} = 1.0 \pm 0.4 \pm 0.3\,. \tag{1}$$

It suggests the strong isospin violation. Here two pion mode is from the isovector ρ meson origin and three pi mode is from the isoscalar ω meson origin. M. Suzuki has estimated the kinematical suppression factor and obtained the production amplitude ratio[7]

$$\left|\frac{A(\rho J/\Psi)}{A(\omega J/\Psi)}\right| = 0.27 \pm 0.02 . \tag{2}$$

Usual size of the isospin symmetry breaking is at most a few %. Thus, we should study the origin of this strong isospin symmetry breaking.

One of the authers (S. T.) has studied the $X(3872)$ using a quark potential model by introducing an extra $(q\bar{q})$ pair to a $c\bar{c}$ system[8] and found a bound state of $q\bar{q}c\bar{c}$ with $J^{PC} = 1^{++}$. The purpose of the present work is to make the situation of the $X(3872)$ clear by studying the effects of the $c\bar{c}$ core state coupling to $D^0\ \overline{D^{0*}}$ and $D^+\ D^{-*}$ molecule states. This approach complements the quark model approach. We have reported the present results at the 19th International IUPAP Conference on Few-Body Problems in Physics (FB19).[9] Since we considered that the overlaps between the participants of the present conference and those of the FB19 were small, we have presented almost same results in this conference. The contents of the present proceedings and those of the FB19[9] are almost same.

2. Wavefunction of $X(3872)$

We consider the $X(3872)$ state is a mixture of the $c\bar{c}$ core state, the $D^0\ \overline{D^{0*}}$ hadronic molecule satate and the $D^+\ D^{-*}$ hadronic molecule states. The observed mass of the $X(3872)$ is (3872.2 ± 0.8) MeV which is about 0.4 MeV above the $D^0\ \overline{D^{0*}}$ threshold (3871.81 ± 0.36) MeV. In order to calculate the wavefunction of the $X(3872)$, we assume that the mass of the $X(3872)$ is about 1 MeV below the $D^0\ \overline{D^{0*}}$ threshold.

The wavefunction of the $X(3872)$ in the center of the mass frame is represented by

$$|X\rangle = c_1\,|c\bar{c}\rangle + c_2\,|D^0\overline{D^{0*}}\rangle + c_3\,|D^+D^{-*}\rangle . \tag{3}$$

The $D^0\ \overline{D^{0*}}$ and $D^+\ D^{-*}$ molecule states are given by

$$|D^0\overline{D^{0*}}\rangle = \int d^3\vec{q}\,\varphi_0(\vec{q})|D^0\overline{D^{0*}}(\vec{q})\rangle , \tag{4}$$

$$|D^+D^{-*}\rangle = \int d^3\vec{q}\,\varphi_+(\vec{q})|D^+D^{-*}(\vec{q})\rangle , \tag{5}$$

where the normalization of the states are

$$\begin{aligned}\langle D^0\overline{D^{0*}}(\vec{q}\,')|D^0\overline{D^{0*}}(\vec{q})\rangle &= \langle D^+D^{-*}(\vec{q}\,')|D^+D^{-*}(\vec{q})\rangle \\ &= \delta^3(\vec{q}\,'-\vec{q}) .\end{aligned} \tag{6}$$

Here $\varphi_0(\vec{q})$ and $\varphi_+(\vec{q})$ are the momentum representation of the wavefunctions of the $D^0\ \overline{D^{0*}}$ and $D^+\ D^{-*}$ hadronic molecule states respectively. The charge conjugation + states are asuumed throughout this paper.

We introduce the monopole form factor type couplings between the $c\bar{c}$ core state and the $D\ \overline{D^*}$ states in the isospin symmetric manner.

$$\langle D^0\overline{D^{0*}}(\vec{q})|V|c\bar{c}\rangle = \langle D^+D^{-*}(\vec{q})|V|c\bar{c}\rangle = \frac{g}{\vec{q}^2+\Lambda^2}\,. \tag{7}$$

We ignore other interactions for simplicity. The Schrödinger equation we have solved is

$$\begin{pmatrix} m_{c\bar{c}}-E & V & V \\ V & m_{D^0}+m_{D^{0*}}+\frac{\hat{p}^2}{2\mu_0}-E & 0 \\ V & 0 & m_{D^+}+m_{D^{-*}}+\frac{\hat{p}^2}{2\mu_+}-E \end{pmatrix} \times \begin{pmatrix} c_1\,|c\bar{c}\rangle \\ c_2\,|D^0\overline{D^{0*}}\rangle \\ c_3\,|D^+D^{-*}\rangle \end{pmatrix} = \begin{pmatrix} 0 \\ 0 \\ 0 \end{pmatrix}, \tag{8}$$

with

$$\frac{1}{\mu_0} = \frac{1}{m_{D^0}} + \frac{1}{m_{D^{0*}}}, \qquad \frac{1}{\mu_+} = \frac{1}{m_{D^+}} + \frac{1}{m_{D^{-*}}}\,. \tag{9}$$

As for the $c\bar{c}$ core state, we consider it is the $J^{PC}=1^{++}$ state and its mass is 3950 MeV from the Godfy and Isgur's result of the quark potential model calculation for the 2 $^3\mathrm{P}_1$ $c\bar{c}$ state.[10]

Let us discuss our numerical results. The parameters of the model are the cutoff Λ and the coupling constant g. We take $\Lambda = 0.3$ GeV and 0.5 GeV which are the typical hadron sizes. The coupling constant g is determined so as to reproduce the observed mass of the $X(3872)$, namely, 3871 MeV. For the case of $\Lambda = 0.3$ (0.5) GeV, we obtained $g = 0.0094$ (0.0185) GeV$^{3/2}$. In the case of the softer cutoff, the coupling is weaker.

The wavefunction we have obtained for the cutoff $\Lambda = 0.3$ GeV is

$$\begin{aligned} |X\rangle &= 0.327\,|c\bar{c}\rangle + 0.882\,|D^0\overline{D^{0*}}\rangle + 0.339\,|D^+D^{-*}\rangle \\ &= 0.327\,|c\bar{c}\rangle + 0.863\,|D\overline{D}^*; I=0\rangle + 0.384\,|D\overline{D}^*; I=1\rangle\,, \end{aligned} \tag{10}$$

and for the cutoff $\Lambda = 0.5$ GeV is

$$\begin{aligned} |X\rangle &= 0.382\,|c\bar{c}\rangle + 0.858\,|D^0\overline{D^{0*}}\rangle + 0.344\,|D^+D^{-*}\rangle \\ &= 0.382\,|c\bar{c}\rangle + 0.850\,|D\overline{D}^*; I=0\rangle + 0.363\,|D\overline{D}^*; I=1\rangle\,. \end{aligned} \tag{11}$$

The cutoff dependence of the admixture of the different componets is small. We obtain the large isospin symmetry breaking and is consistent with the experimental observation Eq. (2).

The calculated coordinate space representation of the wavefunctions of the $D^0\ \overline{D^{0*}}$ and $D^+\ D^{-*}$ hadronic molecule states $r\varphi(r)_0$ and $r\varphi(r)_+$ are shown in Fig. 1. Fig. 1 shows that the $D^0\ \overline{D^{0*}}$ bound state is very widely

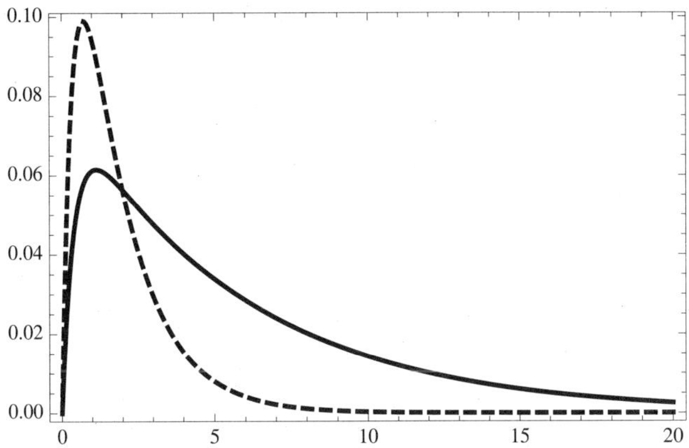

Fig. 1. The solid line shows $r\varphi(r)_0$ and the dashed line shows $r\varphi(r)_+$. The horizontal axis is in units of fm.

spread object and one can safely call it "hadronic molecule". The size of the $D^+\ D^{-*}$ bound state is smaller than that of the $D^0\ \overline{D^{0*}}$ bound state. However it is much bigger than the usual charmonium.

3. Spectrum

In this section, we consider the transition strength $S(E)$ of the weak decay of B meson; $B \to X(3872)K$ and we normalize the $S(E)$ to the production of the 2^3P_1 $c\bar{c}$ state by the weak decay. Then, the $S(E)$ is expressed as follows.

$$S(E) = \frac{-1}{\pi}\mathrm{Im}\langle c\bar{c}|G(E)|c\bar{c}\rangle\,, \tag{12}$$

with the Green's function;

$$G(E) = \frac{1}{E - \hat{H} - i\varepsilon}\,. \tag{13}$$

The state $|c\bar{c}\rangle$ represents the center of mass system of the 2^3P_1 $c\bar{c}$ state with the normalization $\langle c\bar{c}|c\bar{c}\rangle = 1$. Using the free Green's functions and the interaction given in Eq. (7), the Green's function is represented as follows.

$$G(E) = G_1^0 + G_1^0 V G_2^0 V G_1^0 + G_1^0 V G_3^0 V G_1^0 + \cdots\,, \tag{14}$$

$$G_1^0(E) = \frac{1}{E - m_{c\bar{c}} - i\varepsilon}, \tag{15}$$

$$G_2^0(E) = \frac{1}{E - m_{D^0} - m_{D^{0*}} - \frac{\hat{p}^2}{2\mu_0} - i\varepsilon}, \tag{16}$$

$$G_3^0(E) = \frac{1}{E - m_{D^+} - m_{D^{-*}} - \frac{\hat{p}^2}{2\mu_+} - i\varepsilon}. \tag{17}$$

The interaction V gives rise to the attractive force between D and $\overline{D^*}$ effectively and the bound state exists. If the attraction is weak and the bound state is very shallow, then, such state is just the observed X(3872). As mentioned in Sect. 2, we have fixed the strength of the interaction V so as to reproduce the observed $X(3872)$ mass.

On the other hand, the 2^3P_1 $c\bar{c}$ core state becomes resonance. Since such resonance is not observed experimentally, the width of this resonance should be large enough.

The calculated transition strength for the cutoff $\Lambda = 0.3$ GeV is shown in Fig. 2. The $c\bar{c}$ core state of the bare mass of 3950 MeV becomes the

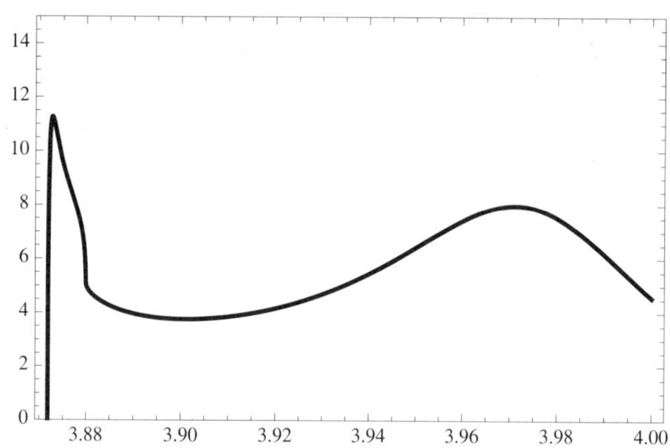

Fig. 2. The calculated transition strength $S(E)$ for the cutoff $\Lambda = 0.3$ GeV. The horizontal axis is in units of GeV. The $S(E)$ is shown above the D^0 $\overline{D^{0*}}$ threshold. The bound state below the D^0 $\overline{D^{0*}}$ threshold, i.e., $X(3872)$ is not shown in figure.

resonance state and its peak position is about 25 MeV shifted up.

We show the transition strength in the case of the cutoff $\Lambda = 0.5$ GeV in Fig. 3. One can find the difference of the shape of the resonance from that

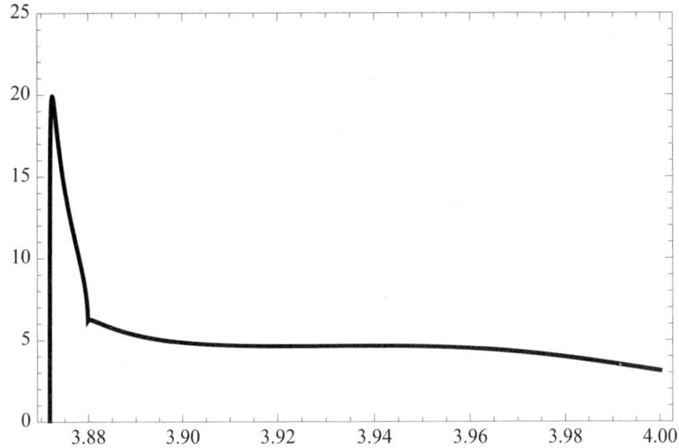

Fig. 3. The calculated transition strength $S(E)$ for the cutoff $\Lambda = 0.5$ GeV. The horizontal axis is in units of GeV. The $S(E)$ is shown above the $D^0\ \overline{D^{0*}}$ threshold. The bound state below the $D^0\ \overline{D^{0*}}$ threshold, i.e., $X(3872)$ is not shown in figure.

for the $\Lambda = 0.3$ GeV case. The mass region of 3950 MeV, the resonance of $J^{PC} = 1^{++}$ has not been observed. Therefore the cutoff $\Lambda = 0.5$ GeV case is more favorable. In the case of the harder cutoff, one needs stronger coupling between the $c\bar{c}$ core state and $D\ \overline{D^*}$ molecule state so as to reproduce the $X(3872)$ mass. Because of this strong coupling, the resonance state around 3950 MeV disappeared.

4. Discussions and Summary

We have studied the structure of the $X(3872)$. We have introduced the interactions between the $c\bar{c}$ core state and the $D\ \overline{D^*}$ molecule state. This interaction effectively gives rise to the attractive force between D and $\overline{D^*}$ and it forms the bound state of the D and $\overline{D^*}$. Since the $c\bar{c}$ core state cannot couple to the charged $D\ \overline{D^*}$ states, no charged partners of the $X(3872)$ exist.

The effect of the isospin symmetry breaking is introduced by the mass difference between the neutral and charged D and D^* mesons. It gives rise to the large isospin mixing in the structure of the $X(3872)$ state and our result is consistent with the experimental observation.

In the wavefunction of the $X(3872)$ we have obtained, there is about 15% of the $c\bar{c}$ core component. It helps up to explain the production rate of the $X(3872)$. If the $X(3872)$ is purely the $D\ \overline{D^*}$ molecule state, the observed production rate of the $X(3872)$ seems to be too large.

Concerning the quark potential model predictions of the charmonium states, the predicted masses are meaningful. However, if the predicted state is above the open charm threshold, one should consider the decay width of that state when you compare the theoretical results with experimental observations.

In order to confirm the present picture of the $X(3872)$, the analyses of the multi-body decay of the $X(3872)$ is important since it gives us the information of the size of the $X(3872)$.

References

1. S. -K. Choi, *et al.*, [Belle Collaboration], *Phys. Rev. Lett.* **91**, 262001 (2003).
2. D. E. Acosta, *et al.*, [CDF II Collaboration], *Phys. Rev. Lett.* **93**, 072001 (2004).
3. V. M. Abazov, *et al.*, [D0 Collaboration], *Phys. Rev. Lett.* **93**, 162002 (2004).
4. B. Aubert, *et al.*, [BABAR Collaboration], *Phys. Rev.* **D71**, 071103 (2005).
5. For a review, see E. S. Swanson, *Phys. Rep.* **429**, 243 (2006).
6. K. Abe, *et al.*, [Belle Collaboration], arXiv:hep-ex/050537.
7. M. Suzuki, *Phys. Rev.* **D72**, 114013 (2005).
8. S. Takeuchi, *Prog. Theor. Phys. Suppl.* **168**, 107 (2007).
9. M. Takizawa and S. Takeuchi, Structure of the X(3872) and its Isospin Symmetry Breaking, to appear in EPJ Web of Conferences.
10. S. Godfrey and N. Isgur, *Phys. Rev.* **D32**, 189 (1985).

EXOTIC NUCLEI WITH OPEN HEAVY FLAVOR MESONS

S. YASUI*

Institute of Particle and Nuclear Studies, High Energy Accelerator Research Organization (KEK), 1-1, Oho, Ibaraki, 305-0801, Japan
**E-mail: yasuis@post.kek.jp*

K. SUDOH

Nishogakusha University, 6-16, Sanbancho, Chiyoda, Tokyo, 102-8336, Japan
E-mail: k-sudoh@nishogakusha-u.ac.jp

The research of exotic nuclei containing charm and bottom flavors is an interesting subject as a possible extension of flavors in strangeness nuclei. We consider the possible existence of exotic charm and bottom nuclei with $\bar{D}$ (B) mesons. With respecting the heavy quark symmetry, we present that the one pion exchange potential with tensor coupling induces an attractive interaction between $\bar{D}$ (B) meson and nucleon. We consider the compound states of a $\bar{D}$ (B) meson and a nucleon with spin-parity $J^P = 1/2^-$ and $3/2^-$ and isospin $I = 0$ and 1. We find only the $J^P = 1/2^-$ and $I = 0$ channel induces a stable bound state, while the others do not. We further discuss the possible bound states of a $\bar{D}$ (B) meson and two nucleons, and discuss the application to nuclei with larger baryon number.

1. Introduction

The research of multi-quark systems is one of the most interesting fields of hadron and nuclear physics. It covers a wide range of physics from exotic hadrons, such as Λ^* with strangeness, X, Y and Z series with charm and Y_{b} with bottom, to exotic nuclei with multi-strangeness, such as Λ-, $\Lambda\Lambda$-, $\Sigma-$, Ξ-hypernuclei and $\bar{K}$-nuclei. The variety of flavors leads us to deep and fruitful understanding about vacuum and matter of QCD. The strangeness has enlarged our knowledge from $\mathrm{SU}(2)_{\mathrm{f}}$ symmetry of up and down quarks to approximate $\mathrm{SU}(3)_{\mathrm{f}}$ symmetry of up, down and strange quarks. It will further promote our understanding of hadrons and nuclei by studies in experimental facilities of J-PARC, GSI and so on.

The heavy flavors of charm and bottom also may provide us new information. The heavy quark masses, $m_{\mathrm{c}} \simeq 1.3$ GeV and $m_{\mathrm{b}} \simeq 4.2$ GeV of charm and bottom quarks, introduce new energy scales which are much larger than the typical energy scale $\Lambda_{\rm QCD} \simeq 0.2$ GeV in light quarks. Such new energy scales enable us to study the hadrons and nuclei within approaches directly based on QCD. For example, the heavy quark effective theory (HQET) gives a systematic expansion by inverse of the quark mass.[1] The HQET can be inverted to heavy meson effective theory in which explicit degrees of freedom are heavy mesons and baryons.[1,2]

The heavy quark effective theory implies the heavy quark symmetry, which induces the mass degeneracy of pseudoscalar and vector mesons of heavy flavors in heavy quark mass limit.[1] For charm, we find the mass difference of D and D^* mesons is $m_{D^*} - m_D \simeq 140$ MeV as observed values in experiments. For bottom, the mass difference is even smaller $m_{B^*} - m_B \simeq 45$ MeV. Because their mass differences are almost equal or less than pion mass, they are regarded as approximately degenerate and to realize the heavy quark symmetry approximately.

The heavy quark symmetry is understood from QCD. In the heavy quark effective theory, the heavy quark is coupled to the magnetic component of the gluon field, which is responsible to the spin flip of the heavy quark, with an order of $1/m_Q$ (m_Q the heavy quark mass). Therefore, the spin flip of the heavy quark is suppressed by $1/m_Q$ and hence it leads to the approximate degeneracy of D (B) and D^* (B^*) mesons. It turns out that both D (B) and D^* (B^*) mesons are relevant degrees of freedom in the dynamics of hadrons. We mention that K and K^* mesons have so large mass difference 392 MeV, and they are not considered to be degenerate conventionally. So K^* is irrelevant as far as only the ground state is concerned.

It would be worthwhile to compare the heavy quark symmetry and the $\mathrm{SU}(4)_\mathrm{f}$ symmetry (Fig. 1). The latter symmetry is a straightforward extension from the $\mathrm{SU}(3)_\mathrm{f}$ symmetry of up, down and strange quarks. The $\mathrm{SU}(3)_\mathrm{f}$ symmetry holds with good approximation for pseudo-scalar meson sector (π, η, K) and vector meson sector (ω, ρ, K^*, ϕ), respectively. However, the masses of D mesons (1864 MeV) and D^* mesons (2008 MeV) are too heavy to be compared to the masses of π, η, K mesons and ω, ρ, K^*, ϕ mesons. Therefore, it seems not natural to regard D mesons belong to the same multiplet of π, η, K mesons, and as well D^* mesons to ω, ρ, K^*, ϕ mesons. Thus, it would be more reasonable to regard D and D^* mesons to obey the heavy quark symmetry rather than the $\mathrm{SU}(4)_\mathrm{f}$ symmetry.

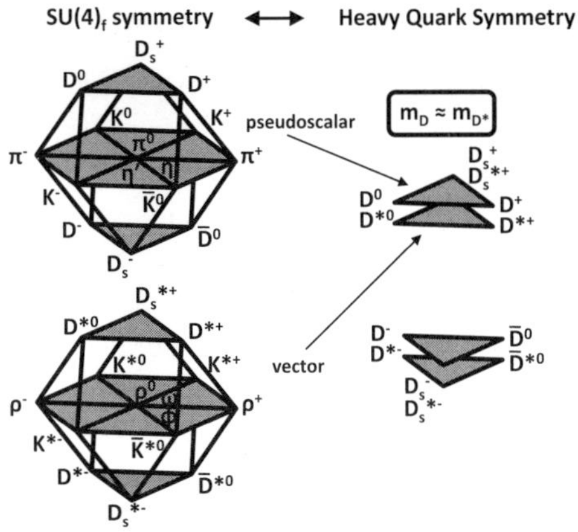

Fig. 1. The meson multiplets of the $SU(4)_f$ symmetry and the heavy quark symmetry.

2. From strangeness nuclei to charm and bottom nuclei

In general we may expect that the different symmetry induces different dynamics and structure of matter. So we consider the heavy quark symmetry of heavy quark systems will give novel dynamics and structure of matter which are qualitatively different from the $SU(3)_f$ symmetry of light quark systems. Here we consider exotic charm and bottom nuclei in which charm and bottom hadrons are included in the normal nuclei.

The charm and bottom nuclei are comparable with strange nuclei (hypernuclei and $\bar{K}$ nuclei) which are now extensively studied in both theoretical and experimental sides.[3,4] We note again that the former has the heavy quark symmetry, while the latter has the $SU(3)_f$ symmetry. Because of the difference of symmetry, we may discover new dynamics and structure in exotic charm and bottom nuclei, which are essentially different from those in strangeness nuclei. Furthermore, as discussed in the introduction, the analysis of charm and bottom hadrons are accessible directly from QCD. Therefore, it will open a deep understanding of nuclei more directly based on QCD. The charm and bottom nuclei are interesting, not only in theoretical studies,[6–8] but also in experimental studies in high-energy hadron facilities such as J-PARC, GSI,[5] and so on. In this paper, we present the recent discussion about the exotic nuclei with containing D (B) mesons with respecting to the heavy quark symmetry.[8]

3. Interactions from heavy quark symmetry

One of the most basic information in the exotic nuclei with D (B) mesons is the interaction between D (B) mesons and nucleons. In many phenomenological studies, the D mesons in nuclei have been discussed by means of the Weinberg-Tomozawa-type contact interaction with $\mathrm{SU}(4)_\mathrm{f}$ flavor symmetry with π, K and D mesons.[7] As discussed in the introduction, however, because the mass scales of charm and bottom are much different from other light flavors, it is expected that the $\mathrm{SU}(4)_\mathrm{f}$ symmetry would not be a good symmetry in nature. Instead of the $\mathrm{SU}(4)_\mathrm{f}$ symmetry, it seems more reasonable to consider the heavy quark symmetry for charm and bottom hadrons. In the latter scheme, both D (B) pseudoscalar mesons and D^* (B^*) vector mesons are fundamental degrees of freedom.

The interaction between the D (B) mesons and nucleons is qualitatively different from that of K mesons and nucleons. To see this difference, let us start from strangeness. We remember the Weinberg-Tomozawa-type contact interaction for K mesons and nucleons. It is well-known that the Weinberg-Tomozawa-type contact interaction can be regarded as exchange of ω and ρ vector mesons. Here, one pion exchange process is absent in principle, because the coupling of $KK\pi$ is forbidden by parity conservation. In heavy flavor case, on the other hand, the couplings of $DD^*\pi$ and $D^*D^*\pi$ ($BB^*\pi$ and $B^*B^*\pi$) exist. Therefore, there is one pion exchange process in the interaction of D (B) mesons and nucleons. This observation indicates that the KN interaction is a short range force characterized with scale of mass of ω and ρ mesons, while the DN (BN) interactions are long range forces characterized with scale of mass of π mesons.

The one pion exchange potential between D (B) mesons and nucleons requires the information of π meson and nucleon vertex, and D (B) meson and nucleon vertex. The former is simply given by

$$\mathcal{L}_{\pi NN} = -\frac{g_{\pi NN}}{m_N} \chi^\dagger_{s'} \vec{\sigma} \cdot \vec{\nabla} \frac{\vec{\pi} \cdot \vec{\tau}}{2} \chi_s, \tag{1}$$

where χ_s is the nucleon field with spin s, m_N is the nucleon mass, and $g^2_{\pi NN}/4\pi = 13.5$ is the coupling constant. The latter is given with help of heavy quark symmetry,

$$\mathcal{L}_{\pi HH} = g\,\mathrm{tr} \bar{H}_a H_b \gamma_\nu \gamma_5 A^\nu_{ba}, \tag{2}$$

where the multiplet field H of P and P^* is defined by

$$H_a = \frac{1 + \not{v}}{2} \left[P^*_{a\mu} \gamma^\mu - P_a \gamma_5 \right], \tag{3}$$

with the velocity v of the mesons.[2] Here we define P as D or B, and P^* as D^* or B^* for short notation. The conjugate field is $\bar{H}_a = \gamma_0 H_a^\dagger \gamma_0$, and the index a denotes up and down flavors. The axial current is given by $A_\mu \simeq \frac{i}{f_\pi}\partial_\mu \mathcal{M}$ with

$$\mathcal{M} = \begin{pmatrix} \frac{\pi^0}{\sqrt{2}} & \pi^+ \\ \pi^- & -\frac{\pi^0}{\sqrt{2}} \end{pmatrix}, \tag{4}$$

where $f_\pi = 135$ MeV is the pion decay constant. The coupling constant $|g| = 0.59$ for πPP^* is determined from the observed decay width $\Gamma = 96$ keV for $D^* \to D\pi$. The coupling of $\pi P^* P^*$, which is difficult to access from experiments, is automatically determined thanks to the heavy quark symmetry. Note that the coupling of πPP does not exist due to the parity conservation. The coupling constant g for πBB^* would be different from one for πDD^* because of $1/m_Q$ corrections with the heavy quark mass m_Q.[9] The recent lattice simulation in the heavy quark limit suggests the closed value adopted above.[10] This would allow us to use the common value for D and B.

4. Exotic bound states

Now we move to discussion about the $P^{(*)}N$ bound states. In this analysis, we consider four states, whose quantum numbers are classified to $J^P = 1/2^-$ and $3/2^-$ with isospin $I = 0$ and 1. The $1/2^-$ states with $I = 0$ and 1 are superpositions of three states, ${}^2S_{1/2}$ for PN, and ${}^2S_{1/2}$ and ${}^4D_{1/2}$ for P^*N with the standard notation ${}^{2S+1}L_J$. The $3/2^-$ states with $I = 0$ and 1 are superpositions of four states, ${}^2D_{3/2}$ for PN, and ${}^4S_{3/2}$, ${}^4D_{3/2}$, and ${}^2D_{3/2}$ for P^*N. A similar analysis has been done in Ref. 11, in which they however do not take into account the mixing between different spin states, ${}^2S_{1/2}$ in PN and P^*N, and ${}^4D_{1/2}$ in P^*N.

With these basis, we explicitly represent the potentials as[8]

$$V_{1/2^-} = \frac{g g_{\pi NN}}{\sqrt{2} m_N f_\pi} \frac{1}{3} \times \begin{pmatrix} 0 & \sqrt{3}C_\mu & -\sqrt{6}T_\mu \\ \sqrt{3}C_\mu & -2C_{m_\pi} & 0 \\ -\sqrt{6}T_\mu & 0 & C_{m_\pi} - 2T_{m_\pi} \end{pmatrix} \vec{\tau}_P \cdot \vec{\tau}_N, \tag{5}$$

$$V_{3/2^-} = \frac{g g_{\pi NN}}{\sqrt{2} m_N f_\pi} \frac{1}{3} \times \begin{pmatrix} 0 & \sqrt{3}T_\mu & -\sqrt{3}T_\mu & \sqrt{3}C_\mu \\ \sqrt{3}T_\mu & C_{m_\pi} & 2T_{m_\pi} & T_{m_\pi} \\ -\sqrt{3}T_\mu & 2T_{m_\pi} & C_{m_\pi} & -T_{m_\pi} \\ \sqrt{3}C_\mu & T_{m_\pi} & -T_{m_\pi} & -2C_{m_\pi} \end{pmatrix} \vec{\tau}_P \cdot \vec{\tau}_N, \tag{6}$$

for $J^P = 1/2^-$ and $3/2^-$, respectively.[a] m_π is the pion mass and $\mu^2 = m_\pi^2 - (m_{P^*} - m_P)^2$ is the modified mass because of the different masses m_{P^*} and m_P for P^* and P. C_m and T_m are defined as

$$C_m = \int \frac{d^3p}{(2\pi)^3} \frac{1}{\vec{q}^{\,2} + m^2} e^{i\vec{q}\cdot\vec{r}} F(\vec{q}; m), \tag{7}$$

$$T_m S_{12}(\hat{r}) = \int \frac{d^3p}{(2\pi)^3} \frac{-\vec{q}^{\,2}}{\vec{q}^{\,2} + m^2} S_{12}(\hat{q}) e^{i\vec{q}\cdot\vec{r}} F(\vec{q}; m), \tag{8}$$

with the tenor operator $S_{12}(\hat{x}) = 3(\vec{\sigma}_1 \cdot \hat{x})(\vec{\sigma}_2 \cdot \hat{x}) - \vec{\sigma}_1 \cdot \vec{\sigma}_2$, and the form factor $F(\vec{q}; m) = (\Lambda_N^2 - m^2)/(\Lambda_N^2 + \vec{q}^{\,2}) \times (\Lambda_P^2 - m^2)/(\Lambda_P^2 + \vec{q}^{\,2})$. Here we adopt $\Lambda_N = 940$ MeV, $\Lambda_D = 1266$ MeV and $\Lambda_B = 1213$ MeV.[8] τ_N and τ_P are isospin operators for nucleon N and meson P.

The kinetic terms are

$$K_{1/2^-} = \mathrm{diag}\left(-\frac{1}{2\tilde{m}_P}\triangle_0, \; -\frac{1}{2\tilde{m}_{P^*}}\triangle_0 + \Delta m_{PP^*}, \; -\frac{1}{2\tilde{m}_{P^*}}\triangle_2 + \Delta m_{PP^*} \right), \tag{9}$$

$$K_{3/2^-} = \mathrm{diag}\left(-\frac{1}{2\tilde{m}_P}\triangle_2, \; -\frac{1}{2\tilde{m}_{P^*}}\triangle_0 + \Delta m_{PP^*}, \; -\frac{1}{2\tilde{m}_{P^*}}\triangle_2 + \Delta m_{PP^*}, \; -\frac{1}{2\tilde{m}_{P^*}}\triangle_2 + \Delta m_{PP^*} \right), \tag{10}$$

[a]It has been pointed out recently that the $P^*N(^2S_{1/2}) - P^*N(^4D_{1/2})$ component in $V_{1/2^-}$ has a numerical error. However the correct value does not change our conclusion qualitatively.[20]

for $J^P = 1/2^-$ and $3/2^-$, respectively. Here we define $\triangle_0 = \partial^2/\partial r^2 + (2/r)\partial/\partial r$ and $\triangle_2 = \triangle_0 + 6/r^2$, $\tilde{m}_{P^{(*)}} = m_N m_{P^{(*)}}/(m_N + m_{P^{(*)}})$, and $\Delta m_{PP^*} = m_{P^*} - m_P$. The eigenvalue equation with the given hamiltonian, $H_{J^P} = K_{J^P} + V_{J^P}$ with $J^P = 1/2^-$ and $3/2^-$, is numerically solved by a variational method. The binding energy is realized as a difference from the threshold $m_N + m_P$.

We find the $\bar{D}N$ and BN bound state solutions for $J^P = 1/2^-$ with $I = 0$.[8] Their binding energies and the relative radii are 1.4 MeV and 9.4 MeV, and 3.8 fm and 1.7 fm, respectively, which are summarized in Table. 1. The wave functions are shown in Fig. 2, where the solid, dashed, and dotted curves represent $PN(^2S_{1/2})$, $P^*N(^2S_{1/2})$, and $P^*N(^4D_{1/2})$, respectively. The narrow (thick) curves show the $\bar{D}N$ (BN) states. We indicate that the $P^*N(^4D_{1/2})$ component is crucially important, though it is smaller than the $PN(^2S_{1/2})$ as shown in Fig. 2. The binding energy is induced mainly by the tensor coupling between $PN(^2S_{1/2})$ and $P^*N(^4D_{1/2})$, which are in off-diagonal terms in $V_{1/2^-}$ in Eq. (5). Indeed, without this tensor coupling, we cannot obtain any bound state. Thus, the PN-P^*N mixing plays an essential role in these systems.

Table 1. The properties of the $\bar{D}N$ and BN bound states for $J^P = 1/2^-$ with $I = 0$.

	$\bar{D}N$	BN
B. E. [MeV]	1.4	9.4
r [fm]	3.8	1.7

We see that the BN state is more deeply bound and compact than the $\bar{D}N$ state. One reason is that the smaller mass splitting between B and B^* strengthens the BN-B^*N mixing. Another reason is that the kinetic energy of B mesons is suppressed simply due to heavy mass. In the heavy quark limit, the complete mass degeneracy of P and P^* induces the ideal PN-P^*N mixing, which would give the maximal binding energy.

It is worth to emphasize that we can not find $\bar{D}N$ and BN bound states for other channels, $J^P = 1/2^-$ with $I = 1$, and $J^P = 3/2^-$ with $I = 0$ or 1. As a result, we conclude that $J^P = 1/2^-$ with $I = 0$ are the most promising channel for detecting stable $\bar{D}N$ and BN bound states.

In experiments, the $\bar{D}N$ and BN bound states would be searched in e^+e^- collisions, or anti-proton beam with deuteron targets.[12] We note there

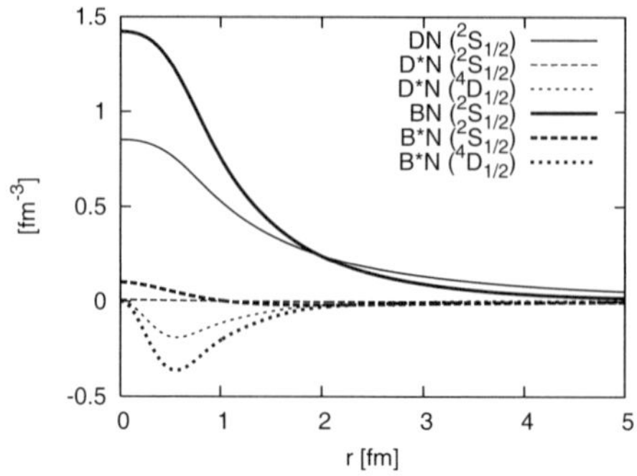

Fig. 2. The wave functions of the $\bar{D}N$ and BN bound states for $J^P = 1/2^-$ with $I = 0$.

is no open decay channel below the threshold, into which the $\bar{D}N$ and BN bound states decay by strong interaction. Therefore, they are stable objects and experimentally well accessible despite of their small binding energies. The $\bar{D}N$ and BN wave functions have the following components,

$$|\bar{D}N\rangle = \left(|D^-p\rangle - |\bar{D}^0 n\rangle\right) + \left(|D^{*-}p\rangle - |\bar{D}^{*0} n\rangle\right), \tag{11}$$

$$|BN\rangle = \left(|B^0 p\rangle - |B^+ n\rangle\right) + \left(|B^{*0} p\rangle - |B^{*+} n\rangle\right), \tag{12}$$

with neglecting the weight of each component. We find that the weak decay processes, $D^-p \to K^+\pi^-\pi^- + p$ and $B^0 p \to D^-\pi^+ + p$, in the first component would be available for reconstruction of the invariant mass.

The $\bar{D}N$ and BN bound states contain five quarks as ingredient. Therefore they can be compared with the KN states as candidate of pentaquarks.[13] It has been theoretically studied that the K meson is not bound with a nucleon by the Weinberg-Tomozawa interaction with chiral symmetry.[14] In contrast, our analysis shows the $\bar{D}$ and B mesons are bound with a nucleon by one pion exchange potential with respecting heavy quark symmetry. Therefore, the $\bar{D}N$ and BN bound states may open another way to search for pentaquark with heavy flavors.[15–17]

5. Toward exotic nuclei

The work on $\bar{D}N$ bound states is extended to exotic nuclei containing $\bar{D}$ mesons. One can expect that the binding energy of the $\bar{D}$ meson in such exotic nuclei becomes larger as baryon number increases. Such exotic nuclei will be investigated in J-PARC and GSI by using anti-proton beams with targets of nuclei.[18] Here we discuss briefly the $\bar{D}NN$ states with baryon number two. The possible states for $I = 1/2$ are classified to

$|\bar{D}(NN)_{0,1}\rangle + |\bar{D}^*(NN)_{1,0}\rangle$ with $J^P = 0^-$ and $|\bar{D}(NN)_{1,0}\rangle + |\bar{D}^*(NN)_{1,0}\rangle + |\bar{D}^*(NN)_{0,1}\rangle$ with $J^P = 1^-$. For $I = 3/2$, they are $|\bar{D}(NN)_{0,1}\rangle$ with $J^P = 0^-$ and $|\bar{D}^*(NN)_{0,1}\rangle$ with $J^P = 1^-$, where subscripts denote the spin and isospin of nucleon pairs. Only the former two states for $I = 1/2$ contain the attractive $\bar{D}N$ pairs for $J^P = 1/2^-$ with $I = 0$. Therefore, one can expect that the $\bar{D}NN$ states, as well as the BNN states, for $J^P = 0^-$ and/or 1^- with $I = 1/2$ would be stable.

It is valuable to compare the $\bar{D}NN$ and BNN states with the K^-pp states ($\bar{K}(NN)_{0,1}$ in our notation).[19] The K^-pp state is considered to have large binding energy 20-70 MeV. However, the K^-pp state decays to πYN by a strong interaction with the decay width 40-70 MeV. On the other hand, the $\bar{D}NN$ and BNN states are stable in the strong decay as discussed above. Therefore, the $\bar{D}NN$ and BNN states may provide more precise information about exotic nuclei. More quantitative analyses including few-body calculation will be presented in coming future.[20]

The charmed and bottom nuclei like $\bar{D}NN$ and BNN states are comparable with the hypothetical existence of the K nuclei, in which K mesons are bound in nuclei.[21] The existence of such nuclei may realize if there is a sufficient number of nucleons around K meson, though KN interaction itself seems not attractive to form KN molecule.[14] We stress again that in our analysis the $\bar{D}$ and B mesons have sufficiently large attraction to form the stable charmed and bottom nuclei beginning with baryon number one.

6. Summary

In summary, bound states of nucleon and $\bar{D}$ and B meson are discussed with respecting the heavy quark symmetry. It is found $\bar{D}N$ and BN bound states with binding energies 1.4 MeV and 9.4 MeV, respectively, for $J^P = 1/2^-$ with $I = 0$, and no bound states in other channels. These states are stable in the strong decay, and can be observed in the weak decay processes $\bar{D}N \to K^+\pi^-\pi^- + p$, and $BN \to D^-\pi^+ + p$. The existence of $\bar{D}N$ and BN bound states would provide an opportunity to probe new exotic states near the thresholds, and open a new way to investigate for exotic nuclei with variety of multi-flavor explored at future hadron facilities such as J-PARC, GSI and so on.

The authors thank A. Dote for fruitful discussions.

References

1. For example, A. V. Manohar and M. B. Wise, Camb. Monogr. Part. Phys. Nucl. Phys. Cosmol. **10** (2000) 1.

2. N. Isgur and M. B. Wise, Phys. Rev. Lett. **66** (1991) 1130; G. Burdman and J. F. Donoghue, Phys. Lett. B **280** (1992) 287; M. B. Wise, Phys. Rev. D **45** (1992) R2188; T. M. Yan, H. Y. Cheng, C. Y. Cheung, G. L. Lin, Y. C. Lin and H. L. Yu, Phys. Rev. D **46** (1992) 1148 [Erratum-ibid. D **55** (1997) 5851]; M. A. Nowak, M. Rho and I. Zahed, Phys. Rev. D **48** (1993) 4370; R. Casalbuoni, A. Deandrea, N. Di Bartolomeo, R. Gatto, F. Feruglio and G. Nardulli, Phys. Rept. **281** (1997) 145; W. A. Bardeen, E. J. Eichten and C. T. Hill, Phys. Rev. D **68** (2003) 054024; T. Matsuki, T. Morii and K. Sudoh, Prog. Theor. Phys. **117** (2007) 1077.
3. Y. Akaishi and T. Yamazaki, Phys. Rev. C **65** (2002) 044005.
4. A. Gal, arXiv:0904.4009 [nucl-th], and references cited therein.
5. M. F. Lutz, B. Pire, O. Scholten and R. Timmermans [The PANDA Collaboration], arXiv:0903.3905 [hep-ex].
6. K. Tsushima, D. H. Lu, A. W. Thomas, K. Saito and R. H. Landau, Phys. Rev. C **59** (1999) 2824; A. Mishra, E. L. Bratkovskaya, J. Schaffner-Bielich, S. Schramm and H. Stocker, Phys. Rev. C **69** (2004) 015202; T. Mizutani and A. Ramos, Phys. Rev. C **74** (2006) 065201.
7. M. F. M. Lutz and E. E. Kolomeitsev, Nucl. Phys. A **730** (2004) 110; J. Hofmann and M. F. M. Lutz, Nucl. Phys. A **763** (2005) 90; D. Gamermann, E. Oset, D. Strottman and M. J. Vicente Vacas, Phys. Rev. D **76** (2007) 074016; J. Haidenbauer, G. Krein, U. G. Meissner and A. Sibirtsev, Eur. Phys. J. A **33** (2007) 107.
8. S. Yasui and K. Sudoh, Phys. Rev. D **80** (2009) 034008.
9. H. Y. Cheng, C. Y. Cheung, G. L. Lin, Y. C. Lin, T. M. Yan and H. L. Yu, Phys. Rev. D **49** (1994) 2490; C. G. Boyd and B. Grinstein, Nucl. Phys. B **442** (1995) 205.
10. H. Ohki, H. Matsufuru and T. Onogi, Phys. Rev. D **77** (2008) 094509.
11. T. D. Cohen, P. M. Hohler and R. F. Lebed, Phys. Rev. D **72** (2005) 074010.
12. A. Sibirtsev, Nucl. Phys. A **680** (2000) 274; J. Haidenbauer, G. Krein, U. G. Meissner and A. Sibirtsev, Eur. Phys. J. A **37** (2008) 55.
13. T. Nakano *et al.* [LEPS Collaboration], Phys. Rev. C **79** (2009) 025210.
14. T. Hyodo, D. Jido and A. Hosaka, Phys. Rev. Lett. **97** (2006) 192002, and references therein.
15. Y. Oh, B. Y. Park and D. P. Min, Phys. Rev. D **50** (1994) 3350; Y. s. Oh and B. Y. Park, Z. Phys. A **359** (1997) 83.
16. S. H. Lee, S. Yasui, W. Liu and C. M. Ko, Eur. Phys. J. C **54** (2008) 259.
17. S. H. Lee and S. Yasui, Eur. Phys. J. C **64** (2009) 283.
18. Y. S. Golubeva, W. Cassing and L. A. Kondratyuk, Eur. Phys. J. A **14** (2002) 255; A. Sibirtsev, K. Tsushima and A. W. Thomas, Eur. Phys. J. A **6** (1999) 351.
19. A. Dote, T. Hyodo and W. Weise, Phys. Rev. C **79** (2009) 014003.
20. K. Sudoh and S. Yasui, in preparation.
21. A.S. Goldhaber, Proceedings of the Second LAMPF II Workshop, edited by H.A. Thiessen, T.S. Bhatia, R.D. Carlini, and N. Hintz, LA-9572-C, Vol. I (1982) 171.

SPECTRUM AND MIXING OF D_s MESONS

AI-LIN ZHANG

Department of Physics, Shanghai University,
Shanghai 200444, China
**E-mail: zhangal@staff.shu.edu.cn*
www.shu.edu.cn

A brief overview of the study of D_s meson is presented. "New" D_s or D_s-like mesons are introduced. The spectrum of D_s mesons is studied and the mixing of D_s mesons is discussed. Interpretation of $D_{sJ}(2632)^+$, $D_{s1}(2700)^\pm$, $D_{sJ}(2860)$ and $D_{sJ}(3040)^+$ is made.

Keywords: Spectrum; Mixing; D_s mesons.

1. Theoretical Study of D_s

A D_s meson consists of a heavy c quark (or antiquark) and a light s antiquark (or quark). Both the heavy quark symmetry and the light quark chiral symmetry approximately apply in D_s mesons. In the past years, D_s mesons have gained great interests. Except for some investigations concentrated on individual D_s, systematically study of D_s mesons has been made in many models. In the relativized quark model,[1–3] both the spectrum and the decay width of D_s have been computed. The spectrum and the decay have been studied in the heavy quark symmetry theory,[4] the relativistic quark model,[5] the chiral quark model,[6] the coupled channels models,[7–9] the lattice theory[10] and the constituent quark model with both the heavy quark symmetry and the light quark chiral symmetry.[11–13] D_s has also been studied in the mass loaded flux tube model[14–16] and some other models.[17–22] Relevant reviews can be found in Refs. 23–26.

There are some significant features for the spectra of D_s mesons. At one hand, the heavy quark symmetry implies that the hyperfine splitting effects of D_s vanish and the spin multiplets of D_s mesons degenerate. For the $j^P = \frac{1}{2}^-$ ground state, there exists a doublet: $J^P = 0^-(\,^1S_0)$ and $J^P = 1^-(\,^3S_1)$. For the $j^P = \frac{1}{2}^+$ or the $j^P = \frac{3}{2}^+$ first excited states, there exist two doublets: $J^P = (0^+,\ 1^+)$ and $J^P = (1^+,\ 2^+)$. On the other hand, the light quark chiral symmetry indicates that there exists a parity degeneracy in the multiplets. In this case, the $j^P = \frac{1}{2}^-$ multiplet degenerate in mass with the $j^P = \frac{1}{2}^+$ multiplet. That is to say, the $J^P = (0^+,\ 1^+)$ states degenerate in mass with the $J^P = (0^-,\ 1^-)$ states. In the real world, both the heavy quark symmetry and the light quark chiral symmetry are broken. Accordingly, there are mass splittings among these degenerated multiplets.

The study of D_s mesons is usually made in two kinds of classification schemes of the D_s mesons. One is in the nonrelativistic $n^{2S+1}L_J$ scheme, another is in the heavy quark symmetric nj^P scheme.[15] The general relations between the $^{2S+1}L_J$ eigenstates and the j^P eigenstates (C-G coefficients) can be found in Ref. 27. Some explicit expressions of the mixing between P-wave and D-wave multiplets are:

$$
\begin{aligned}
|\frac{1}{2}^+\rangle &= \sqrt{\frac{2}{3}}\,|^3P_1\rangle - \sqrt{\frac{1}{3}}\,|^1P_1\rangle, \\
|\frac{3}{2}^+\rangle &= \sqrt{\frac{1}{3}}\,|^3P_1\rangle + \sqrt{\frac{2}{3}}\,|^1P_1\rangle. \\
|\frac{3}{2}^-\rangle &= \sqrt{\frac{3}{5}}\,|^3D_2\rangle - \sqrt{\frac{2}{5}}\,|^1D_2\rangle, \\
|\frac{5}{2}^-\rangle &= \sqrt{\frac{2}{5}}\,|^3D_2\rangle + \sqrt{\frac{3}{5}}\,|^1D_2\rangle.
\end{aligned} \tag{1}
$$

In Table. 1, the predicted spectra of D_s in the $n^{2S+1}L_J$ scheme is presented.[1] In Table. 2, the predicted spectra of D_s in the nj^P scheme in Refs.[4,5] is presented. In these two tables, the candidates are our inputs according to existing literatures.

As well known, a physical state may not be the $^{2S+1}L_J$ or the j^P eigenstate, it may be a mixing state of these eigenstates. For the P-wave 1^+ D_s, there are two different 1^+ D_s (denoted as $|1^+\rangle$ and $|1^+\rangle'$, respectively) which are the mixing between $|^3P_1\rangle$ and $|^1P_1\rangle$, or the mixing between $|\frac{1}{2}^+\rangle$ and $|\frac{3}{2}^+\rangle$. Therefore,

$$
\begin{aligned}
|1^+\rangle &= a|^3P_1\rangle + b|^1P_1\rangle, \\
|1^+\rangle' &= c|^3P_1\rangle + d|^1P_1\rangle. \\
|1^+\rangle &= e|\frac{1}{2}^+\rangle + f|\frac{3}{2}^+\rangle, \\
|1^+\rangle' &= g|\frac{1}{2}^+\rangle + h|\frac{3}{2}^+\rangle.
\end{aligned}
$$

The coefficients a, b, c, d, e, f, g and h determine the mixing angles. The

Table 1. Spectrum of D_s mesons (GeV).

Candidates	J^P	$n^{2S+1}L_J$	Ref.[1]
$D_s^{\pm}(1969)$	0^-	1^1S_0	1.98
$D_s^{\star\pm}(2112)^0$	1^-	1^3S_1	2.13
$D_{s0}^{\star}(2317)^{\pm}$	0^+	1^3P_0	2.48
$D_{s1}(2536)^{\pm}$	1^+	1^3P_1	2.57
$D_{s1}(2460)^{\pm}$	1^+	1^1P_1	2.53
$D_{s2}(2573)^{\pm}$	2^+	1^3P_2	2.59
$D_{sJ}(2632)$?	1^-	1^3D_1	2.90
?	2^-	1^3D_2	-
?	2^-	1^1D_2	-
$D_{sJ}(2860)$	3^-	1^3D_3	2.92
?	0^-	2^1S_0	2.67
$D_{s1}(2700)^{\pm}$	1^-	2^3S_1	2.73

Table 2. Spectrum of D_s mesons (GeV).

Candidates	J^P	nj^P	Ref.[4]	Ref.[5]
$D_s^{\pm}(1969)$	0^-	$1\frac{1}{2}^-$	1.981	1.965
$D_s^{\star\pm}(2112)^0$	1^-	$1\frac{1}{2}^-$	2.111	2.113
$D_{s0}^{\star}(2317)^{\pm}$	0^+	$1\frac{1}{2}^+$	2.508	2.487
$D_{s1}(2536)^{\pm}$	1^+	$1\frac{3}{2}^+$	2.515	2.535
$D_{s1}(2460)^{\pm}$	1^+	$1\frac{1}{2}^+$	2.569	2.605
$D_{s2}(2573)^{\pm}$	2^+	$1\frac{3}{2}^+$	2.56	2.581
$D_{sJ}(2632)$?	1^-	$1\frac{3}{2}^-$	-	2.913
?	2^-	$1\frac{5}{2}^-$	-	2.900
?	2^-	$1\frac{3}{2}^-$	-	2.953
$D_{sJ}(2860)$	3^-	$1\frac{5}{2}^-$	-	2.925
?	0^-	$2\frac{1}{2}^-$	2.670	2.700
$D_{s1}(2700)^{\pm}$	1^-	$2\frac{1}{2}^-$	2.716	2.806
?	0^+	$2\frac{1}{2}^+$	-	3.067
?	1^+	$2\frac{3}{2}^+$	-	3.114
$D_{sJ}(3040)^+$	1^+	$2\frac{1}{2}^+$	-	3.165
?	2^+	$2\frac{3}{2}^+$	-	3.157

case is similar for the D-wave 2^- D_s. For the 1^- D_s, the mixing between $|1^3D_1\rangle$ and $|2^3S_1\rangle$ was also proposed.[28]

The ideal mixing angle of the 1^+ D_s is $\theta = 35.3°$. The mixing angle $\theta \approx 38°$ was predicted for the 1^+ D_s[2] while $\phi \approx -(55 \pm 5)°$ was predicted for this multiplet.[22]

Obviously, for some J^P states, there may be other kinds of mixing. Only when the mixing mechanism and the mixing effects are clear, can people has the right picture of the classification of D_s mesons. Unfortunately, people's knowledge of mixing is very poor. How to detect the mixing information of the physical states through their spectroscopy, production and decay deserve more exploration.

The dynamics of the strong decay of mesons is not very clear, therefore the descriptions of the strong decay are model dependent. In the $^3P_0$ model,[29] the elementary process is described by the creation of a $q\bar{q}$ pair with the quantum numbers of the vacuum, $J^{PC} = 0^{++}$, in the final state. In the relativized quark model,[1] mesons' decay proceeds through a single quark transition via the emission of a pseudoscalar meson. In the chiral quark model,[6] the transition is mediated by an effective interaction. There are some other models not mentioned here. It is natural that the predictions of the decay widths of D_s mesons may be different.

The $nj^P = 1\,\frac{1}{2}^+$ P-wave D_s was predicted to have a large decay width, while the $nj^P = 1\,\frac{3}{2}^+$ P-wave D_s was predicted to have a small width.[2,6] The predictions of the decay width of the D-wave D_s are controversial. The widths of some decay channels for the $nj^P = 1\,\frac{3}{2}^-$ were predicted to have large differences.[3,6] The $nj^P = 1\,\frac{5}{2}^-$ was predicted to have a small width,[6] while it was predicted to have a large width in Ref. 3. The predictions of the decay widths for the radially excited D_s are also different.

2. "New" D_s and Interpretation of Them

2.1. *"New" D_s*

So far, there seems no confusion for the charmed mesons $c\bar{q}$, but the case is different for the charmed strange mesons $c\bar{s}$. Some "new" D_s or D_s-like mesons have been reported in the past years.

$D^\star_{s0}(2317)^\pm$

$D^\star_{s0}(2317)^\pm$ was first observed by BaBar,[30] and subsequently confirmed by CLEO[31] and BELLE.[32] This state has the mass 40 MeV below the DK threshold and has a narrow width $\Gamma < 3.8$ MeV.[33] It was interpreted as the conventional $0^+(\frac{1}{2}^+)$ D_s,[30,34] the DK or $D\pi$ molecules,[35] the baryonium[36,37] or the coupled channels effects.[38] Now $D^\star_{s0}(2317)^\pm$ is popularly assumed the conventional 0^+ D_s with some new features.

$D_{s1}(2460)^\pm$

$D_{s1}(2460)^\pm$ was first reported by CLEO,[31] and subsequently confirmed by BELLE[32] and BaBar.[39] Its mass is about 50 MeV below the $D^\star K$ threshold and its full width $\Gamma < 3.5$ MeV.[33] This state has also a lower mass compared to most theoretical predictions and has a very narrow full width. It was interpreted as the conventional $1^+(\frac{1}{2}^+)$ D_s[31] or the baryonium.[36,37]

$D_{s1}(2700)^\pm$

$D_{s1}(2700)^\pm$ was first observed by Belle.[40,41] It is also reported by Babar,[42] but the significance of the signal was not stated. $D^\star_{s1}(2710)^+$ was recently observed by BaBar.[43] $D_{s1}(2700)^\pm$ decays into D^0K^+ and $D^\star K$ channels. This state has mass 2690 ± 7 MeV and full width $\Gamma = 110 \pm 27$ MeV.[33] It was interpreted as the conventional $1^-(2^3S_1)$ D_s[15,16,28] or the conventional $1^-(1^3D_1)$ D_s.[44]

$D_{sJ}(2860)$

$D_{sJ}(2860)$ was first reported by BaBar,[42] and recently observed again by BaBar[43] in another channel. This state decays into D^0K^+, $D^+K^0_s$ and $D^\star K$ channels. It was interpreted as the conventional $0^+(2^3P_0)$ D_s[9,44] or the conventional $3^-(1^3D_3)$ D_s.[15,16,44,45] This state has not yet been confirmed by any other experiment.

$D_{sJ}(3040)^+$

Very recently, $D^\star_{sJ}(3040)^+$ was observed by BaBar.[43] This state decays into $D^\star K$ channel, while not observed in DK channel. Its decay properties and the angular analysis suggest its unnatural parity $J^P = 0^-, 1^+, 2^-, \cdots$. In Refs. 16,46, this state was interpreted as the conventional $1^+(2\frac{1}{2}^+)$ D_s.

$D_{sJ}(2632)^+$

$D_{sJ}(2632)^+$ is a surprisingly narrow D_s reported by SELEX some years ago.[47] Its mass is about 274 MeV and 116 MeV above the D^0K^+ and the $D_s\eta$ threshold, respectively. This state has an exotic relative branching ratio $\Gamma(D^0K^+)/\Gamma(D_s^+\eta) = 0.16 \pm 0.06$. That is to say, its decay favors the $D_s\eta$ mode over the DK mode though the two channels share the same quark flavors and phase space. So far, this state has not been observed by any other experiment. It was interpreted as the conventional $1^-(2^3S_1)$ D_s[48] or the four-quark state.[49,50] In our opinion, $D_{sJ}(2632)^+$ is very possibly the conventional 1^- $(1\ ^3D_1)$ D_s[16] if it is confirmed by experiment.

2.2. *Interpretation of $D_{sJ}(2632)^+$, $D_{s1}(2700)^\pm$, $D_{sJ}(2860)$ and $D_{sJ}(3040)^+$*

In terms of the Regge trajectories on (L, M^2)-plots[51,52] and on (n, M^2)-plots,[53] the classification of some D_s mesons was performed.[54] In the semi-classic flux tube model,[55,56] the lowest orbitally excited D_s mesons were systematically studied and classified.[15,16] Based on these investigations, we obtained the interpretation of $D_{sJ}(2632)^+$, $D_{s1}(2700)^\pm$, $D_{sJ}(2860)$ and $D_{sJ}(3040)^+$ as follows.

If $D_{sJ}(2632)^+$ exists, it seems not the radially excited 1^- (2^3S_1) D_s, it is very possibly the orbitally excited 1^- (1^3D_1) D_s.

$D_{s1}(2700)^\pm$ is very possibly the radially excited $1^-(2^3S_1)$ D_s.

If $D_{sJ}(2860)$ and its decay modes are confirmed by much more experiments, it must be the orbitally excited 3^- (1^3D_3) D_s.

If $D_{sJ}(3040)^+$ exists, it is very likely the radially excited 1^+ $(2\frac{1}{2}^+)$ D_s.

If the arrangements of $D_{s1}(2700)^\pm$ and $D_{sJ}(2860)$ are confirmed, partners of them should exist.[57] Partners of the D-wave 2^- D_{s2} will have masses $\approx 2700 - 2800$ MeV, and partners of the radially excited 0^- D'_s and 1^+ D'_{s1} will have masses ≈ 2600 MeV and ≈ 2970 MeV, respectively. Obviously, higher excited D_s have lower masses compared with most existing predictions.

3. Conclusions and Discussions

In the paper, a brief overview of the study of D_s meson (especially existing study of the spectrum and the decay) is presented. The main properties and existing explanation of the "New" D_s or D_s-like mesons are introduced. Interpretation of $D_{sJ}(2632)^+$, $D_{s1}(2700)^\pm$, $D_{sJ}(2860)$ and $D_{sJ}(3040)^+$ is made.

D_s mesons seem to have different features from D mesons. Whether it is the truth deserve more exploration and more experiments are required. If D_s mesons really have different features from D mesons, the reasons have to be found out.

It seems that the P-wave D_s mesons have been established. However, whether there is unclear dynamics responsible for the lower masses of $D^\star_{s0}(2317)^\pm$ and $D_{s1}(2460)^\pm$ has to be studied. The mixing mechanism and the mixing effects to physical states deserve more study.

More and more D_s mesons have been reported, whether there is supernumerary excited D_s is not clear. People has to answer where the orbitally excited D-wave and the radially excited D_s mesons are. It seems that the higher excited D_s have lower masses. If the higher excited D_s really have lower masses, the dynamics or the reason has to be found out.

When a "new" meson with some new properties is reported by experiment, exotic explanations outside normal meson models are usually made. Are exotic explanations outside normal meson models really necessary?

Acknowledgments

This work is supported by the National Natural Science Foundation of China under the grant: 10775093. The author is grateful to the organizers of "Hadron and Nuclear Physics" workshop at RCNP, Osaka University, Japan for the local support.

References

1. S. Godfrey and N.Isgur, *Phys. Rev.* **D32**, 189 (1985).
2. S. Godfrey and R. Kokoski, *Phys. Rev.* **D43**, 1679 (1991).
3. F. E. Close and E. S. Swanson, *Phys. Rev.* **D72**, 094004 (2005).
4. E. J. Eichten, C. T. Hill and C. Quigg, *Phys. Rev. Lett.* **71**, 4116 (1993).
5. D. Ebert, V. O. Galkin and R. N. Faustov, *Phys. Rev.* **D57**, 5663 (1998); Erratum-ibid. **D59**, 019902 (1999).
6. M. Di Pierro and E. Eichten, *Phys. Rev.* **D64**, 114004 (2001).
7. Eef van Beveren and George Rupp, *Phys. Rev. Lett.* **91**, 012003 (2003).
8. Yu. A. Simonov and J. A. Tjon, *Phys. Rev.* **D70**, 114013 (2004).
9. Eef van Beveren and George Rupp, *Phys. Rev. Lett.* **97**, 202001 (2006).
10. J. Hein, S. Collins, C. T. Davies, A.A. Khan, H. Newton, C. Morningstar, J. Shigemitsu and J. Sloan, *Phys. Rev.* **D62**, 074503 (2000).
11. M.A. Nowak, M. Rho and I. Zahed , *Phys. Rev.* **D48**, 4730 (1993).
12. W.A. Bardeen and C.T. Hill, *Phys. Rev.* **D49**, 409 (1994).
13. W.A. Bardeen, E.J. Eichten and C.T. Hill, *Phys. Rev.* **D68**, 054024 (2003).
14. T.J. Allen, T. Coleman, M.G. Olsson and S. Veleli, *Phys. Rev.* **D69**, 074010 (2004).
15. Hong-Yun Shan and Ailin Zhang, *Chin. Phys.* **C34**, 16 (2010).
16. Bing Chen, Deng-Xia Wang and Ailin Zhang, *Phys. Rev.* **D80**, 071502(R) (2009).
17. T. Matsuki and T. Morii, *Phys. Rev.* **D56**, 5646 (1997).
18. J. L. Goity and W. Roberts, *Phys. Rev.* **D60**, 034001 (1999).
19. Yu. S. Kalashnikova, A. V. Nefediev and Yu. A. Simonov, *Phys. Rev.* **D64**, 014037 (2001).
20. J. Erdmenger, N. Evans and J. Grosse, *JHEP* **0701**, 098 (2007).
21. T. Matsuki, T. Morii, and K. Sudoh, *Eur. Phys. J.* **A31**, 701 (2007).
22. Xian-Hui Zhong and Qiang Zhao, *Phys. Rev.* **D78**, 014029 (2008).
23. P. Colangelo, F. De Fazio and R. Ferrandes, *Mod. Phys. Lett.* **A19**, 2083 (2004).
24. E.S. Swanson, *Phys. Rept.* **429**, 243 (2006).
25. J. L. Rosner, *J. Phys.* **G34**, S127 (2007).
26. Shi-Lin Zhu, *Int. J. Mod. Phys.* **E17**, 28 (2008).
27. R. N. Cahn and J. D. Jackson, *Phys. Rev.* **D68**, 037502 (2003)
28. F. E. Close, C. E. Thomas, Olga Lakhina and E. S. Swanson, *Phys. Lett.* **B647**, 159 (2007).
29. A. Le Yaouanc, L. Oliver, O. Pene and J. C. Raynal, *Phys. Rev.* **D8**, 2223 (1973); **D11**, 1272 (1975).

30. B. Aubert *et al.*, BaBar Collaboration, *Phys. Rev. Lett.* **90**, 242001 (2003).
31. D. Besson *et al.*, CLEO Collaboration, *Phys. Rev.* **D68**, 032002 (2003).
32. K. Abe *et al.*, Belle Collaboration, *Phys. Rev. Lett.* **92**, 012002 (2004).
33. C. Amsler *et al.*, Particle Data Group, *Phys. Lett.* **B667**, 1 (2008).
34. W. A. Bardeen, E. J. Eichten and C. T. Hill, em Phys. Rev. **D68**, 054024 (2003).
35. T. Barnes, F. E. Close and H. J. Lipkin, *Phys. Rev.* **D 68**, 054006 (2003).
36. V. Dmitrasinovic, *Phys. Rev.* **D70**, 096011 (2004).
37. Yu-Qi Chen and Xue-Qian Li, *Phys. Rev. Lett.* **93**, 232001 (2004).
38. Eef van Beveren and George Rupp, *Phys. Rev. Lett.* **91**, 012003 (2003).
39. B. Aubert *et al.*, *Phys. Rev.* **D69**, 031101 (2004).
40. K. Abe, *et al.*, Belle Collaboration, hep-ex/0608031.
41. J. Brodzicka *et al.*, Belle Collaboration, *Phys. Rev. Lett.* **100**, 092001 (2008).
42. B. Aubert, *et al.*, BaBar Collaboration, *Phys. Rev. Lett.* **97**, 222001 (2006).
43. B. Aubert, *et al.*, BaBar Collaboration, *Phys. Rev.* **D80**, 092003 (2009).
44. Bo Zhang, Xiang Liu, Wei-Zhen Deng and Shi-Lin Zhu, *Eur. Phys. J.* **C50**, 617 (2007).
45. P. Colangelo, F. De. Fazio and S. Nicotri, *Phys. lett.* **B642**, 48 (2006).
46. Zhi-Feng Sun and Xiang Liu, *Phys. Rev.* **D80**, 074037 (2009).
47. A. Evdokimov *et al.*, SELEX Collaboration, *Phys. Rev. Lett.* **93**, 242001 (2004).
48. Kuang-Ta Chao, *Phys. Lett.* **B599**, 43 (2004).
49. L. Maiani, F. Piccinini, A. D. Polosa and V. Riquer, *Phys. Rev.* **D70**, 054009 (2004).
50. Y. R. Liu, Shi-Lin Zhu, Y. B. Dai and C. Liu, *Phys. Rev.* **D70**, 094009 (2004).
51. G. Chew and S. Frautschi, *Phys. Rev. Lett.* **8**, 41 (1962).
52. S. Filipponi, G. Pancheri and Y. Srivastava, *Phys. Rev. Lett.* **80**, 1838 (1998).
53. A. V. Anisovich, V. V. Anisovich, and A. V. Sarantsev, *Phys. Rev.* **D62**, 051502(R) (2000).
54. Ailin Zhang, *Phys. Rev.* **D72**, 017902 (2005).
55. D. Lacourse and M. G. Olsson, *Phys. Rev.* **D39**, 2751 (1989).
56. Alexander Selem and Frank Wilczek, Ringberg 2005, New trends in HERA physics, 337-356; hep-ph/0602128.
57. Ailin Zhang, arXiv: 0904.2453.

LIGHT PSEUDOSCALAR MESON AND HEAVY MESON SCATTERING LENGTHS

YAN-RUI LIU*

Department of Physics, Tokyo Institute of Technology, Tokyo 152-8551 Japan and Institute of High Energy Physics, P.O. Box 918-4, Beijing 100049, China
**E-mail: yrliu@th.phys.titech.ac.jp*

XIANG LIU

School of Physical Science and Technology, Lanzhou University, Lanzhou 730000, China

Using heavy meson chiral perturbation theory, we have calculated systematically the pion (π, K, η) and heavy pseudoscalar meson S-wave scattering lengths up to $\mathcal{O}(p^3)$ in the heavy quark symmetry limit. From a lattice simulation, we estimate the unknown low-energy constants and predict all the other unmeasured scattering lengths. We found six channels are attractive, including the isoscalar DK channel. The DK interaction may help to lower the mass of the "bare" charm-strange scalar state in the quark model through the couple-channel effect.

1. Introduction

To understand exotic resonances, strange enhancements or any anomalous structures observed experimentally, various configurations have been proposed. In addition to the conventional quark model and the exotic state assumptions like multiquark, hybrid, or glueball, the mixing between these two pictures or final state interaction is probably more important to interpret the observed phenomena. We have to study the mass, decay width, scattering or production properties. Many new exotic heavy quark mesons, such as $D_{sJ}(2317)$, $D_{sJ}(2460)$, and the increasing XYZ states have been observed. They provide new chances to understand the nature of the strong interaction.

Most of these exotic states are near-threshold mesons. A natural question is: are they molecules? In fact, one understands none of the molecular interpretations has been confirmed, both from the experimental side and from the theoretical side. The bound state problem and the scattering problem are crucial to understand this question. In this contribution, we will present one study on the scattering problem. What we consider is an S wave system composed of a light pseudoscalar meson and a heavy pseudoscalar meson, which is related with $D_{sJ}(2317)$.

$D_{sJ}(2317)$ was observed in the invariant mass distribution of $D_s\pi^0$ in the e^+e^- annihilation process.[1–3] Its mass is about 50 MeV below DK threshold and its width is very narrow. This state was expected by the quark model but with a higher mass. The quantum is $J^P = 0^+$. It has been proposed as a chiral partner of D_s, P-wave excited states of $D_s(D_s^*)$, couple-channel effects between the $c\bar{s}$ state and DK continuum, conventional $c\bar{s}$ states, four-quark states etc. In fact, when one considers the large contribution of the S-wave DK continuum, the mass of $D_{sJ}^*(2317)$ agrees well with the experimental value, which is indicated by the coupled-channel effect[4] and the QCD sum rule (QSR) approach.[5] Therefore, $D_{sJ}^*(2317)$ is very probably a conventional $c\bar{s}$ state with $J^P = 0^+$. Since $D_{sJ}(2317)$ strongly couples to the DK channel, the study of the DK interaction is very interesting. Here we present our study on the scattering lengths in the pion-heavy meson channels in the heavy meson chiral perturbation theory (HMχPT),[6,7] where the pion denotes π, K or η while the heavy mesons are the pseudoscalar charmed or bottom mesons.

The scattering length is an important observable, which encodes the information of the underlying interaction. From the relation between the cross section σ and the scattering length a, $\sigma = 4\pi a^2$, one may define this threshold parameter through

$$T_{th} = 8\pi(1 + \frac{m}{M})a, \tag{1}$$

where T_{th} is the threshold T-matrix and $m(M)$ is the mass of the light (heavy) meson. For the sign convention, we choose one consistent with the definition

$$\lim_{k\to\infty} k \cot \delta_0 = 1/a, \tag{2}$$

where δ_0 is the S wave phase shift and k is the relative momentum. Therefore $a > 0$ indicates an attractive interaction while $a < 0$ indicates a repulsive interaction. The present study is helpful to test HMχPT, find out interesting channels, and make predictions.

2. Framework

When the masses of the light quarks go to zero, QCD has an approximate SU(3) chiral symmetry. This symmetry is spontaneously broken into SU(3) vector symmetry due to the non-vanishing quark condensate. As a result, eight Goldstone bosons (π, K, η) appear. Because the quark mass actually does not vanish, these mesons get their masses. Based on this fact, chiral perturbation theory was developed, where the Lagrangians are organized with the order of the momentum and the quark mass

$$\mathcal{L} = \mathcal{L}[p^2, m_q] + \mathcal{L}[p^4, m_q^2, p^2 m_q] + \cdots \tag{3}$$

When one applies this theory to scattering processes, the T matrices are also organized with the chiral order $T = T^2 + T^4 + \cdots$. The Weinberg's power counting rule gives the relation between the chiral order of the T-matrix and number of the loops and the Lagrangians. As an extended theory, one usually use the (heavy) baryon chiral perturbation theory to study the meson-baryon systems. A similar power counting rule exists in the heavy baryon chiral perturbation theory (HBχPT).

On the other hand, QCD manifests its heavy quark symmetry when the mass of the heavy quark goes to infinity. When one considers the meson-heavy meson systems, HMχPT is a useful tool. In the heavy quark symmetry limit, the recoil order corrections are neglected. One may use a power counting rule similar to that in HBχPT to include the chiral corrections order by order. We will adopt the HMχPT formalism to calculate the scattering lengths up to $\mathcal{O}(p^3)$. Such a study may be helpful to test the convergence of HMχPT in the scattering processes.

In the heavy quark limit, the heavy mesons form degenerate doublets: $(0^-, 1^-)$, $(0^+, 1^+)$, etc. One usually uses a superfield

$$H = \frac{1+\not{v}}{2}\left(P_\mu^* \gamma^\mu + \delta P \gamma_5\right) \tag{4}$$

to denote the ground state meson doublet, where $\bar{H} = \gamma^0 H^\dagger \gamma^0$, $P = (D^0, D^+, D_s^+)$, $P_\mu^* = (D^{0*}, D^{+*}, D_s^{+*})_\mu$, or $P = (B^-, \bar{B}^0, \bar{B}_s^0)$, $P_\mu^* = (B^{-*}, \bar{B}^{0*}, \bar{B}_s^{0*})_\mu$, v is the velocity of the heavy meson and δ is a relative phase between the pseudoscalar meson and its vector partner. For our study at thresholds, we need the following chiral Lagrangians,[8]

$$\mathcal{L}^{(2)}_{\phi\phi} = f^2 \mathrm{tr}(u_\mu u^\mu + \frac{\chi_+}{4}), \tag{5}$$

$$\mathcal{L}^{(1)}_{H\phi} = -\langle (i\partial_0 H)\bar{H}\rangle + \langle H\Gamma_0 \bar{H}\rangle + g\langle H u_\mu \gamma^\mu \gamma_5 \bar{H}\rangle, \tag{6}$$

$$\mathcal{L}^{(2)}_{H\phi} = c_0\langle H\bar{H}\rangle \mathrm{tr}(\chi_+) - c_1\langle H\chi_+\bar{H}\rangle \\ -c_2\langle H\bar{H}\rangle \mathrm{tr}(u_0 u_0) - c_3\langle H u_0 u_0 \bar{H}\rangle, \tag{7}$$

$$\mathcal{L}^{(3)}_{H\phi} = \kappa\langle H[\chi_-, u_0]\bar{H}\rangle, \tag{8}$$

where

$$u_\mu = \tfrac{i}{2}\{\xi^\dagger, \partial_\mu \xi\}, \quad \Gamma_\mu = \tfrac{i}{2}[\xi^\dagger, \partial_\mu \xi], \quad \xi = \exp(i\phi/2f), \tag{9}$$

$$\chi_\pm = \xi^\dagger \chi \xi^\dagger \pm \xi \chi \xi, \quad \chi = \mathrm{diag}(m_\pi^2, m_\pi^2, 2m_K^2 - m_\pi^2), \tag{10}$$

$$\phi = \sqrt{2}\begin{pmatrix} \frac{\pi^0}{\sqrt{2}} + \frac{\eta}{\sqrt{6}} & \pi^+ & K^+ \\ \pi^- & -\frac{\pi^0}{\sqrt{2}} + \frac{\eta}{\sqrt{6}} & K^0 \\ K^- & \bar{K}^0 & -\frac{2}{\sqrt{6}}\eta \end{pmatrix}, \quad f \approx 92.4\ \mathrm{MeV}. \tag{11}$$

In the Lagrangians, tr(...) and $\langle ... \rangle$ denote the trace in flavor and Dirac space, respectively. The low energy constants (LECs) κ, c_0, c_1, c_2, c_3 will be determined from a lattice measurement.

3. The calculation

At leading order, one may easily derive the T matrices from the Lagrangian $\mathcal{L}^{(1)}_{H\phi}$. They are Weinberg-Tomozawa terms

$$T^{(1)}_{DK} = 0, \quad T^{(0)}_{DK} = \tfrac{2m_K}{f_K^2}, \quad T^{(1)}_{D\bar{K}} = -\tfrac{m_K}{f_K^2}, \quad T^{(0)}_{D\bar{K}} = \tfrac{m_K}{f_K^2}, \\ T_{D_sK} = -\tfrac{m_K}{f_K^2}, \quad T_{D_s\bar{K}} = \tfrac{m_K}{f_K^2}, \quad T^{(3/2)}_{D\pi} = -\tfrac{m_\pi}{f_\pi^2}, \quad T^{(1/2)}_{D\pi} = \tfrac{2m_\pi}{f_\pi^2}, \\ T_{D_s\pi} = T_{D\eta} = T_{D_s\eta} = 0, \tag{12}$$

with $K = (K^+, K^0)^T$, $\bar{K} = (\bar{K}^0, K^-)^T$. The superscripts of T-matrices denote the total isospin. Here we have replaced f with the renormalized decay constants f_π, f_K and f_η. The resulting $\mathcal{O}(p^3)$ corrections will be taken into account later.

It is convenient to define two combinations of LECs at the next-leading order

$$C_1 = 8c_0 - 4c_1 + 2c_2 + c_3, \\ C_0 = 8c_0 + 4c_1 + 2c_2 - c_3. \tag{13}$$

From $\mathcal{L}_{H\phi}^{(2)}$, we get the following T-matrices

$$
\begin{aligned}
T_{DK}^{(1)} &= \tfrac{m_K^2}{2f_K^2}(C_1+C_0), \quad T_{DK}^{(0)} = \tfrac{m_K^2}{f_K^2}C_1, \\
T_{D\bar{K}}^{(1)} = \tfrac{m_K^2}{f_K^2}C_1, \quad T_{D\bar{K}}^{(0)} &= \tfrac{m_K^2}{f_K^2}C_0, \quad T_{D_sK} = \tfrac{m_K^2}{f_K^2}C_1, \quad T_{D_s\bar{K}} = \tfrac{m_K^2}{f_K^2}C_1, \\
T_{D\pi}^{(3/2)} = \tfrac{m_\pi^2}{f_\pi^2}C_1, \quad T_{D\pi}^{(1/2)} &= \tfrac{m_\pi^2}{f_\pi^2}C_1, \quad T_{D_s\pi} = \tfrac{m_\pi^2}{2f_\pi^2}(C_1+C_0) \\
T_{D\eta} &= \tfrac{1}{3f_\eta^2}\left[(2C_1+C_0)m_\eta^2 + 4c_1(m_\eta^2 - m_\pi^2)\right], \\
T_{D_s\eta} &= \tfrac{1}{6f_\eta^2}\left[(7C_1-C_0)m_\eta^2 - 16c_1(m_\eta^2 - m_\pi^2)\right]. \qquad (14)
\end{aligned}
$$

The effects due to the decay constant renormalization are beyond the order we are considering.

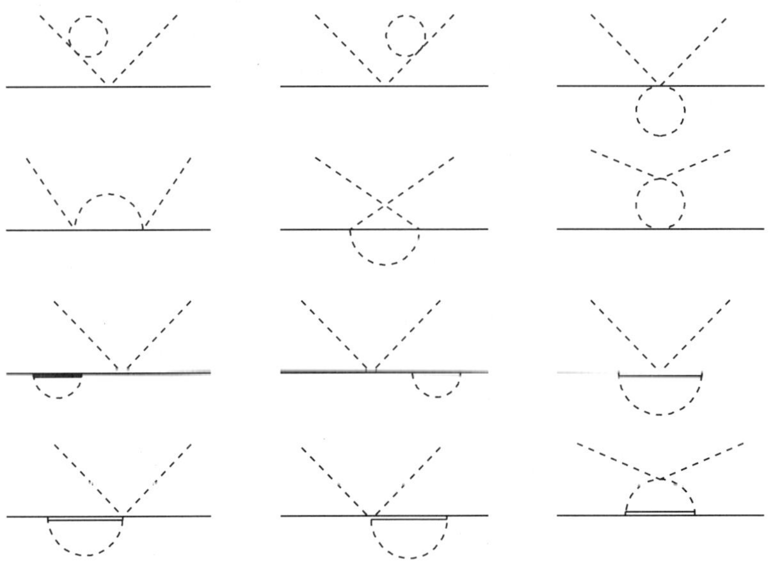

Fig. 1. Non-vanishing loop diagrams in the calculation of meson-heavy meson scattering lengths to the third chiral order in HMχPT. Dashed lines represent Goldstone bosons while solid (double) lines represent pseudoscalar (vector) heavy mesons. The fourth diagram generates imaginary parts for kaon-heavy meson and eta-heavy meson scattering lengths.

At the third order, we have the non-vanishing corrections from the loop diagrams in Fig. 1 and the contributions from the counter terms $\mathcal{L}_{H\phi}^{(3)}$. One example for the T matrices[8] is

$$T_{DK}^{(0)} = \frac{m_K^2}{8\pi^2 f_K^4}\left\{3m_K\left(1-\ln\frac{|m_K|}{\lambda}-\ln\frac{m_\eta}{\lambda}\right)\right.$$
$$\left.-3\sqrt{m_\eta^2-m_K^2}\arccos\frac{-m_K}{m_\eta}+\frac{1}{6}g^2\pi\left(7m_\eta+\frac{6m_\pi^2}{m_\eta+m_\pi}\right)\right\}$$
$$+16\kappa^r\frac{m_K^3}{f_K^2}, \tag{15}$$

where $\kappa^r = \kappa^r(\lambda) = \kappa - \frac{3}{4}\frac{L}{f^2}$ is the renormalized LEC with $L = \frac{\lambda^{d-4}}{16\pi^2}[\frac{1}{d-4} + \frac{1}{2}(\gamma_E - 1 - \ln 4\pi)]$. Here we have used the dimensional regularization and minimal subtraction scheme when deriving the analytical expressions. We retain the chiral logarithm terms. $\lambda = 4\pi f_\pi \sim 1$ GeV denotes the scale of chiral symmetry breaking. In the T-matrices, the corrections proportional to g^2 come from intermediate vector meson contributions (the last six diagrams depicted in Fig 1) and the divergences from these diagrams cancel out. The fourth diagram generates imaginary parts for $T_{DK}^{(1)}$, $T_{D_s\bar{K}}$, $T_{D\eta}$ and $T_{D_s\eta}$.

The derived threshold T-matrices satisfy the crossing symmetry:

$$T_{D\bar{K}}^{(1)} = \frac{1}{2}[T_{DK}^{(1)} + T_{DK}^{(0)}]_{m_K\to -m_K},\quad T_{D\bar{K}}^{(0)} = \frac{1}{2}[3T_{DK}^{(1)} - T_{DK}^{(0)}]_{m_K\to -m_K}, \tag{16}$$
$$T_{D_s\bar{K}} = [T_{D_sK}]_{m_K\to -m_K}, \tag{17}$$

and the SU(3) relations ($m_u = m_d = m_s$)

$$T_{DK}^{(1)} = T_{D_s\pi} = \frac{1}{2}[T_{D\bar{K}}^{(1)} + T_{D\bar{K}}^{(0)}],\quad T_{D_sK} = T_{D\pi}^{(3/2)} = T_{D\bar{K}}^{(1)}, \tag{18}$$
$$T_{D_s\bar{K}} = \frac{1}{3}[T_{D\pi}^{(3/2)} + 2T_{D\pi}^{(1/2)}] = \frac{1}{2}[T_{DK}^{(1)} + T_{DK}^{(0)}], \tag{19}$$
$$T_{D_s\pi} + T_{D_s\eta} = \frac{1}{3}[2T_{D\pi}^{(3/2)} + T_{D\pi}^{(1/2)}] + T_{D\eta}. \tag{20}$$

These relations are very useful to cross-check the results.

4. Numerical results and discussions

There are several works on these scattering lengths in the literature. The S-wave DK scattering length was predicted to be 5 ± 1 GeV^{-1} in a unitarization model.[9] The S-wave scattering lengths $a_{B\pi}^{(1/2)} = 0.26(26)m_\pi^{-1}$, $a_{D\pi}^{(1/2)} = 0.29(4)m_\pi^{-1}$ were extracted from a lattice calculations of the scalar form factors in the semileptonic decays.[10] Recently, a lattice study on the scattering lengths of the light hadrons with the charmed mesons and charmonia were performed in full QCD.[11] Unfortunately their pion masses are still quite large. The scattering lengths of heavy mesons and Goldstone bosons were discussed to the next leading order in chiral perturbation theory recently.[12]

Our numerical calculation depends on four parameters: c_1, C_1, C_0 and κ^r. We extract c_1 from the mass splitting between heavy mesons within the same doublet. The obtained value is $c_1 = 0.47\ \text{GeV}^{-1}$. We use a recent lattice simulation[11] to estimate other parameters. The resulting values are $C_1 = -0.63\ \text{GeV}^{-1}$, $C_0 = 5.64\ \text{GeV}^{-1}$ and $\kappa^r = -0.33\ \text{GeV}^{-2}$ from $a_{D\pi}^{(3/2)}$, a_{D_sK} and $a_{D_s\pi}$. We present our results in Table 1 where several scattering lengths existing in the literature were given.

Table 1. Threshold T-matrices for the elastic scattering of pions and charmed pseudoscalar mesons in unit of fm with the scale $\lambda = 4\pi f_\pi$. For the results from a unitarized method, one may consult Ref. 12.

	$\mathcal{O}(p)$	$\mathcal{O}(p^2)$	$\mathcal{O}(p^3)$	Total	Scattering lengths	Other results
$T_{DK}^{(1)}$	0	9.4	$-1.5+5.6i$	$7.9+5.6i$	$0.25+0.17i$	
$T_{DK}^{(0)}$	15.3	-2.4	0.1	13.0	0.41	0.98 ± 0.20[9]
$T_{D\bar{K}}^{(1)}$	-7.6	-2.4	-1.2	-11.2	-0.35	
$T_{D\bar{K}}^{(0)}$	7.6	21.2	2.5	31.4	0.99	
T_{D_sK}	-7.6	-2.4	0.3	-9.7	-0.31 (input)	
$T_{D_s\bar{K}}$	7.6	-2.4	$-1.5+8.3i$	$3.8+8.3i$	$0.12+0.27i$	
$T_{D\pi}^{(3/2)}$	-3.2	-0.3	-0.8	-4.3	-0.16 (input)	
$T_{D\pi}^{(1/2)}$	6.5	-0.3	0.3	6.4	0.24	0.41 ± 0.06[10]
$T_{D_s\pi}$	0.0	1.1	-1.1	0.0	0.00 (input)	
$T_{D\eta}$	0.0	9.9	$1.2+5.5i$	$11.0+5.5i$	$0.34+0.17i$	
$T_{D_s\eta}$	0.0	-13.8	$0.5+11.1i$	$-13.3+11.1i$	$-0.41+0.35i$	

The heavy quark symmetry works better for bottom systems. The heavy quark flavor symmetry tells us that we may still use the above determined values for C_1, C_0 and κ^r to estimate the scattering lengths in this case. The calculated results do not differ significantly from the charm case.[8] Larger difference will appear after including the recoil corrections. We focus mainly on charmed systems in the following discussions.

From Table 1, we note that $a_{DK}^{(1)}$, $a_{DK}^{(0)}$, $a_{D\bar{K}}^{(0)}$, $a_{D\pi}^{(1/2)}$ and $a_{D\eta}$ are all positive, which indicates that the interactions for these channels are all attractive. The chiral expansion converges well in four channels: isoscalar DK, isovector $D\bar{K}$, D_sK and $D_s\bar{K}$. The convergence of $T_{DK}^{(1)}$, $T_{D\eta}$ and $T_{D_s\eta}$ will be manifest when the $\mathcal{O}(p^4)$ corrections are included.

For the isoscalar $D-K$ channel, the attraction is relatively strong. However, one cannot know whether the DK interaction is strong enough to form a bound state just through the scattering length. The exploration of the phase shifts for the elastic DK scattering is helpful to clarify this question. Ref. 13 presents such an interesting study in a chiral quark model. The phase shift does not exceed 40° and a DK molecular state seems to be impossible. Nevertheless, the attractive DK interaction may help to lower the mass of the near-threshold $(\bar{c}s)$ meson through the coupled channel effects.[5]

The result for the isoscalar $D - \bar{K}$ channel is very interesting. Since no quark pair annihilation occurs in this channel, the relatively large scattering length implies that the possibility to form a four-quark resonance state is not excluded. If the strong attraction is confirmed in future studies, this channel is certainly worthwhile further exploration. However, a critical problem of convergence exists in this channel. The contribution from the second order is very large. Future accurate determination of LECs is probably helpful to diminish this contribution.

The $D-\eta$ channel is also interesting. Twenty years ago, the formation of η-mesic nucleus[14] was proposed based on the observation that η is neutral and the strong force between η and N is attractive. Later the possibility of the η-hypernuclei was also discussed.[15] The bound state of a D meson and $^{208}P_b$ was studied in Ref. 16 while the DN bound state was explored in Ref. 17. Now, with the observation that both $D\eta$ and ηN are attractive, one may guess the formation of a $\eta - D-$heavy nucleus state is also possible. However, such a bound state would be very difficult to be detected experimentally.[18]

Contrary to the $D - \eta$ channel, the scattering length $a_{D_s\eta}$ is negative, which indicates the interaction is repulsive. In order to get a positive $a_{D_s\eta}$ and nearly vanishing $a_{D_s\pi} \sim 0$, one requires $C_1 > 1.4$ GeV^{-1} and $C_0 < 3.6$ GeV^{-1}.

Future improvements may be made for the present calculation. We have used the recent lattice results in full QCD to estimate the LECs where the lattice pion mass is large.[11] Due to the lack of data, the deduced LECs have large uncertainties. Since no $J^P = 0^+$ resonance close to the $D\pi$ threshold has been reported, this channel is ideal to determine the LECs. To test HMχPT in the scattering problem, the experimental determination of pion-heavy meson scattering lengths is strongly expected. On the other hand, further lattice simulations can provide more reliable inputs. We hope our chiral corrections may be useful when performing chiral extrapolations. One may also improve the analytical expressions by including $1/M$ corrections and extending the calculation to higher orders in future exploration. In addition, the effects of the nearby 0^+ resonances may affect the predictions.

One can consider them using the non-perturbative methods.

In short summary, we have performed a systematic calculation of the light pseudoscalar meson heavy meson scattering lengths to the third chiral order in a heavy meson chiral perturbation theory in the heavy quark limit. We found four channels converge well, four channels will manifest the convergence at higher orders and three do not converge. Since DK channel is attractive, the coupling of $c\bar{s}$ and DK continuum may lower the mass of $D_s(2317)$ to its physical mass through the coupled channel effects.

References

1. B. Aubert et al., BaBar Collaboration, Phys. Rev. Lett. **90**, 242001 (2003).
2. P. Krokovny et al., Belle Collaboration, Phys. Rev. Lett. **91**, 262002 (2003).
3. D. Besson et al., CLEO Collaboration, Phys. Rev. **D 68**, 032002 (2003).
4. E. van Beveren and G. Rupp, Phys. Rev. Lett. **91**, 012003 (2003); E. van Beveren and G. Rupp, Eur. Phys. J. **C 32**, 493 (2004).
5. Y.B. Dai, X.Q. Li, S.L. Zhu and Y.B. Zuo, Eur. Phys. J. **C 55**, 249 (2008), arXiv: `hep-ph/0610327`.
6. T.M. Yan, H.Y. Cheng, C.Y. Cheung, G.L. Lin, Y.C. Lin, H.L. Yu, Phys. Rev. **D 46**, 1148 (1992); M.B. Wise, Phys. Rev. **D 45**, R2188 (1992); P. Cho, Nucl. Phys. **B 396**, 183 (1993).
7. R. Casalbuoni, A. Deandrea, N.Di Bartolomeo, R. Gatto, F. Feruglio, G. Nardulli, Phys. Rept. **281**, 145 (1997).
8. Y.R. Liu, X. Liu, S.L. Zhu, Phys. Rev. D **79**, 094026 (2009), arXiv: 0904.1770 [hep-ph].
9. E. van Beveren, G. Rupp, AIP Conf. Proc. **687**, 86 (2003).
10. J.M. Flynn and J. Nieves, Phys. Rev. **D 75**, 074024 (2007).
11. L.M. Liu, H.W. Lin, K. Orginos, arXiv: 0810.5412 [hep-lat].
12. F.K. Guo, C. Hanhart, and U.-G. Meissner, arXiv: 0901.1597 [hep-ph].
13. Dan Zhang, Qiao-Yan Zhao, Qiu-Yang Zhang, Chin. Phys. Lett. 26, 091201 (2009), arXiv: 0905.1804 [nucl-th].
14. Q. Haider and L.C. Liu, Phys. Lett. **B 172**, 257 (1986); L.C. Liu and Q. Haider, Phys. Rev. **C 34**, 1845 (1986); arXiv: 0902.4248 [nucl-th].
15. V.V. Abaev and B.M.K. Nefkens, Phys. Rev. **C 53**, 385 (1996).
16. K.Tsushima, D.H. Lu, A.W. Thomas, K. Saito, and R.H. Landau, Phys. Rev. **C 59**, 2824 (1999).
17. S. Yasui and K. Sudoh, Phys. Rev. D 80, 034008 (2009).
18. K. Saito, K. Tsushima, A.W. Thomas, Prog. Part. Nucl. Phys. **58**, 1 (2007).

PHOTOPRODUCTION OF $\Lambda(1405)$ AT SPring-8/LEPS

M. NIIYAMA (FOR THE LEPS COLLABORATION)

The Institute of Physical and Chemical Research, Wako, Saitama

Photoproduction of the $\Lambda(1405)$ has been studied at SPring-8/LEPS in order to understand its production mechanisms and the internal structure. The ratios of $\Lambda(1405)$ production to $\Sigma^0(1385)$ production were measured in two photon energy ranges: near the production threshold ($1.5 < E_\gamma < 2.0$ GeV) and far from it ($2.0 < E_\gamma < 2.4$ GeV). The observed ratio decreases in the higher photon energy region, which may suggest different production mechanisms and form factors for these hyperon resonances. In addition to production cross sections, the measurement of the $KN\Lambda^*$ and $K^*N\Lambda^*$ coupling using the photon beam asymmetry of $\Lambda(1405)$ production is underway. The status of the analysis will be described.

Keywords: hyperon, photoproduction

1. Introduction

Recently, the photoproduction of Λ and Σ^0 hyperons off protons has been studied with high statistics data,[1–4] mainly motivated by missing resonances which could couple to the KY channels.[5] The authors of Ref. 5 also predicted nucleon resonances which couple to the KY^* channels, $K\Sigma(1385)$ and $K\Lambda(1405)$. In addition to searching for missing resonances, it is of interest to study the internal structure of hyperon resonances. The $\Sigma^0(1385)$ hyperon is firmly established as a q^3 baryon. On the contrary, the internal structure of the $\Lambda(1405)$ hyperon is ambiguous. In the quark model, $\Lambda(1405)$ is assigned as a p-wave q^3 baryon.[7] However, it is also widely discussed as a candidate for a meson-baryon molecular state[8,9] or a $q^4\bar{q}$ pentaquark baryon.[10] An overview of the study of the $\Lambda(1405)$ before the year 1998 can be found in Ref. 11. Recently, Jido *et al.* predicted the second pole of $\Lambda(1405)$ with higher mass and narrow width which strongly couple with $\bar{K}N$ channel using chiral Lagrangian and coupled unitary method.[12] The second pole of $\Lambda(1405)$ is important to understand the internal structure of the $\Lambda(1405)$, low energy $\bar{K}N$ interaction, and kaonic nucleus reported by FINUDA group.[13] Hyodo *et al.* pointed out that the second pole is enhanced in $\gamma p \to K^+(890)\Lambda(1405)$ reaction in which K^- exchange is dominant.[14] In order to investigate the second pole of $\Lambda(1405)$ and the structure of $\Lambda(1405)$ with the photon beam, the knowledge of $KN\Lambda^*$ and $K^*N\Lambda^*$ coupling is important. In $\gamma p \to K^+\Lambda(1405)$ reaction, both of K^- and K^{*-} exchange contribute, and the strength of each contribution can be investigated from the photon beam asymmetry.

The cross section of hyperon photoproduction depends on its production mechanism and the form factor of the hyperon. Photoproduction of $\Sigma^0(1385)$ has been measured by the CLAS Collaboration at TJNAF.[15] A theoretical calculation by Oh *et al.* using an effective Lagrangian was then compared with the data and the contributions from nucleon resonances were discussed.[16] Experimentally, the $\Lambda(1405)$ has been studied in meson-induced and proton-induced reactions so far. However, understanding of the photoproduction of $\Lambda(1405)$ is very limited because of the lack of experimental data. We measured photoproduction cross section of $\Lambda(1405)$ and $\Sigma^0(1385)$ using 1.5-2.4 GeV photons at SPring-8/LEPS. This is the first measurement of the cross section of $\Lambda(1405)$ photoproduction.

2. $\Lambda(1405)$ photoproduction at SPring-8/LEPS

In this experiment, forward going K^+'s from the $\gamma p \to K^+X$ reaction were detected in the LEPS spectrometer. In order to distinguish $\Lambda(1405)$ and $\Sigma^0(1385)$, a time projection chamber (TPC) was used together with the LEPS spectrometer to facilitate the detection of the decay products of these hyperon resonances. The main decay mode of $\Sigma^0(1385)$ is the $\Lambda\pi$ channel. On the other hand, $\Lambda(1405)$ is prohibited from decaying into a $\Lambda\pi$ pair due to the isospin conservation. Thus, $\Sigma^0(1385)$ production can be identified by detecting a Λ hyperon. Both of $\Lambda(1405)$ and $\Sigma^0(1385)$ decay into the $\Sigma\pi$ channel, but the contamination of $\Sigma^0(1385)$ can be estimated using the yield obtained from its $\Lambda\pi$ decay mode and known branching ratio of $\Lambda\pi$ and $\Sigma\pi$ decay of $\Sigma^0(1385)$. Fig. 1 shows missing mass spectra of $\gamma p \to K^+X$ reaction for events with $K^+\Sigma\pi$ final state. The experimental data are shown as closed circles. The data were fitted with spectra determined by MC simulation of $K^+\Lambda(1405)$, $K^+\Sigma^0(1385)$, $K^+\Lambda(1520)$, nonresonant $(K^+\Sigma\pi)$ and $K^{*0}\Sigma^+$ production. The solid histograms show fit results, where the strengths of $\Sigma^0(1385)$ production were obtained from its $\Lambda\pi$ decay mode, and those of the other production were obtained by fitting. The solid lines, open circles, dashed lines and dot-dashed line show spectra of $K^+\Lambda(1405)$, $K^+\Sigma^0(1385)$, nonresonant $(K^+\Sigma\pi)$ and $K^{*0}\Sigma^+$ production, respectively. One can see the contribution of $\Lambda(1405)$ is significant in lower photon energy region. On the other hand, in higher photon energy region, the contribution $\Lambda(1405)$ is unclear.[17]

The production ratios of $\Lambda(1405)$ to $\Sigma^0(1385)$ were obtained in two photon energy ranges, and the scattering angle of the K^+ in the center-of-mass frame, $\cos(\theta_K)$, was required to be $0.8 < \cos(\theta_K) < 1.0$. The production ratios of $\Lambda(1405)$ to $\Sigma^0(1385)$ were obtained as $\Lambda(1405)/\Sigma(1385) = 0.54$ and 0.084 for $1.5 < E_\gamma < 2.0$ GeV and $2.0 < E_\gamma < 2.4$ GeV, respectively. The absolute values of the production cross sections were obtained from high statistics liquid hydrogen data with the input of production ratios of $\Lambda(1405)$ / $\Sigma^0(1385)$ obtained above. The differential cross sections of production were found to be $d\sigma/d(\cos\theta) \sim 0.43$ μb, and 0.072 μb for $1.5 < E_\gamma < 2.0$ GeV and $2.0 < E_\gamma < 2.4$ GeV, respectively. Those of $\Sigma^0(1385)$ production were $d\sigma/d(\cos\theta) \sim 0.80$ μb and 0.87 μb for $1.5 < E_\gamma < 2.0$ GeV and $2.0 < E_\gamma < 2.4$ GeV, respectively.

The photon beam asymmetry is sensitive to the electric and magnetic coupling of exchanged particles in t channel reaction, and we can investigate K^-, K^{*-} exchange mechanisms and $KN\Lambda^*$, $K^*N\Lambda^*$ couplings in $\gamma p \to K^+\Lambda(1405)$ reaction. In order to measure the photon beam asymmetry, three times more data were obtained with a liquid hydrogen target. The photon beam asymmetries were obtained for ground state Λ and Σ^0 production reactions which were consistent with our previous measurement.[4] The analysis for photon beam asymmetries for $\Lambda(1405)$ and $\Sigma^0(1385)$ production is underway, and the results will be shown in near future.

3. Summary and prospects

The production cross sections of $\Sigma^0(1385)$ are almost constant with respect to the photon energy, and are consistent with a theoretical prediction of ~ 0.8 μb by Oh *et al.*[16] in both photon energy ranges. In contrast, the cross section of $\Lambda(1405)$ photoproduction is slightly different with theoretical calculations. Nacher *et al.* predicted the cross section of ~ 0.1 μb at $E_\gamma = 1.7$ GeV using chiral Lagrangian and coupled unitary method. Nam *et al* obtained the cross section as $0.05 \sim 0.1$ μb at $1.6 < E_\gamma < 2.0$ GeV using effective Lagrangian. The production cross section of $\Lambda(1405)$ obtained by this work is ~ 0.4 μb in the lower photon energy region and is larger than these theoretical values. Recently, D. Jido and Y. Kanada-En'yo proposed a nucleon like molecular state consists of K^+K^-p which decays into $K^+\Lambda(1405)$.[18] Such a nucleon resonance can contribute to raise the production cross section. However, in view of our limited statistics, further data are needed for more quantitative discussions. We already took data with three times more statistics to study photoproduction of Λ(1405) at SPring-8/LEPS in the year 2008. A precise measurement of the production cross section and photon beam asymmetry will be done in near future.

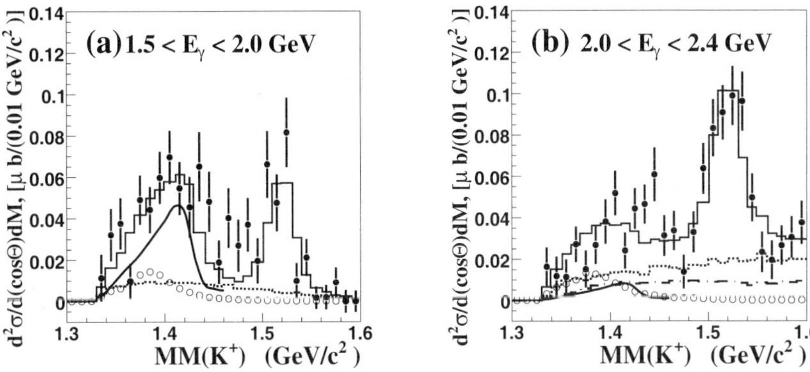

Fig. 1. Missing mass of the $\gamma p \to K^+X$ reaction in data set (II) in two photon energy ranges: (a) $1.5 < E_\gamma < 2.0$ GeV and (b) $2.0 < E_\gamma < 2.4$ GeV. The experimental data are shown as closed circles. The data were fitted with spectra determined by MC simulation of $K^+\Lambda(1405)$, $K^+\Sigma^0(1385)$, $K^+\Lambda(1520)$, nonresonant $(K^+\Sigma\pi)$ and $K^{*0}\Sigma^+$ production. The solid histograms show fit results. The solid lines, open circles, dashed lines and dot-dashed line show spectra of $K^+\Lambda(1405)$, $K^+\Sigma^0(1385)$, nonresonant $(K^+\Sigma\pi)$ and $K^{*0}\Sigma^+$ production, respectively.

References

1. M.Q. Tran *et al.*, Phys. Lett. B445, 20 (1998).
2. J.W.C. McNabb *et al.*, Phys. Rev. C69, 042201 (2004).
3. K.-H. Glander *et al.*, Eur. Phys. J. A19, 251 (2004).
4. M. Sumihama *et al.*, Phys. Rev. C73, 035214 (2006).
5. S. Capstick and W. Roberts, Phys. Rev. D58, 074011 (1998).
6. W.-M. Yao *et al.* [Particle Data Group 2006], J. Phys. G 33, 1 (2006).
7. N. Isgur and G. Karl, Phys. Rev. D18, 4187 (1978).
8. N. Kaiser, P.B. Siegel and W. Weise, Nucl. Phys. A594, 325 (1995).
9. J.C. Nacher, E. Oset, H. Toki and A. Ramos, Phys. Lett. B455, 55 (1999).
10. T. Inoue, Nucl. Phys. A790, 530 (2007).
11. R.H. Dalitz, Euro. Phys. J C15, 748 (2000).
12. D. Jido, J. A. Oller, E. Oset, A. Ramos, and U. -G, Meissner, Nucl. Phys. A755, 669 (2005).
13. M. Agnello et al., Phys. Rev. Lett. 94 (2005).
14. T. Hyodo, A. Hosaka, M.J. Vicente Vacas, E. Oset, Phys. Lett. B593, 75 (2004).

15. L. Guo and D.P. Weygand, hep-ex/0601010, in NSTAR05 Proceedings.
16. Y. Oh, C.M. Ko and K. Nakayama, Phys. Rev. C 77, 045204 (2008).
17. M. Niiyama *et al.*, Phys. Rev. C78, 035202 (2008).
18. D. Jido and Y.K. En'yo, Phys. Rev. C78, 035203 (2008).

THE INVESTIGATION OF STRANGENESS PHOTOPRODUCTION IN THE THRESHOLD REGION AT LNS-TOHOKU

THE NKS/NKS2 COLLABORATION

M. KANETA*, B. BECKFORD, M. EJIMA, T. FUJII, Y. FUJII, T. FUJIBAYASHI, T. GOGAMI, K. FUTATSUKAWA, O. HASHIMOTO, K. HOSOMI, K. HIROSE†, A. IGUCHI, S. KAMEOKA, H. KANDA, H. KATO, D. KAWAMA, T. KAWASAKI, C. KIMURA, S. KIYOKAWA, T. KOIKE, T. KON, Y. MA, K. MAEDA, N. MARUYAMA, A. MATSUMURA, Y. MIYAGI, Y. MIURA, K. MIWA, S.N. NAKAMURA, H. NOMURA, A. OKUYAMA, A. OHTANI, M. SATO, A. SHICHIJO, K. SHIROTORI, T. TAKAHASHI‡, H. TAMURA, N. TANIYA, H. TSUBOTA, K. TSUKADA§, N. TERADA, M. UKAI¶, D. UCHIDA, T. WATANABE, T.O. YAMAMOTO, H. YAMAUCHI

Department of Physics, Tohoku University,
Sendai, Miyagi 980-8578, Japan
**E-mail: kaneta@lambda.phys.tohoku.ac.jp*

T. ISHIKAWA, T. KINOSHITA, H. MIYAHARA, T. NAKABAYASHI, H. SHIMIZU, K. SUZUKI, T. TAMAE, T. TERASAWA, H. YAMAZAKI

Laboratory of Nuclear Science, Tohoku University,
Sendai, Miyagi 980-0826, Japan

Y.C. HAN, T.S. WANG

School of Nuclear Science and Technology, Lanzou University,
Lanzhou 730000, China

A. SASAKI

Department of Electrical and Electronic Engineering, Akita University,
Akita, Akita 010-8502, Japan

†Now at LNS, Tohoku University, Sendai, Miyagi 980-0826, Japan
‡Now at Inst. of Particle, Nuclear Studies, KEK, Tsukuba, Ibaraki 305-0801, Japan
§Now at RIKEN, Wako, Saitama 351-0198, Japan
¶Now at Gifu University, Gifu, Gifu 501-1193, Japan

O. KONNO

Department of Electrical Engineering, Ichinoseki National College of Technology, Ichinoseki, Iwate 021-8511, Japan

P. BYDŽOVSKÝ, M. SOTONA

Nuclear Physics Institute, CS-25068 Řež, Czech Republic

The strangeness photoproduction process has been intensively studied based on the high-quality data of the charged kaon channel, $\gamma + p \rightarrow K^+ + \Lambda(\Sigma^0)$. On the other hand, there was no reliable data for the neutral kaon channel $\gamma+n \rightarrow K^0+\Lambda(\Sigma^0)$. The theoretical investigations suffer seriously from the lack of the data. A Substantial effort has been made to measure the $\gamma+n \rightarrow K^0+\Lambda$ process in the $K^0_S \rightarrow \pi^+\pi^-$ decay channel, using a liquid deuterium target and a tagged photon beam (E_γ = 0.8-1.1 GeV) in the threshold region at Laboratory of Nuclear Science (LNS), Tohoku University.

We have taken exploratory data quite successfully with use of Neutral Kaon Spectrometer (NKS) at LNS-Tohoku in 2003 and 2004. The data is compared theoretical models and it indicates a hint that the K^0 differential cross section has a backward peak in the energy region. The second generation of the experiment, named NKS2, is designed to extend the NKS experiment by considerably upgrading the original neutral kaon spectrometer, fully replacing the spectrometer magnet, tracking detectors and all the trigger counters. The new spectrometer NKS2 has significantly larger acceptance for neutral kaons compared with NKS, particularly covering forward angles and much better invariant mass resolution. The estimated acceptance of NKS2 is three (ten) times larger for $K^0_S(\Lambda)$ than that of NKS. The spectrometer is newly constructed and installed at LNS, Tohoku University in 2005. The deuterium target data was taken with the tagged photon beam in 2006-2007.

We will report recent results of NKS2 in this paper. Additionally, a status of the upgrade project that gives us larger acceptance and capability of $K^0+\Lambda$ coincidence measurement will be repsented.

Keywords: Strangeness; Photoproduction

1. Physics Motivation

The strangeness photoproduction is mainly studied by K^+ channel. Table 1 shows some collaboration names and facilities for the study of strangeness production by the electro-magnetic interaction. On the other hand, the K^0 production study was rarely carried out because of experimental difficulty. There were almost no data for the $\gamma+n$ channels though many high statistics data are available for $\gamma + p$ channels at a time the study began.

Many theoretical studies are phenomenological and are based on the K^+ data of $\gamma + p$ reactions. Thus, it is natural that available theoretical

Table 1. Experiments and facilities for strangeness production study by electro-magnetic interaction.

Facility / Detector(Collaboration)	Country	Beam species	Energy [GeV]
ELSA / SAPHIR[3]	Germany	γ	~ 2.6
JLab / CLAS[4]	U.S.A.	γ, e	0.5 - 6.0
JLab / HMS,SOS	U.S.A.	γ, e	0.5 - 6.0
SPring-8 / LEPS[5]	Japan	$\vec{\gamma}$	1.5 - 2.4
LNS / NKS	Japan	γ	0.5 - 1.1

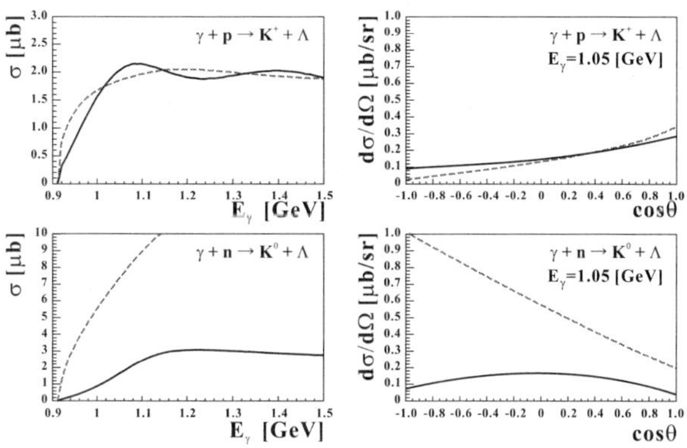

Fig. 1. Model calculation of total cross section (left) and differential cross section (right). Top figures are for the $K^+ + \Lambda$ photoproduction. Bottom figures are for the neutral channel. Solid lines are from Kaon-MAID[1] model. Dash lines are from Saclay-Lyon A (SLA)[2] model.

models predict the excitation functions quite similarly in the $\gamma + p$ channel. Two typical calculations based on the isobar models, Kaon-MAID[1] and Saclay-Lyon A[2] (SLA), are compared in Fig. 1. The figures are total cross section as a function of photon energy and differential cross section as a function of cosine of the Kaon agnle with respect to the beam axis for the $\gamma + p \to K^+ + \Lambda$ and $\gamma + n \to K^0 + \Lambda$ channels. There is large discrepancies in the $\gamma + n$ channel, and it is strongly suggested that measurements of the $\gamma + n \to K^0 + \Lambda$ channels by detecting neutral kaons will provide key information in the investigation on the strangeness production by the electromagnetic interaction.

2. Experimental Setup

The experiment is carried out at the second experimental hall of LNS-Tohoku. The electron beam is accelerated up to 150 MeV in LINAC and

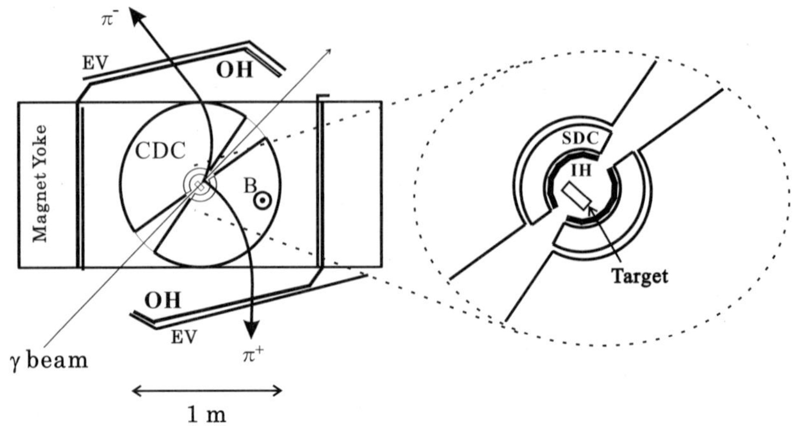

Fig. 2. The schematic view of NKS. This is a slice of the spectrometer on the beam plane. See Ref. 7 about the NKS detectors in detail.

injected to Stretcher-Booster (STB) ring. The maximum electron energy is 1.2 GeV in STB. Bremsstrahlung photons were produced by the internal radiator and tagged by the scintillation counters.[6] The energy range of the tagged photon is 0.9 to 1.1 GeV with 1.2 GeV electron. The photon energy is suitable to study the kaon photo-production near the threshold (915 MeV for $\gamma + n \to K^0 + \Lambda$). The detail information of the tagging system can be found elsewhere.[6]

The NKS detector is shown in Fig. 2. Liquid deuterium target cell is placed at the center of the dipole magnet and surrounded by detectors: Inner Hodoscope (IH), Straw Drift Chamber (SDC), Cylindrical Drift Chamber (CDC), and Outer Hodoscope (OH). Electron Veto (EV) counters are put on the beam plane to reject the electron on the trigger level. A typical time-of-flight (TOF) resolution is about 510 ps and allows the separation of pion from proton. NKS CDC lacks an acceptance in forward region to avoid high-rate signals from electron photon conversion. This resulted in a smaller acceptance and limited coverage of all kinematic region.

NKS2 is the second generation experiment for the strangeness photoproduction. A schematic view of NKS2 is shown in Fig. 3. The design concept is as follows. CDC covers forward region to provide a larger acceptance and kinematic region than NKS. Stereo wire implemented in CDC to increase momentum resolution in vertical direction. The dipole magnet is larger than NKS, resulting more precise momentum resolution due to longer trajectory of charged particle. Previously, the scintillation light was led by fiber optics

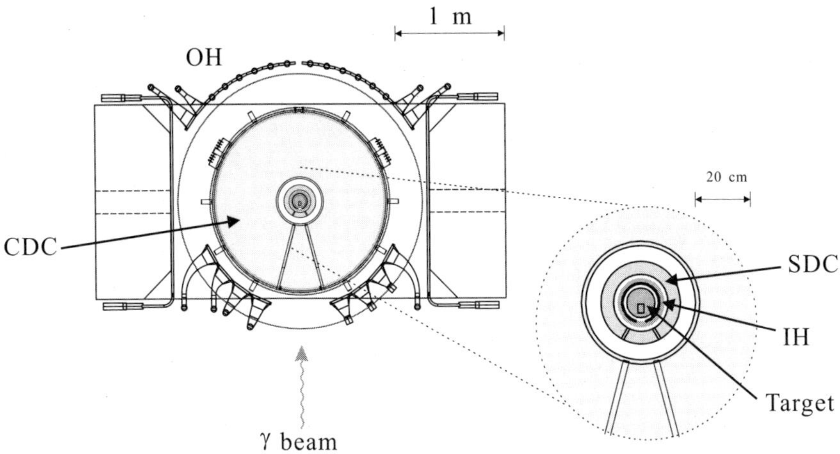

Fig. 3. The schematic view of NKS2. This is a slice of the spectrometer on the beam plane.

to the PMT's at NKS IH and a part of OH. A higher degree of $\pi/K/p$ separation using TOF and momentum was desired. Therefore, a fine-mesh dynode PMT for IH was used to work in the magnetic field (0.42 T). NKS OH scintillator placed in the magnet are connected by fiber optics. It was cause of large timing resolution, and then the scintillators place in horizontal and connected to PMT's by normal acrylic light guide in NKS2.

Fine-mesh PMT's are also incorporated for a part of OH. The TOF resolution of NKS2 was determined to be less than 350 ps. A 3 σ separation of π and p is attainable until 0.7 GeV/c.

3. Results from NKS

The data taking of NKS is done in 2003-2004 and the total number of photons is 4.6×10^{12}. The results from NKS is published in Ref. 7. Please see the reference for the analysis method in detail.

The data is compared with two theoretical models, Kaon-MAID and SLA, and one phenomenological parametrization PH1 (Legendre polynomials)[7] in Fig. 4. A parameter of an opposite sign is used in PH2, giving the inverse angular distribution in the CM system.

Only K^0_S were tagged in the γ+d reaction. There is a contribution of $K^0\Sigma$ production but it is negligible in E_γ=0.9-1.0 GeV due to the different threshold energy. The model is for the elementary reaction and thus we add the effect of Fermi motion of neutron to the theoretical calculation. Also

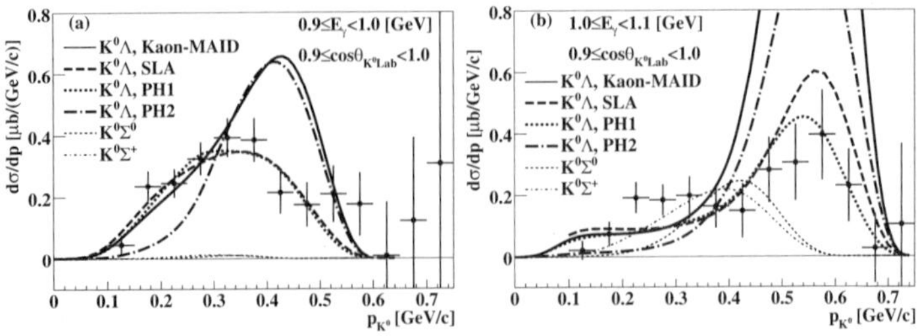

Fig. 4. The differential cross section of K^0 as a function of the kaon momentum. Those figures are from Ref. 7. Left (right) figure is 0.9 to 1.0 (1.0 to 1.1) GeV of photon energy. Only statistical error is shown in the plots. The data is compared with some models. The $K^0\Sigma^{0(+)}$ lines are from Kaon-MAID models.

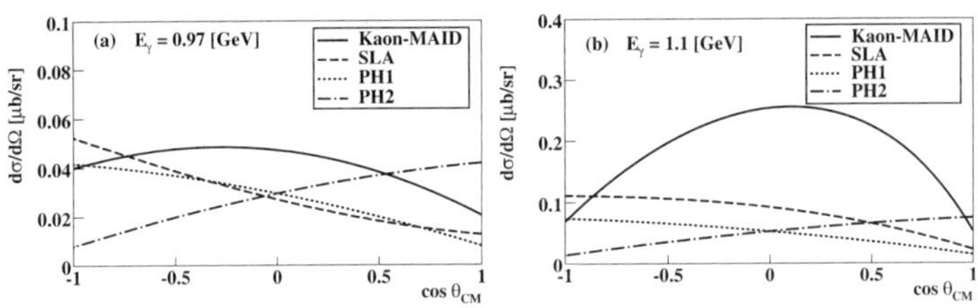

Fig. 5. The angle dependence of the K^0 cross section calculated by models. Those figures are from Ref. 7.

a non-relativistic Bonn OBEPQ deuteron wave function from Ref. 8 was included.

The data prefer SLA model rather than Kaon-MAID in both $E_\gamma = 0.9$-1.0 and 1.0-1.1 GeV region. In Fig. 5, differential cross-section is shown as a function of the kaon angle in CM system. SLA and PH1 show backward peak. NKS data shows in the region of $0.9 < \cos\theta_{K^0Lab} < 1.0$. To discuss the angular dependence, it is necessary to have large acceptance and is the reason why we started the second generation experiment NKS2.

4. Recent Results from NKS2

The data taking of NKS2 carried out in 2006-2007 using liquid deuterium target. Total number of photons is 3.0×10^{12}.

The same analysis method with NKS is applied to obtain the K^0 cross section. Preliminary results of K^0 is shown in Fig. 6. The data is separated in 4 bin in the K^0 momentum region (0.0 to 1.0 GeV/c). The results are from about 30% of total data.

The results from NKS is shown in the same plot but it is only in $\cos\theta_{K^0 Lab}$ = 0.9 - 1.0. The dash line is a prediction of SLA model with the best fit parameter of NKS data.

There is a contribution from $K^0\Sigma$ production as you see in Fig. 4, but it is small in E_γ=0.9-1.0 GeV. The NKS2 acceptance is expanded in a region of $\cos\theta_{K^0 Lab}$ <0.9. There is capability to discuss an angular dependence of K^0 cross section directly.

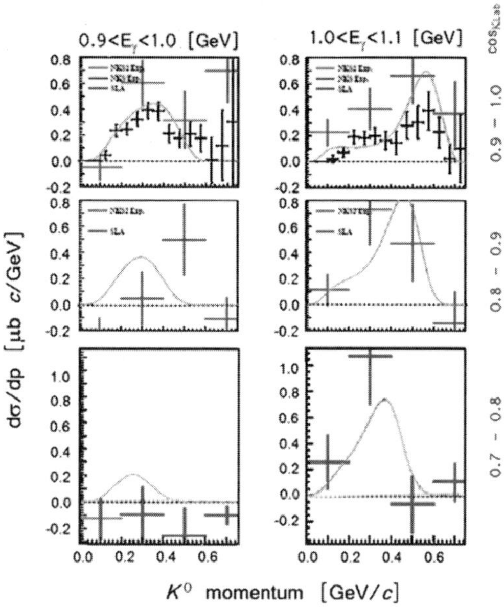

Fig. 6. K^0 cross-section of photo-production on deuterium target from NKS2. The number of photons used in the figure is about 30% of total data recorded.

Figure 7 shows preliminary Λ differential cross-section as a function of momentum for three angle bins. The number of events are the same with K^0 analysis. The data includes two reactions: $\gamma+n \rightarrow K^0+\Lambda$ and $\gamma+p \rightarrow K^+ + \Lambda$.

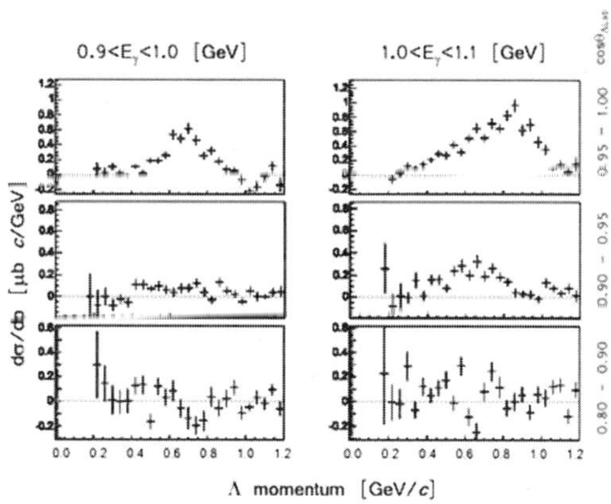

Fig. 7. Λ cross-section of photo-production on deuterium target from NKS2. The number of photons used in the figure is about 30% of total data recorded.

The large NKS2 acceptance has a capability of $K^0+\Lambda$ coincidence measurement that provides us information of reaction plane. A complete system of the kinematics can be obtained and then the data to a model without Fermi motion correction can be compared. Additionally, we will have a reaction plane information from $K+\Lambda$ event. That information give us a capability of Λ polarization information.

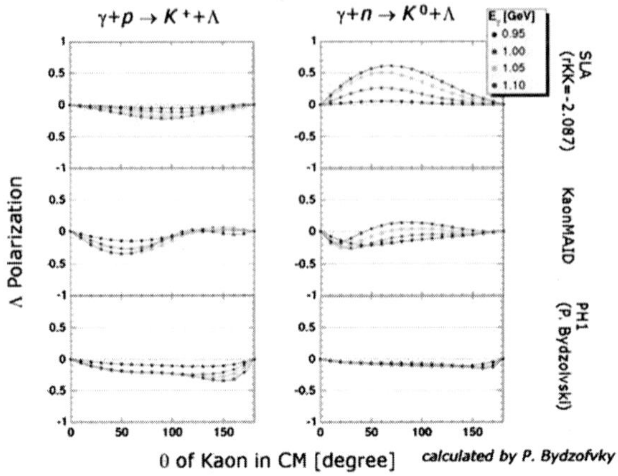

Fig. 8. Model predictions of Λ self polarization as a function of Kaon angle with respect to the beam axis in CM system. The calcuration is for E_γ = 0.95, 1.00, 1.05, and 1.10 GeV. The models used in the prediction are SLA, KaonMAID and PH1 (a phenomenological model by P. Bydzolvski).

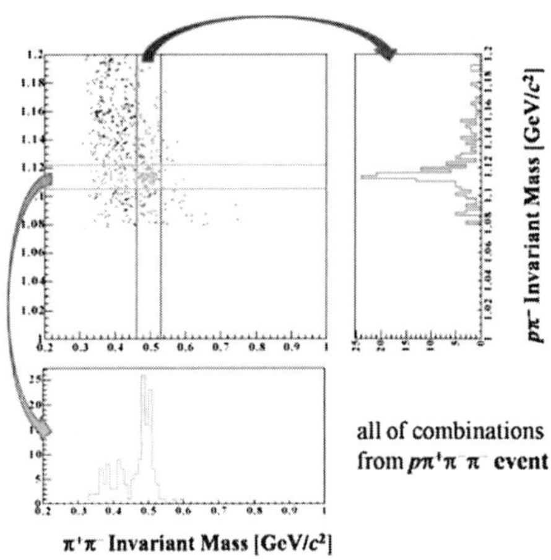

Fig. 9. A scatter plot of $\pi^+\pi^-$ invariant mass and $p\pi^-$ invariant mass from $p\pi^+\pi-\pi^-$ track event. The particle idefication of pion/proton is done by Time-Of-Flight and memntum. We selected $p\pi^+\pi-\pi^-$ events from 4 track event. The all of data recorded is used for the analysis.

Figure 8 shows model predictions of self polarization of Λ. The polarization shows different dependeces in K^+ and K^0 channels. The measurement will open a new window of strangeness photoproduction study.

The scatter plot in Fig. 9 is a correlation of two invariant masses. The horizontal axis is the invariant mass of $\pi^+\pi^-$ pair and the vertical one is from $p\pi^-$ pair. The bands around K^0 and Λ mass are also shown in the scatter plot. After selected Λ (K^0) band, the region is plojected to $\pi^+\pi^-$ ($p\pi^-$) invariant mass axis. As the results, we can see a clear peak of K^0 (Λ). The number of $K^0+\Lambda$ events are about 90 from all of data recorded. It is clear that we need to expand the acceptance of NKS2 to have large amount of statistics for the $K^0 + \Lambda$ coincidence envet.

5. Detector Upgrade Project

A larger acceptance for the $K + \Lambda$ coincidence measurement was needed for three (or four) track analysis. In the NKS2 tracking, it is required that all of charged particle should pass through CDC because it is only the detector of 3D tracking Therefore, we need to have a 3D tracker inside of CDC to increase the acceptance.

It is estimated a gain of K^0 and Λ yield with the new 3D tracker. The gain of the coincidence measurement is about 8 times of the NKS2, assuming four tracks can be reconstructed by the new detector independently. We decided to adopt a drift chamber as new inner detector (named Vertex Drift Chamber (VDC)) from a viewpoint of material thickness.

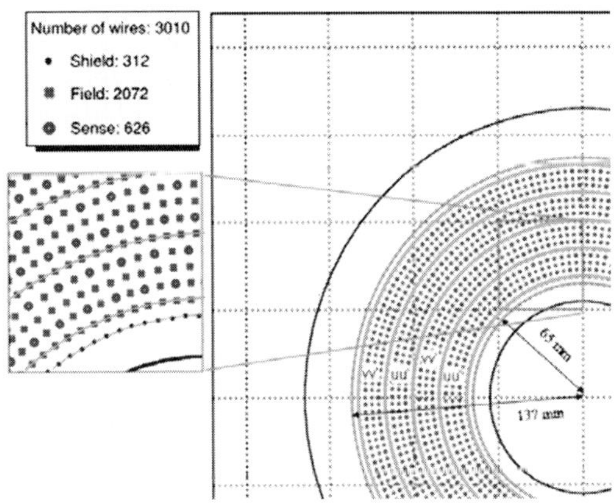

Fig. 10. A Schematic view of the vertex drift chamber.

VDC has 8 layers and its cell shape is trapezoid (the ratio of two directions is less than 10%). Only stereo wire is used and no axial wire due to space problem. Figure 10 shows wire positions at the end-cap of VDC. Every two layers, the stereo angle becomes opposite direction. The stereo angle is increasing with the radius (6 to 11 degree).

VDC commissioning and trigger study runs are held in 2008 and 2009. An event display from the test is shown in Fig. 11. Chamber cells that have a hit is shown on the beam plane. A zigzag hit pattern is observed on the event display and it means that the two tracks tilted to the beam plane, and is a candidate of a hadronic event.

6. Summary

We report the recent results of the neutral kaon photoproduction by NKS and NKS2 at LNS, Tohoku. The first measurement of neutral kaon via $\gamma + n$

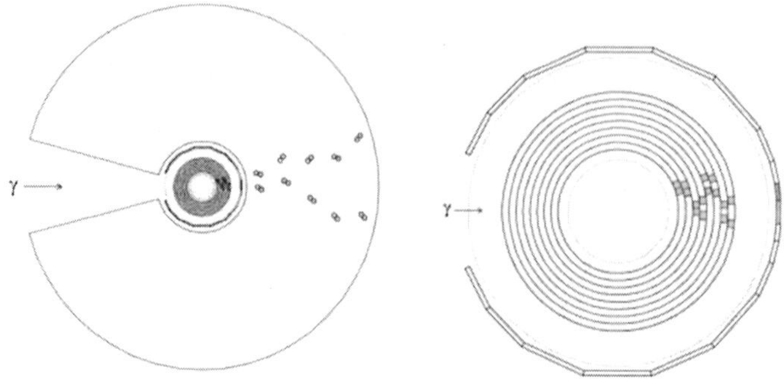

Fig. 11. An event display from a beam test. Left figure shows VDC, IH and CDC and right one is zoomed in VDC and IH. Valid hits are shown by solid rectangle on VDC and IH. They are shown by open circles on CDC.

reaction near the threshold by NKS. NKS2 covers larger acceptance over all kinematic regions and has a capability to discuss angular dependence of cross-section. The upgrade project of NKS2 is ongoing. We will have the solid angle expanded and about 8 times acceptance of NKS2 for the $K + \Lambda$ coincidence measurement. The performance is tested by commissioning run.

Acknowledgments

This work is supported by Grant-in-Aid for Scientific Research from Japan Society for the Promotion of Science (JSPS) and The Ministry of Education, Culture, Sports, Science and Technology(MEXT), Nos. 09304028, 12002001, 14740150, and 16GS0201 and by the Grant Agency of the Czech Republic under Grant 202/05/2142 and 202/08/0984.

References

1. F. X. Lee, T. Mart, C. Bennhold and L. E. Wright, *Nucl. Phys.* **A695**, 237 (2001).
2. T. Mizutani, C. Fayard, G. H. Lamot and B. Saghai, *Phys. Rev.* **C58**, 75 (1998).
3. W. J. Schwille *et al.*, *Nucl. Instrum. Meth.* **A344**, 470 (1994).
4. B. A. Mecking *et al.*, *Nucl. Instrum. Meth.* **A503**, 513 (2003).
5. T. Nakano *et al.*, *Nucl. Phys.* **A684**, 71 (2001).
6. H. Yamazaki *et al.*, *Nucl. Instrum. Meth.* **A536**, 70 (2005).
7. K. Tsukada *et al.*, *Phys. Rev.* **C78**, p. 014001 (2008).
8. R. Machleidt, K. Holinde and C. Elster, *Phys. Rept.* **149**, 1 (1987).

COMPLETE POLARIZATION EXPERIMENTS WITH THE POLARIZED HD TARGET AT SPring-8

T. OHTA[1*], S. BOUCHIGNY[2], J.P. DIDELEZ[2], M. FUJIWARA[1], K. FUKUDA[1], H. KOHRI[1], T. HOTTA[1] T. KUNIMATSU[1] C. MORISAKI[1], G. ROUILLE[2], M. TANAKA[3], K. UEDA[1], M. URAKI[1], M. UTSURO[1], S.Y. WANG[4,5] and M. YOSOI[1].

[1]*Research Center for Nuclear Physics (RCNP), Ibaraki, Osaka 567-0047, Japan*
[2]*IN2P3, Institute de Physique Nucleaire, F-91406 Orsay, France*
[3]*Kobe Tokiwa University, Kobe 654-0838, Japan*
[4]*Institute of Physics, Academia Sinica, Taipei, Taiwan 11529, Republic of China*
[5]*Department of Physics, National Kaohsiung Normal University, Kaohsiung 824, Taiwan*
**E-mail: takeshi@rcnp.osaka-u.ac.jp*

We plan to have hadron photoproduction experiments by using polarized photon beams at the laser-electron photon facility (LEPS) at SPring-8. We develope a HD (Hydrogen-Deuteride) target as a polarized target for future experiments using both the polarized photon beams and the polarized target since 2005.[1] The HD polarized target will be prodeced at RCNP (Osaka University) and installed in the LEPS facility. Such a double polarization experiment is expected to give important information to investigate the nucleon hidden structure and hadron photoproduction dynamics. We report on the present status of the developement of the polarzed HD target at RCNP.

Keywords: strangeness, photoproduction, polarized target

1. Physics motivation

1.1. *Strangeness content in the proton and neutron*

It is generally accepted that the low-energy properties of nucleon are well described in terms of three constituent u and d quarks. Therefore, recent experimental results are very surprising. Expriments from the lepton deep inelastic scattering indicate that there may be non-negligible strange quark content in the nucleon and that the strange quarks give 10-20% contributions to the nucleon spin.[2,3] After these results, there are many experiments[4–9] were perfomed, but some of them show the different conclusions

(~0%). This problem is actively discussed now. This discrepancy should be solved by providing new experimental information on the $s\bar{s}$-quark content of the nucleon.

1.2. *ϕ-meson photoproduction and double polarization asymmetry*

This $\gamma p \rightarrow \phi p$ reaction is dominated by the diffractive production within the vector-meson-dominance model (VDM) through Pomeron exchange as shown in Fig. 1(a). Conventional meson exchanges such as one-pion exchange (OPE) as shown in Fig. 1(b) are also contributed. In the t channel, however OPE are strongly suppressed by the OZI rule. If the proton has the $s\bar{s}$-quark content, the $s\bar{s}$ knockout and uud knockout processes may occur in the ϕ-meson photoproduction as shown in Fig. 1(c, d). In case of the ϕ-meson photoproduction on the neutron, the same processes are possible, while the sign of the interference term of the π-exchange and η-exchange in opposite to that for the proton.

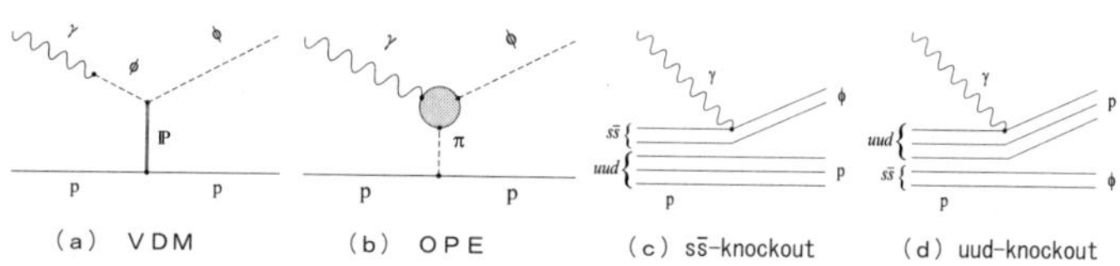

Fig. 1. Feynman diagram of ϕ-meson photoproduction

Titov *et al*[10] pointed out the double polarization measurement becomes very sensitive to the $s\bar{s}$ contents in the proton. They found that the study of cross section in the small ϕ meson scattering angle region is not sensitive to the $s\bar{s}$-quark content in the nucleon. The cross section of VDM is ten times lager than that of the $s\bar{s}$ knockout, as shown in Fig. 2, where the strangeness admixture is assumed to be 1%. But the beam-target asymmetry (C_{zz}^{BT}) for the $s\bar{s}$ direct knockout ϕ meson photoproduction (Fig. 2(c)) is very sensitive to the $s\bar{s}$-quark content in the nucleon (Fig. 3).[10]

The interference between the VDM model amplitude and the $s\bar{s}$ knockout amplitude gives a distinct contribution to the asymmetry even at the small ϕ meson angles. The asymmetry C_{zz}^{BT} is defined as

$$C_{zz}^{BT} = \frac{\mathrm{d}\sigma(\overset{\rightarrow}{\rightarrow}) - \mathrm{d}\sigma(\overset{\leftarrow}{\rightarrow})}{\mathrm{d}\sigma(\overset{\rightarrow}{\rightarrow}) + \mathrm{d}\sigma(\overset{\leftarrow}{\rightarrow})} = \frac{\mathrm{d}\sigma(\frac{3}{2}) - \mathrm{d}\sigma(\frac{1}{2})}{\mathrm{d}\sigma(\frac{3}{2}) + \mathrm{d}\sigma(\frac{1}{2})} \tag{1}$$

where dσ represents $d\sigma/dt$, and $\frac{1}{2}$ and $\frac{3}{2}$ denote the sum of the initial proton and photon helicities. The polarized target enables the measurement of double polarization asymmetries in ϕ meson photoproduction to study the strangeness content in the proton and neutron.

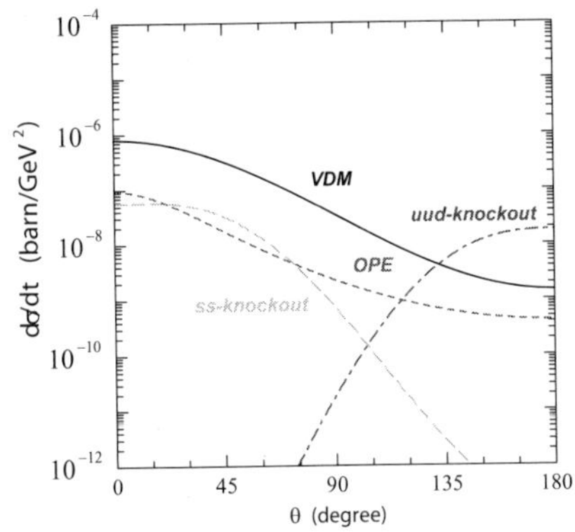

Fig. 2. The unpolarized ϕ-meson photoproduction cross section $d\sigma/dt$ at E_γ=2.0 GeV. The solid, dotted, dashed, and dot-dashed lines give the cross sections of VDM, OPE, $s\bar{s}$ knockout, and uud knockout with the strangeness admixture of 1%.[10]

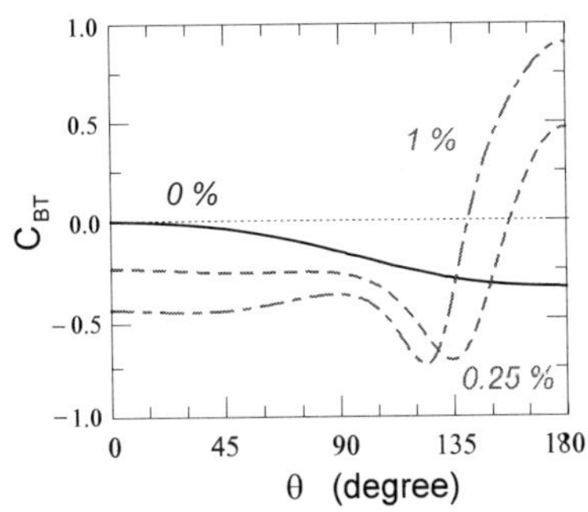

Fig. 3. The double polarization asymmetry C_{zz}^{BT} for the ϕ-meson photoproduction at Eγ=2.0 GeV, strangeness quark content is assumed to be 0% (solid), 0.25% (dashed), and 1% (dot-dashed).[10]

2. Characteristics of the polarized HD target

The HD is a hetero nuclear molecule which is consists of hydrogen and deuterium. Both the proton with spin 1/2 and the deuteron with spin 1 can be polarized and independently reversible.[11] In order to achieve high polarizations of proton and deuteron, we employ the static method using "brute force" at low temperature ($\sim$ 10 mK) and high magnetic field (17 Tesla). The relaxation time of about one year is achievable by keeping the HD target under the experimental condition (300 mK and 1 Tesla). The HD is an idealistic polarized targets for the LEPS experiments, although thin aluminum wires (at most 20% in weight) is included in the target cell to insure the cooling efficiency and it takes a few months for freezing polarization.

From the Boltzmann law, the polarization degree is only related to magnetic field (B) and temperature (T) as the following equations.

$$P_H = \tanh\left(\frac{\mu_p B}{kT}\right) \quad (for\ proton) \tag{2}$$

$$P_D = \frac{4\tanh\left(\frac{\mu_d B}{2kT}\right)}{3+\tanh^2\left(\frac{\mu_d B}{2kT}\right)} \quad (for\ deuteron) \tag{3}$$

Where the magnetic moments of proton and deuteron are $\mu_p = 1.40917 \times 10^{-26}[J/T]$, $\mu_d = 4.33059 \times 10^{-27}[J/T]$, and k is the Boltzmann constant. From this equation, the proton polarization is about 85%, and the deuteron polarization is about 24% at B=17 T and T=14 mK, as shown in Fig. 4.

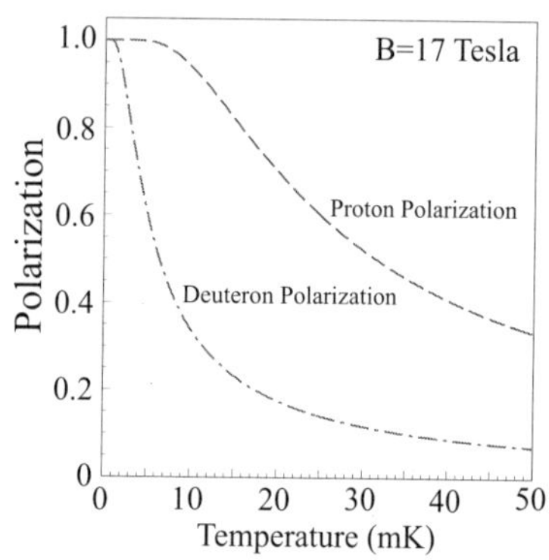

Fig. 4. Polarizations for proton and deuteron as a function of temperature at 17 T. Under the condition of 14 mK and 17 T, These polarization reach 85% for the proton (dashed) and 24% for the deuteron (dot-dashed), respectively.

Although the polarization of the deuteron reaches 24% at 14 mK and 17 T, the polarization of 60% for the deuteron can be obtained by transferring the proton polarization to the deuteron using the dynamics nuclear polarization method.

3. The Production of the polarized HD target

We have tried to proceduce a polarized HD target and to measure the polarization of the HD at RCNP for the first time in the beginning of 2009. Polarization was measured using the NMR method with continuous RF wave by sweeping the magnetic field. The polarization desired from the NMR signal was 41.4±3.1%, which is about a half of the caluculated value at 14 mK and 17 T on the basis of the Boltzmann law. The relaxation time of HD was estimated 106±16 days.

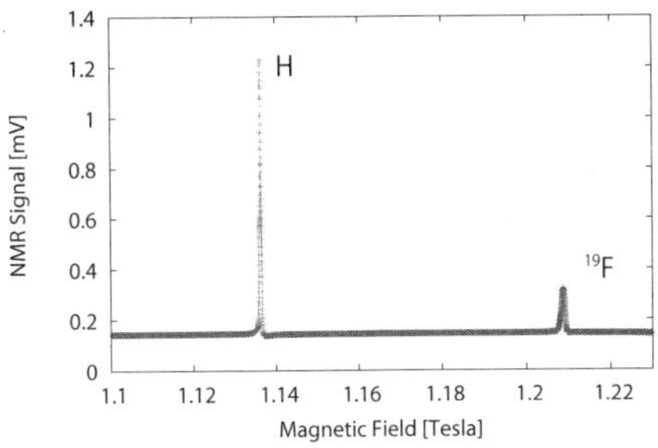

Fig. 5. NMR signal observed for H and F. The left side signal is for hydrogen. The right side signal is for ^{19}F included in cell.

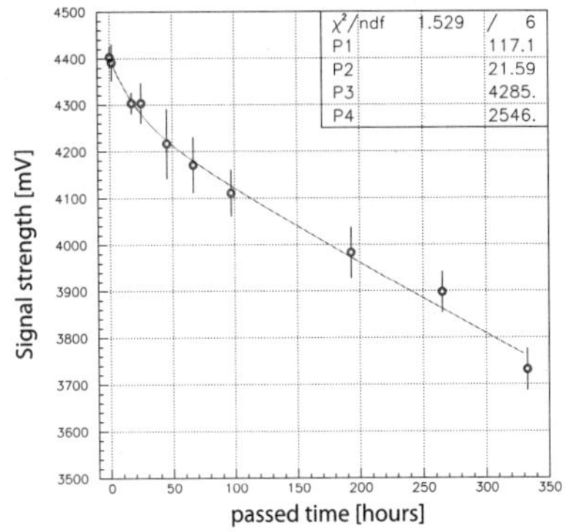

Fig. 6. The relaxation time was estimated 106±16 days by fitting to the life time equation of polarization. Signal strength was decreasing exponentially for two weeks at 1 T and at 300 mK that is the same as the experimental condition.

4. Future plan

During these 5 years, we have carried out a variety of tests for the cryogenic-devices and for the developed the NMR system. We proceed the test for transportation of the polarized HD target from RCNP to SPring-8. We plan to make the first of the experiment with the polarized HD target in 2010.

References

1. M. Fujiwara *et al.*, Photoproduction Experiment with Polarized HD Target at SPring-8, LEPS/RCNP proposal (2003).
2. D. Adams *et al.*, Phys. Lett. **B329**, 399 (1994).
3. K. Abe *et al.*, Phys. Rev. Lett. **74**, 346 (1995).
4. L.A. Ahrens *etal.*, Phys. Rev. **D 35**, 785 (1987).
5. J.F. Donoghue and C.R. Nappi, Phys. Lett. **B168**, 105 (1986).
6. A. Bertin *et al.*, Phys. Lett. **B388**, 450 (1996).
7. D.S. Armstrong *et al.*, Phys. Rev. Lett. **95**, 092001 (2005).
8. A. Acha *etal.*, Phys. Rev. Lett. **98**, 032301 (2007).
9. A. Airapetian *et al.*, Phys. Rev. Lett. **92**, 012005 (2004).
10. A.I. Titov, Y. Oh, and S.N. Yang, Phys. Rev. Lett. **79**, 1634 (1997).
11. A. Honig, Phys. Rev. Lett. **19**, 1009 (1967).

BARYON ANTIDECUPLET IN THE CHIRAL QUARK-SOLITON MODEL

HYUN-CHUL KIM

Department of Physics, Inha University, Incheon 402-751, Korea
**E-mail: hchkim@inha.ac.kr*

GHIL-SEOK YANG and KLAUS GOEKE

Institut für Theoretische Physik II, Ruhr-Universität Bochum, D-44780 Bochum, Germany

TIM LEDWIG

Institut für Kernphysik, Universität Mainz, D-55099 Mainz, Germany

In the present talk, we review a series of recent investigations on the structure of the baryon antidecuplet, in particular, of the Θ^+ baryon within the framework of the chiral quark-soliton model. Taking into account the $1/N_c$ rotational and linear m_s corrections, we calculate the mass splittings of the SU(3) baryons, and the vector and axial-vector transition form factors. We discuss the results in the context of recent experimental data. We also briefly show very recent results of the electromagnetic mass splittings of the baryon antidecuplet in a "model-independent" approach, which will provide a basic setting for the *physical* mass splittings of the SU(3) baryons.

Keywords: Baryon antidecuplet, Θ^+ baryon, the chiral quark-soliton model

1. Introduction

Even though many people have become dubious about the existence of the pentaquark baryon Θ^+[1–3] after the null results of a series of the CLAS experiments,[4–7] several crucial questions related to the Θ^+ remain unanswered and even seem to be evaded. Based on these experimental results, regardless of the existence of the Θ^+ or not, one can come to the following three main conclusions. Firstly, the Θ^+ was elusive in almost all high-energy experiments,[8–16] whereas its evidence was reported by many experiments[17–24] which were mainly conducted at relatively low energies. It was not answered or explained whether these positive results were just experimental noises or the real evidences of the Θ^+. However, we can conclude at least that the finding of the Θ^+ seems to be reaction-dependent.

Secondly, The decay width of the Θ^+, if it exists, is very small ($\Gamma_\Theta <$ 1 MeV) according to the DIANA results.[17,25] It implies that the $KN\Theta$ coupling should be tiny. This may provide an indirect and partial clue to the first question.

Lastly, the total cross section of the Θ^+ photoproduction is small.[4–7,26] It indicates that the $K^*N\Theta$ coupling constant should be also very small. Thus, understanding this smallness of the $KN\Theta$ and $K^*N\Theta$ coupling constants theoretically may shed light on the present cloudy situation for the existence of the Θ^+.

In this talk, we want to briefly review recent investigations on the $KN\Theta$ and $K^*N\Theta$ coupling constants within the framework of the self-consistent chiral quark-soliton model(χQSM).[27,28] We will also mention a very recent work on the mass splittings of the SU(3) baryons in a "model-independent" approach, with electromagnetic mass splittings considered.[29] The latter study provides the most general and systematic analysis on the structure of the SU(3) baryons with a given frmaework of the χQSM.

The present talk consists of two parts. In the first part (Section 2), we review briefly the general formalism of the χQSM to show how to calculate the vector and axial-vector form factors. We present the results and discuss them. We also predict the decay width of the Θ^+. We also discuss the connection of the present reported results to the original work by DPP. In the second part (Section 3), we explain how to incorporate the effects of isospin-symmetry breaking arising from both electromagnetic interactions and hadronic current quark mass differences. We will close the talk, summarizing and giving outlook in the last section.

2. $KN\Theta$ and $K^*N\Theta$ coupling constants, and the decay width of the Θ^+

The Θ^+-to-neutron transition matrix elements of the vector and axial-vector currents can be parametrized six real transition form factors $F_i^{n\Theta}$ and $G_i^{n\Theta}$ as follows:

$$\begin{aligned}
&\langle \Theta(p')|\bar{s}\gamma^\mu u|n(p)\rangle \\
&= \bar{u}_\Theta(\boldsymbol{p}')\left[F_1^{n\Theta}(Q^2)\gamma^\mu + \frac{F_2^{n\Theta}(Q^2)i\sigma^{\mu\nu}q_\nu}{M_\Theta + M_n} + \frac{F_3^{n\Theta}(Q^2)q^\mu}{M_\Theta + M_n}\right]u_n(\boldsymbol{p}), \quad (1) \\
&\langle \Theta(p')|\bar{s}\gamma^\mu\gamma_5 u|n(p)\rangle \\
&= \bar{u}_\Theta(\boldsymbol{p}')\left[G_1^{n\Theta}(Q^2)\gamma^\mu + G_2^{n\Theta}(Q^2)q^\mu + G_3^{n\Theta}(Q^2)P^\mu\right]\gamma^5\, u_n(\boldsymbol{p}), \quad (2)
\end{aligned}$$

where the $u_{\Theta(n)}$ denotes the spinor of the Θ^+ (neutron) with the corresponding mass $M_{\Theta(n)}$. The Q^2 stands for the momentum transfer $Q^2 = -q^2 = -(p'-p)^2$ and P represents the total momentum $P = p' + p$. $F_i^{n\Theta}$ and $G_i^{n\Theta}$ stand for real transition form factors that will be related to the strong coupling constants for the K^* and K with the help of the vector-meson dominance (VMD)[30,31] and Goldberger-Treiman relation.

In the VMD, the vector-transition current can be expressed as the K^* current by the current field identity (CFI):

$$V^\mu(x) \;=\; \bar{s}(x)\gamma^\mu u(x) = \frac{m_{K^*}^2}{f_{K^*}} K^{*\mu}(x)\,, \tag{3}$$

where m_{K^*} and f_{K^*} denote, respectively, the mass of the K^* meson, $m_{K^*} =$ 892 MeV, and decay constant defined as

$$f_{K^*}^2 = \frac{m_{K^*}^2}{m_\rho^2} f_\rho^2. \tag{4}$$

The decay constant f_ρ for the rho meson can be determined as

$$f_\rho^2 = \frac{4\pi\alpha^2 m_\rho}{3\Gamma_{\rho^0\to e^+e^-}}, \tag{5}$$

where α denotes the electromagnetic fine-structure constant. The f_{K^*} is determined by the experimental data for ρ-meson $m_\rho = 770$ MeV and $\Gamma_{\rho^0\to e^+e^-} = (7.02 \pm 0.11)$ keV,[32] for which we obtain the values $f_\rho \approx 4.96$ and $f_{K^*} \approx 5.71$.

Making use of the CFI, we can express the strong vector and tensor coupling constants for the $K^*N\Theta$ vertex in terms of the transition form factors:

$$g_{K^*n\Theta} \;=\; f_{K^*}F_1^{\Theta n}(0), \;\; f_{K^*n\Theta} \;=\; f_{K^*}\,F_2^{\Theta n}(0)\,. \tag{6}$$

Using the generalized Goldberger-Treiman relation, we can express the strong coupling constant $g_{Kn\Theta}$ for the $KN\Theta$ vertex as follows

$$g_{Kn\Theta} = \frac{G_1^{\Theta n}(0)\,(M_\Theta + M_n)}{2f_K}, \tag{7}$$

where $f_K \approx 1.2 f_\pi$ stands for the kaon decay constant.

The vector and axial-vector transition form factors in Eqs.(1,2) have been evaluated within the self-consistent χQSM.[27,28] The model is featured by the following effective low-energy partition function with quark fields ψ with the number of colors N_c and the pseudo-Goldstone boson field $U(x)$ in Euclidean space:

$$\mathcal{Z}_{\chi\mathrm{QSM}} = \int \mathcal{D}\psi\mathcal{D}\psi^\dagger\mathcal{D}U \exp\left[-\int d^4x\psi^\dagger iD(U)\psi\right]$$

$$= \int \mathcal{D}U \exp(-\mathcal{S}_{\text{eff}}[U]), \tag{8}$$

$$\mathcal{S}_{\text{eff}}(U) = -N_c \text{Tr} \ln iD(U), \tag{9}$$

where

$$\begin{aligned} D(U) &= \gamma^4(i\partial\!\!\!/ - \hat{m} - MU^{\gamma_5}) \\ &= -i\partial_4 + h(U) - \delta m, \end{aligned} \tag{10}$$

$$\begin{aligned} \delta m &= \frac{m_s - \bar{m}}{3}\gamma^4\mathbf{1} + \frac{\bar{m} - m_s}{\sqrt{3}}\gamma^4\lambda^8 \\ &= M_1\gamma^4\mathbf{1} + M_8\gamma^4\lambda^8. \end{aligned} \tag{11}$$

The current quark mass matrix is defined as $\hat{m} = \text{diag}(\bar{m}, \bar{m}, m_s) = \bar{m} + \delta m$. The $\bar{m}$ denotes the average of the up and down current quark masses with isospin symmetry assumed. The M represents the constituent quark mass of which the best value for the numerical results is $M = 420$ MeV. The $\mathbf{1}$ denotes the singlet flavor matrix diag$(1, 1, 1)$. The pseudo-Goldstone boson field U^{γ_5} is defined as

$$U^{\gamma_5} = \exp(i\gamma_5\lambda^a\pi^a) = \frac{1+\gamma_5}{2}U + \frac{1-\gamma_5}{2}U^\dagger \tag{12}$$

with $U = \exp(i\lambda^a\pi^a)$. For the quantization, we consider here Witten's embedding of SU(2) soliton into SU(3):

$$U_{\text{SU(3)}} = \begin{pmatrix} U_{\text{SU(2)}} & 0 \\ 0 & 1 \end{pmatrix} \tag{13}$$

with the SU(2) hedgehog chiral field

$$U_{\text{SU(2)}} = \exp[i\gamma_5\hat{\boldsymbol{n}}\cdot\boldsymbol{\tau}P(r)], \tag{14}$$

Here, the $P(r)$ denotes the profile function of the chiral soliton $U_{\text{SU(2)}}$. We refer to Refs. 33,34 as to how one can compute the form factors within the χQSM.

The results for the $K^*N\Theta$ and $KN\Theta$ coupling constants are given as[27,28]

$$g_{K^*N\Theta} = 0.81, \quad f_{K^*N\Theta} = 0.84, \quad g_{KN\Theta} = 0.83. \tag{15}$$

with the effects of SU(3) symmetry breaking taken into account. We want to mention that the leading order and $1/N_c$ corrections are almost canceled, so that the m_s corrections become significant. Moreover, the vector coupling constant $g_{K^*n\Theta}$ vanishes in exact SU(3) symmetry due to the generalized

Ademollo-Gatto theorem[a], which will be explained below. The value of $g_{K^*n\Theta}$ with SU(3) symmetry breaking comes solely from the wavefunction corrections. Using the value of $g_{Kn\Theta} = 0.83$. we immediately obtain the decay width of the Θ^+ as $\Gamma(\Theta \to NK) = 0.71$ MeV which is in qualitative agreement with the data of the DIANA collaboration.[25]

Diakonov et al.[1] estimated the decay width of the Θ^+: $\Gamma_\Theta \approx 15$ MeV, based on the experimental data for the πNN coupling constant and

$$g_{\pi NN} \approx \frac{7}{10}\left(G_0 + \frac{1}{2}G_1\right) \approx 13.3, \tag{16}$$

where terms proportional to G_2 and $c_{\overline{10}}$ were neglected. However, the coupling constant $g_{Kn\Theta}$ is proportional to $G_0 - G_1$, so that it is not possible to determine it by the $g_{\pi NN}$ only. Thus, DPP[1] have taken the old results from the χQSM calculations.[35,36] Ledwig et al.[28] uses the same formalism as Refs. 35,36 but several parts have been further elaborated and the ratio G_1/G_0 are found to be 0.68.

Had DPP[1] used the ratio $G_1/G_0 = 0.68$ instead of $G_1/G_0 = 0.4$, the decay width would have turned out to be $\Gamma(\Theta NK) = 3.4$ MeV, which is much smaller than the value $\Gamma(\Theta NK) \approx 15$ MeV published in Ref. 1, whereas $\Gamma(\Delta N\pi)$ would remain unchanged. Even though we consider the criticism of Ref.38 with $G_1/G_0 = 0.68$, one will get the decay width of the Θ which is also smaller than predicted in Ref.1. Thus, we want to point out that the predicted physics in Ref.1 by using the χQSM is therefore unchanged.

3. The electromagnetic mass splittings of the baryon antidecuplet

We now turn to the electromagnetic (EM) mass splittings of the baryon antidecuplet. The motivation is not to compute the EM mass splittings themselves but to provide a basic setting for the prediction of the SU(3) baryon mass splittings in a systematic way with minimum uncertainties.[39]

Considering the effects of SU(3) symmetry breaking, we can parametrize

[a]We want to mention that the generalized Ademollo-Gatto therem was first done by M.V. Polyakov.

the masses of the SU(3) baryons in general as follows:

$$(M_8)_H = \overline{M}_8 + \begin{bmatrix} -3x + y \text{ for } N \\ -x \quad \text{for } \Lambda \\ x \quad \text{for } \Sigma \\ 2x - y \text{ for } \Xi \end{bmatrix}, \tag{17}$$

$$(M_{10})_H = \overline{M}_{10} + \begin{bmatrix} -z \text{ for } \Delta \\ 0 \text{ for } \Sigma^* \\ z \text{ for } \Xi^* \\ 2z \text{ for } \Omega^- \end{bmatrix}, \tag{18}$$

$$(M_{\overline{10}})_H = \overline{M}_{\overline{10}} + \begin{bmatrix} -2v \text{ for } \Theta^+ \\ -v \text{ for } N^* \\ 0 \text{ for } \Sigma_{\overline{10}} \\ v \text{ for } \Xi_{3/2} \end{bmatrix}, \tag{19}$$

where M_8, M_{10}, and $M_{\overline{10}}$ stand for the mass matrices corresponding to the baryon octet, decuplet, and antidecuplet, respectively. The $\overline{M}_8$, $\overline{M}_{10}$, and $\overline{M}_{\overline{10}}$ are the corresponding *center* values. The parameters x and y , z, and v are originated from the flavor SU(3) symmetry breaking (m_8). They will be fixed by using the experimental data. The contributions from SU(3) symmetry breaking are presented in the bases of (N, Λ, Σ, Ξ) for the octet, (Δ, Σ^*, Ξ^*, Ω^-) for the decuplet, and (Θ^+, N^*, $\Sigma_{\overline{10}}$, $\Xi_{3/2}$) for the antidecuplet. Equations (17,18,19) are identified as the well-known Gell-Mann-Okubo formula.[40,41]

We now consider the effects of isospin symmetry breaking ($m_3 \neq 0$). We can parametrize them in the same way as we have done in the case of SU(3) symmetry breaking, so that we obtain for the contributions of isospin symmetry breaking

$$\begin{aligned} (M_8)_H^{\text{iso}} &= aT_3 + bYT_3, \\ (M_{10})_H^{\text{iso}} &= rT_3, \\ (M_{\overline{10}})_H^{\text{iso}} &= sT_3, \end{aligned} \tag{20}$$

where the T_3 and the Y stand for the third component of the isospin and the hypercharge, respectively. Thus, the hadronic parts of the SU(3) baryon masses can be written as

$$M_8 = \overline{M}_8 + \begin{bmatrix} (a+b)T_3 \\ 0 \\ aT_3 \\ (a-b)T_3 \end{bmatrix} + \begin{bmatrix} -3x + y \\ -x \\ x \\ 2x - y \end{bmatrix} \text{ for } \begin{bmatrix} N \\ \Lambda \\ \Sigma \\ \Xi \end{bmatrix}, \tag{21}$$

$$M_{10} = \overline{M}_{10} + \begin{bmatrix} rT_3 \\ rT_3 \\ rT_3 \\ 0 \end{bmatrix} + \begin{bmatrix} -z \\ 0 \\ z \\ 2z \end{bmatrix} \text{ for } \begin{bmatrix} \Delta \\ \Sigma^* \\ \Xi^* \\ \Omega^- \end{bmatrix}, \tag{22}$$

$$M_{\overline{10}} = \overline{M}_{\overline{10}} + \begin{bmatrix} 0 \\ sT_3 \\ sT_3 \\ sT_3 \end{bmatrix} + \begin{bmatrix} -2v \\ -v \\ 0 \\ v \end{bmatrix} \text{ for } \begin{bmatrix} \Theta^+ \\ N^* \\ \Sigma_{\overline{10}} \\ \Xi_{3/2} \end{bmatrix}, \tag{23}$$

Note that, however, the baryon masses given in Eqs.(21-23) do not contain the EM corrections. Thus, in order to get the *physical* baryon masses, we have to take into account the EM corrections. We will briefly show in the following how to analyze these EM corrections to the baryon masses.

The EM corrections to the baryon masses can be derived from the baryonic two-point correlation functions of the EM current J_μ:

$$M_B^{\mathrm{EM}} = \langle B|J_\mu(x)J^\mu(0)|B\rangle = \langle B|\mathcal{O}^{\mathrm{EM}}|B\rangle, \tag{24}$$

where J^μ is defined as

$$J^\mu(x) = e\bar{\psi}(x)\gamma_\mu\hat{Q}\psi(x) \tag{25}$$

with the electric charge e and the quark charge operator $\hat{Q}$ defined as the Gell-Mann-Nishijima relation

$$\hat{Q} = \frac{1}{2}\left(\lambda_3 + \frac{1}{\sqrt{3}}\lambda_8\right). \tag{26}$$

Since the EM current is taken as an octet operator, we can express the product of the EM currents in terms of irreducible operators. Because of Bose symmetry, we have only the singlet (**1**), the octet (**8**), and the eikosiheptaplet (**27**) irreducible operators for the $\mathcal{O}_{\mathrm{EM}}$. The matrix element (24) can be parametrized by four different collective operators. These four parameters can be determined by using the empirical data[42] as shown in Table 1. The results listed in the third column of Table 1 are the reproduced values within the present scheme.

In Table 2, we list the final results of the EM mass splittings of the baryon antidceuplet. As mentioned before, we do not intend to predict the

Table 1. EM mass differences for the baryon octet in units of MeV. The numbers with asterisks are taken as input.

	Refs. 42	Reproduced
$(M_p - M_n)_{\mathrm{EM}}$	$0.76 \pm 0.30^*$	0.74 ± 0.12
$(M_{\Sigma^+} - M_{\Sigma^-})_{\mathrm{EM}}$	$-0.17 \pm 0.30^*$	-0.15 ± 0.25
$(M_{\Xi^0} - M_{\Xi^-})_{\mathrm{EM}}$	$-0.86 \pm 0.30^*$	-0.88 ± 0.19

EM mass splittings themselves, since there are no experimental data for them and we do not expect that they would be measured in near future. However, having obtained the EM mass splittings, we can proceed to predict the *physical* mass splittngs of the SU(3) baryons. The corresponding works will appear elsewhere.

Table 2. Electromagnetic mass differences of the baryon anti-decuplet in units of MeV.

$(\Delta M_{B_{\overline{10}}})_{\mathrm{EM}}$	This work
$(M_{p^*} - M_{n^*})_{\mathrm{EM}}$	-0.77 ± 0.07
$\left(M_{\Sigma^+_{\overline{10}}} - M_{\Sigma^0_{\overline{10}}}\right)_{\mathrm{EM}}$	-0.77 ± 0.07
$\left(M_{\Sigma^0_{\overline{10}}} - M_{\Sigma^-_{\overline{10}}}\right)_{\mathrm{EM}}$	0.15 ± 0.07
$\left(M_{\Sigma^+_{\overline{10}}} + M_{\Sigma^-_{\overline{10}}} - 2M_{\Sigma^0_{\overline{10}}}\right)_{\mathrm{EM}}$	-0.92 ± 0.07
$\left(M_{\Xi^+_{3/2}} - M_{\Xi^0_{3/2}}\right)_{\mathrm{EM}}$	-0.77 ± 0.07
$\left(M_{\Xi^0_{3/2}} - M_{\Xi^-_{3/2}}\right)_{\mathrm{EM}}$	0.15 ± 0.07
$\left(M_{\Xi^-_{3/2}} - M_{\Xi^{--}_{3/2}}\right)_{\mathrm{EM}}$	1.06 ± 0.11
$\left(M_{\Xi^+_{3/2}} - M_{\Xi^-_{3/2}}\right)_{\mathrm{EM}}$	0.62 ± 0.12
$\left(M_{\Xi^0_{3/2}} - M_{\Xi^{--}_{3/2}}\right)_{\mathrm{EM}}$	1.02 ± 0.19
$\left(M_{\Xi^+_{3/2}} - M_{\Xi^{--}_{3/2}}\right)_{\mathrm{EM}}$	0.44 ± 0.20
$\left(M_{\Sigma^+_{\overline{10}}} - M_{\Sigma^-_{\overline{10}}}\right)_{\mathrm{EM}}$	-0.62 ± 0.12

4. Summary and outlook

In the present talk, we have reviewed recent works[27,28] on the $KN\Theta$ and $K^*N\Theta$ coupling constants based on the chiral quark-soliton model. The results are summarized as follows: the K^* vector and tensor coupling constants for the Θ^+: $g_{K^*N\Theta} = 0.74 - 0.87$ and $f_{K^*N\Theta} = 0.53 - 1.16$, and $\Gamma(\Theta \to NK) = 0.71$ MeV with the $KN\Theta$ coupling constant $g_{Kn\Theta} = 0.83$. We also discussed the connection of the presently reported results to the original work by Diakonov et al.[1] If the present result had used in their work, they would have obtained $\Gamma(\Theta \to NK) = 3.4$ MeV. These improvements even solidify the predicted physics in Ref. 1 regardless of the criticism given by Jaffe.[38]

We finally have discussed the recent investigation on the electromagnetic mass splittings for the baryon antidecuplet. The results will be used for deriving all the masses of the SU(3) baryons, which will appear elsewhere.

Acknowledgments

HChK would like to thank the organizers of HNP09 for the possibility to give a talk and for the hospitality during his stay at Osaka University. The authors are grateful to Maxim V. Polyakov for the crucial discussion and criticism. The present work is supported by Basic Science Research Program through the National Research Foundation of Korea (NRF) funded by the Ministry of Education, Science and Technology (grant number: 2009-0073101).

References

1. D. Diakonov, V. Petrov, and M. V. Polyakov, Z. Phys. A **359**, 305 (1997).
2. M. Praszałowicz, in: Workshop on Skyrmions and Anomalies, M.Jezabek and M. Praszałowicz (Eds.), World Scientific, 1987, p.112. Phys. Lett. B 575, 234 (2003).
3. T. Nakano *et al.* [LEPS Collaboration], Phys. Rev. Lett. **91**, 012002 (2003).
4. M. Battaglieri *et al.* [CLAS Collaboration], arXiv:hep-ex/0510061.
5. B. McKinnon *et al.* [CLAS Collaboration], Phys. Rev. Lett. **96**, 212001 (2006).
6. S. Niccolai *et al.* [CLAS Collaboration], Phys. Rev. Lett. **97**, 032001 (2006).
7. R. De Vita *et al.* [CLAS Collaboration], Phys. Rev. D **74**, 032001 (2006).
8. J. Z. Bai and [BES Collaboration], Phys. Rev. D**70**, 012004 (2004).
9. B. Aubert and [BABAR Collaboration], hep-ex/0408064.
10. K. Abe and [BELLE Collaboration], hep-ex/0409010.
11. S.R. Armstrong, Nucl. Phys. Proc. Suppl.**142**, 364 (2005).
12. S. Schael and [ALEPH Collaboration], Phys. Lett. B**599**, 1 (2004).
13. I. Abt and [HERA-B Collaboration], Phys. Rev. Lett. **93**, 212003 (2004).
14. Yu.M. Antipov and [SPHINX Collaboration], Eur. Phys. J. A**21**, 455 (2004).
15. M.J. Longo and [HyperCP Collaboration], Phys. Rev. D**70**, 111101 (2004).
16. D.O. Litvintsev and [CDF Collaboration] , Nucl. Phys. Proc. Suppl. **142**, 374 (2005).

17. V. V. Barmin *et al.* [DIANA Collaboration], Phys. Atom. Nucl. **66**, 1715 (2003).
18. S. Stepanyan and [CLAS Collaboration], Phys. ReV. Lett. **91**, 252001 (2003).
19. V. Kubarovsky and [CLAS Collaboration], Phys. ReV. Lett. **92**, 032001 (2004).
20. A. Airapetian and [HERMES Collaboration], Phys. Lett. B**585**, 213 (2004).
21. S. Chekanov and [ZEUS Collaboration] , Phys. Lett. B**591**, 7 (2004).
22. A.E. Asratyan and A.G. Dolgolenko and M.A. Kubantsev, Phys. Atom. Nucl. **67**, 682 (2004).
23. A. Aleev and [SVD Collaboration] , Phys. Atom. Nucl. **68**, 974 (2005).
24. M. Abdel-Bary and [COSY-TOF Collaboration] , "Phys. Lett. B**595**, 127 (2004).
25. V. V. Barmin *et al.* [DIANA Collaboration], Phys. Atom. Nucl. **70**, 35 (2007).
26. T. Nakano et al. Phys. Rev. C **79**, 025210 (2009).
27. T. Ledwig, H. -Ch. Kim and K. Goeke, Nucl. Phys. A **811**, 353 (2008).
28. T. Ledwig, H. -Ch. Kim and K. Goeke, Phys. Rev. D **78**, 054005 (2008).
29. Gh.-S. Yang, H.-Ch. Kim, and M.V. Polyakov, in preparation (2010).
30. J. J. Sakurai, Annals Phys. **11**, 1 (1960); "Currents and Mesons", (The University of Chicago Press, Chicago, 1969).
31. R. P. Feynman, "Photon-Hadron Interactions," (W.A. Benjamin, Inc. Reading, MA, 1972).
32. C. Amsler et al. Phys. Lett. B **667**, 1 (2008).
33. H.-Ch. Kim, A. Blotz, M.V. Polyakov and K. Goeke, Phys. Rev. D **53**, 4013 (1996).
34. C.V. Christov, A.Z. Gorski, K. Goeke and P.V. Pobylitsa, Nucl. Phys. A **592**, 513 (1995).
35. C. V. Christov, A. Blotz, K. Goeke, P. Pobylitsa, V. Petrov, M. Wakamatsu and T. Watabe, Phys. Lett. B **325**, 467 (1994).
36. A. Blotz, M. Praszałowicz and K. Goeke, Phys. Rev. D **53**, 485 (1996).
37. A. Silva, H.-Ch. Kim, D. Urbano and K. Goeke, Phys. Rev. D **72**, 094011 (2005).
38. R.L. Jaffe, Eur. Phys. J. C**35**, 221 (2004).
39. Gh.-S. Yang, H.-Ch. Kim, and M.V. Polyakov, in preparation (2010).
40. M. Gell-Mann, Phys. Rev. **125** (1962) 1067.
41. S. Okubo, Prog. Theor. Phys. **27** (1962) 949.
42. J. Gasser and H. Leutwyler, Phys. Rept. **87**, 77 (1982).

STRUCTURE OF THE σ MESON AND THE SOFTENING

T. HYODO*

Department of Physics, Tokyo Institute of Technology,
Meguro, Tokyo 152-8551, Japan
**E-mail: hyodo@th.phys.titech.ac.jp*

D. JIDO

Yukawa Institute for Theoretical Physics, Kyoto University,
Kyoto 606–8502, Japan

T. KUNIHIRO

Department of Physics, Kyoto University,
Kyoto 606–8502, Japan

We study the structure of the σ meson, the lowest-lying resonance of the $\pi\pi$ scattering in the scalar-isoscalar channel, through the softening phenomena associated with the partial restoration of chiral symmetry. We build dynamical chiral models to describe the $\pi\pi$ scattering amplitude, in which the σ meson is described either as the chiral partner of the pion or as the dynamically generated resonance through the $\pi\pi$ attraction. It is shown that the internal structure is reflected in the softening phenomena; the softening pattern of the dynamically generated σ meson is qualitatively different from the behavior of the chiral partner of the pion. On the other hand, in the symmetry restoration limit, the dynamically generated σ meson behaves similarly to the chiral partner.

Keywords: σ meson, chiral dynamics, softening, hadronic molecule

1. Introduction

The study of the structure of hadron resonances is one of the central issues in modern hadron spectroscopy. Among others, the structure of the σ meson has been intensively studied, since the σ meson is considered to play an important role in various aspects of hadron and nuclear physics. For instance, since the scalar-isoscalar excitation of QCD vacuum can be regarded as the amplitude fluctuation of the chiral order parameter $\langle\bar{q}q\rangle$, the nature of the σ meson is crucial to understand the dynamical chiral symmetry breaking in QCD.[1] While the existence of the scalar-isoscalar resonance in the $\pi\pi$ scattering is established by recent analyses (see e.g. Ref. 2), its internal structure is still controversial. Thus, it is our aim here to investigate the structure of the σ meson.

There are many proposals for the structure of the σ meson based on various effective models: collective $\bar{q}q$ excitation,[1,3,4] tetraquark structure with strong diquark correlation,[5] dynamically generated $\pi\pi$ resonance,[6] and so on. Conventionally, the validity of certain structure has been tested by comparing the model prediction with experimental data, for instance, mass spectrum and the decay properties.

It is also possible to investigate the structure of the resonance from the response to the change of the internal/external parameter of the models. For instance, the study of the N_c scaling is successful to disentangle the $\bar{q}q$ and other structures for meson resonances.[7] Following this philosophy, we would like to study the spectral change of the σ meson when the chiral symmetry is partially restored, in order to discriminate the different internal structures.

In association with chiral symmetry restoration, the softening of the σ meson has been discussed.[1,4,8] In the linear realization of chiral symmetry, the σ meson forms a chiral four-vector together with the pion, and hence they are chiral partners. The partial restoration of chiral symmetry induces the softening of the σ spectrum, which results in the enhancement of the $\pi\pi$ cross section in the scalar-isoscalar channel near threshold. It was shown later that the threshold enhancement in the $I = J = 0$ channel takes place also in the nonlinear realization of chiral symmetry without the bare σ field[9,10] where the σ meson is expressed as a dynamically generated resonance from the attractive $\pi\pi$ interaction. Although similar threshold enhancement of the cross section is observed in both cases, the mechanism which causes the softening is quite different. Based on these observations, we demonstrate that the softening phenomena reflects the structure of the σ meson, paying attention to the nature of the s-wave resonance.

When the symmetry is completely restored, the chiral partners emerge as the pair of particles with a degenerate mass. If we regard the mass degeneracy as the condition for the chiral partner, we can extend the notion of the chiral partner for the dynamically generated σ meson in the nonlinear realization. We will study the structure of the $\pi\pi$ scattering amplitude of our model in the restoration limit, to discuss the chiral partner of the pion.

2. Formulation

Here we describe the $\pi\pi$ scattering amplitude with the σ resonance in the $I = J = 0$ channel. We consider the low energy behavior of the amplitude based on chiral effective Lagrangian, and then introduce the unitarity condition to extend the applicability of the model to the resonance energy region.

We start from the Lagrangian of two-flavor linear sigma model to derive the $\pi\pi$ scattering amplitudes:

$$\mathcal{L} =\frac{1}{4}\mathrm{Tr}\left[\partial M\partial M^{\dagger} - \mu^2 MM^{\dagger} - \frac{2\lambda}{4!}(MM^{\dagger})^2 + h(M + M^{\dagger})\right], \qquad (1)$$

where $M = \sigma + i\boldsymbol{\tau}\cdot\boldsymbol{\pi}$. In this Lagrangian, the σ meson is treated as the chiral partner of the pion. For negative μ^2, chiral symmetry is spontaneously broken and three parameters in the Lagrangian μ, λ, and h are related to the chiral condensate (pion decay constant) $\langle\sigma\rangle = f_\pi$, the mass of the pion m_π, and the mass of the σ meson m_σ in the mean-field level.

Crossing symmetry enables us to express the general $\pi\pi$ scattering amplitude as $T_{\mathrm{tree}}(s,t,u) = A(s,t,u)\delta_{ab}\delta_{cd} + A(t,s,u)\delta_{ac}\delta_{bd} + A(u,t,s)\delta_{ad}\delta_{bc}$, where the invariant amplitude A is given, from the Lagrangian (1) at tree level, by[9]

$$A(s) =\frac{s-m_\pi^2}{\langle\sigma\rangle^2} - \frac{(s-m_\pi^2)^2}{\langle\sigma\rangle^2}\frac{1}{s-m_\sigma^2}.$$

In this expression, the first (second) term can be regarded as the leading (higher) order contribution in the chiral perturbation theory. The coefficient of the leading order term is fixed by the low energy theorem, while the low energy constant for the higher order terms is not constrained by the symmetry and should be determined by experiments. Thus, we introduce a parameter x to express the general amplitude

$$A(s;x) =\frac{s-m_\pi^2}{\langle\sigma\rangle^2} - x\frac{(s-m_\pi^2)^2}{\langle\sigma\rangle^2}\frac{1}{s-m_\sigma^2}. \qquad (2)$$

By choosing $x = 1$, we recover the result of the original Lagrangian (1). If we take $x = 0$, then we are left with the leading order interaction, which also corresponds to the heavy m_σ limit. In this way, we can smoothly connect the original linear sigma model ($x = 1$) and the leading order chiral perturbation theory without the bare σ field ($x = 0$). Projecting Eq. (2) onto the $I = J = 0$ channel, we obtain the tree-level amplitude for the $\pi\pi$

scattering as

$$T_{\text{tree}}(s;x) = \frac{m_\sigma^2 - m_\pi^2}{\langle\sigma\rangle^2}\left[\frac{2s - m_\pi^2}{m_\sigma^2 - m_\pi^2}(1-x) - 5x \right.$$
$$\left. - 3x\frac{m_\sigma^2 - m_\pi^2}{s - m_\sigma^2} - 2x\frac{m_\sigma^2 - m_\pi^2}{s - 4m_\pi^2}\ln\left(\frac{m_\sigma^2}{m_\sigma^2 + s - 4m_\pi^2}\right)\right], \quad (3)$$

in the center-of-mass frame.

Next we consider the unitarity condition Im $T^{-1}(s) = -\Theta(s)/2$ for $s > 4m_\pi^2$, with the two-body phase space function $\Theta(s) = (16\pi)^{-1}\sqrt{1 - 4m_\pi^2/s}$. Based on the N/D method, we write down the general expression of the unitary scattering amplitude $T(s;x)$. Matching the chiral interaction $T_{\text{tree}}(s;x)$ with the loop expansion of the full amplitude $T(s;x)$, we obtain the amplitude which is consistent with both chiral low energy theorem and unitarity as[11]

$$T(s;x) = \frac{1}{T_{\text{tree}}^{-1}(s;x) + G(s)},$$

$$G(s) = \frac{1}{2}\frac{1}{(4\pi)^2}\left\{a(\mu) + \ln\frac{m_\pi^2}{\mu^2} + \sqrt{1 - \frac{4m_\pi^2}{s}}\left[\ln\frac{\sqrt{1 - \frac{4m_\pi^2}{s}} + 1}{\sqrt{1 - \frac{4m_\pi^2}{s}} - 1}\right]\right\},$$

where $a(\mu)$ is the subtraction constant at the subtraction point μ. We determine the subtraction constant by excluding the nontrivial CDD pole (states which does not originate in the two-body dynamics) in the amplitude[12]

$$G(s) = 0 \quad \text{at} \quad s = m_\pi^2, \quad (4)$$

which leads to $a(m_\pi) = -\pi/\sqrt{3}$. With this subtraction constant, the full scattering amplitude T reduces into the tree level one T_{tree} at $s = m_\pi^2$.

The full amplitude $T(s;x)$ corresponds to the nonperturbative resummation of the s-channel loop diagrams up to infinite order. For the $x = 1$ case, the bare σ pole in the Lagrangian acquire the finite width through the coupling to the $\pi\pi$ state, and the full scattering amplitude exhibits a pole in the complex energy plane. On the other hand, for the $x = 0$ case without the bare σ pole, a resonance can be dynamically generated as the pole of the amplitude, if the two-body interaction is sufficiently attractive. Indeed, it is shown that the resummation of the leading order interaction generates the σ meson dynamically.[13] In the following we compare the properties of these σ states: one originating in the chiral partner of the pion in the linear σ model ($x = 1$), and another generated dynamically from the attractive $\pi\pi$ interaction ($x = 0$).

3. Chiral symmetry restoration

3.1. *Prescription for the symmetry restoration*

Now we introduce the effect of chiral symmetry restoration through the modification of the model parameters. It is known that the chiral condensate should decrease with the chiral symmetry restoration, so we parametrize the condensate by

$$\langle\sigma\rangle = \Phi\langle\sigma\rangle_0,$$

where $\langle\sigma\rangle_0$ is the condensate in vacuum and we vary the parameter Φ from one to zero to express the symmetry restoration. The NJL model indicates that m_π hardly changes as symmetry restoration,[1] so we assume that it is a constant:

$$m_\pi = \text{const.}$$

The bare mass of the σ should be degenerated with pion when the symmetry is restored. This can be achieved by the mean-field relation of Eq. (1)

$$m_\sigma = \sqrt{\lambda\frac{\langle\sigma\rangle^2}{3} + m_\pi^2},$$

with λ and m_π being fixed.

3.2. *Behavior of the amplitude in the restoration limit*

It is instructive to study the behavior of the $\pi\pi$ scattering amplitude in the limit $\langle\sigma\rangle \to 0$. Here we analytically derive the pole of the amplitude in the restoration limit. In subsection 4.2, we will numerically demonstrate that the obtained result corresponds to the asymptotic behavior of the σ pole for $\langle\sigma\rangle \to 0$.

We first consider the $x = 1$ case, where the σ meson is the chiral partner of the pion. To make the $\langle\sigma\rangle$ dependence in m_σ explicit, we rewrite the tree level amplitude as

$$\begin{aligned} T_{\text{tree}}(s;1) = &-\frac{5\lambda}{3} - \frac{\lambda^2\langle\sigma\rangle^2}{3}\frac{1}{s - m_\pi^2 - \frac{\lambda}{3}\langle\sigma\rangle^2} \\ &- \frac{2\lambda^2\langle\sigma\rangle^2}{9}\frac{1}{s-4m_\pi^2}\ln\frac{m_\pi^2 + \frac{\lambda}{3}\langle\sigma\rangle^2}{s - 3m_\pi^2 + \frac{\lambda}{3}\langle\sigma\rangle^2}. \end{aligned}$$

The second term represents the bare pole of the σ meson. As $\langle\sigma\rangle \to 0$, the pole mass of this term decreases and finally it coincides with the pion mass. Because of the renormalization condition (4), the tree-level amplitude $T_{\text{tree}}(s;1)$ coincides with the full amplitude $T(s;1)$ at $s = m_\pi^2$, so the full amplitude $T(s;1)$ also has a pole as in the same way with $T_{\rm tree}(s;1)$. Approximating the amplitude by the Breit-Wigner form around the pole,

$$T(s;1) \sim \frac{g^2}{s - M_{\rm pole}^2},$$

we extract the mass of the state $M_{\rm pole}$ and the coupling to the scattering state g. In the present case, we find

$$g \to 0, \quad M_{\rm pole} \to m_\pi \quad \text{for} \quad \langle\sigma\rangle \to 0.$$

This is what we anticipate for the properties of the chiral partner; the mass degeneracy with the pion and the vanishing of the coupling constant to the $\pi\pi$ state.

Next we consider the $x = 0$ case without the bare σ pole. In this case, $\langle\sigma\rangle$ dependence of the tree-level amplitude (3) exclusively stems from the overall factor,

$$T_{\rm tree}(s;x) \propto \frac{1}{\langle\sigma\rangle}.$$

Taking the restoration limit $\langle\sigma\rangle \to 0$, this term diverges, and therefore the full amplitude is solely determined by the loop function $G(s)$:

$$T(s;x) = \frac{1}{T_{\rm tree}^{-1}(s,x) + G(s)} \to \frac{1}{G(s)} \quad \text{for} \quad \langle\sigma\rangle \to 0.$$

Thus, the pole of the amplitude in the restoration limit is given by the zero of $G(s)$. The present renormalization scheme requires $G(s) = 0$ for $s = m_\pi^2$, which indicates the existence of a pole at $\sqrt{s} = m_\pi$ in the σ channel. The coupling constant g can be obtained by calculating the residue of the pole:

$$g^2|_{\langle\sigma\rangle\to 0} = (4\pi)^2 \left(\frac{1}{2} - \frac{\pi}{3\sqrt{3}}\right)^{-1} m_\pi^2. \tag{5}$$

Thus, for the dynamically generated σ meson, the amplitude has a pole at the pion mass with the coupling constant which is proportional to m_π for $\langle\sigma\rangle \to 0$.

This result has an interesting implication for the chiral partner. Strictly speaking, the notion of the "chiral partner" is defined only in the chiral limit ($m_\pi \to 0$), where the SU(2)×SU(2) symmetry is exact in the Wigner phase. In the chiral limit, Eq. (5) indicates $g^2|_{\langle\sigma\rangle\to 0} = 0$, so the asymptotic value of the mass and coupling constant of the dynamically generated σ meson is exactly the same with the chiral partner case. Namely, the dynamically generated σ meson behaves as if it is the chiral partner of the pion, for $m_\pi \to 0$.

4. Numerical study for the softening phenomena

4.1. *Structure of the σ meson in vacuum*

We first show the description of the scattering amplitude in vacuum. We choose canonical values of the parameters as $\langle\sigma\rangle_0 = 93$ MeV, $m_\pi = 140$ MeV, and $m_\sigma = 550$ MeV. By taking the parameter $x = 1$, the σ meson is described as the chiral partner of the pion (chiral σ), and we refer to this case as "model A". Choosing $x = 0$, we obtain the dynamically generated σ meson in the amplitude. This case is called as "model B".

To check the agreement with the physical amplitude, we study the properties of these models without symmetry restoration. We calculate the scattering length $a = \frac{1}{32\pi}T(4m_\pi^2)$ (in units of m_π^{-1}) in these models. The results are shown in Table 1, together with the pole position of the amplitude in the complex energy plane. We find a qualitative agreement with the recent analyses of experimental data,[2,14] $a_{\rm exp} \sim 0.216$ and $z = 441 - 272i$ MeV.

4.2. *Softening of the σ meson*

We then study the variation of the scattering amplitude in the σ channel along with partial restoration of chiral symmetry. We plot the reduced cross section $\bar{\sigma} = |T|^2/s$ and the trace of the pole position as functions of the total center-of-mass energy $\sqrt{s}$, by changing the parameter Φ from 1 to 0.

The invariant mass spectra and the pole trajectory of model A are shown in Fig. 1, where the softening of σ is clearly observed. As the symmetry is restored, the σ pole moves toward the $\pi\pi$ threshold and the spectrum gets narrow and enhanced around threshold. In this case, since the σ meson is treated as the chiral partner of pion, the softening phenomena is driven by the decrease of the bare mass of the σ and its consequence of the reduction of the phase space for the decay.[8] In the limit $\langle\sigma\rangle \to 0$, the pole approaches the mass of the pion, as indicated by the analysis in section 3.2.

We show the results of model B in Fig. 2. In this case, although the threshold enhancement takes place, the change of the spectrum as well as the trace of the pole are qualitatively different from those of model A.

Table 1. Properties of the models: value of parameter x, possible origin of the pole, the scattering length a in units of m_π^{-1}, and the pole position of the amplitude in vacuum.

	x	origin	a $[m_\pi^{-1}]$	pole position [MeV]
model A	1	chiral partner	0.244	$423 - 126i$
model B	0	dynamically generated	0.174	$364 - 356i$

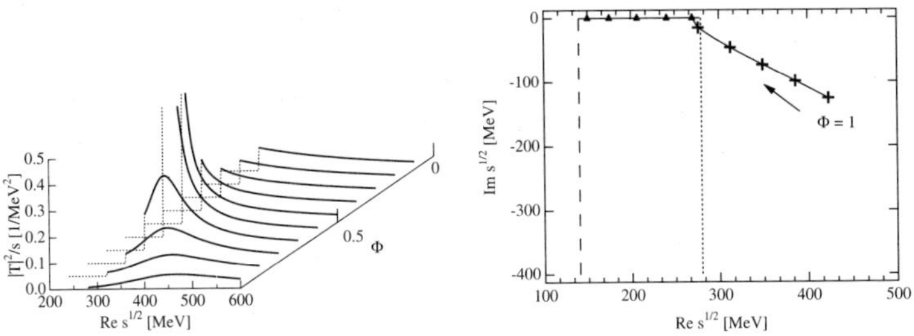

Fig. 1. Mass spectra of the σ meson (left) and the trace of the pole positions (right) in model A ($x = 1$). The poles on the first (second) Riemann sheet are plotted by triangles (crosses). Symbols are marked with each 0.1 step of Φ. Arrow indicates the direction of the movement of the pole as the parameter Φ is decreased from 1 to 0. Dotted (dashed) line represents the energy corresponds to the threshold (mass of pion).

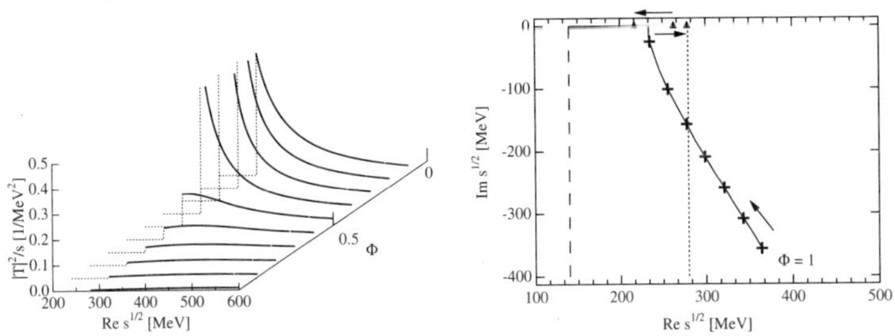

Fig. 2. Mass spectra of the σ meson (left) and the trace of the pole positions (right) in model B ($x = 0$). The poles on the first (second) Riemann sheet are plotted by triangles (crosses). Symbols are marked with each 0.1 step of Φ. Arrows indicate the direction of the movement of the pole as the parameter Φ is decreased from 1 to 0. Dotted (dashed) line represents the energy corresponds to the threshold (mass of pion).

Especially, the pole in the complex energy plane goes to the subthreshold energy region, keeping the width finite. This is a peculiar feature of the dynamically generated σ meson. As a consequence, the strong enhancement of the cross section occurs at much later stage of the symmetry restoration, compared with the model A.

Let us discuss how this structure appears in model B. The mechanism of the threshold enhancement in the nonlinear σ model has been studied in Ref. 9; the partial restoration of chiral symmetry induces the reduction of the pion decay constant. Since the low energy interaction is proportional to $1/\langle\sigma\rangle^2$, the symmetry restoration effectively increases the attractive interaction. Thus, the dynamically generated σ resonance will eventually turns into a $\pi\pi$ bound state when the interaction becomes sufficiently attractive.

At this stage, however, it is important to recall that the σ is in s-wave. In this case, when the attraction is increased, the resonance first becomes the *virtual state* which is the pole on the second Riemann sheet of the energy plane but lies below the threshold.[15] The peculiar pole trajectory in Fig. 2 is due to the appearance of the virtual state.

It is worth mentioning that the finite pion mass is important for the appearance of the virtual state. If $m_\pi = 0$, the $\pi\pi$ threshold lies at $\sqrt{s} = 0$, so there is no region where the virtual state appears. Indeed, the virtual state was not seen in the analysis of the dynamically generated σ meson in Ref. 10, studied in the chiral limit. In addition, the s-wave nature of the σ is also essential for the virtual state. Therefore, the behavior of the ρ meson in p-wave amplitude will not exhibit such a structure (see also Ref. 16).

As we further restore the symmetry, the σ meson becomes the bound state. In the limit $\langle\sigma\rangle \to 0$, the pole moves toward the pion mass. This is again in agreement with the result in section 3.2: the appearance of the pole at the pion mass in the restoration limit.

5. Summary

We have studied the properties of the σ meson in the $\pi\pi$ scattering associated with the restoration of chiral symmetry. Two models are constructed based on chiral low energy interaction and unitarity of the scattering amplitude: one describes the σ meson as the chiral partner of the pion, and the other treats the σ meson as dynamically generated resonance.

For the dynamically generated σ meson, we find the qualitative difference from the chiral partner σ in the softening behavior, namely, the movement of the pole of the amplitude and its consequence of the change of the spectrum. The difference stems from the mechanism which drives the softening, and it is the appearance of the virtual state that leads to the distinct behavior of the dynamically generated σ.

We also study the asymptotic properties of the σ pole in the restoration limit. For the σ meson as the chiral partner, we find that the mass of the σ pole approaches the pion mass and the coupling to the $\pi\pi$ state vanishes. For the dynamically generated σ, we also find the mass degeneracy with the pion, and the coupling strength vanishes in the chiral limit $m_\pi \to 0$. Namely, the behavior of the dynamically generated σ pole in the restoration limit is essentially the same with what we expect for the chiral partner. This is a nontrivial result which urge us to speculate the possibility of the dynamically generated σ meson as the chiral partner of the pion.

In this way, through the comparison of two models, we draw two conclusions: (i) with partial restoration of chiral symmetry, the difference of the internal structures is reflected in the spectral change of the σ channel, and (ii) in the symmetry restoration limit, the difference of the structure is reduced and we obtain essentially the same behavior of the σ pole for $m_\pi \to 0$. More comprehensive analysis, including the case with the σ meson as the CDD pole, is now underway.[17]

Acknowledgments

The authors are grateful to M. Oka and Y. Kanada-En'yo for useful discussion. This work is partly supported by the Global Center of Excellence Program by MEXT, Japan through the Nanoscience and Quantum Physics Project of the Tokyo Institute of Technology, the Grant-in-Aid for Scientific Research from MEXT and JSPS (Nos. 21840026, 20540273, 2054026, and 20028004), and the Grant-in-Aid for the Global COE Program "The Next Generation of Physics, Spun from Universality and Emergence" from MEXT of Japan. This work was done in part under the Yukawa International Program for Quark-hadron Sciences (YIPQS).

References

1. T. Hatsuda and T. Kunihiro, *Phys. Rept.* **247**, 221 (1994).
2. I. Caprini, G. Colangelo and H. Leutwyler, *Phys. Rev. Lett.* **96**, 132001 (2006).
3. Y. Nambu and G. Jona-Lasinio, *Phys. Rev.* **122**, 345 (1961); *Phys. Rev.* **124**, 246 (1961).
4. T. Hatsuda and T. Kunihiro, *Prog. Theor. Phys.* **74**, 765 (1985); *Phys. Lett.* B **185**, 304 (1987).
5. R. L. Jaffe, *Phys. Rev.* **D15**, 267 (1977).
6. J. A. Oller, E. Oset and J. R. Pelaez, *Phys. Rev. Lett.* **80**, 3452 (1998).
7. J. R. Pelaez, *Phys. Rev. Lett.* **92**, 102001 (2004).
8. T. Hatsuda, T. Kunihiro and H. Shimizu, *Phys. Rev. Lett.* **82**, 2840 (1999).
9. D. Jido, T. Hatsuda and T. Kunihiro, *Phys. Rev.* **D63**, 011901 (2001).
10. K. Yokokawa, T. Hatsuda, A. Hayashigaki and T. Kunihiro, *Phys. Rev.* **C66**, 022201 (2002).
11. J. A. Oller and E. Oset, *Phys. Rev.* **D60**, 074023 (1999).
12. T. Hyodo, D. Jido and A. Hosaka, *Phys. Rev.* **C78**, 025203 (2008).
13. J. A. Oller and E. Oset, *Nucl. Phys.* **A620**, 438 (1997).
14. S. Pislak *et al.*, *Phys. Rev.* **D67**, 072004 (2003).
15. A. Bohm, *Quantum Mechanics: Foundations and Applications* (Springer, New York, 2001).
16. C. Hanhart, J. R. Pelaez and G. Rios, *Phys. Rev. Lett.* **100**, 152001 (2008).
17. T. Hyodo, D. Jido and T. Kunihiro, *in preparation*.

LIGHT SCALAR MESON $\sigma(600)$ AND ITS CONTINUUM

HUA-XING CHEN

Department of Physics and State Key Laboratory of Nuclear Physics and Technology, Peking University, Beijing 100871, China, and Research Center for Nuclear Physics, Osaka University, Ibaraki, Osaka 567–0047, Japan
E-mail: hxchen@rcnp.osaka-u.ac.jp

ATSUSHI HOSAKA

Research Center for Nuclear Physics, Osaka University, Ibaraki, Osaka 567–0047, Japan
E-mail: hosaka@rcnp.osaka-u.ac.jp

HIROSHI TOKI

Research Center for Nuclear Physics, Osaka University, Ibaraki, Osaka 567–0047, Japan
E-mail: toki@rcnp.osaka-u.ac.jp

SHI-LIN ZHU

Department of Physics and State Key Laboratory of Nuclear Physics and Technology, Peking University, Beijing 100871, China
E-mail: zhusl@pku.edu.cn

We study the sigma meson in the QCD sum rule. We use the linear combination of five independent currents. We find that for some cases the Borel mass stability and threshold value stability are both weak, while for some cases not. To know the reason for this we ought to consider the π-π continuum contribution in the phenomenological side of the QCD sum rule. Then we study the stability of the extracted mass against the Borel mass and the threshold value and find the $\sigma(600)$ mass at 530 MeV $\pm$ 40 MeV.

Keywords: scalar meson, tetraquark, QCD sum rule.

The light scalar mesons have been subject to intensive discussions for many years.[1–3] The expected members are $\sigma(600)$, $\kappa(800)$, $f_0(980)$ and $a_0(980)$ of flavor SU(3) nonet. The existence of $\sigma(600)$ which is denoted as $f_0(600)$ in the Particle Data Group has been confirmed also by a model independent theoretical analysis.[4–7] Yet, their nature is not fully understood.[8] We have studied them by using the method of QCD sum rule.[9] Now we want to improve our previous QCD sum rule analysis. In the QCD sum rule, we calculate matrix elements from the QCD (OPE) and relate them to observables by using dispersion relations. Under suitable assumptions, the QCD sum rule has proven to be a very powerful and successful non-perturbative method in the past decades,[10,11] and it has been applied to the study of tetraquarks by many authors.[12–14]

To start this work, first we show the five independent interpolating fields which can couple to $\sigma(600)$. They were obtained in Ref. 9:

$$\begin{aligned}
S_3^\sigma &= (u_a^T C\gamma_5 d_b)(\bar{u}_a\gamma_5 C\bar{d}_b^T - \bar{u}_b\gamma_5 C\bar{d}_a^T)\,,\\
V_3^\sigma &= (u_a^T C\gamma_\mu\gamma_5 d_b)(\bar{u}_a\gamma^\mu\gamma_5 C\bar{d}_b^T - \bar{u}_b\gamma^\mu\gamma_5 C\bar{d}_a^T)\,,\\
T_6^\sigma &= (u_a^T C\sigma_{\mu\nu} d_b)(\bar{u}_a\sigma^{\mu\nu} C\bar{d}_b^T + \bar{u}_b\sigma^{\mu\nu} C\bar{d}_a^T)\,,\\
A_6^\sigma &= (u_a^T C\gamma_\mu d_b)(\bar{u}_a\gamma^\mu C\bar{d}_b^T + \bar{u}_b\gamma^\mu C\bar{d}_a^T)\,,\\
P_3^\sigma &= (u_a^T C d_b)(\bar{u}_a C\bar{d}_b^T - \bar{u}_b C\bar{d}_a^T)\,,
\end{aligned} \tag{1}$$

where the sum over repeated indices (μ, ν, $\cdots$ for Dirac, and $a, b, \cdots$ for color indices) is taken. Either plus or minus sign in the second parentheses ensures that the diquarks form the antisymmetric combination in the flavor space. The currents S, V, T, A and P are constructed by scalar, vector, tensor, axial-vector, pseudoscalar diquark and antidiquark fields, respectively. The subscripts 3 and 6 show that the diquarks (antidiquark) are combined into the color representation $\mathbf{\bar{3}_c}$ and $\mathbf{6_c}$ ($\mathbf{3_c}$ or $\mathbf{\bar{6}_c}$), respectively. (See Ref. 9 for more discussion about these interpolating fields.)

We have used the single currents and linear combination of two single currents to perform the QCD sum rule analysis.[4]In this work we use general currents by taking linear combinations of these five currents:

$$\eta = t_1 e^{i\theta_1} S_3^\sigma + t_2 e^{i\theta_2} V_3^\sigma + t_3 e^{i\theta_3} T_6^\sigma + t_4 e^{i\theta_4} A_6^\sigma + t_5 e^{i\theta_5} P_3^\sigma\,,$$

where t_i and θ_i are ten mixing parameters, whose linear combination describes the full space of currents which can couple to $\sigma(600)$. We can not determine them in advance and therefore we choose them randomly for the study of the QCD sum rule.

Then we calculate the OPE up to dimension eight. By randomly choosing the ten mixed parameters of Eq. (2), we perform the QCD sum rule analysis. We test as many as fifty sets of numbers. Some of them lead to negative spectral densities in the low energy region of interest, which should be, however, positive from their definition. Except the negative cases, there are fifteen sets which lead to positive spectral densities. We use them to perform the QCD sum rule analysis, and find that the Borel mass stability is usually weak. However, the threshold value stability is sometimes weak, while sometimes not so weak. Many effects contribute to the mass dependence on the threshold value, but for $\sigma(600)$ the π-π continuum contribution is probably the dominant one.

we take the spectral density of the π-π continuum as

$$\rho^C(s) = a(t_i, \theta_i) s^2 \sqrt{1 - \frac{4m_\pi^2}{s}}\,, \tag{2}$$

and add it in the phenomenological side and perform the QCD sum rule analysis. After adding the continuum contribution, the threshold value stability becomes much better. We show the mass of the case (13) in Fig. 1 as a functions of s_0, which are the lower three curves. At the same time, we also show the mass calculated without $\rho^C(s)$, which are the upper three curves. We see that the stability of the threshold value is significantly improved.

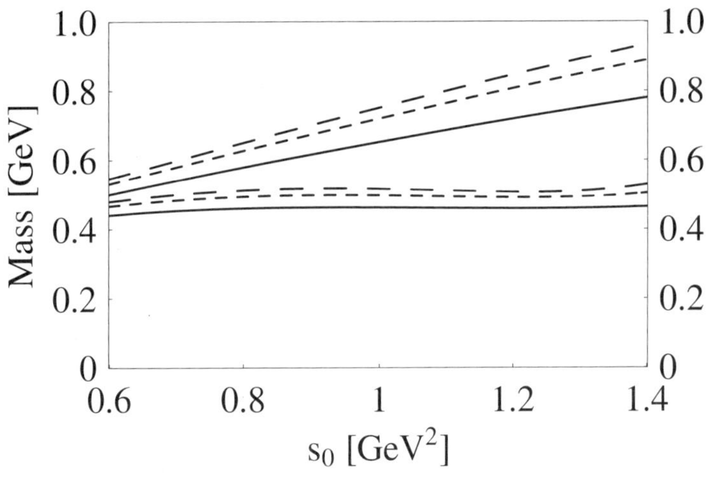

Fig. 1. The extracted mass of $\sigma(600)$ as a tetraquark state calculated by the mixed current η, as functions of the threshold value s_0. We choose the parameter set (13) as an example. The solid, short-dashed and long-dashed curves are obtained by setting $M_B^2 = 0.5$, 1 and 2 GeV^2, respectively. The upper three curves are obtained without adding the contribution of the π-π continuum, while the lower three curves are obtained after adding the contribution of the π-π continuum.

We made the same analysis for all the other cases. In summary, we have studied the light scalar meson $\sigma(600)$ in the QCD sum rule. We have used general tetraquark currents which are linear combinations of five independent current operators. We have introduced a contribution from the π-π continuum, and obtained a good threshold value stability. In our working region 0.8 GeV^2 $< s_0 <$ 1.2 GeV^2 and 1 GeV^2 $< M_B^2 <$ 2 GeV^2, we find that all the cases lead to a mass within the region 495 MeV$\sim$ 570 MeV. From this mass range, the mass of $\sigma(600)$ is extracted to be 530 MeV $\pm$ 40 MeV. Very interesting observation is that the mass increases slightly with the Borel mass. We have made a model calculation by taking the Gaussian width of 200 MeV centered at 550 MeV and try to make a sum rule analysis. We see a similar increase trend as seen in the QCD sum rule analysis.

References

1. C. Amsler and N. A. Tornqvist, Phys. Rept. **389**, 61 (2004).
2. D. V. Bugg, Phys. Rept. **397**, 257 (2004).
3. E. Klempt and A. Zaitsev, Phys. Rept. **454**, 1 (2007).
4. I. Caprini, G. Colangelo and H. Leutwyler, Phys. Rev. Lett. **96**, 132001 (2006).
5. T. Hatsuda and T. Kunihiro, Phys. Rept. **247**, 221 (1994).
6. J. A. Oller and E. Oset, Nucl. Phys. A **620**, 438 (1997) [Erratum-ibid. A **652**, 407 (1999)].
7. J. Sugiyama, T. Nakamura, N. Ishii, T. Nishikawa and M. Oka, Phys. Rev. D **76**, 114010 (2007).
8. C. Amsler *et al.* [Particle Data Group], Phys. Lett. B **667**, 1 (2008).
9. H. X. Chen, A. Hosaka and S. L. Zhu, Phys. Rev. D **76**, 094025 (2007).
10. M. A. Shifman, A. I. Vainshtein and V. I. Zakharov, Nucl. Phys. B **147**, 385 (1979).
11. L. J. Reinders, H. Rubinstein and S. Yazaki, Phys. Rept. **127**, 1 (1985).
12. M. E. Bracco, A. Lozea, R. D. Matheus, F. S. Navarra and M. Nielsen, Phys. Lett. B **624**, 217 (2005).
13. S. Narison, Phys. Rev. D **73**, 114024 (2006).
14. H. J. Lee and N. I. Kochelev, Phys. Lett. B **642**, 358 (2006).

QUARK POTENTIALS ON SEVERAL COLOR CHANNELS IN THE LATTICE QCD

T. SAITO*

Integrated Information Center, Kochi University, Kochi, 780-8520, Japan

Y. NAKAGAWA

Research Institute for Information Science and Education, Hiroshima University, Higashi-Hiroshima 739-8521, Japan

A. NAKAMURA

Research Institute for Information Science and Education, Hiroshima University, Higashi-Hiroshima 739-8521, Japan

H. TOKI

Research Institute for Nuclear Physics, Osaka University, Ibaraki, Osaka 567-0047, Japan

We report a numerical study of two-quark potentials depending on the color-irreducible representation of the $SU(3)$ group by using a quenched-lattice QCD simulation with the Coulomb gauge in the confinement phase. In addition, we display preliminary numerical results for color-decomposed three-quark potentials.

Keywords: Lattice QCD, Coulomb gauge, color interactions

1. Introduction

The recent Belle experiment has reported the existence of the new exotic particle $Z(4430)^+$, which is expected to be a 4-quark state made of $u\bar{d}c\bar{c}$ with an electric charge.[1] This might be a good candidate for investigating the internal structure of hadrons, providing a great hint for the color confinement and chiral symmetry dynamics. Besides other candidates for new exotics have been also reported in Ref. 2. However, it is difficult to theoretically understand a composition of those exotics in the simple quark model classification.

In order to deal with the virtual gluons and quarks dynamics properly, our study addresses a lattice calculation of the quark potential inside hadrons depending on the color-irreducible representations. The previous lattice computations mainly concentrate on the Wilson loop potential (gauge-invariantly defined); exactly speaking, this is not a color singlet state but a mixed state of color-singlet and color-octet between quark and antiquark; but, its behavior coincides with the phenomenological Cornell potential very well.[3] Therefore, this means that the color-octet contribution is insignificant for the basic mesonic systems. However, through the discovery of the exotic hadrons, it becomes very important to known the long-distance properties of *color-non-singlet forces*, which would construct a more fundamental combination such as a color-antitriplet diquark and color-octet molecular state (maybe broaden), etc., finally forming a color-singlet observable. In addition, $Z(4430)^+$ including some light quarks is a more extended object than the usual $c\bar{c}$ mesons with the radius $R \sim 0.2$ fm, over the range of which it should be thought non-perturbatively.

Those color-dependent forces are calculable on the lattice when using a Polyakov line correlator (PLC) under the classification of the color-group irreducible representation to some quark components. In practice, this approach has been applied to the quark-gluon plasma (QGP) physics in the lattice QCD simulations. Above the phase transition temperature, the quark potentials on each color channel are color-screened due to a heat-bath medium and they are altered to a Yukawa-type potential with a non-zero color-Debye mass; the screened potential depends qualitatively on the quadratic Casimir factors.[4–6] In contrast, there are a few studies for a numerical simulation along this line in the confinement phase,[7,8] since the quark-antiquark interaction increases linearly with the distances of quark separations; this would make a signal of the PLC functions worse below the QGP critical temperature.

Besides, since the dynamics of color confinement remains an unsolved problem, the treatment of the confining potentials is a hard task. However, we can achieve to obtain useful data applicable to the model study such as a potential model, if using the Coulomb gauge QCD framework. This theory is suitable for the Hamiltonian formulation.[9] Therefore one can employ variational analyses with no indefinite metric; a color-Coulomb instantaneous potential that depends on the quadratic Casimir factor can be defined and causes a strong confining force which is responsible for color confinement. Furthermore the infrared singularity results from the accumulation of Fadeev-Popov (FP) ghost operator of QCD. This is so called Gribov-

Zwanziger confinement scenario,[10,11] which has been extensively confirmed in the lattice computations recently.[12–16]

In this study, we calculate the color forces in the confinement region by using the Coulomb gauge QCD theory with the PLC on each (irreducible) color-channel in the $SU(3)$ group. We show that the some results for the two-quark state; namely, color-singlet, color-octet, color-antitriplet and color-sextet parts are addressed. We also report the preliminary calculation of the colored-three quark potentials.

2. Formulation

2.1. *Hamiltonian of Coulomb gauge QCD*

The Hamiltonian of QCD in the Coulomb gauge[10,11] can be given by

$$H = \frac{1}{2}\int d^3x((E_i^{tr}(\vec{x}))^2 + B_i^2(\vec{x})) + \frac{1}{2}\int d^3x d^3y(\rho(\vec{x})\mathcal{V}(\vec{x},\vec{y})\rho(\vec{y})), \tag{1}$$

where E_i^{tr}, B_i and ρ are the transverse electric field, the transverse magnetic field and the color-charge density, respectively. The function $\mathcal{V}$ in the second term is made by the FP operator in the spatial direction, $M = -\vec{D}\vec{\partial} = -(\vec{\partial}^2 + g\vec{A}\times\vec{\partial})$,

$$\mathcal{V}(\vec{x},\vec{y}) = \int d^3z \left[M(\vec{x},\vec{z})^{-1}(-\vec{\partial}^2_{(\vec{z})})M(\vec{z},\vec{y})^{-1}\right]. \tag{2}$$

From the partition function with the Hamiltonian Eq. (1), one can evaluate the time-time gluon propagator composed of the following two parts:

$$g^2\langle A_0(x)A_0(y)\rangle = g^2 D_{00}(x-y) = V(x-y) + P(x-y), \tag{3}$$

where

$$V(x-y) = g^2\langle\mathcal{V}(\vec{x},\vec{y})\rangle\delta(x_4 - y_4). \tag{4}$$

Equation (4) is the instantaneous color-Coulomb potential, which is responsible for color confinement and their string tension is produced from the lowest infrared mode of the FP operators.[17,18]

2.2. *Polyakov line correlators on several color channels*

We employ a partial-length Polyakov line (PPL) which has been developed in Refs. 12 and 13 as

$$L(\vec{x},T) = \prod_{t=1}^{T} U_0(\vec{x},t), \quad T = 1, 2, \cdots, N_t. \tag{5}$$

A PPL correlator in the color-singlet channel is given by

$$G_1(R,T) = \frac{1}{3}\left\langle \mathrm{Tr}[L(R,T)L^\dagger(0,T)]\right\rangle, \tag{6}$$

where R stands for $|\vec{x}|$. From Eq. (6) one evaluates a color-singlet potential on a lattice,

$$V(R,T) = \log\left[\frac{G_1(R,T)}{G_1(R,T+a)}\right]. \tag{7}$$

For the smallest temporal-lattice extension, i.e., $T=0$, we define

$$V(R,0) = -\log[G_1(R,1)]. \tag{8}$$

Here $V(R,0)$ in the Coulomb gauge is the color-Coulomb instantaneous potential $V_{coul}(R)$. We here have Zwanziger's inequality, $V_{phys}(R) \leq V_{coul}(R)$ 19, where $V_{phys}(R)$ is the physical (Wilson loop) potential.

One can make every correlator function for other color channels by the combination of the PLC under the irreducible color-$SU(3)$ representation ($3\otimes\bar{3} = 1\oplus 8$ for $q\bar{q}$ and $3\otimes 3 = 6\oplus\bar{3}$ for qq).

3. Numerical results

The quark potentials between two quarks on each color channel are shown in Fig. 1. These are calculated by the link-link correlator giving the instantaneous potential. Color-singlet and color-antitriplet channels yield attractive confining forces for the large distances; on the other hands, the color-octet and color-sextet channels are repulsions. Furthermore, we plotted the ratio of those forces in Fig. 2. This result is consistent with the behavior expected from the quadratic Casimir factors.

The three-quark potentials can be also formulated in terms of the PLC; for example, in the color singlet qqq state (from the decomposition of $3\otimes 3\otimes 3 = 1\oplus 8\oplus 8^{'}\oplus 10$ for qqq), we have the following formula:

$$\begin{aligned} e^{-V_{qqq}} \sim\ & [27\mathrm{Tr}(L_1)\mathrm{Tr}(L_2)\mathrm{Tr}(L_3) \\ & - 9\mathrm{Tr}(L_1)\mathrm{Tr}(L_2L_3) - 9\mathrm{Tr}(L_2)\mathrm{Tr}(L_3L_1) - 9\mathrm{Tr}(L_3)\mathrm{Tr}(L_1L_2) \\ & + 3\mathrm{Tr}(L_1L_2L_3) + 3\mathrm{Tr}(L_1L_3L_2)]\,. \end{aligned} \tag{9}$$

Some numerical calculations in the QGP phase have been reported in Ref. 20 and the closed Wilson-loop three-quark potential has been also studied in Ref. 21.

Here we display the preliminary results for the color-singlet and color-octet instantaneous potentials in Figs. 3 and 4, in which we use the Yacobi coordinates to simply plotting data. The quadratic Casimirs are $C^1_{qqq} = -2$

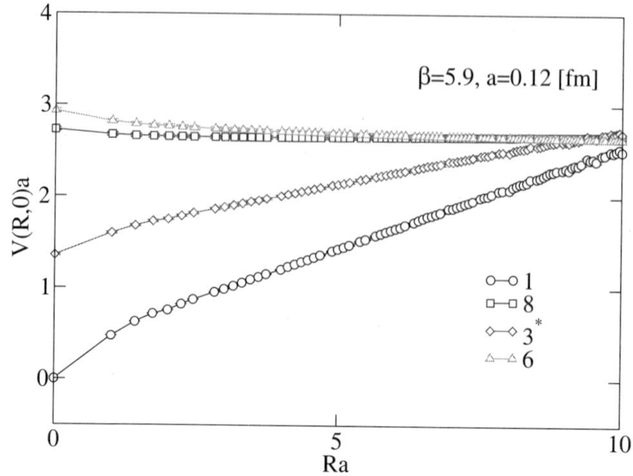

Fig. 1. Quark potentials for two-quark state on the instantaneous part ($\beta = 5.9$, $a = 0.12$fm).

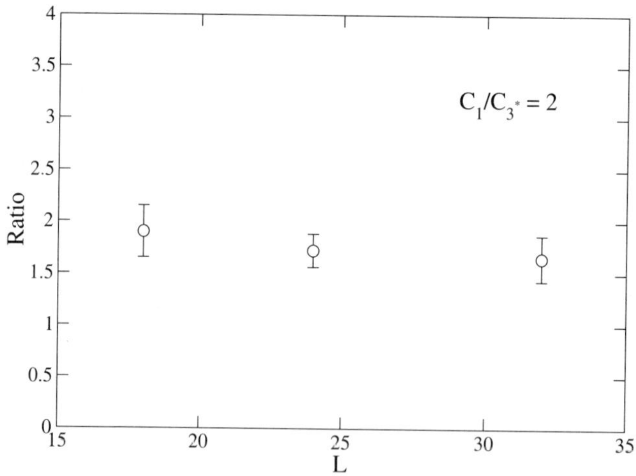

Fig. 2. Ratio of the string tensions on the singlet $q\bar{q}$ and antitriplet qq as the lattice volume size increases.

and $C^{8}_{qqq} = -1/2$ respectively. Here the numerical results obtained also show that they are attraction; more details of this calculation for the other color-channels will come out elsewhere.

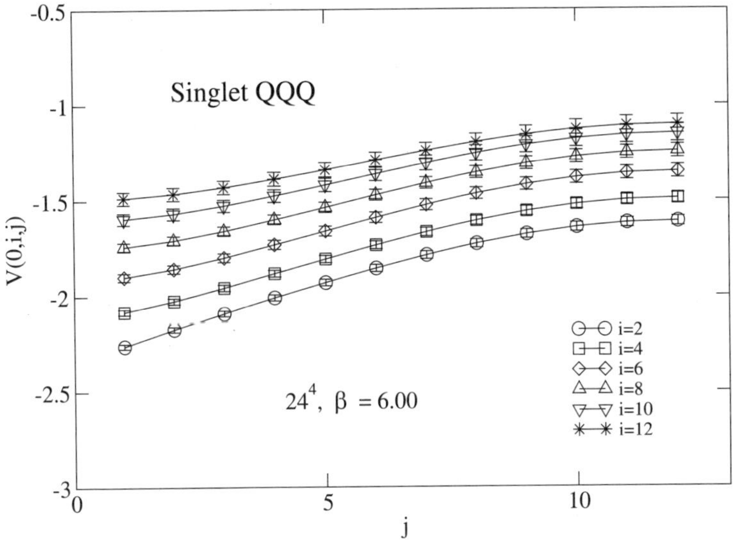

Fig. 3. The three-quark potential ($V(0,i,j)$) in the color-singlet channel; here we employ Yacobi coordinate when plotting.

4. Summary

We have studied the color-dependent forces among quarks in the confinement region by using a quenched-lattice QCD simulation with the Coulomb gauge fixing. We confirm that for two-quark system the color-singlet and color-antitriplet (diquark) channels yield the attractive forces, while the color-octet and color-sextet channels are repulsions. Moreover, we show the preliminary numerical results for the color forces made of three quarks. For the instantaneous part, those forces seem to depend on the quadratic Casimir factor; however, it is also necessary to calculate the full-length Polyakov line correlator so as to account for the vacuum polarization effect for this simple conclusions, although the color-singlet potential is consistent with the Wilson loop (Cornell) potential.[12–14]

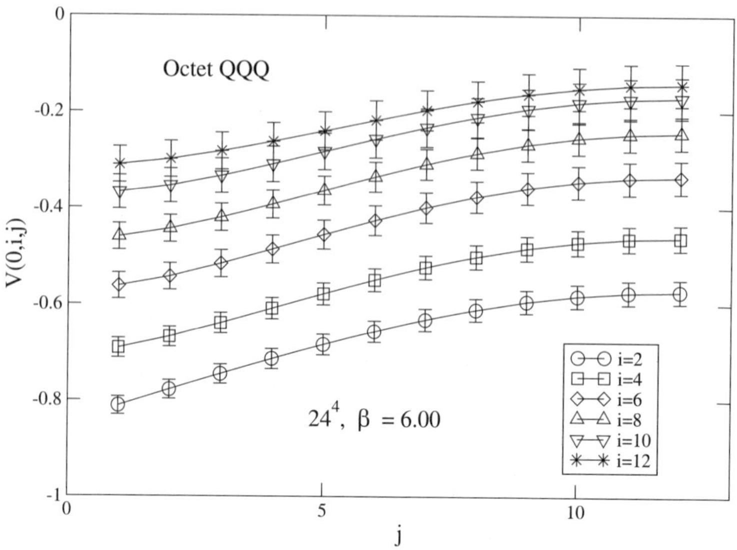

Fig. 4. The three-quark potential ($V(0,i,j)$) in the color-octet channel; here we employ Yacobi coordinate when plotting.

References

1. Belle Collaboration, S.-K. Choi et al., Phys. Rev. Lett. 100, 142001 (2008).
2. E. S. Swanson, Phys. Rep. 429, 243 (2006).
3. G. S. Bali, Phys. Rept. 343, 1 (2001).
4. A. Nakamura and T. Saito, Prog. Theor. Phys. 111 (2004) 733-743; Prog. Theor. Phys. 112 (2004) 183-188.
5. Y. Maezawa, N. Ukita, S. Aoki, S. Ejiri, T. Hatsuda, N. Ishii, and K. Kanaya, Phys. Rev. D75 (2007) 074501.
6. M. Döring, K. Hübner, O. Kaczmarek, F. Karsch, Phys. Rev. D75 (2007) 054504.
7. A. Nakamura, T. Saito, Phys. Lett. B621 (2005) 171-175.
8. Y. Nakagawa, A. Nakamura, T. Saito, H. Toki, Phys. Rev. D77, 034015 (2008).
9. A. P. Szczepaniak, E. S. Swanson, Phys. Rev. D65 (2001), 025012.
10. V. N. Gribov, Nucl. Phys. B139, 1 (1978).
11. D. Zwanziger, Nucl. Phys. B518, 237 (1998).
12. J. Greensite and Š. Olejník, Phys. Rev. D 67, 094503 (2003).
13. J. Greensite, Š. Olejník and D. Zwanziger, Phys. Rev. D 69, 074506 (2004).
14. A. Nakamura, T. Saito, Prog. Theor. Phys. 115, 189 (2006).
15. Y. Nakagawa, A. Nakamura, T. Saito, H. Toki, D. Zwanziger, Phys. Rev. D 73, 094504 (2006).
16. Y. Nakagawa, A. Voigt, E.-M. Ilgenfritz, M. Müller-Preussker, A. Nakamura, T. Saito, A. Sternbeck, H. Toki, Phys.Rev.D79,114504 (2009).
17. Y. Nakagawa, A. Nakamura, T. Saito, H. Toki, Phys. Rev. D 75, 014508

(2007).
18. Y. Nakagawa, A. Nakamura, T. Saito, H. Toki, Phys. Rev. D 81, 054509 (2010).
19. D. Zwanziger, Phys. Rev. Lett. 90, 102001 (2003).
20. K. Huebner, O. Kaczmarek, O. Vogt, Phys. Rev. D 77,074504 (2008).
21. T. T. Takahashi1, H. Matsufuru1, Y. Nemoto, and H. Suganuma, Phys. Rev. Lett. 86, 18 (2001); Phys. Rev. D 65, 114509 (2002).

CHIRAL MAGNETIC EFFECT (CME) AT LOW TEMPERATURE FROM INSTANTON VACUUM

SEUNG-IL NAM

Department of Physics,
Chung-Yuan Christian University,
Chung-Li 32023, Taiwan
E-mail: sinam@cycu.edu.tw
www.phys.cycu.edu.tw/∼hep

In this talk, we report our present work on the chiral magnetic effect (CME) under a strong magnetic field $\boldsymbol{B} = B_0\hat{x}_3$ at low temperature. To this end, we use the instanton vacuum with the finite instanton-number fluctuation Δ, which relates to the nontrivial topological charge Q_t. We compute the vacuum expectation values of the local chiral density $\langle\rho_\chi\rangle$, chiral charge density $\langle n_\chi\rangle$ and induced electromagnetic current $\langle j_\mu\rangle$. We observed that the longitudinal EM current is much larger than the transverse one, $|j_\perp/j_\parallel| \sim Q_t$, and the $\langle n_\chi\rangle$ equals to the $|\langle j_{3,4}\rangle|$. It also turns out that the CME becomes insensitive to the magnetic field as T increases, since the instanton effect decreases.

Keywords: chiral magnetic effect, P- and CP-violations, instanton vacuum configuration

1. Introduction

Recently, in Refs. 1,2, it was suggested that the nontrivial topological charge of QCD vacuum, Q_t can be probed by measuring asymmetric electric currents, due to an event-by-event P- and CP-violation, in the non-central heavy-ion collision experiments such as RHIC, FAIR, and LHC. Their main idea is that unequal numbers of the chirally left- and right-handed quarks, $N_L - N_R \neq 0$ proportional to the Q_t in terms of the axial Ward-Takahashi identity,[3] can produce the asymmetric electric current under a strong magnetic field B. Interestingly enough, a possible experimental evidence was already reported by STAR collaboration at RHIC.[4] Moreover, in Ref. 5, using the lattice QCD simulation, the electromagnetic (EM) current fluctuation as well as local chirality were explored at finite T, resulting in that the fluctuations become insensitive to the external magnetic field as T increases. In Ref. 6, it was also shown that the electric current, induced by the external magnetic field with the nontrivial $Q_{\rm t}$, reads

$$\boldsymbol{j} = -\frac{\mu_\chi}{2\pi^2}\boldsymbol{B}, \tag{1}$$

where the μ_χ represents the chiral chemical potential and contains the information on the $Q_{\rm t}$ inside of it. In fact, this is a consequence of the EM axial anomaly.[7]

In the present report, based on our previous work,[8] we investigate the CME at low T, employing the instanton vacuum configuration with the nontrivial $Q_{\rm t}$, which relates to the finite instanton-number fluctuation, i.e. the number difference between the instantons and anti-instantons in the grand canonical ensemble, $\Delta \equiv N_I - N_{\bar{I}} \neq 0$ resulting in the CP-violation. To this end, we first write the effective action derived from the instanton vacuum configuration as a functional of Δ.[9] Using generic functional external-source and linear Schwinger methods, we compute the relevant physical quantities, the vacuum expectation values of the local chiral density $\langle\rho_\chi\rangle$, chiral charge density $\langle n_\chi\rangle$, EM current $\langle j_\mu\rangle$, induced by the external magnetic field. In order to consider the T-dependence of the relevant quantities, we also employ the Harrington-Shepard caloron to obtain the T-dependent constituent-quark mass. Fianlly, the standard fermionic Matsubara formula is taken into account for the anti-perioidc sum over the Euclidean time, i.e. T. We consider all the relevant physical quantities up to $\mathcal{O}(\Delta^2) \sim \mathcal{O}(Q_{\rm t}^2)$.

2. Effective action from the instanton vacuum

In this Section, we briefly introduce a P- and CP-violating effective action $\mathcal{S}_{\rm eff}$, derived by Diakonov *et al.* from the instanton vacuum configuration in the large N_c limit at zero temperature ($T = 0$).[9] Employing a dilute grand canonical ensemble of the (anti)instantons with a finite instanton-number fluctuation, $\Delta \equiv N_+ - N_- \neq 0$, which corresponds to a CP-violating vacuum, but a fixed total number of the pseudo-particles $N_+ + N_- = N$, the $\mathcal{S}_{\rm eff}$ can be written in momentum space with Euclidean metric as follows:

$$\begin{aligned}\mathcal{S}_{\rm eff} &= \mathcal{C} + \frac{N_+}{V}\ln\lambda_+ + \frac{N_-}{V}\ln\lambda_- - \frac{mN_c}{4\pi^2\bar{\rho}^2}(\lambda_+ + \lambda_-) \\ &- N_c\int\frac{d^4k}{(2\pi)^4}\mathrm{Tr}_\gamma \ln\left[\frac{\not{k} - \frac{i}{2}[\lambda_+(1+\gamma_5) + \lambda_-(1-\gamma_5)]F^2(k)}{\not{k} - im}\right],\end{aligned} \tag{2}$$

From the effective action, we can obtain the following two self-consistent (saddle-point) equations with respect to the $\lambda_\pm$:

$$\lambda_\pm \frac{N_\pm}{V} = \frac{(1\pm\delta)M_0 m N_c}{4\pi^2\bar{\rho}^2} + N_c \int \frac{d^4k}{(2\pi)^4} \mathrm{Tr}_\gamma \left[\frac{\frac{1}{2}(1\pm\gamma_5)(1+\delta\gamma_5)^2 M^2}{k^2+(1+\delta\gamma_5)^2 M^2}\right], \quad (3)$$

where the $\lambda_\pm$ is approximated as $M_0(1\pm\delta)$ in the last line of Eq. (3) with account of the fact $\Delta \ll N$ in the thermodynamic limit.[9] The momentum-dependent constituent-quark mass is defined as $M(k) = M_0 F^2(k)$.[10] By adding and subtracting the instanton (+) and anti-instanton (−) contributions in Eq. (3), we can obtain an expression for δ as a function of relevant parameters:

$$\delta = \left(\frac{2\pi^2\bar{\rho}^2}{mM_0N_c}\right)\frac{\Delta}{V}. \quad (4)$$

Taking into account all the ingredients discussed so far, finally, we can write the relevant effective action with $\Delta \neq 0$ for further investigations:

$$\mathcal{S}_{\rm eff} = -\int \frac{d^4k}{(2\pi)^4} \mathrm{Tr}_{c,f,\gamma} \ln \left[\frac{\not{k} - i(1+\delta\gamma_5)M(k)}{\not{k} - im}\right], \quad (5)$$

where the $\mathrm{Tr}_{c,f,\gamma}$ denotes the trace over color, flavor and Lorentz indices.

According to the axial Ward-Takahashi identity,[3] the $Q_{\rm t}$ is proportional to the number difference between the chirally left- and right-handed quarks, $Q_{\rm t} \propto N_R - N_L$. Hence, the nontrivial $Q_{\rm t}$ indicates the chirality flip. Note that, similarly, if a chirally left-handed quark is scattered from an instanton to an anti-instanton, the quark helicity is flipped to the right-handed one, and vice versa. This means that the nonzero Δ results in $N_R - N_L \neq 0$. In this way, the $Q_{\rm t}$ can be considered to be proportional to the Δ: $Q_{\rm t} \sim \Delta$.[10,11] The δ in Eq. (4) can be rewritten in terms of a real and small parameter ϵ, which satisfies the condition $|\epsilon| \leq 1$, for convenience as follows:

$$\delta = \left(\frac{2\pi^2\bar{\rho}^2}{mM_0N_c}\right)\frac{\epsilon N}{V}, \quad (6)$$

3. The chiral magnetic effect in the magnetic field

In this Section, we compute relevant physical quantities for the CME in the presence of the external magnetic field, using the effective action given in the last Section. They are the vacuum expectation values (v.e.v.) of the local chiral density ρ_χ, chiral charge density n_χ, and electromagnetic (EM) current j_μ, which are defined as follows:

$$\rho_\chi(x) = iq^\dagger(x)\gamma_5 q(x), \quad n_\chi(x) = iq^\dagger(x)\gamma_4\gamma_5 q(x), \quad j_\mu(x) = iq^\dagger(x)\gamma_\mu q(x). \quad (7)$$

The ρ_χ and n_χ represent the strength of the parity breaking in a system, whereas the j_μ here the EM current induced by the external magnetic field. It is very convenient to compute these quantities employing the effective action with corresponding pseudoscalar ($\mathcal{P}$), axial vector ($\mathcal{A}$), and vector ($\mathcal{V}$) external sources. Hence, the effective action can be rewritten as:

$$\mathcal{S}_{\text{eff}} = -\int \frac{d^4k}{(2\pi)^4} \mathrm{Tr}\ln\left[\frac{\not{K} - i(1+\delta\gamma_5)M(K) + i\delta\gamma_5\mathcal{P} + \delta\gamma_4\gamma_5\mathcal{A}_4 + \gamma_\mu\mathcal{V}_\mu}{\not{K} - im + i\delta\gamma_5\mathcal{P} + \delta\gamma_4\gamma_5\mathcal{A}_4 + \gamma_\mu\mathcal{V}_\mu}\right], \tag{8}$$

where $\mathrm{Tr} \equiv \mathrm{Tr}_{c,f,\gamma}$ with $N_c = 3$ and $N_f = 2$. The K_μ indicates the covariant quark momentum, gauged by the photon field as $k_\mu + A_\mu$. Note that we have written the external sources $\mathcal{P}$ and $\mathcal{A}_4$, which correspond to the ρ_χ and n_χ, to be proportional to the δ, since they signal the nontrivial topological charge Q_t as the parity-breaking quark mass $\delta\gamma_5 M$ does.

First, we calculate the v.e.v. of the chiral density, using the standard functional method as follows:

$$\langle\rho_\chi\rangle = -i\delta N_c N_f \int \frac{d^4k}{(2\pi)^4} \mathrm{Tr}_\gamma\left\{\left[\frac{1}{\not{K} - i(1+\delta\gamma_5)M(K)} - \frac{1}{\not{K} - im}\right]\gamma_5\right\}, \tag{9}$$

where we have performed the trace of over the color and flavor indices. Since we are interested in the response of the nonperturbative vacuum to the external magnetic field for the nonzero topological charge Q_t, we only collect the terms proportional to the δ and the photon field strength tensor $F_{\mu\nu}$. In order to perform the trace over the Lorentz index in the r.h.s. of Eq. (9) under the external EM field, we employ the linear Schwinger method.[12,13] According to that, the quark propagator in the presence of the instanton background can be written up to $\mathcal{O}(F_{\mu\nu})$, which is equivalent up to $\mathcal{O}(B_0)$, as follows:

$$\begin{aligned} S(k,A) = \frac{1}{\not{K} - i(1+\delta\gamma_5)M(K)} &\approx \frac{\not{k} + \not{A} + i(1+\delta\gamma_5)\left[M + \frac{1}{2}\bar{M}(\sigma\cdot F)\right]}{k^2 + (1+\delta^2)M^2} \\ &\times \left[1 - \frac{\tilde{M}(\sigma\cdot F) + i\left(1 - \frac{\delta}{4}\gamma_5\right)\hat{M}(k)\gamma_\mu K_\nu F_{\mu\nu} - 2i\delta M\gamma_5\not{K}}{k^2 + (1+\delta^2)M^2}\right]. \end{aligned} \tag{10}$$

Here, $\sigma\cdot F = \sigma_{\mu\nu}F_{\mu\nu}$. We define relevant functions related to the momentum-dependent quark mass:

$$M = M_0\left(\frac{2}{2+k^2\bar{\rho}^2}\right)^2, \quad \bar{M} = -\frac{8M_0\bar{\rho}^2}{(2+k^2\bar{\rho}^2)^3}, \quad \tilde{M} = \frac{1}{2} + M\bar{M}, \quad \hat{M} = 4i\bar{M}. \tag{11}$$

After a straightforward trace manipulation, one arrives at a simple expression for the chiral density:

$$\langle \rho_\chi \rangle_{F,\delta} = 2\delta^2 \mathcal{F}_a B_0^2, \tag{12}$$

where the subscripts F and δ in the l.h.s. of Eq. (12) indicate the fact that we picked up only the terms proportional to the $F_{\mu\nu}$ and δ as explained above. In deriving Eq. (12), to make the problem easy, we assumed a static external EM field, which is assigned for example as in Ref. 5:

$$A_\mu = A_\mu^{\rm cl} + A_\mu^{\rm fluc} = \left(-\frac{B_0}{2}x_2, \frac{B_0}{2}x_1, 0, 0\right) + (a_1, a_2, a_3, a_4), \tag{13}$$

where the external EM field consists of the classical ($A_\mu^{\rm cl}$) in the symmetry gauge and fluctuation ($A_\mu^{\rm fluc}$) parts. We take the $a_{1\sim4}$ as small and constant EM potentials. Among the EM field strength tensors, $F_{12} = B_0$ and its dual remain finite, while others disappear. The relevant integral $\mathcal{F}_a$ in Eq. (12) reads:

$$\mathcal{F}_a = -4N_c N_f \int \frac{d^4k}{(2\pi)^4} \frac{\bar{M}\left(M\bar{M} + \frac{1}{2}\right)}{[k^2 + (1+\delta^2)M^2]^2}, \tag{14}$$

As understood from Eq. (12), the parity breaking of the vacuum becomes enhanced quadratically with respect to the δ and B_0.

Similarly to the ρ_χ, we can compute the v.e.v. of the chiral charge density as follows:

$$\langle n_\chi \rangle = -\delta N_c N_f \int \frac{d^4k}{(2\pi)^4} {\rm Tr}_\gamma \left\{ \left[\frac{1}{\not{K} - i(1+\delta\gamma_5)M(K)} - \frac{1}{\not{K} - im} \right] \gamma_4\gamma_5 \right\}. \tag{15}$$

By solving Eq. (15) and picking up relevant terms, one is lead to a compact expression for it as follows:

$$\langle n_\chi \rangle_{F,\delta} = \delta \mathcal{F}_b \epsilon_{4\nu\rho\sigma}(iA_\sigma)F_{\nu\rho} = 2\delta \mathcal{F}_b (iA_3) B_0, \tag{16}$$

where we have used the fact that only the $\tilde{F}_{34} = B_0$ is nonzero in our symmetric-gauge EM field. From Eq. (16), we observe that the chiral charge density increases linearly with respect to the δ as well as the B_0. The integral $\mathcal{F}_b$ is written as:

$$\mathcal{F}_b = -4N_c N_f \int \frac{d^4k}{(2\pi)^4} \frac{M\bar{M}}{[k^2 + (1+\delta^2)M^2]^2}. \tag{17}$$

Finally, we attempt to compute the v.e.v. of the EM current, induced by the external magnetic field in the presence of the nontrivial topological charge:

$$\langle j_\mu \rangle = -N_c N_f \int \frac{d^4k}{(2\pi)^4} {\rm Tr}_\gamma \left\{ \left[\frac{1}{\not{K} - i(1+\delta\gamma_5)M(K)} - \frac{1}{\not{K} - im} \right] \gamma_\mu \right\}. \tag{18}$$

Expectedly, we can obtain a very similar expression for the $\langle j_\mu\rangle_{F,\delta}$ to that for the v.e.v. of the chiral charge density:

$$\langle j_\mu\rangle_{F,\delta} = -[3\delta^2 F_{\mu\nu} + 2\delta\tilde{F}_{\mu\nu}](iA_\nu)\mathcal{F}_b, \tag{19}$$

where the definition of the $\mathcal{F}_b$ is the same with that given in Eq. (17). Considering Eq. (19) and assuming finite values for $A_{1\sim4}$, we can write the induced EM currents for $Q_{\rm t} \neq 0$ separately for its transverse ($\perp$) and longitudinal ($\|$) components:

$$\begin{aligned}\langle j_{1,2}\rangle_{F,\delta} &\equiv j_\perp = \mp 3\delta^2\mathcal{F}_b(iA_{2,1})B_0,\\ \langle j_{3,4}\rangle_{F,\delta} &\equiv j_\| = \mp 2\delta\mathcal{F}_b(iA_{4,3})B_0.\end{aligned} \tag{20}$$

The $j_\|$ increases linearly as the B_0 grows, which is the indication of the CME as in Eq. (1). Equating the transverse and longitudinal components, we have the ratio as follows:

$$\left|\frac{j_\perp}{j_\|}\right| = \frac{3}{2}\frac{|A_{2,1}|}{|A_{4,3}|}\delta. \tag{21}$$

From Eq. (21), if we assume that $|A_{1,2}| \sim |A_{3,4}|$ along the $\hat{x}_3$-direction, in other words, the strengths of the fluctuations $a_{1\sim4}$ in Eq. (13) are almost the same to each other, it is obvious that the transverse components of the induced EM current are much smaller than those of the longitudinal ones, $|j_\perp/j_\|| \ll 1$, due to the fact that $\delta \ll 1$. This tendency is just consistent with that given in the recent lattice QCD simulation.[5]

Considering the general expression for the induced EM current in Eq. (1), the induced EM current along the $\hat{x}_3$-direction in Eq. (20) can be rewritten as

$$\langle j_3\rangle_{F,\delta} = -\frac{1}{2\pi^2}[4\pi^2\delta\mathcal{F}_b(iA_4)]B_0, \tag{22}$$

and gives the following expression for the chiral chemical potential,

$$\mu_\chi = 4\pi^2\mathcal{F}_b(i\delta A_4). \tag{23}$$

Comparing Eqs. (16) and (20), it can be easily shown that the v.e.v. of the chiral charge density is the same with the third and fourth components of the induced EM current in the leading contributions in δ:

$$\langle n_\chi\rangle_{F,\delta} = \mp\langle j_{3,4}\rangle_{F,\delta} \approx \frac{1}{2\pi^2}(i\delta A_{4,3})B_0 = \frac{1}{2\pi^2}\mu_\chi B_0, \tag{24}$$

where we have used the result from Eq. (23). Eq. (24) for A_3 is consistent with that given in Refs. 6,14 for a homogeneous magnetic field.

4. Chiral magnetic effect at low temperature

To investigate the physical quantities in hand at low but finite temperature ($T \lesssim T_c^\chi$), we want to discuss briefly how to modify the instanton variables, $\bar{\rho}$ and $\bar{R}$ at finite T. We will follow our previous work[15] and Refs. 16,17 to this end. We write the instanton distribution function at finite T with the HS caloron as follows:

$$d(\rho, T) = \underbrace{C_{N_c}\,\Lambda^b_{\mathrm{RS}}\,\hat{\beta}^{N_c}}_{\mathcal{C}}\,\rho^{b-5}\exp\left[-(A_{N_c}T^2 + \bar{\beta}\gamma n\bar{\rho}^2)\rho^2\right]. \tag{25}$$

Using the instanton distribution function in Eq. (25), we can compute the average value of the instanton size, $\bar{\rho}^2$ straightforwardly as follows:[11]

$$\bar{\rho}^2(T) = \frac{\int d\rho\,\rho^2 d(\rho,T)}{\int d\rho\, d(\rho,T)} = \frac{\left[A^2_{N_c}T^4 + 4\nu\bar{\beta}\gamma n\right]^{\frac{1}{2}} - A_{N_c}T^2}{2\bar{\beta}\gamma n}, \tag{26}$$

where $\nu = (b-4)/2$. Substituting Eq. (26) into Eq. (25), the distribution function can be evaluated further as:

$$d(\rho, T) = \mathcal{C}\,\rho^{b-5}\exp\left[-\mathcal{M}(T)\rho^2\right],\quad \mathcal{M}(T) = \frac{1}{2}A_{N_c}T^2 + \left[\frac{1}{4}A^2_{N_c}T^4 + \nu\bar{\beta}\gamma n\right]^{\frac{1}{2}}. \tag{27}$$

Finally, in order to estimate the T-dependence of the constituent-quark mass M_0, it is necessary to consider the normalized distribution function, defined as follows:

$$d_N(\rho, T) = \frac{d(\rho,T)}{\int d\rho\, d(\rho,T)} = \frac{\rho^{b-5}\mathcal{M}^\nu(T)\exp\left[-\mathcal{M}(T)\rho^2\right]}{\Gamma(\nu)}. \tag{28}$$

Now, we want to employ the large-N_c limit to simplify the expression of $d_N(\rho, T)$. Since the parameter b becomes infinity as $N_c \to \infty$, and the same for ν. In this limit, as understood from Eq. (28), $d_N(\rho, T)$ can be approximated as a δ-function:[9]

$$\lim_{N_c\to\infty} d_N(\rho, T) = \delta[\rho - \bar{\rho}(T)]. \tag{29}$$

Considering the constituent-quark mass is represented by[9]

$$M_0 \propto \sqrt{n}\int d\rho\,\rho^2\delta[\rho - \bar{\rho}(T)] = \sqrt{n(T)}\,\bar{\rho}^2(T), \tag{30}$$

we can modify the M_0 as a function of T as follows:

$$M_0 \to M_0\left[\frac{\sqrt{n(T)}\,\bar{\rho}^2(T)}{\sqrt{n(0)}\,\bar{\rho}^2(0)}\right] \equiv M_0(T) \tag{31}$$

where we will use $M_0 \approx 350$ MeV as done for zero T.

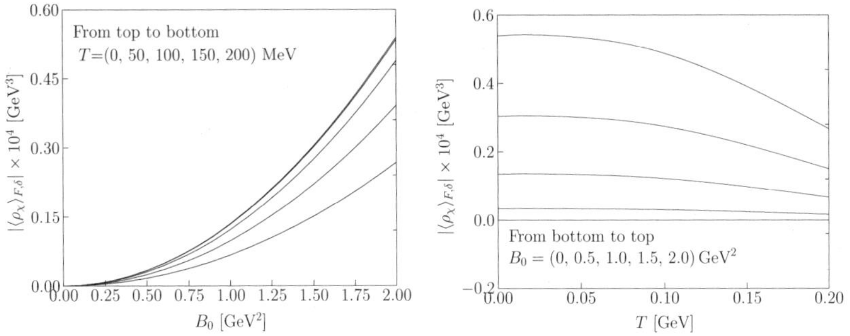

Fig. 1. Chiral local density, $|\langle\rho_\chi\rangle_{F,\delta}| \times 10^4$, as functions of B_0 (left) and T (right) for $\epsilon = 10^{-3}$. The five curves in each panel correspond to those for $T = (0, 50, 100, 150, 200)$ MeV (left, from top to bottom) and $B_0 = (0, 0.5, 1.0, 1.5, 2.0)\ \mathrm{GeV}^2$ (right, from bottom to top).

5. Numerical results and discussions

First, we present the numerical results for the local chiral density $|\langle\rho_\chi\rangle_{F,\delta}|$ in Eq. (12) as a function of B_0 in the left panel of Fig. 1. There, we depict them for different temperatures, T = (0, 50, 100, 150, 200) MeV, separately. As a trial, we choose $\epsilon = 10^{-3}$, which gives $\delta \approx 0.0136$ in Eq. (6). In other words, the P- and CP-violation effects are about 1% in comparison to non-violating quantities. Moreover, since the $\langle\rho_\chi\rangle$ is almost linearly proportional to ϵ^2, one can easily estimate it for different ϵ (or δ) values. Obviously as shown in the figure, the $|\langle\rho_\chi\rangle|$ grows rapidly with respect to B_0, manifesting the CME. As T increases, the strength of the curve decreases. At the same time, the slope of the curve, $\partial|\langle\rho_\chi\rangle_{F,\delta}|/\partial B_0$, gets smaller. This tendency means that the $|\langle\rho_\chi\rangle_{F,\delta}|$ becomes insensitive to the external magnetic field as T gets higher, and the same for the CME. The reason for this tendency can be understood by the decreasing instanton effect.

In the right panel of Fig. 1, we show the $|\langle\rho_\chi\rangle_{F,\delta}|$ as a function of T for different strengths of the magnetic fields, B_0 = (0, 0.5, 1.0, 1.5, 2.0) GeV^2. As shown there, the strength of a curve depends on that of the magnetic field. As T increases, the $|\langle\rho_\chi\rangle_{F,\delta}|$ becomes small, indicating that the instanton effect is reduced. In other words, decreasing tunneling effect corresponding to $Q_t \to 0$. Thus, unless there is another mechanism to make the Q_t nontrivial, such as the sphaleron, the $|\langle\rho_\chi\rangle_{F,\delta}|$ as well as the CME decrease monotonically with respect to T for a finite B_0 value. Interestingly, the $|\langle\rho_\chi\rangle_{F,\delta}|$ decreases much faster for the larger B_0 value, since the T-dependence of the quantity becomes more obvious and strengthened.

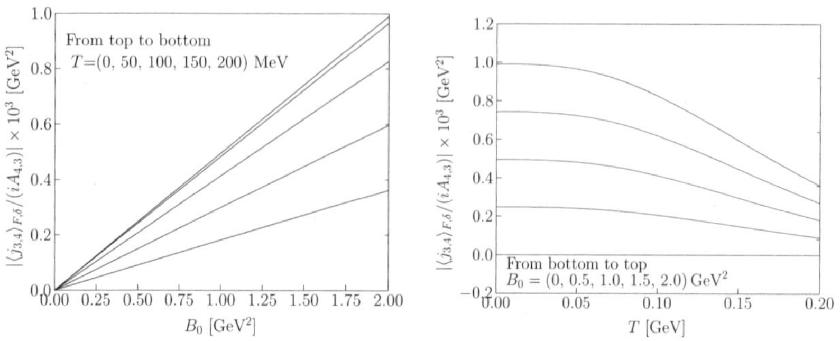

Fig. 2. Induced EM current indicating the CME, $|\langle j_{3,4}/(iA_{4,3})\rangle_{F,\delta}| \times 10^3$, as functions of B_0 (left) and T (right) for $\epsilon = 10^{-3}$. The five curves in each panel correspond to those for $T = (0, 50, 100, 150, 200)$ MeV (left, from top to bottom) and $B_0 = (0, 0.5, 1.0, 1.5, 2.0)\,\text{GeV}^2$ (right, from bottom to top).

In the left panel of Fig. 2, we draw $|\langle j_{3,4}/(iA_{4,3})\rangle_{F,\delta}|$ as functions of B_0 for $T =$ (0, 50, 100, 150, 200) MeV, separately. The curves are increasing linearly with respect to B_0 as expected, indicating the CME. Being similar to the local chiral density, the $|\langle j_{3,4}/(iA_{4,3})\rangle_{F,\delta}|$ becomes insensitive to the magnetic field, as T increases. Hence, we can conclude that the signal of the CME is weakened with respect to T at a certain value of B_0.

6. Summary and conclusion

In the present report, we have investigated the chiral magnetic effect (CME). We have observed the followings:

- The transverse EM current induced by the external magnetic field is far smaller than that of the longitudinal one, $j_{\parallel} \gg j_{\perp}$, and their ratio is proportional to δ, $|j_{\perp}/j_{\parallel}| \propto \delta \sim Q_t$.[5]
- We obtain a relation such that the strength of the chiral charge density is the same with that of the induced EM current in the $\hat{x}_3$- and $\hat{x}_4$-directions: $\langle n_\chi\rangle = |\langle j_{3,4}\rangle|$ as long as the Lorentz invariance remains unbroken.[6,7]
- The correct relation between the fourth component of the EM field and the the chiral chemical potential is found: $i\delta A_4 = \mu_\chi$.
- The CME, represented by the local chiral density and induced EM current, becomes insensitive to the external magnetic field as T increases. This can be understood by that the instanton, i.e. tunneling effect gets diminished with respect to T: $Q_t \to 0$.[5]

- The local chiral density and induced EM current as functions of T decrease faster for the larger external magnetic field, since the decreasing instanton effect is strengthened.

For more details on the present report, one can refer to Ref. 8.

Acknowledgment

This report is prepared for the proceeding for the international workshop *Hadron and Nuclear Physics* (HNP2009), 16∼19 Nov 2009, Osaka, Japan. The author appreciate the hospitality during his stay at RCNP, Osaka University, Japan, and the financial support from A. Hosaka of RCNP. He also thanks K. Fukushima, M. M. Musakhanov, and C. W. Kao for fruitful discussions. This work was supported by the NSC96-2112-M033-003-MY3 from the National Science Council (NSC) of Taiwan.

References

1. D. Kharzeev, Phys. Lett. B **633**, 260 (2006).
2. S. A. Voloshin, Phys. Rev. C **70**, 057901 (2004).
3. D. E. Kharzeev *et al.*, Nucl. Phys. A **803**, 227 (2008).
4. S. A. Voloshin [STAR Collaboration], arXiv:0806.0029 [nucl-ex].
5. P. V. Buividovich *et al.*, arXiv:0907.0494 [hep-lat].
6. K. Fukushima *et al.*, Phys. Rev. D **78**, 074033 (2008).
7. E. D'Hoker and J. Goldstone, Phys. Lett. B **158**, 429 (1985).
8. S. i. Nam, arXiv:0911.0509 [hep-ph], accepted for publication in Phys. Rev. D.
9. D. Diakonov, M. V. Polyakov and C. Weiss, Nucl. Phys. B **461**, 539 (1996).
10. D. Diakonov, Prog. Part. Nucl. Phys. **51**, 173 (2003).
11. T. Schafer and E. V. Shuryak, Rev. Mod. Phys. **70**, 323 (1998).
12. J. S. Schwinger, Phys. Rev. **82**, 664 (1951).
13. J. F. Nieves and P. B. Pal, Phys. Rev. D **73**, 105003 (2006)
14. N. Sadooghi and A. Jafari Salim, Phys. Rev. D **74**, 085032 (2006).
15. S. i. Nam, arXiv:0905.3609 [hep-ph].
16. B. J. Harrington and H. K. Shepard, Nucl. Phys. B **124**, 409 (1977).
17. D. Diakonov and A. D. Mirlin, Phys. Lett. B **203**, 299 (1988).

WALL VORTICES IN THE MULTI-COMPONENT DUAL SUPERCONDUCTOR THEORY

P. M. ZHANG[1,2,3], C. K. WANG[3,4], L. P. ZOU[3,4], D. J. JIA[5] and X. Y. WANG[6]

[1]*Research Center for Hadron and CSR Physics, Lanzhou University & Institute of Modern Physics of CAS, Lanzhou, China*

[2]*Center of Theoretical Nuclear Physics, National Laboratory of Heavy Ion Accelerator, Lanzhou, China*

[3]*Institute of Modern Physics, Chinese Academy of Sciences, Lanzhou, China*

[4]*Graduate University of Chinese Academy of Sciences, Beijing 100049, China*

[5] *College of Physics and Electronic Engineering, Northwest Normal University, Lanzhou, China*

[6] *School of Science, Lanzhou University of Technology, Lanzhou, China*

In terms of the decomposition of gauge potential, the two-component dual superconductor theory was obtained based on the $SU(2)$ Yang-Mills theory. We investigate the wall vortices and knots in the multi-component dual superconductor theory.

1. Introduction

The mechanism of dual superconductivity[1–4] has been proposed for many years to tackle the problem of confinement in pure Yang-Mills theories, in which the QCD vacuum is regarded as a superconductor of chromomagnetic charges corresponding to the confinement of chromoelectric charges. Due to its clear picture to understand the color confinement, the dual superconductivity is regarded as one of the most promising effective models of QCD. Even more, the dynamical chiral symmetry breaking could be demonstrated well according to color monopole condensate.[5–9]

In the dual superconductor theory, various hadronic states (meson, baryon and glueball etc.) are just the objects with different shape (rod, star, or rings) where the quarks together with gluons are confined in the form of hadronic color-electric flux tube. Furthermore, the flux-tube model can be applied for the study of scattering properties of hadrons.[10] The dual superconductor theory can be easily applied to the string picture of hadrons also.

Cho decomposition,[11–13] and the following Faddeev-Niemi connection[14–17] have been developed to find out the effective degree of freedom in order to explore the problem of confinement. In this approach, monopoles arise naturally with the new variable $\hat{n}$, which is a general local color direction. And without special gauge fixing, the gluon fields could be decomposed into the background part and dynamical part. The decomposition theory of gauge potential is a powerful tool not only in the particle physics,[18–22] also in the condensed matter physics[23–25] and other fields.[26,27] It has been suggested that the two-component dual superconductor theory could be achieved with the decomposition of gauge potential.[28]

Recently, wall vortex was proposed in the Abelian-Higgs model in order to understand the bag model.[29–32] The promising result is wall vortices are stable with their huge flux, which against the situation in the ordinary superconductivity. The wall vortex provide a natural boundary for the Higgs vacuum and Columb vacuum, and confine the chromoelectric charge inside according to the dual superconductivity theory. However, since dual superconductor theory of QCD is usually regarded as the multi-component theory,[33–35] it is necessary to check the wall vortex in the multi-component dual Ginzburg-Landau theory.

In this paper, we outline the two-component dual superconductor theory according to the Faddeev-Niemi connection and Cho decomposition in Sec II. Then the wall vortices are investigated and the solutions are presented in Sec III. In Sec IV, we discuss knots which are related with the glueball. The conclusion are presented in the last section.

2. Two-component dual superconductor theory

Dual superconducting scenario is formulated by the dual Ginzburg-Landau (GL) theory[36] as a phenomenological theory for the nonperturbative QCD. In order to obtain the dual GL theory, usually people consider Abelian gauge fixing (selecting the off-diagonal gauge degrees of freedom) and Abelian projection[37] (dropping off-diagonal gauge degrees of freedom). The Abelian dominance for the nonperturbative quantities was conjectured,[38] which reduces the $SU(3)$ gauge theory into the $[U(1)]^2$ Abelian gauge theory with additional two types of color-magnetic monopole currents.

However, it is believed that the decomposition of gauge potential could investigate the dual superconducting picture more naturally, including Abelian dominance and monopole condensate.[39,40] In the following we outline the two-component dual GL theory based on the idea.

2.1. *Niemi's approach*

Employing a novel decomposition of gauge potential,[16] Niemi obtained two-component dual GL theory[28] based on the pure $SU(2)$ Yang-Mills theory.

In general, the $SU(2)$ gauge potential could be separated into the Cartan part $A_\mu^3 = A_\mu$ and off-diagonal part

$$W_\mu^\pm = A_\mu^1 \pm iA_\mu^2. \tag{1}$$

Introducing

$$D_\mu^\pm = \partial_\mu \pm igA_\mu, \quad D_\mu^{ij} = \delta^{ij}\partial_\mu - g\varepsilon^{ij}A_\mu$$

one can find

$$-\frac{1}{4}(F_{\mu\nu}^a)^2 = -\frac{1}{4}(F_{\mu\nu} + G_{\mu\nu})^2 - \frac{1}{2}F_{\mu\nu}G_{\mu\nu} + \frac{1}{2}|D_\mu^+ W_\nu|^2 - \frac{1}{2}|D_\mu^+ W_\mu|^2 \tag{2}$$

where

$$F_{\mu\nu} = \partial_\mu A_\nu - \partial_\nu A_\mu, \quad G_{\mu\nu} = \frac{ig}{2}(W_\mu W_\nu^* - W_\nu W_\mu^*)$$

and

$$(F_{\mu\nu}^3)^2 = (F_{\mu\nu} + G_{\mu\nu})^2 + 2G_{\mu\nu}F_{\mu\nu}.$$

With the decomposition proposed by Faddeev and Niemi[16,17]

$$W_\mu = \psi_1 e_\mu + \psi_2^* e_\mu^*,$$

one can obtain

$$\begin{aligned}|(\partial_\mu + igA_\mu^3)W_\nu|^2 &= |(D_\mu^+ + i\Gamma_\mu)\psi_1|^2 + |(D_\mu^- + i\Gamma_\mu)\psi_2|^2 \\ &\quad + \rho^2|(\partial_\mu - i\Gamma_\mu)e_\nu|^2 + \psi_1\psi_2(\partial_\mu e_\nu)^2 + \psi_1^*\psi_2^*(\partial_\mu e_\nu^*)^2\end{aligned}$$

where

$$\rho^2 = (A_\mu^1)^2 + (A_\mu^2)^2 = W_\mu W_\mu^*, \quad \Gamma_\mu = ie_\nu\partial_\mu e_\nu^*.$$

With above results, Niemi proposed that the e_μ part can be removed totally, only the scalar field $\psi_{1,2}$ are left. Then, the two-component GL theory was obtained

$$\begin{aligned}\mathcal{L} = &-\frac{1}{4}F_{\mu\nu}^2 + |(\partial_\mu + igA_\mu)\psi_1|^2 + |(\partial_\mu - igA_\mu)\psi_2|^2 \\ &- \lambda(|\psi_1|^2 - |\psi_2|^2)^2 + \gamma(\psi_1^*\psi_2 + \psi_2^*\psi_1)\end{aligned} \tag{3}$$

where the last term is the Josephsen coupling term.

2.2. *Cho decomposition*

The $SU(2)$ gauge potential can be expressed with a local color direction $\hat{n}$

$$\vec{A}_\mu = A_\mu \hat{n} - \frac{1}{g}\hat{n} \times \partial_\mu \hat{n} + \vec{X}_\mu = \hat{A}_\mu + \vec{X}_\mu, \tag{4}$$

where $A_\mu = \hat{n} \cdot \vec{A}_\mu$ is the "electric" potential and $\hat{n} \cdot \vec{X}_\mu = 0$. The restricted potential $\hat{A}_\mu$ is defined with

$$\hat{D}_\mu \hat{n} = \partial_\mu \hat{n} + g\hat{A}_\mu \times \hat{n} = 0 \tag{5}$$

where $\hat{A}_\mu$ by itself describes an $SU(2)$ connection which enjoys the full $SU(2)$ gauge degrees of freedom. $\hat{A}_\mu$ actually has a dual structure

$$\hat{F}_{\mu\nu} = (F_{\mu\nu} + H_{\mu\nu})\hat{n},$$

$$F_{\mu\nu} = \partial_\mu A_\nu - \partial_\nu A_\mu, \quad H_{\mu\nu} = -\frac{1}{g}\hat{n} \cdot (\partial_\mu \hat{n} \times \partial_\nu \hat{n}) = \partial_\mu C_\nu - \partial_\nu C_\mu \tag{6}$$

where C_μ is the "magnetic" potential.

Since $\hat{A}_\mu$ is already a $SU(2)$ connection, the restricted $SU(2)$ Yang-Mills-Higgs (YMH) theory can be constructed as following[41]

$$\mathcal{L} = -|\hat{D}_\mu \Phi|^2 + V(\Phi, \Phi^\dagger) - \frac{1}{4}\hat{F}^2_{\mu\nu} \tag{7}$$

where

$$\hat{D}_\mu \Phi = (\partial_\mu + \frac{g}{2i}\vec{\sigma} \cdot \hat{A}_\mu)\Phi, \quad V(\Phi, \Phi^\dagger) = \mu^2 \Phi^\dagger \Phi - \frac{\lambda}{2}(\Phi^\dagger \Phi)^2. \tag{8}$$

Introduce two orthonormal doublets ξ and η which form a basis and $\Phi = \phi_1 \xi + \phi_2 \eta$, the above theory is reduced to

$$\mathcal{L} = -\frac{1}{4}\mathcal{F}^2_{\mu\nu} - \sum_{i=1}^{2} \frac{1}{2}|D_\mu \phi_i|^2 - V(|\phi_i|)$$

with $U(1)$ potential $\mathcal{A}_\mu$

$$\mathcal{A}_\mu = A_\mu + C_\mu, \quad D_\mu = \partial_\mu + ig\mathcal{A}_\mu$$

which is the two-component dual GL theory.

3. Wall vortices in the two-component dual GL theory

In this part, we consider the wall vortices for the large flux limit in the following two-component GL theory

$$\mathcal{L} = -|(\partial_\mu + igA_\mu)\phi|^2 + \mu\phi^\dagger\phi - \frac{\lambda}{2}(\phi^\dagger\phi)^2 - \frac{1}{4}F_{\mu\nu}^2. \tag{9}$$

With the ansatz for vortices

$$\phi = \frac{1}{\sqrt{2}}\rho\begin{pmatrix}\cos\frac{f(r)}{2}\exp(-in\varphi)\\ \sin\frac{f(r)}{2}\end{pmatrix}, \quad A_\mu = \frac{n}{g}A(r)\partial_\mu\varphi, \tag{10}$$

the Hamiltonian becomes

$$H = \frac{1}{2}\dot{\rho}^2 + \frac{1}{8}\rho^2(\dot{f}^2 + \frac{n^2}{r^2}\sin^2 f) + \frac{n^2}{2r^2}\rho^2(A - \frac{\cos f + 1}{2})^2 + \frac{n^2\dot{A}^2}{2g^2r^2} + \frac{\lambda}{8}(\rho^2 - \rho_0^2)^2 \tag{11}$$

where $\rho_0 = \sqrt{2\mu/\lambda}$.

Under the boundary condition

$$\rho(0) = 0,\ \ \rho(\infty) = \rho_0;\ \ f(0) = \pi,\ \ f(\infty) = 0;\ \ A(0) = 0,\ \ A(\infty) = 1,$$

we could find the normal vortex solutions of above system,[43] the corresponding topological charge is

$$Q = \oint A_\mu dx^\mu = \frac{4\pi n}{g}.$$

When $\lambda = g^2$, there are self-dual solutions

$$\begin{aligned} \dot{\rho} \pm \frac{n}{r}(A - \cos^2\frac{f}{2})\rho &= 0 \\ \dot{f} \pm \frac{n}{r}\sin f &= 0 \\ H \pm \frac{\sqrt{\lambda}}{2}(\rho^2 - \rho_0^2) &= 0. \end{aligned} \tag{12}$$

For the large-n, the system could be solved explicitly almost everywhere [a]

$$\rho(r) = 1, \quad f(r) = 2ArcTan(Cr^{-n}), \quad A(r) = \frac{1}{1 + C^2r^{-2n}} \tag{13}$$

where C is an integral constant. For the large flux limit ($n \to \infty$), we have

$$\rho(r) = 1,\ \ A(r) = \begin{cases} 0, & r < 1 \\ 1, & r > 1 \end{cases},\ \ f(r) = \begin{cases} \pi, & r < 1 \\ 0, & r > 1 \end{cases} \tag{14}$$

Above solutions of wall vortices are shown with different n in Fig. 1 and Fig. 2. With these solutions, it is obvious that the wall vortices appear when the $n \to$

[a] at the point $r = 1$, there is singularity.

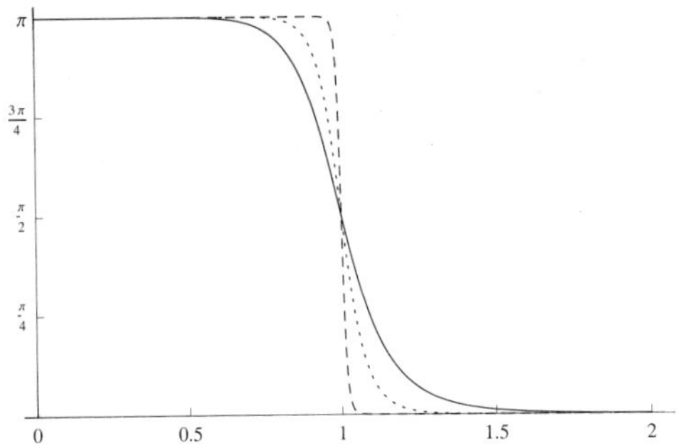

Fig. 1. $f(r)$ with different n ($C = 1$): $n = 10$ (solid line); $n = 20$ (dotted line); $n = 100$ (dashed line)

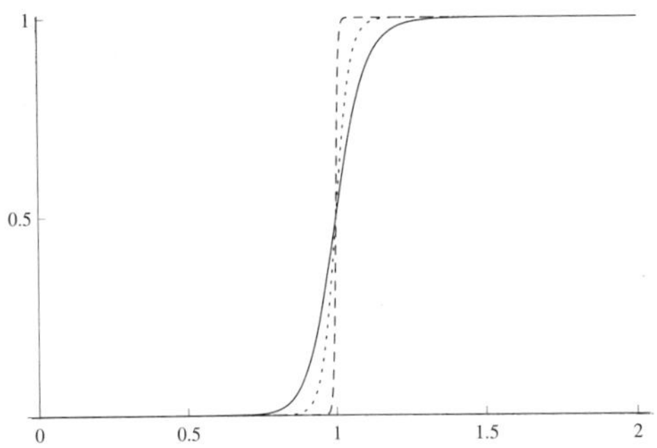

Fig. 2. $A(r)$ with different n ($C = 1$): $n = 10$ (solid line); $n = 20$ (dotted line); $n = 100$ (dashed line)

∞. Therefore, we conclude that the wall vortices exist in the two-component GL model.

At last, since it is argued that the $SU(N)$ QCD theory corresponds to the $U(1)^{N-1}$ GL theory, the multi-component theory should be considered also. Here we explain that the the similar exotic objects could be obtained in the three-component GL theory. The three-component GL model is

$$\mathcal{L} = -\frac{1}{4}F_{\mu\nu}^2 - \frac{1}{2}\sum_{i=1,3}|D_\mu\phi_i|^2 - V(|\phi_i|). \tag{15}$$

Considering the ferromagnetic phase

$$\zeta = e^{i\theta}U\begin{pmatrix}1\\0\\0\end{pmatrix} = e^{i(\theta-\tau)}\begin{pmatrix}e^{-i\alpha}\cos^2\frac{\beta}{2}\\ \sqrt{2}\cos\frac{\beta}{2}\sin\frac{\beta}{2}\\ e^{i\alpha}\sin^2\frac{\beta}{2}\end{pmatrix}$$

which means the spin $< F >= \zeta_a^* F_{ab}\zeta_b$ satisfied $< F >^2 = 1$.

The model is reduced to

$$F = \int dr[\frac{1}{2}(\partial_i\rho)^2 + \frac{\rho^2}{4}(\partial_i\hat{n})^2 - \mu\rho^2 + \frac{\rho^4}{2}(c_0 + c_2) + \frac{\rho^2}{4}B_i^2 + \frac{1}{4g^2}\left((\partial_i B_j - \partial_j B_i) - \hat{n}\cdot(\partial_i\hat{n}\times\partial_j\hat{n})\right)^2]$$

where

$$\hat{n} = \zeta^\dagger\vec{\sigma}\zeta = (\sin\beta\cos\alpha, \sin\beta\sin\alpha, \cos\beta)^T, \quad B_i = A_i - \frac{i}{g}\zeta^\dagger\partial_i\zeta$$

which is equivalent to the two-component GL model.

Therefore, wall vortices exist in the three-component GL theory also.

4. Knot in the Two-component dual GL theory

Firstly, with the Lagrangian (9), we show the two-component GL theory is equivalent to generalized Faddeev-Niemi model.[42]

Let's introduce

$$\phi = \frac{1}{\sqrt{2}}\rho\xi, \quad \xi^\dagger\xi = 1,$$

$$\vec{n} = \xi^\dagger\vec{\sigma}\xi, \quad \vec{n}^2 = 1,$$

and

$$C_\mu = \frac{2i}{g}\xi^\dagger\partial_\mu\xi$$

then the Hamiltonian becomes

$$H = \frac{1}{2}(\partial_\mu\rho)^2 + \frac{1}{2}g^2\rho^2B_\mu^2 + \frac{1}{8}\rho^2(\partial_\mu\vec{n})^2 + \frac{1}{4}\left[G_{\mu\nu} - \frac{1}{2g}\vec{n}\cdot(\partial_\mu\vec{n}\times\partial_\nu\vec{n})\right]^2 + V(\rho) \quad (16)$$

where

$$G_{\mu\nu} = \partial_\mu B_\nu - \partial_\nu B_\mu, \quad B_\mu = A_\mu - C_\mu/2$$

which is the generalized Faddeev-Niemi model.

As we know, the generalized Faddeev-Niemi model admits a 3-dimension finite energy object, knot labeled by Hopf Number[44]

$$Q_{CS} = -\frac{1}{8\pi^2}\int \varepsilon^{ijk}A_iF_{jk}dx^3 = mn$$

where m means the times of flux passing through the disk of the ring and n means the times of flux passing along the ring.

Here, the knot can be regarded as a topological glueball of the twisted chromoelectric flux ring.[17,21,45] It is made of two real (physical) chromoelectric flux rings linked together, whose knot quantum number is given by the linking number of two flux rings.

5. Conclusion

In terms of decomposition theory of gauge potential, the two-component dual superconductor theory could be obtained from $SU(2)$ pure Yang-Mills theory or restricted Yang-Mills-Higgs theory. Wall vortices are obtained in the two-component Ginzburg-Landau theory. Furthermore, we show that the three-component Ginzburg-Landau theory could be reduced to two-component case in the ferromagnetic phase, which means wall vortices exist in the three-component system also. At last, knot, as a topological stable solution of multi-component dual superconductor theory, can be understood as the glueball.

Acknowledgments

P. M. Zhang wishes to thank A. J. Niemi, Y. M. Cho, and H. Toki for the illuminating discussions. This work was supported in part by NSFC (Grant No. 10604024), West Light Foundation of The Chinese Academy of Sciences, SRF for ROCS, SEM, Center of Theoretical Nuclear Physics, National Laboratory of Heavy Ion Accelerator of Lanzhou and National Program on Key Basic Research Project (Grant No. 2010CB832902).

References

1. Y. Nambu, Phys. Rev. D10, 4262 (1974).
2. S. Mandelstam, Phys. Rept. 23C, 245 (1976).
3. G. 't Hooft, Nucl. Phys. B 138, 1 (1978).
4. Georges Ripka, Dual Superconductor Models of Color Confinement, (Lecture Notes in Physics 639, Springer Verlag 2005).
5. H. Suganuma, S. Sasaki and H. Toki, Nucl. Phys. B435, 207 (1995).
6. S. Sasaki, H. Suganuma and H. Toki, Prog. Theor. Phys. 94, 373 (1995).
7. S. Sasaki, H. Suganuma and H. Toki, Phys. Lett. B387, 145 (1996).
8. H. Toki, S. Sasaki, H. Ichie, H. Monden and H. Sugamuma, Jour. Korean Phys. Soc (Proc. Suppl.) 29, S346 (1996).
9. H. Toki and H. Suganuma, Progr. Part. Nucl. Phys. 45, S397 (2000).
10. N. Isgur and J. Paton, Phys. Lett. 124B, 247 (1983).
11. Y. M. Cho, Phys. Rev. D 21, 1080 (1980).
12. Y. M. Cho, Phys. Rev. Lett. 46, 302 (1981).
13. Y. S. Duan, M. L. Ge, Sinica Sci. 11, 1072 (1979) (in Chinese).
14. L. D. Faddeev, A. J. Niemi, Phys. Rev. Lett. 82, 1624 (1999).
15. L. D. Faddeev, A. J. Niemi, Phys. Lett. B 449, 214 (1999).
16. L. D. Faddeev, A.J. Niemi, Phys. Lett. B 525, 195 (2002).
17. L. D. Faddeev and A. J. Niemi, Nucl. Phys. B776, 38 (2007).
18. S. Shabanov, Phys. Lett. B 458, 322 (1999).
19. Y. M. Cho, H. W. Lee and D. G. Pak, Phys. Lett. B 525, 347 (2002).
20. Y. M. Cho, M. L. Walker and D. G. Pak, J. High Energy Phys. 05, 073 (2004).

21. Y. M. Cho and D. G. Pak, Phys. Lett. B 632, 745 (2006).
22. L. E. Oxman, JHEP 12, 089 (2008).
23. Y. S. Duan, H. Zhang and S. Li, Phys. Rev. B58, 125 (1998).
24. P. M. Zhang, Y. S. Duan, X. Liu and C. K. Zhong, Ann. Phys. 313, 16 (2004).
25. P. M. Zhang, Y. S. Duan and H. Zhang, Physica A 370, 245 (2006).
26. Y. S. Duan and X. H. Meng, J. Math. Phys. 34, 1149 (1993).
27. P. M. Zhang, Y. S. Duan and L. M. Cao, Mod. Phys. Lett. A18, 2587 (2003); P. M. Zhang, Y. S. Duan and J. R. Ren, ibid. 19, 745 (2004).
28. A. J. Niemi, JHEP 0408 (2004) 035.
29. S. Bolognesi, Nucl. Phys. B 730, 127 (2005).
30. S. Bolognesi, Nucl. Phys. B 730, 150 (2005).
31. S. Bolognesi and S. B. Gudnason, Nucl. Phys. B 741, 1 (2006).
32. S. Bolognesi, Nucl. Phys. B 752, 93 (2006).
33. Y. Koma, H. Suganuma, H. Toki, Phys. Rev. D 60, 074024 (1999).
34. Y. Koma, H. Toki, Phys. Rev. D 62, 054027 (2000).
35. Y. Koma, M. Koma, D. Ebert and H. Toki, Nucl. Phys. B648, 189 (2003).
36. T. Suzuki, Prog. Theor. Phys. 80, 929 (1988); S. Maedan and T. Suzuki, ibid. 81, 229 (1989).
37. G. 't Hooft, Nucl. Phys. B 190, 455 (1981).
38. Z. F. Ezawa and A. Iwazaki, Phys. Rev. D 25, 2681 (1982); D 26, 631 (1982).
39. Y. M. Cho, Phys. Rev. D 62, 074009 (2000).
40. Y. M. Cho and D. G. Pak, Phys. Rev. D 65, 074027 (2002).
41. Y. M. Cho, Phys. Rev. B 72, 212526 (2005); Y. M. Cho and P. M. Zhang, ibid. 73, 180506 (2006).
42. L. D. Faddeev and A. Niemi, Nature 387, 58 (1997).
43. Y. M. Cho and P. M. Zhang, Eur. Phys. J. B 65, 155 (2008).
44. E. Babaev, L. Faddeev, A. Niemi, Phys. Rev. B 65, 100512 (2002).
45. Y. M. Cho, B. S. Park and P. M. Zhang, Int. J. Mod. Phys. A 23, 267 (2008).

BARYONS AND BARYONIC MATTER IN HOLOGRAPHIC QCD

KANABU NAWA[1], HIDEO SUGANUMA[2], ATSUSHI HOSAKA[1] and TORU KOJO[3]

[1] *Research Center for Nuclear Physics (RCNP), Osaka University, Osaka, Japan*
[2] *Department of Physics, Kyoto University, Kyoto, Japan*
[3] *RIKEN BNL Research Center, Brookhaven National Laboratory, NY, USA*

We study baryons and baryonic matter in holographic QCD with D4/D8/$\overline{\text{D8}}$ multi-D brane system in type IIA superstring theory. Baryons are described as "Skyrmions" of a four-dimensional meson effective action which is derived from the five-dimensional Yang-Mills theory formulated in holographic QCD. We compare our results with those in the hidden local symmetry approach on the flat space-time, and show that the curved extra-dimension largely changes the roles of (axial) vector mesons for the baryons.

Keywords: AdS/CFT, Gauge/Gravity duality, Skyrmions, Finite Density QCD

1. Review of holography

Holography is a new concept of superstring theory proposed by Maldacena in 1997[1] as a duality between two theories having different spatial dimensions. The essential object is "D_p brane" as a $(p+1)$-dimensional membrane in 10-dimensional space-time.[2] Actually, D_p brane appears as a "soliton" of fundamental strings (F-strings), on which open strings attach their ends with the Dirichlet boundaries. The fluctuation modes of open strings gives the gauge fields $A_{\mu=0\sim p}$, so that $(p+1)$-dimensional *gauge theory* appears on the surface of D_p brane (see Fig. 1, Left). Moreover, due to the "open-closed duality" of the F-strings, D_p brane can be seen as the source of the closed strings, which gives the metric g_{MN} $(M, N = 0 \sim p+1)$, or graviton, around D_p brane. In this sense, $\{(p+1)+1\}$-dimensional *supergravity* appears around D_p brane (see Fig. 1, Right). Holography indicates that $(p+1)$-dimensional gauge theory and $\{(p+1)+1\}$-dimensional supergravity are equivalent with each other through D_p brane, often called the "Gauge/Gravity duality" (see Fig. 2).

In fact, holography provides the strong-weak duality (S-duality) between the gauge theory and the supergravity; their coupling strengths are inversely

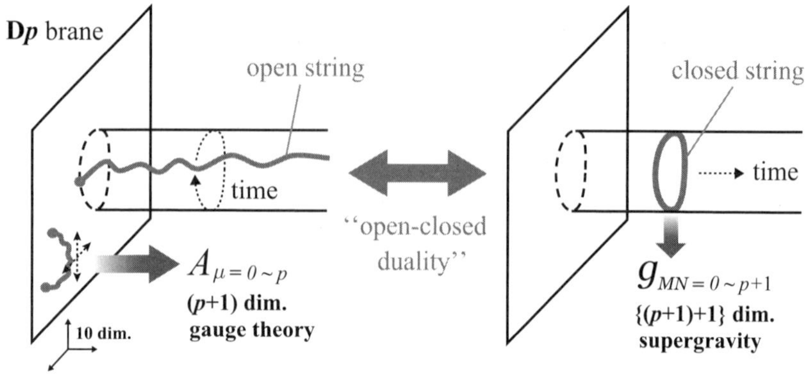

Fig. 1. Interaction diagram between D_p brane and F-strings. Fluctuation modes of open strings induce the $(p+1)$-dimensional gauge theory with A_μ ($\mu = 0 \sim p$) on the surface of D_p brane (Left). Due to the "open-closed duality" of F-strings, D_p brane can be seen as the source of closed strings, inducing the $\{(p+1)+1\}$-dimensional supergravity around D_p brane with metric g_{MN} $(M, N = 0 \sim p+1)$ (Right).

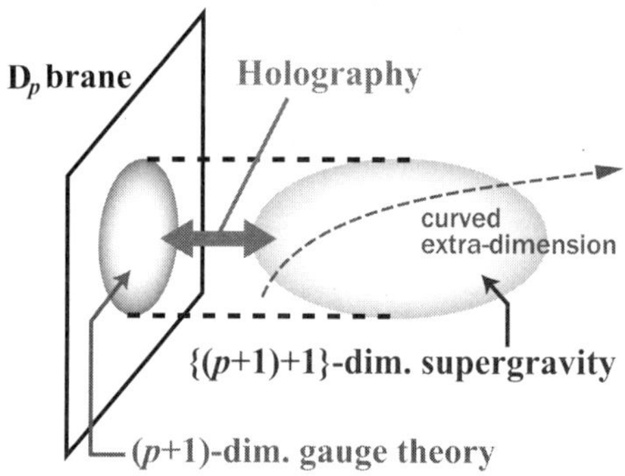

Fig. 2. Schematic figure of holographic duality between $(p+1)$-dimensional gauge theory and $\{(p+1)+1\}$-dimensional supergravity through D_p brane.

related with each other. Therefore, the non-perturbative (*strong-coupling*) aspects of one side can be analyzed from the perturbative (*weak-coupling*) aspects of the other "dual" side.

2. Baryons and baryonic matter in holographic QCD

If QCD as $SU(3)_c$ gauge theory with quarks and gluons can be constructed on a special configuration of D branes, its *non-perturbative* aspects can be analyzed from the *tree-level* dual supergravity. This strategy is called "holographic QCD". In 2005, Sakai and Sugimoto succeeded in constructing

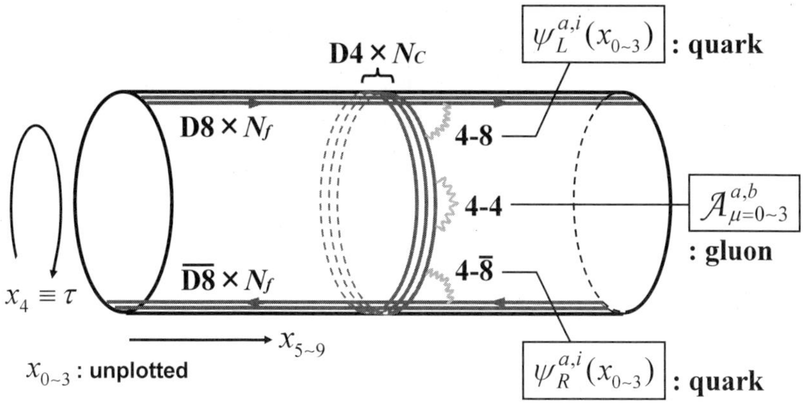

Fig. 3. Sakai-Sugimoto model with D4/D8/$\overline{\text{D8}}$ multi-D brane system in type IIA superstring theory. Flat coordinates $x_{0\sim3}$ are not plotted. Fluctuation modes of 4-4, 4-8, and 4-$\bar{8}$ open strings give gluons and quarks with $SU(3)_c$ symmetry and chiral symmetry.

the massless QCD on the D4/D8/$\overline{\text{D8}}$ multi-D brane system in type IIA superstring theory, called the Sakai-Sugimoto model[3] (see Fig. 3). In this model, many of the *meson* phenomenologies can be successfully derived from the classical supergravity description of the D branes. However, such classical supergravity is found to be dual with strong-coupling "large-N_c" QCD. In the large-N_c limit, QCD reduces to a theory of weakly-interacting mesons,[4] while baryons appear as their soliton solutions.[5] In this sense, the baryon description becomes non-trivial in the large-N_c holographic model.

In our study, we describe the baryon as a chiral soliton, i.e., "Skyrmion" in holographic QCD.[6,7] We find a stable soliton solution in the hedgehog configuration with the mass $M_{\rm B} \simeq 834$MeV and the root-mean-square mass radius $\sqrt{\langle r^2 \rangle} \simeq 0.37$fm. We find that, due to the heavy ρ meson field in the core region of the baryon, the baryon size tends to shrink by $\sim 10\%$ (see Fig. 4, Left). This result is different from those in the flat space phenomenology like the hidden local symmetry (HLS),[8,9] where the ρ meson field gives a strong repulsion for the baryon (see Fig. 4, Right). In fact, the meson masses come from the "oscillation energy" of the meson wave function in the extra-fifth dimension in the holographic approach. Therefore, a *curvature* in the extra-dimension is essential to give the *discrete* mass spectra for mesons in four-dimensional space-time as in QCD (see Fig. 5). Such curvature now modifies the meson effective action, which can largely changes the roles of ρ meson field for the baryon as shown in Fig. 4.[10]

We also investigate the baryonic matter in holographic QCD by treating

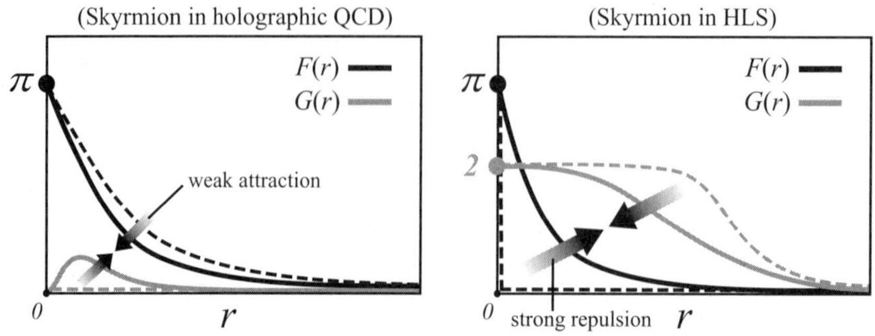

Fig. 4. Skyrmions in holographic QCD (Left) and in hidden local symmetry (HLS) (Right). $F(r)$ and $G(r)$ are hedgehog configurations of π and ρ meson fields as $\pi_a(x) = \hat{x}_a F(r)$ and $\rho_{ia}(x) = \varepsilon_{iab}\hat{x}_b G(r)/r$, respectively. Dashed lines correspond to the configurations without the interaction between π and ρ mesons. $G(r)$ has different boundaries at $r = 0$, i.e., for the HLS approach as $G(0) = 2$, and for holographic QCD as $G(0) = 0$ due to the curved extra-dimension.

the system of single brane-induced Skyrmion on a three-dimensional closed manifold S^3 with a radius R, which models the baryonic matter on $\boldsymbol{R}^3$ with a baryon-number density $\rho_{\rm B} = 1/(2\pi^2 R^3)$ (see Fig. 6). The density dependence of baryon mass and root-mean-square mass radius are discussed as a function of R.[11] In Fig. 7, we show the density dependence of total energy density of single baryon by changing the value of R, where one can see the "delocalization" of the baryon in high density phase. The critical radius into uniform phase is found as $R_{\rm crit} \simeq 1.19$ with ρ meson field (Fig. 7, Left) and $R_{\rm crit} = \sqrt{2}$ without ρ meson field (Fig. 7, Right). In fact, ρ meson field gives a slight shrinkage of baryon size,[6] so that the smaller radius R of S^3, i.e., the larger baryon-number density is needed for the delocalization. By taking the inputs; $f_\pi = 92.4$MeV and $m_\rho = 776$MeV, we get the critical densities as $\rho_{\rm B} \simeq 7.12\rho_0$ (Fig. 7, Left) and $\rho_{\rm B} \simeq 4.26\rho_0$ (Fig. 7, Right) with the normal nuclear density $\rho_0 \simeq 0.17{\rm fm}^{-3}$.

We can also discuss the QCD phase transitions with their order parameters. For example, the delocalization phase transition of baryon can be seen by the spatial fluctuation of energy density around the uniform distribution (Fig. 8). In fact, such delocalization can be related with the QCD deconfinement.[12] Furthermore, we can also estimate the chiral-symmetry restoration by taking the spatially-averaged chiral condensate as the global order parameter of bulk hadronic matter (Fig. 9). Such condensate is related with the global order parameter of ferromagnetic material with magnetic domain-walls. Here it is noted that, within the Skyrme model, deconfine-

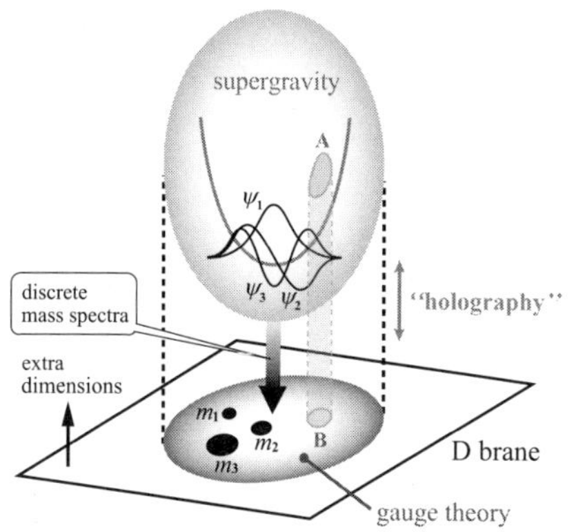

Fig. 5. Schematic figure of holographic duality between gauge theory as QCD on D brane and supergravity around D brane. Masses of mesons $m_{n=1,2,3,\cdots}$ are given by the oscillation of meson wave functions $\psi_n(z)$ in the extra-dimension z.[6] Therefore, the curvature in the extra-dimension around D brane is essential to give the *discrete* meson mass spectra as QCD on D brane. Then, regions A and B have different curvatures, and the gauge symmetry on A will be distorted on B except for the gauge symmetry of QCD.

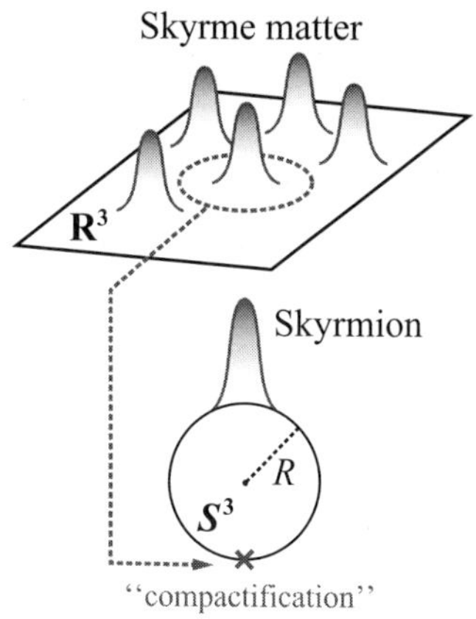

Fig. 6. Skyrme matter on a flat space $\mathbf{R}^3$, and single Skyrmion on a three-dimensional closed manifold S^3. These two can be related through the compactification of a unit cell of the Skyrme matter on $\mathbf{R}^3$ into S^3 with a radius R. Interactions between Skyrmions are simulated by the curvature of S^3. Total baryon-number density is realized on S^3 as $\rho_{\mathrm{B}} = 1/(2\pi^2 R^3)$, so that the decrease of R models the increase of the total baryon-number density in the medium on $\mathbf{R}^3$.

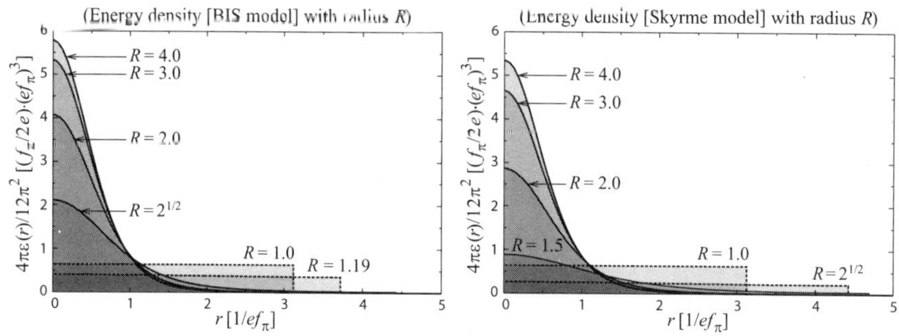

Fig. 7. Energy density of single baryon with ρ meson field (Left) and without ρ meson field (Right) in holographic QCD. Dashed lines correspond to the uniform configurations.

ment and chiral-symmetry restoration can be closely related through the classical configuration of meson fields.

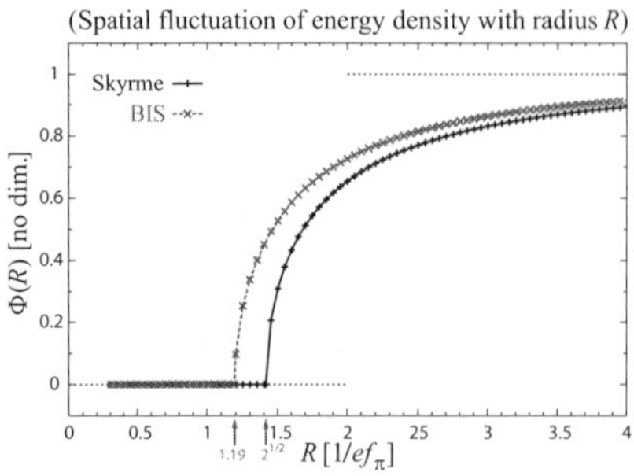

Fig. 8. Order parameter of delocalization: spatial fluctuation $\Phi(R)$ of energy density of single baryon with ρ meson field (BIS) and without ρ meson field (Skyrme) for radius R.

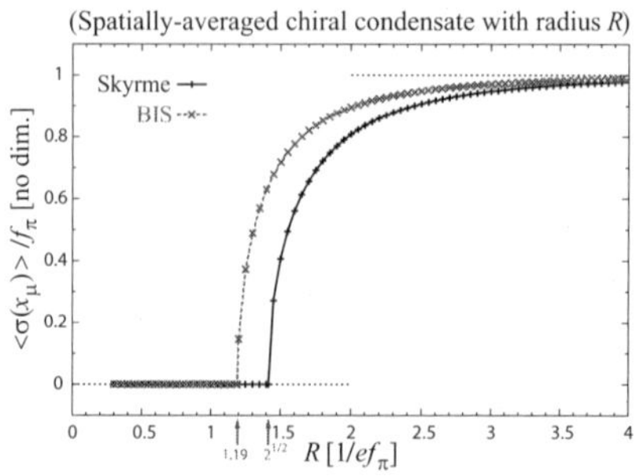

Fig. 9. Order parameter of chiral symmetry restoration: spatially-averaged chiral condensate $\langle\sigma(x_\mu)\rangle/f_\pi$ with ρ meson field (BIS) and without ρ meson field (Skyrme) for radius R.

3. Summary and outlook

We have studied baryons[6,7] and baryonic matter[11] in holographic QCD with D4/D8/$\overline{\text{D8}}$ multi-D brane system in type IIA superstring theory. We have found that holographic QCD has a stable hedgehog soliton solution as a baryon, which is basically composed by the large amplitude pion field including ρ meson field in the interior region. By comparing our results with those in the flat-space phenomenology like the hidden local symmetry, we have shown that the roles of ρ meson field for the baryon can be largely changed due to the curved extra-dimension.[10]

Owing to the "S-duality" of holography, the *non-perturbative* natures in the gravitational side, e.g., the Hawking temperature[13] and the black hole formation[14] can be analyzed from the dual *tree-level* gauge theory side. Furthermore, the high-T_c superconductor in the *strong-coupling* fermionic system like the cuprate or the organic material can also be analyzed from the dual *classical* supergravity side, which is called the "holographic superconductor".[15] We hope that holography will continuously provide powerful connections between many areas of physics for the future.

References

1. J.M. Maldacena, Adv. Thoer. Math. Phys. **2**, 231 (1998).
2. J. Polchinski, Phys. Rev. Lett. **75**, 4724 (1995).
3. T. Sakai and S. Sugimoto, Prog. Theor. Phys. **113**, 843, (2005); **114**, 1083 (2005).
4. G. 'tHooft, Nucl. Phys. **B 72**, 461 (1974); **B 75**, 461 (1974).
5. E. Witten, Nucl. Phys. B **160**, 57 (1979).

6. K. Nawa, H. Suganuma and T. Kojo, Phys. Rev. D**75**, 086003 (2007).
7. K. Nawa, H. Suganuma and T. Kojo, Prog. Theor. Phys. Suppl. **168**, 231 (2007); Mod. Phys. Lett. A **23**, 2364 (2008); Prog. Theor. Phys. Suppl. **174**, 347 (2008); H. Suganuma, K. Nawa and T. Kojo, Nucl. Phys. B (Proc. Suppl.) **186**, 248 (2009).
8. M. Bando, T. Kugo and K. Yamawaki, Prog. Theor. Phys. **73**, 1541 (1985); Phys. Rept. **164**, 217 (1988).
9. Y. Igarashi, M. Johmura, A. Kobayashi, H. Otsu, T. Sato and S. Sawada, Nucl. Phys. B **259**, 721 (1985); T. Fujiwara, Y. Igarashi, A. Kobayashi, H. Otsu, T. Sato and S. Sawada, Prog. Theor. Phys. **74**, 128 (1985).
10. K. Nawa, A. Hosaka and H. Suganuma, Phys. Rev. D**79**, 126005 (2009).
11. K. Nawa, H. Suganuma and T. Kojo, Phys. Rev. D**79**, 026005 (2009).
12. N.S. Manton and P.J. Ruback, Phys. Lett. B**181**, 137 (1986); A. D. Jackson, A. Wirzba and L. Castillejo, Nucl. Phys. A**486**, 634 (1988); H. Forkel, A.D. Jackson, M. Rho, C. Weiss, A. Wirzba and H. Bang, Nucl. Phys. A**504**, 818 (1989).
13. E. Witten, Adv. Theor. Math. Phys. **2**, 505 (1998); S. Mukherji and S. S. Pal, JHEP **0205**, 026 (2002).
14. L. A.-Gaume, C. Gomez, and M. A. Vazquez-Mozo, Phys. Lett. B **649**, 478 (2007).
15. S. A. Hartnoll, C. P. Herzog, and G. T. Horowitz, Phys. Rev. Lett. **101**, 031601 (2008).

MESONS AND NUCLEONS FROM HOLOGRAPHIC QCD

ULUGBEK YAKHSHIEV* and HYUN-CHUL KIM

Department of Physics, Inha University, Incheon 402-751, Korea
**E-mail: yakhshiev@inha.ac.kr*

YOUNGMAN KIM

Asia Pasific Center for Theoretical Physics and Department of Physics
Pohang University of Science and Technology, Pohang, Gyeongbuk 790-784, Korea

We present in this talk a recent investigation on a unified approach within the framework of a hard-wall model of AdS/QCD. We first study a theoretical inconsistency in existing models. In order to remove this inconsistency, we propose a unified approach in which the mesons and the nucleons are treated on the same footing, the same infrared (IR) cutoff being employed in both fermionic and bosonic sectors. We also suggest a possible way of improving the model by introducing a five-dimensional anomalous dimension.

Keywords: AdS/QCD, hadrons

1. Introduction

The AdS/QCD model has been extensively used in describing a wide variety of phenomena in particle and nuclear physics and even in condensed matter physics. The model[1–3] is constructed as a 5D holographic dual of QCD based on the general wisdom of AdS/CFT,[4–6] the 5D gauge coupling being identified by matching the two-point vector correlation functions in deep Euclidean region. Although the AdS/QCD model is based on an *ad hoc* ground, it reflects some of most important features of gauge/gravity duality. Moreover, it is rather successful in describing properties of mesons (see, for example, the following recent review[7]).

QCD is not a conformal theory, in particular, in the low-energy region, so one should also incorporate this property in constructing an effective AdS/QCD model. Consequently, different models have been developed. In Refs. 1–3, the size of the extra dimension (also known as the compactification scale) was fixed at the point that corresponds approximately to the

QCD scale parameter Λ_{QCD}, i.e., an infrared (IR) cutoff parameter was explicitly introduced. It is often considered as the confinement scale that also breaks sharply the conformal invariance. These AdS/QCD models are called the *hard-wall model*. On the contrary, there is an alternative approach called a *soft-wall model* in which the conformal invariance is broken smoothly by introducing the dilaton-like field in the 5D AdS space.[8–10]

While both approaches describe meson properties relatively well, one confronts a serious problem in the baryonic sector. Since there is no theoretical reason that the confinement scale for the baryon should be the same as that for the meson, different values of the IR cutoff for the baryon have been introduced.[11–16] Actually, the main reason for that is due to the fact that it is impossible to reproduce the meson and baryon properties, in particular, excited nucleon states,[14] with the same value of the IR cutoff used. However, when one calculates the meson-nucleon coupling constants, a serious inconsistency arises.[17] In order to determine the coupling constants consistently, one must use the *same* IR cutoff. Otherwise, one cannot fully consider whole information on the meson and nucleon wavefunctions. This is the motivation of the present investigation and has been studied in our recent work.[18] In the present talk, we will briefly review how to resolve this inconsistency and will put forward possible methods to describe the meson and baryon on an equal footing.

2. The effective action of a hard-wall model with holographic mesons and nucleons

We start from a hard-wall model for mesons developed in Refs. 1,2 and study its applications to nucleons.[14,18] The model has a geometry of 5D AdS

$$ds^2 = g_{MN}dx^M dx^N = \frac{1}{z^2}(\eta_{\mu\nu}dx^\mu dx^\nu - dz^2)\,, \tag{1}$$

where $\eta_{\mu\nu}$ stands for the 4D Minkowski metric: $\eta_{\mu\nu} = \mathrm{diag}(1, -1, -1, -1)$. The 5D AdS space is compactified by two different boundary conditions, i.e. the IR boundary at $z = z_m$ and the UV one at $z = \epsilon \to 0$. Considering the global chiral symmetry $\mathrm{SU}(2)_L \times \mathrm{SU}(2)_R$ of QCD, we need to introduce 5D local gauge fields A_L and A_R of which the values at $z = 0$ play a role of external sources for $\mathrm{SU}(2)_L$ and $\mathrm{SU}(2)_R$ currents respectively. Since chiral symmetry is known to be broken to $\mathrm{SU}(2)_V$ spontaneously as well as explicitly, we introduce a bi-fundamental field X with respect to the local gauge symmetry $\mathrm{SU}(2)_L \times \mathrm{SU}(2)_R$, in order to realize the spontaneous and explicit breakings of chiral symmetry in the AdS side. Considering these two, we can construct the bi-fundamental 5D bulk scalar field X in terms of the current quark mass m_q and the quark condensate σ

$$X_0(z) = \langle X \rangle = \frac{1}{2}(m_q z + \sigma z^3) \tag{2}$$

with isospin symmetry assumed.

The 5D gauge action in AdS space with the scalar bulk field and the vector field can be expressed as

$$\begin{aligned} S_M &= \int d^4x \int dz\, \mathcal{L}_M\,, \\ \mathcal{L}_M &= \frac{1}{z^5}\,\mathrm{Tr}\left[|DX|^2 + 3|X|^2 - \frac{1}{2g_5^2}(F_L^2 + F_R^2)\right], \end{aligned} \tag{3}$$

where covariant derivative and field strength tensors are defined as $DX = \partial X - iA_L X + iXA_R$ and $F_{L,R}^{MN} = \partial^M A_{L,R} - \partial^N A_{L,R} - i[A_{L,R}^M, A_{L,R}^N]$. The 5D gauge coupling g_5 is fixed by matching the 5D vector correlation function to that from the operator product expansion (OPE): $g_5^2 = 12\pi^2/N_c$. The 5D mass of the bulk gauge field $A_{L,R}$ is determined by the relation $m_5^2 = (\Delta - p)(\Delta + p - 4)$[5,6] where Δ denotes the canonical dimension of the corresponding operator with spin p and turns out to be $m_5^2 = 0$ because of gauge symmetry. The effective action (3) describes the mesonic sector[1,2] completely apart from exotic mesons.[19]

To consider baryons in the flavor-two ($N_F = 2$) sector, one needs to introduce a bulk Dirac field corresponding to the nucleon at the boundary.[14,18] This hard-wall model was applied to describe the neutron electric dipole moment[20] and holographic nuclear matter.[21] In this model, the nucleons are the massless chiral isospin doublets (p_L, n_L) and (p_R, n_R) which satisfy the 't Hooft anomaly matching. Then the spontaneous breakdown of chiral symmetry induces a chirally symmetric mass term for nucleons

$$\mathcal{L}_{\chi SB} \sim -M_N \begin{pmatrix} \bar{p}_L \\ \bar{n}_L \end{pmatrix} \Sigma\, (p_R, n_R) + \text{h.c.}, \tag{4}$$

where $\Sigma = \exp(2i\pi^a\tau^a/f_\pi)$ is the nonlinear pseudo-Goldstone boson field that transforms as $\Sigma \to U_L \Sigma U_R^\dagger$ under $\mathrm{SU}(2)_L \times \mathrm{SU}(2)_R$. The τ^a and f_π represent the SU(2) Pauli matrices and the pion decay constant, respectively. Thus, we have to consider the following mass term in the AdS side

$$\mathcal{L}_{\mathrm{I}} = -g \begin{pmatrix} \bar{p}_L \\ \bar{n}_L \end{pmatrix} X\, (p_R, n_R) + \text{h.c.}, \tag{5}$$

where g denotes the mass coupling (or Yukawa coupling) between X and nucleon fields, which is usually fitted by reproducing the nucleon mass

$M_N = 940$ MeV. In this regard, we can introduce two 5D Dirac spinors N_1 and N_2 of which the Kaluza-Klein (KK) modes should include the excitations of the massless chiral nucleons (p_L, n_L) and (p_R, n_R), respectively. By this requirement, one can fix the IR boundary conditions for N_1 and N_2 at $z = z_m$.

Note that the 5D spinors $N_{1,2}$ do not have chirality. However, one can resolve this problem in such a way that the 4D chirality is encoded in the sign of the 5D Dirac mass term. For a positive 5D mass, only the right-handed component of the 5D spnior remains near the UV boundary $z \to 0$, which plays the role of a source for the left-handed chiral operator in 4 dimension. It is vice versa for a negative 5D mass. The 5D mass for the $(d+1)$ bulk dimensional spinor is determined by the AdS/CFT expression

$$(m_5)^2 = \left(\Delta - \frac{d}{2}\right)^2 \tag{6}$$

and turns out to be $m_5 = 5/2$. However, since QCD does not have conformal symmetry in the low-energy regime, the 5D mass might acquire an anomalous dimension due to a 5D renormalization flow. Though it is not known how to derive it, we will introduce some anomalous dimension to see its effects on the spectrum of the nucleon.

Summarizing all these facts, we are led to the 5D gauge action for the nucleons

$$\begin{aligned} S_N &= \int d^4x \int dz \, \frac{1}{z^5} \, \mathrm{Tr}\left[\mathcal{L}_\mathrm{K} + \mathcal{L}_\mathrm{I}\right], \\ \mathcal{L}_\mathrm{K} &= i\bar{N}_1\Gamma^M\nabla_M N_1 + i\bar{N}_2\Gamma^M\nabla_M N_2 - \frac{5}{2}\bar{N}_1 N_2 + \frac{5}{2}\bar{N}_2 N_1 \\ \mathcal{L}_\mathrm{I} &= -g\left[\bar{N}_1 X N_2 + \bar{N}_2 X^\dagger N_1\right], \end{aligned} \tag{7}$$

where

$$\nabla_M = \partial_M + \frac{i}{4}\,\omega_M^{AB}\Gamma_{AB} - iA_M^L \,. \tag{8}$$

The non-vanishing components of the spin connection are $w_M^{5A} = -w_M^{A5} = \delta_M^A/z$ and $\Gamma_{AB} = \frac{1}{2i}[\Gamma^A, \Gamma^B]$ are the Lorentz generators for spinors. The Γ matrices are related to the ordinary γ matrices as $\Gamma^M = (\gamma^\mu, -i\gamma_5)$.

The more details of the present approach can be found in Ref. 18.

3. Results and discussions

Before presenting the results of this work, we note the most of input parameters of the model such as m_q, σ and z_m are quite well fitted in the mesonic sector.[1] Hence, we have only one free parameter g to reproduce the data in the baryonic sector. However, the IR cutoff z_m in the baryonic sector, which is often interpreted as a scale of the confinement, takes different values from those in the mesonic sector. Actually, Ref. 14 performed two different fittings of these parameters. In the first fitting of Ref. 14, the z_m and the σ were fixed in the mesonic sector, and the g is fitted to the nucleon mass. In the second fitting, the z_m and the g were taken respectively to be $(205\,\mathrm{MeV})^{-1}$ and 14.4 such that the masses of the nucleon and the Roper resonance $N(1440)$ were reproduced. Since there is no reason for a nucleon to have the same scale of the confinement as that for a meson, this might be an acceptable argument as far as one treats mesons and baryons separately. However, there is one caveat. When it comes to some observables such as the meson-baryon coupling constants, we need to treat the mesons and baryons on the same footing and require inevitably a common z_m. Otherwise, we are not able to consider whole information on both mesons and baryons. Moreover, a model uncertainty brings on by the mass coupling g. Thus, in the present section, we will carry out the numerical analysis very carefully, keeping in mind all these facts.

We first take different values of the z_m from those in the mesonic sector and try to fit the data as was done in Ref. 14. In this case, σ is defined as $\sigma = 4\sqrt{2}(g_5 z_m^3)^{-1}$. In general, one can examine two different limits of the mass coupling g (see Ref. 18): In the limit of the small mass coupling, there are three free parameters m_q, g and z_m. All other parameters can be related to z_m. On the other hand, in the limit of the strong mass coupling, the g can be fixed by some condition,[18] which leaves only two free parameters. Obviously, the dependence on the current quark mass m_q must be tiny because of its smallness, so we can simply neglect it. In this case, we have only one free parameter.

The results of the calculations are listed in table 1. In the upper part of the table, we present the results in the limit of the strong mass coupling. They are more or less the same as those obtained in Ref. 14. For comparison, we list the results for the small mass coupling in the lower part of table 1 where the mass coupling g is chosen to be 6. While the spectrum of the nucleon seems to be qualitatively well reproduced, that of the ρ meson is fairly underestimated in comparison with the experimental data. In the case of the strong mass coupling, the situation becomes even worse. One can note however, as shown in Ref. 18, the ordering of the nucleon-parity states are correctly reproduced for $0 < g < g_{\mathrm{crit}}$.

The results listed in table 1 indicate that it is not possible to reproduce

Table 1. The results of the spectra of the nucleon and the ρ meson. In the limit of the small mass coupling, there are three free parameters m_q, g and z_m, while in the limit of the strong mass coupling, the g is fixed near its critical value.[18] All dimensional quantities are expressed in units of MeV. The asterisks indicate input parameters. The parameter σ is defined as $\sigma = 4\sqrt{2}(g_5 z_m^3)^{-1}$. The 5D nucleon mass m_5 is given by the AdS/CFT, Eq. (6).

m_5	z_m^{-1}	$\sigma^{1/3}$	m_q	g	$(p,n)^+$ (939)	N^+ (1440)	N^- (1535)	ρ (776)	ρ (1475)
				The limit of the strong mass coupling					
5/2	130.9*	126.4	0	15.9	940	1336	1367	315	723
5/2	129.7*	125.2	3*	15.7	940	1328	1358	312	716
5/2	126.0*	121.7	10*	15.4	940	1305	1332	303	696
				The case of the small mass coupling					
5/2	147.2*	142.1	0	6.0*	940	1440	1457	354	813
5/2	147.0*	141.9	3*	6.0*	940	1439	1456	354	812
5/2	146.3*	141.3	10*	6.0*	940	1435	1451	352	808

the spectra of the ρ meson and the nucleon at the same time.[a] As an attempt to improve the above-presented results, we want to introduce an anomalous dimension of the 5D nucleon mass while the 5D mass of the bulk vector field does not acquire any anomalous dimension because of the gauge symmetry.

The results are drawn in figures 1 and 2. One can see that, when anomalous dimension is set to zero (i.e. $m_5 = 5/2$ is fixed), the experimental data is badly reproduced. The inclusion of an additional parameter (i.e. considering m_5 as a free parameter) improves the output data well but leads to larger values of the possible anomalous dimension.

Note that here the nucleon mass is not used as an input. Varying the value of m_5, we try to fit the spectrum of the nucleon. We present the results from two different parameter sets called model A and model B. In this analysis, we take the values of the z_m and σ from Ref. 1. Note that the ρ meson mass is used as an input in model A, while model B corresponds to the global fitting done in Ref. 1. The 5D nucleon mass is varied in the range of $0 \le m_5 \le 5/2$, its anomalous dimension being considered as mentioned before. As shown in figures 1 and 2, the best result is obtained with $m_5 = 0$. Though the absolute value of the nucleon mass turn out to be overestimated in contrast to the previous analysis presented in table 1, qualitatively it is well reproduced within 30 % while ρ meson mass is fitted

[a]Note that in the present work we do not aim at the fine-tuning of the parameters to reproduce the experimental data. The output data in baryonic sector is quite stable for changes in σ.

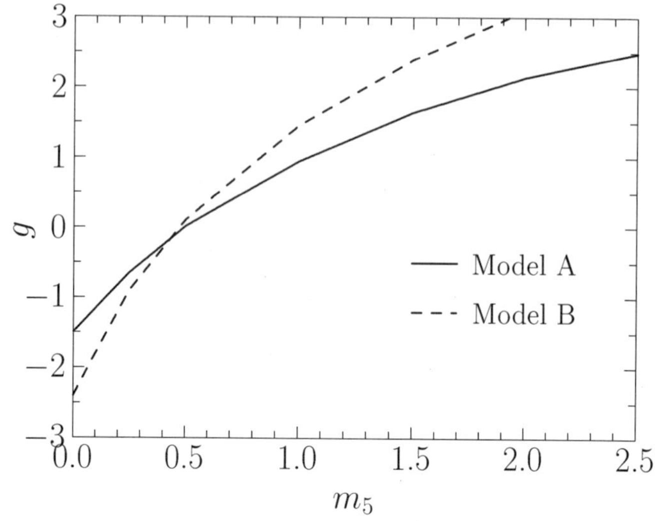

Fig. 1. Mass coupling g dependence on the renormalized 5D mass m_5. The parameters for models A (solid curve) and B (dashed one) are taken from mesonic sector (see Ref.1).

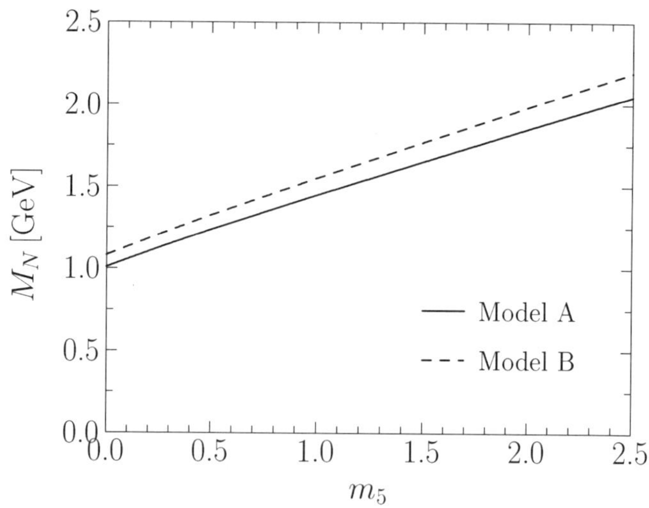

Fig. 2. The mass of the lowest-lying nucleon (in GeV) as a function of the renormalized 5D mass m_5. The notations are same as in fig.1.

around its experimental value.

We are now in a position to include meson-baryon coupling constants in the present numerical analysis. We will consider here the πNN and the ρNN coupling constants in addition to the ρ meson and the nucleon spectra. One has to keep in mind that in order to calculate the meson-baryon coupling constants it is essential to use the same z_m for the mesonic and baryonic sectors. Otherwise, it is not possible to keep whole information on the wavefunctions. Thus, it is of utmost importance to compute all observables with the same set of parameters. We perform a global fitting procedure to obtain the results listed in table 2. Note that we consider here the chiral limit ($m_q = 0$), since its effects on the results are rather tiny.[b] We assume also that the 5D nucleon mass acquires a large anomalous

Table 2. The results of the spectra of the nucleon and the ρ meson and the πNN and ρNN coupling constants. The parameters z_m, σ, and g are found by the global fitting procedure. The anomalous dimension of the 5D nucleon mass is chosen in such a way that the 5D mass vanishes. All other definitions are the same as in table 1.

z_m^{-1}	$\sigma^{1/3}$	g	$(p,n)^+$ (939)	N^+ (1440)	N^- (1535)	ρ (776)	ρ (1475)	$g_{\pi NN}$ (13.1)	$g_{\rho NN}$ (2.4)
285*	256*	-2.0*	890	1791	1797	685	1573	1.65	1.39
285*	237*	-2.0*	890	1790	1796	685	1573	1.76	1.39
285*	256*	-8.0*	930	1826	1856	685	1573	4.89	1.34
285*	237*	-8.0*	920	1817	1843	685	1573	5.12	1.35
285*	227*	-9.6*	930	1826	1856	685	1573	6.12	1.34
280*	252*	-2.0*	874	1760	1765	673	1546	1.65	1.39
280*	233*	-2.0*	874	1759	1764	673	1546	1.76	1.39

dimension so that it may vanish, i.e., $m_5 = 0$. The best fit is obtained with the parameters fitted as follows: $z_m = (285\,\text{MeV})^{-1}$, $\sigma = (227\,\text{MeV})^3$, and $g = -9.6$. The masses of the ground-state nucleon and the ρ meson are in good agreement with the experimental data. Moreover, those of the excited states are qualitatively well reproduced within $10 - 20\,\%$. However, the coupling constants are in general about $50\,\%$ underestimated. We mention that in Ref. 17 the dependence of the meson-baryon coupling constants on the z_m was investigated without considering hadron spectra but the results for the coupling constants are more or less in the same level as in the present work.

[b]Note that in the chiral limit, the nucleon mass is different from experiments, $M_n \simeq$ 939 MeV. For instance, $M_n \simeq 882$ MeV in the chiral limit.[22]

4. Summary and outlook

We have investigated the mesons and the nucleons in a unified approach, based on a hard-wall model of AdS/QCD.[1,14] In order to study the nucleon spectrum, we have developed an approximated method in which the effective potential can be expanded. The method of this approximation was shown to work very well. In particular, the correct ordering of the nucleon parity states was *analytically* as well as *explicitly* shown in this method.[18]

We then have performed several numerical analyzes, varying the model parameters such as the IR cutoff z_m, the quark condensate σ, and the mass coupling (or Yukawa coupling) g. In order to improve the results of the nucleon and the ρ meson spectra on an equal footing, we have introduced an anomalous dimension of the 5D nucleon mass. We found that the zero 5D nucleon mass, $\Delta = 2$, produces the best fit.

We have proceeded to compute the πNN and ρNN coupling constants with the same IR cutoff z_m used. This is crucial in order to keep whole information about the wavefunctions. We carried out the global fitting procedure in which we obtained the best fit with the values of the parameters: $z_m = (285\,\mathrm{MeV})^{-1}$, $\sigma = (227\,\mathrm{MeV})^3$, and $g = -9.6$. The mass spectra of the nucleon and the ρ meson are in relatively good agreement with the experimental data within $10 - 20\,\%$, whereas the πNN and ρNN coupling constants underestimated by about $50\,\%$.

Last but not least, we want to mention the following significant point. While the present AdS/QCD model for the baryon respects some important physics such as spontaneous chiral symmetry breaking, it still suffers from a serious flaw. Any construction of AdS/QCD models must satisfy the UV matching with QCD. However, the present framework of the AdS/QCD model for the baryon does not match the OPE results of QCD. It indicates that there is still a very important component missing in the present version of the model, that is, the correct surface terms in the 5D effective action are missing. The corresponding investigation is under way.[25]

Acknowledgments

U.T. Yakhshiev would like to thank the organizers of HNP09 for the possibility to give a talk and for the hospitality during his stay at Osaka University. The present work is supported by Basic Science Research Program through the National Research Foundation of Korea (NRF) funded by the Ministry of Education, Science and Technology (grant number: 2009-0073101). Y. Kim acknowledges the Max Planck Society(MPG), the Korea

Ministry of Education, Science, Technology(MEST), Gyeongsangbuk-Do and Pohang City for the support of the Independent Junior Research Group at the Asia Pacific Center for Theoretical Physics(APCTP).

References

1. J. Erlich, E. Katz, D. T. Son, M. A. Stephanov, *Phys. Rev. Lett.* **95**, 261602 (2005).
2. L. Da Rold, A. Pomarol, *Nucl. Phys.* **B 721**, 79 (2005).
3. L. Da Rold, A. Pomarol, *JHEP* **0601**, 157 (2006).
4. J. M. Maldacena, *Adv. Theor. Math. Phys.* **2**, 231 (1998) [*Int. J. Theor. Phys.* **38**, 1113 (1999)].
5. S. S. Gubser, I. R. Klebanov, A. M. Polyakov, *Phys. Lett. B* **428**, 105 (1998).
6. E. Witten, *Adv. Theor. Math. Phys.* **2**, 253 (1998).
7. J. Erdmenger, N. Evans, I. Kirsch, E. Threlfall, *Eur. Phys. J. A* **35**, 81 (2008).
8. B. Batell, T. Gherghetta, *Phys. Rev. D* **78**, 026002 (2008).
9. B. Batell, T. Gherghetta, D. Sword, *Phys. Rev. D* **78**, 116011 (2008).
10. O. Andreev, *Phys. Rev. D* **73**, 107901 (2006).
11. R. Contino, A. Pomarol, *JHEP* **0411**, 058 (2004).
12. T. Gherghetta, A. Pomarol, *Nucl. Phys. B* **586**, 141 (2000).
13. W. Mück, K. S. Viswanathan, *Phys. Rev. D* **58**, 106006 (1998).
14. D. K. Hong, T. Inami, H. U. Yee, *Phys. Lett. B* **646**, 165 (2007).
15. G. E. Arutyunov, S. A. Frolov, *Nucl. Phys. B* **544**, 576 (1999).
16. M. Henningson, K. Sfetsos, *Phys. Lett. B* **431**, 63 (1998).
17. N. Maru, M. Tachibana, *Eur. Phys. J. C* **63**, 123 (2009).
18. H.-Ch. Kim, Y. Kim, U. T. Yakhshiev, *JHEP* **0911**, 034 (2009).
19. H.-Ch. Kim, Y. Kim, *JHEP* **0901**, 034 (2009).
20. D. K. Hong, H.-Ch. Kim, S. Siwach, H. U. Yee, *JHEP* **0711**, 036 (2007).
21. Y. Kim, C. H. Lee, H. U. Yee, *Phys. Rev. D* **77**, 085030 (2008).
22. M. Procura, B.U. Musch, T. Wollenweber, T.R. Hemmert, W. Weise, *Phys. Rev. D* **73**, 114510 (2006).
23. Y. Kim, P. Ko, X.-H. Wu, *JHEP* **0806**, 094 (2008).
24. N. Evans, A. Tedder, *Phys. Lett. B* **642**, 546 (2006).
25. H.-Ch. Kim, Y. Kim, I. J. Shin, and U. T. Yakhshiev, *in preparation* (2010).

MESON PROPERTIES IN NUCLEAR MEDIUM

S. HIRENZAKI*, N. IKENO and H. NAGAHIRO

Department of Physics, Nara Women's University,
Nara, 630-8506, Japan
**E-mail: zaki@cc.nara-wu.ac.jp*

J. YAMAGATA-SEKIHARA

Departamento de Física Teórica and IFIC,
Centro Mixto Universidad de Valencia-CSIC, Valencia, 46071, Spain

We discuss the interests and difficulties of the studies of the meson-nucleus bound systems. As recent topics, we briefly explain the research activities of the deeply bound pionic atoms, the $\eta'(958)$ mesic nuclei, and the ϕ mesic nuclei.

Keywords: Mesic Atoms, Mesic Nuclei, Medium effects.

1. Introduction

In the contemporary hadron-nuclear physics, meson-nucleus systems are considered to be very interesting objects in the following two points. First, they are *strongly interacting exotic many body systems* which will be the new frontier of the nuclear physics. Second, meson-nucleus systems provide appropriate laboratories for the studies of *hadron properties and aspects of symmetries of the strong interaction at finite density* which are significantly important to explore symmetry breaking and restoration patterns of QCD.[1] In this context, the studies of meson-nucleus bound states such as mesic atoms and mesic nuclei are extremely interesting in the contemporary hadron physics even they have been already studied for decades.[2] In addition, the measurements of light vector mesons spectra in nucleus as the invariant mass of lepton pairs are also interesting and provide new information.[3,4]

So far, the properties of various mesons in nucleus have been studied both theoretically and experimentally. In this contribution, we would like to pay attention to the pionic atoms, the $\eta'(958)$ mesic nuclei, and the ϕ mesic nuclei as some of the recent active topics in this field.

2. Deeply bound pionic atoms

The studies of structure and formation of the deeply bound pionic atoms, which can not be observed in X-ray experiments, were started about 20 years ago.[5,6] So far, we have confirmed (i) the existence of the quasi-stable bound states even for the deepest $1s$ level in ^{208}Pb,[5,6] (ii) the (d,^{3}He) spectroscopy is a powerful tool for the deeply bound state observations,[7] and (iii) the deeply bound states provide the new information on the pion-nucleus optical potentials.[8,9] And we can deduce the information on chiral symmetry restoration in nucleus from the deeply bound pionic atom spectroscopy.[10–12]

Recently, a new (d,^{3}He) experiment for the formation of the deeply bound pionic atoms was proposed.[13,14] In this experiment, the expected energy resolution is much better than previous data[10] and the non-yrast $2s$ state is expected to be observed firstly. We consider the effective nuclear densities probed by various pionic states in connection with so-called Seki-Masutani relation.[15] The part of our results are given in Ref. 16.

3. $\eta'(958)$ nuclei and $U_A(1)$ anomaly

Since the huge η' mass is believed to have a close relation to $U_A(1)$ anomaly,[17–19] we think it must be very interesting to study the formation reaction of the η' meson in nucleus and discuss the possibility to deduce the η' properties, especially mass shift, at finite density. The effects of the anomaly in medium have never been understood well, hence the η' mass in the medium should provide us important information on the $U_A(1)$ anomaly effects.

The theoretical calculation based on NJL model showed that the mass of η' could reduced by around 150 MeV at normal nuclear density.[20] The formation of the slow η' in nucleus was reported in Refs. 21,22. We show here one of the calculated results in Fig. 1. Here, the η'-nucleus optical potential $U(r)$ is assumed to be, $U(r) = (V_0 + iW_0)\dfrac{\rho(r)}{\rho_0}$, where $\rho(r)$ is the nuclear density distribution and ρ_0 denotes the nuclear saturation density. The calculated results with $(V_0, W_0) = (0, -5\text{ MeV})$ and $(-100\text{ MeV}, -5\text{ MeV})$ are shown in Fig. 1 as functions of the excited energy defined as $E_{\text{ex}} = m_{\eta'} - B_{\eta'} + [S_p(j_p) - S_p(p_{3/2})]$, where $B_{\eta'}$ is the η' binding energy and S_p the proton separation energy. The η' production threshold energy E_0 is indicated in the figure by the vertical solid lines. Once we can observe these kinds of spectra in experiments, we could deduce the in-medium η' mass information as the potential depth in the equation of motion.

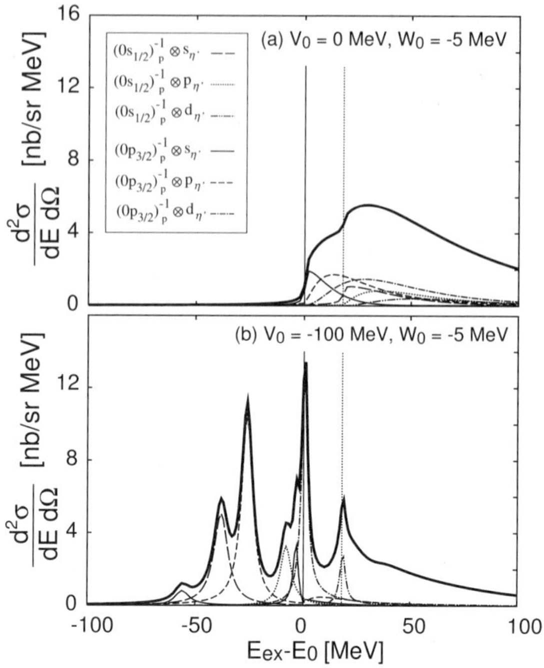

Fig. 1. The calculated spectra of the $^{12}\mathrm{C}(\gamma,p)^{11}\mathrm{B} \otimes \eta'$ reaction at $E_\gamma = 3$ GeV are shown as functions of the excited energy E_{ex} defined in the text. E_0 is the η' production threshold energy. The η'-nucleus optical potential depths are (a) $V_0 = 0$, $W_0 = -5$ MeV and (b) $V_0 = -100$ MeV, $W_0 = -5$ MeV. The total spectra are shown by the thick solid lines, and the dominant contributions of subcomponents are shown by dotted and dashed lines, as indicated in the figure. The vertical lines indicate the η' production threshold energy with the ground $p_{3/2}$ proton-hole configuration (solid line) and the excited $s_{1/2}$ proton-hole configuration (dotted line) in the final states. The figure is taken from Ref. 21.

4. ϕ mesic nuclei and OZI rule

The properties of ϕ meson in nucleus are expected to have close connection to OZI rule and the strange quark component in nucleon due to the $\bar{s}s$ nature of ϕ meson. So far, in-medium properties of ϕ meson have been studied theoretically using effective models, which depend on the K and $\bar{K}$ meson properties in medium because of the strong $\phi \to K\bar{K}$ coupling. The study of QCD sum rule[23] and the data taken at KEK[4] suggested 3% mass reduction of ϕ at the normal nuclear density, while the ϕ meson selfenergy calculated in Refs. 24,25 indicated a significantly smaller attractive potential for ϕ. In addition, the results of attenuation experiment in Ref. 26 are different from those in Ref. 4. Thus, the properties of ϕ in nuclear medium are not clear yet at present and still controversial.

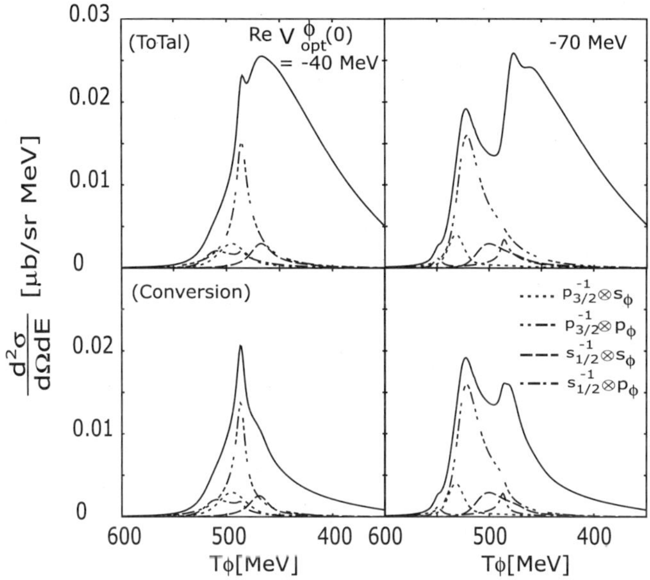

Fig. 2. Calculated spectra for the ϕ-nucleus systems formation in Ref. 27 plotted as functions of the emitted ϕ-meson energy in $^{12}C(\bar{p},\phi)$ reaction at $p_{\bar{p}} = 1.3$ GeV/c. Total cross section (upper panels) and conversion parts (lower panels) are calculated by the Green's function method[28] with deep potentials obtained by scaling the theoretical potential[25] as Re $V^{\phi}_{\rm opt}(E = m_\phi) = -40$ and -70 MeV at nuclear center with keeping the imaginary part unmodified. The dashed and dotted lines indicate the contributions from different subcomponent $[\ell_j^{-1} \otimes \ell_\phi]$ as shown in the figure.

We show the calculated total spectra[27] in Fig. 2 with assuming the potential depth Re $V^{\phi}_{\rm opt}(E = m_\phi)$ at nuclear center to be -40 and -70 MeV. The conversion spectra[28] are also shown in the lower panels. The potential depth of -40 MeV is consider to be consistent to the data[4] within the error. In this case, we may be able to see a tiny spike in the spectra as the indication of the attractive interaction in the total spectrum. In case of Re $V^{\phi}_{\rm opt}(E = m_\phi) = -70$ MeV, we can expect clear signals of the bound state.

5. Discussions and conclusions

In this report, we introduce some recent interesting activities on meson-nucleus bound systems. We hope our theoretical results stimulate other research activities and develop this field.

Finally, we would like make to some comments on the formation reactions of meson-nucleus systems. We consider in this report the missing mass method, which was successfully used to observe the deeply bound pionic states. However, it will have difficulties to be applied to the formation of heavy meson states in nucleus. We think it will be more difficult to achieve necessary energy resolution, which is typically $\sim\hbar\omega$ in nucleus, since we need more energetic incident particles for heavier meson production. There is also another difficulty due to overlapping of the contributions from different nuclear excitations coupling to the mesic states with different energies in the missing mass spectra.

On the other hand, we need to notice that the most of the mesons decay outside the nucleus in the invariant mass methods like in Ref. 4 for mesons with small decay widths to leptons and the spectra from these mesons have no information on in-medium meson properties. In addition, it is necessary to confirm quantitatively that the target nucleus still remain as the nucleus of enough spatial dimension in the final state, since the incident proton energy in Ref. 4 is 12 GeV which is high enough to break nucleus even after producing a ϕ meson. A proposed experiment will make use of the even higher energy proton beam in JPARC.[29]

Recently, several coincident measurements have been considered where the emitted particles due to meson decay and/or meson absorption in nucleus are observed in coincident with the energetic forward particles.[30–33] In these coincident methods, we can expect some advantages; (i) by selecting the events with highly energetic forward particles, we can guarantee the existence of the nucleus in the final states in the meson production processes, and guarantee to have the only contribution of meson decay/absorption inside the nucleus for the subthreshold meson production kinematics, and (ii) by considering the cross sections as the functions of the total energy (not of the invariant mass) of the emitted particles from the meson decay/absorption in the nucleus, we can avoid the overlapping of the contributions of mesic states with the different energies coupling to the different nuclear excitations.

However, we also have some disadvantages unfortunately. The dominant difficulty is the reduction of the reaction rate. For example, the branching fraction of the ϕ meson decaying into e^+e^- pair in vacuum is around

3.0×10^{-4} which should be multiplied to the formation spectra in inclusive missing mass spectroscopies to estimate the coincident measurements of e^+e^- pair creation. If we consider the strong absorption effects to ϕ, the fraction will be even worse like 5.3×10^{-5} estimated by assuming the strong width of ϕ in nucleus to be 20 MeV. To consider the decay processes of ϕ into the channels with hadrons is another possibility to expect larger branching fraction. In this case, we need careful treatment of the final state interaction effects for the emitted hadrons.

References

1. T. Hatsuda and T. Kunihiro, Phys. Rep. **247**, 221 (1994), and references therein.
2. C. J. Batty, E. Friedman and A. Gal, Phys. Rep. **287**, 385 (1997), E. Friedman and A. Gal, Phys. Rept. **452**, 89 (2007).
3. R. S. Hayano and T. Hatsuda, submitted to Rev. Mod. Phys. , arXiv:0812.1702 [nucl-ex].
4. R. Muto, et al., Phys. Rev. Lett. **98**, 042501(07).
5. H. Toki and T. Yamazaki, Phys. Lett. B **213** (1988) 129.
6. H. Toki, S. Hirenzaki, T. Yamazaki and R. S. Hayano, Nucl. Phys. A **501** (1989) 653.
7. S. Hirenzaki, H. Toki and T. Yamazaki, Phys. Rev. C **44** (1991) 2472.
8. H. Gilg *et al.*, Phys. Rev. C **62** (2000) 025201.
9. K. Itahashi *et al.*, Phys. Rev. C **62** (2000) 025202.
10. K. Suzuki *et al.* , Phys. Rev. Lett. 92 (2004) 072302.
11. E. E. Kolomeitsev, N. Kaiser and W. Weise, Phys. Rev. Lett. 90 (2003) 092501.
12. D. Jido, T. Hatsuda and T. Kunihiro, Phys. Lett. B670 (2008) 109.
13. K. Itahashi *et al.*, "Precision Spectroscopy of Pionic Atoms in (d,^{3}He) Nuclear Reactions" ; Exp. proporsal NP0702-RIBF-027 for RIBF, Dec. (2006).
14. K. Itahashi *et al.*, "Spectroscopy of Pionic Atom in ^{122}Sn(d,^{3}He) Nuclear Reaction" ; Exp. proporsal NP0802-RIBF-054 for RIBF, Jan. (2008).
15. R. Seki and K. Masutani, Phys. Rev. C27 (1983) 2799.
16. N. Ikeno *et al.*, in this proceedings.
17. T. Kunihiro and T. Hatsuda, Phys. Lett. B **206**, 385 (1988); **206**, 278(E) (1988).
18. V. Bernard, R. L. Jaffe, and U. -G. Meissner, Nucl. Phys. **B308**, 753 (1988).
19. T. Kunihiro, Phys. Lett. B **219**, 363 (1989).
20. M. Buballa, Phys. Rep. 407, 205 (2005).
21. H. Nagahiro and S. Hirenzaki, Phys. Rev. Lett. **94** (2005) 232503 [arXiv:hep-ph/0412072].
22. H. Nagahiro, M. Takizawa and S. Hirenzaki, Phys. Rev. C **74** (2006) 045203 [arXiv:nucl-th/0606052].
23. T. Hatsuda and S. H. Lee, Phys. Rev. C **46**, 34 (1992).
24. F. Klingl, T. Waas and W. Weise, Phys. Lett. B **431** (1998) 254.

25. D. Cabrera, M. J. Vicente Vacas, Phys. Rev. C **67**, 045203 (2003).
26. T. Ishikawa *et al.*, Phys. Lett. B **608** (2005) 215.
27. J. Yamagata-Sekihara, D. Cabrera, M. J. V. Vacas and S. Hirenzaki, submitted to Prog. Theor. Phys. , arXiv:1001.2235 [nucl-th].
28. O. Morimatsu and K. Yazaki, Nucl. Phys. **A435**, 727 (1985); **A483**, 493 (1988).
29. S. Yokkaichi *et al.*, Proposal P16, J-PARC.
30. P. Bühler *et al.*, Proposal P29, J-PARC.
31. J. Yamagata-Sekihara, D. Jido, H. Nagahiro and S. Hirenzaki, Phys. Rev. C **80** (2009) 045204 [arXiv:0812.4359 [nucl-th]].
32. K. Ozawa and R. S. Hayano, letter of intent for J- PARC, "Combined measurements of nuclear ω bound state and ω mass modification in $p(\pi, n)\omega$ reaction".
33. M. Kaskulov, H. Nagahiro, S. Hirenzaki and E. Oset, Phys. Rev. C **75**, 064616 (2007) [arXiv:nucl-th/0610085].

FORMATION OF η-MESIC NUCLEI BY (π,N) REACTION AT J-PARC

H. NAGAHIRO*, S. HIRENZAKI

Department of Physics, Nara Womens' University,
Nara, 630-8506, Japan
**E-mail: nagahiro@cc.nara-wu.ac.jp*

D. JIDO

Yukawa Institute for Theoretical Physics, Kyoto University,
Kyoto 606-8502, Japan

We study the formations of of the η-nucleus systems in the (π, N) reactions with nuclear targets, which can be performed at existing and/or forthcoming facilities, including the Japan Proton Accelerator Research Complex, to investigate the η-nucleus interaction. Based on the $N^*(1535)$ dominance in the ηN system, the η-mesic nuclei are suitable systems for the study of in-medium properties of the $N^*(1535)$ baryon resonance, such as reduction of the mass difference of N and N^* in the nuclear medium, which affects the level structure of the η and N^*-*hole* modes. We find that clear information on the in-medium N^*- and η-nucleus interactions can be obtained through the formation spectra of the η-mesic nuclei.

Keywords: mesons in nuclei; η meson; partial restoration of chiral symmetry

1. Introduction

The study of meson-nucleus bound systems is one of the important subjects in nuclear physics. The detailed investigations of the structure of the bound states provide us quantitative information on the hadron-nucleus interactions. So far, the structure of atomic states of pion, kaon, Σ^-, and $\bar{p}$ have been successfully observed and investigated comprehensively both in theoretical and experimental points of view.[1] One of the remarkable developments in experimental aspects is the establishment of the (d,^{3}He) spectroscopy for the formation of deeply bound pionic atoms with recoil-free kinematics.[2–5] It opens new possibilities of the formation of other hadron-nucleus bound systems.[6–9]

Bound states of the η meson in nuclei were predicted first by Haider and Liu.[10] After that, many works were devoted to studies of the structure of the bound states, the formation reactions of the η-mesic nuclei and in-medium properties of the η meson.[6–9,11–18] Especially, the η meson in the nuclear medium has been recently investigated in the aspect of chiral symmetry.[6–9,13–17]

The first experimental search of the η bound states in nuclei[19] was performed in the (π^+, p) reaction with several nuclear targets in finite momentum transfer to aim to observe narrow states as predicted in Ref. 10, and the result turned out to be negative. In this study, we revisit the (π, N) reactions with the recoil-free kinematics for the formation of the η-mesic nuclei in order to get clearer information on the level structure of the η and N^*-*hole* modes and in-medium properties of $N^*(1535)$ in the viewpoint of the chiral symmetry for baryons. We will find that the appropriate kinetic energy of the injecting pion in this reaction can be attained by the Japan Proton Accelerator Research Complex (J-PARC) facility. The detailed discussions are given in Refs. 8,9,17 and 18.

2. Level crossing of the η-meson and N^*-hole modes

We briefly review the interesting feature of the η meson in the nuclear medium.[17] The in-medium η propagator is given by,

$$D_\eta(\omega, k; \rho)^{-1} = \omega^2 - k^2 - m_\eta^2 - \Pi_\eta(\omega, k; \rho), \tag{1}$$

where ω and k denote the energy and momentum of the η meson, m_η is its mass, and Π_η denotes the η self-energy in the nuclear medium. Thanks to the strong coupling of the ηN system to the $N^*(1535)$ resonance, we can evaluate the η self-energy by using the N^* dominance hypothesis. Considering the lowest N^*-nucleon-hole excitation, we obtain the η self-energy in small η momentum[8,17] as

$$\Pi_\eta(\omega, k; \rho) = \frac{g_\eta^2 \rho}{\omega + m_N^*(\rho) - m_{N^*}^*(\rho) + i\Gamma_{N^*}(\omega, \rho)/2} + \text{(crossed term)}. \tag{2}$$

Here, g_η is the coupling constant of the s-wave $\eta N N^*$ vertex and can be determined as $g_\eta \simeq 2.0$ to reproduce the in-vacuum partial width $\Gamma_{N^* \to \eta N} \simeq 75$ MeV at tree level. $m_N^*(\rho)$ and $m_{N^*}^*(\rho)$ are effective masses (free mass plus medium self-energy) of N and N^* in the nuclear medium with density ρ, respectively.

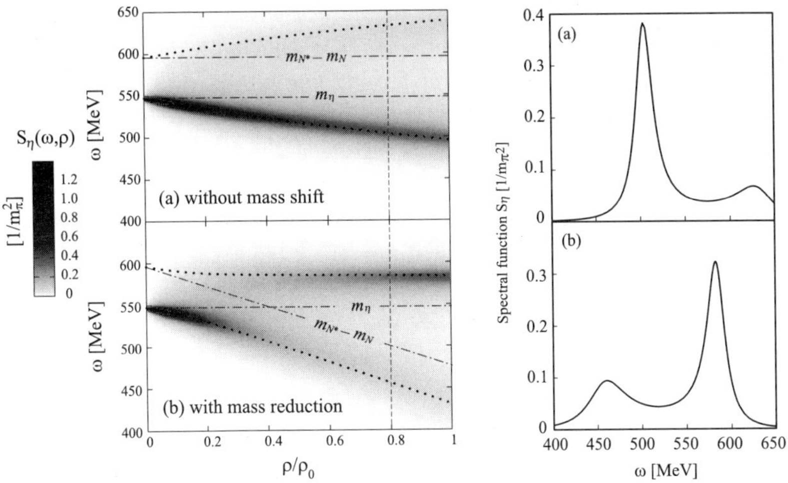

Fig. 1. (left panel) Contour maps of the η meson spectral density in nuclear matter in Eq. (3) as functions of the baryon density and the η energy, assuming (a) the N and N^* masses not to change in medium and (b) 20% mass gap reduction of N and N^* at normal nuclear density ρ_0. (right panel) Spectral functions of the η meson in nuclear matter as functions of the η energy at $\rho/\rho_0 = 0.8$ (indicated by the vertical dashed lines in the left panel).

The η propagator (1) with the self-energy (2) has two poles with a positive real part in the complex energy plane in each density. These poles represent the η meson and N^*-*hole* modes in the nuclear medium.[13,14,17] Corresponding to these poles, the η spectral density S_η given by

$$S_\eta(\omega,\rho) = -\frac{1}{\pi}\mathrm{Im}(D_\eta(\omega, k=0;\rho)) \tag{3}$$

has two peaks in a function of real energy at a certain density as shown in Fig. 1 (right panel). We also show the contour maps of the η spectral density as functions of baryon density and η energy in Fig. 1 (left panel), where the real parts of the pole positions are indicated by the dotted lines. Figure 1(a) (left) shows the strength of two branches in the case that the effective masses of N and N^* do not change in medium. In this case, the two branches slightly come away from each other for higher ρ as a result of level repulsion, and the strength of the lower mode is always larger than the upper mode as also shown explicitly in the right panel of Fig. 1(a). The similar behavior of the η spectral function based on the chiral unitary approach were also reported in Refs. 13,14, where the reduction of the mass gap between N and N^* is very small. In contrast, in the case that the mass gap becomes smaller in the nuclear medium, the behavior of the η spectral density significantly changes. Suppose that the mass gap of N and N^* linearly decreases by 20% at ρ_0, the level crossing between the two branches takes place around $\rho \sim 0.4\rho_0$ as shown in Fig. 1(b). As a consequence, the strength of the upper mode becomes stronger due to the level mixing, and the lower mode shifts downwards considerably as the density increases.

A possible source of the mass gap reduction is the partial restoration of chiral symmetry in the nuclear medium. If $N^*(1535)$ is a chiral partner of nucleon, the N and N^* mass difference should decrease as chiral symmetry is being restored. In this study, we use the following parameterization of the mass gap reduction based on the chiral doublet model:[20–22]

$$m^*_{N^*}(\rho) - m^*_N(\rho) = \left(1 - C\frac{\rho}{\rho_0}\right)(m_{N^*} - m_N), \tag{4}$$

where m_N and m_{N^*} are the N and N^* masses in free space, respectively. Here the parameter C represents the strength of the chiral restoration at the normal nucleon saturation density ρ_0, and its empirical value around 0.2. Figures 1(b) correspond to the case with $C = 0.2$ in the chiral doublet model.

These characteristic phenomena caused by the level crossing can be a signal of the reduction of the N and N^* mass gap, which supports the partial restoration of the chiral symmetry in the nuclear medium.

3. η-nucleus optical potentials

As a consequence of the level crossing, the in-medium η self-energy has a strong energy dependence. In addition, this mass gap reduction as density increases gives also a strong density dependence on the η self-energy and then on the η optical potential as pointed out in Ref. 8. The detailed discussions the η-nucleus optical potentials we used are described in Refs. 8,9,16,17 and 18.

In order to describe the in-medium properties of N^*, we use two kinds of the chiral models, which are based on distinct physical pictures of N^*. One is the chiral doublet model,[20–22] in which N^* is regarded as the chiral partner of the nucleon. The other is the chiral unitary model, in which N^* is introduced as a dynamically generated resonance in the coupled channel meson-baryon scattering.[14,15]

In the first approach, N^* is introduced as a particle with a large width and appears in an effective Lagrangian together with the nucleon field in

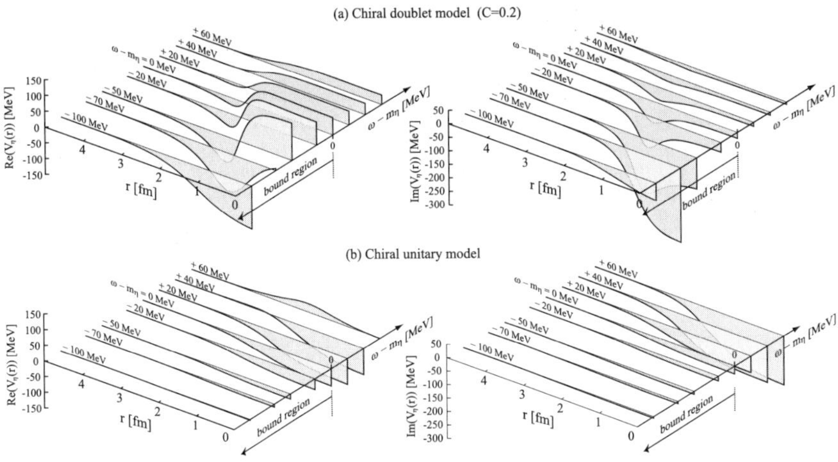

Fig. 2. The η-nucleus optical potentials with (a) the chiral doublet model ($C = 0.2$) and (b) the chiral unitary model as functions of the radial coordinate r for η energies $\omega - m_\eta = -100, -70, -50, -20, 0, +20, +40, +60$ MeV. Left and right figures show the real and imaginary parts of the optical potentials V_η, respectively.

linear realization of chiral symmetry. The η-nucleus optical potential

$$V_\eta(\omega, r) \equiv \frac{\Pi_\eta(\omega, k=0; \rho(r))}{2\mu}$$
$$= \frac{g_\eta^2}{2\mu} \frac{\rho(r)}{\omega + m_N^*(\rho(r)) - m_{N^*}^*(\rho(r)) + i\Gamma_{N^*}(\omega, \rho(r))/2} + (\text{crossed term}). \quad (5)$$

is sensitive to the mass difference of N and N^* in nuclei. Especially, the sign of the real part of the η optical potential changes when the mass gap of N^* and N becomes smaller than the η energy ω.[8,9,16] This means that the attractive η-nucleus interaction at low densities can turn to be repulsive depending on the values of the mass gap and the η energy. This feature can be seen in Fig. 2(a), where we plot the η-nucleus optical potentials as functions of radial coordinate r at η energies. For instance, at the η threshold $\omega - m_\eta = 0$, the optical potential has a repulsive core inside the nucleus and an attractive pocket on the surface. Furthermore, this mass reduction yields also the strong energy dependence of the η-nucleus optical potential as shown in Fig. 2(a). At $\omega - m_\eta \simeq -100$ MeV, the real part of the optical potential is about 140 MeV attractive while that of the η threshold $\omega - m_\eta = 0$ is 55 MeV repulsive at $r = 0$.

In the second approach to describe the $N^*(1535)$ resonance, which is the chiral unitary model, the optical potential has quite different features from the previous case. In this model, it was found that N^* has a dominant component of the $K\Sigma$ channel.[23,24] Since the Σ hyperon is free from the Pauli blocking in the nuclear medium, only a tiny change of the mass gap is expected in the nuclear medium.[13,14] Therefore, the η optical potential is attractive in the bound energy region, $\omega \leq m_\eta$, as shown in Fig. 2(b), and has weaker energy dependence compared with that of the chiral doublet model with $C = 0.2$.

4. Formation spectra of η-mesic nuclei by (π^+,p) reaction

In Fig. 3 we show the momentum transfer as a function of the kinetic energy of incident pion. We find the recoilless condition can be satisfied in the energy range expected at J-PARC project. We should stress here that it is important to set the emitted proton angle being zero degree otherwise the momentum transfer should be large and then many subcomponents give finite contribution to expected spectra.

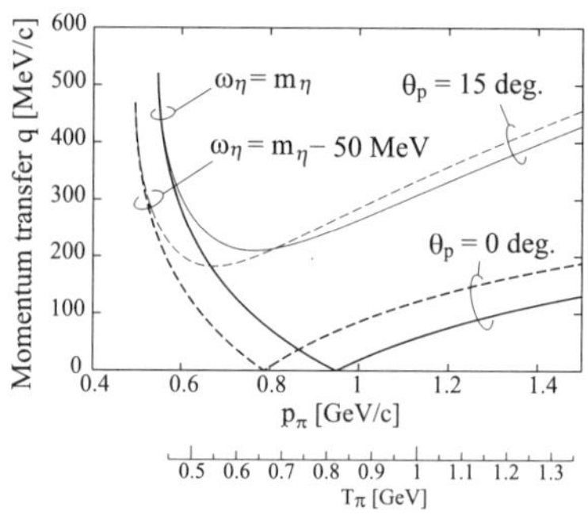

Fig. 3. Momentum transfers at η energies $\omega_\eta = m_\eta$ and $m_\eta - 50$ MeV as functions of the incident pion momentum p_π and also kinetic energy T_π. θ_p denotes the emitted proton angle in Laboratory frame.

In Fig. 4, we show the calculated ${}^{12}\mathrm{C}(\pi^+,p){}^{11}\mathrm{C}\otimes\eta$ cross sections for the formation of the η-${}^{11}\mathrm{C}$ system with the chiral doublet model potential for $C = 0.2$ (upper panels in Fig. 4) and that of the chiral unitary model (lower panels in Fig. 4). The incident pion kinetic energy T_π is 820 MeV corresponding to the recoilless condition at the η threshold. (right panels)

As reported in Refs. 9,16, in the chiral unitary model case, we find a bound state peak in the subcomponent $(0s_{1/2})_n^{-1}\otimes s_\eta$ around $E_{\mathrm{ex}} - E_0 \sim 10$ – 15 MeV ($\omega_\eta - m_\eta \sim -10$ – -5 MeV) as indicated in Fig. 4(b). The

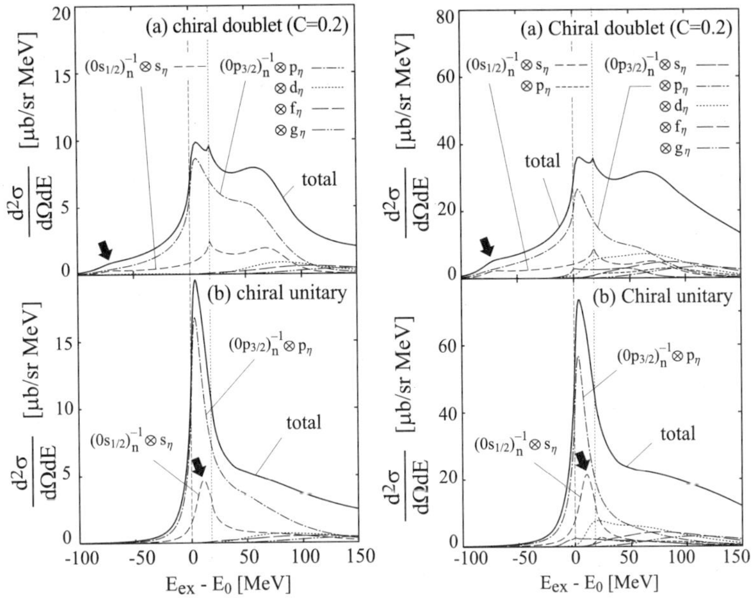

Fig. 4. Calculated spectra of the $^{12}C(\pi^+,p)^{11}C\otimes\eta$ reaction at (1) $T_\pi = 820$ MeV and (2) $T_\pi = 650$ MeV and the emitted proton angle $\theta_p = 0°$ as functions of the excited energy E_{ex}. E_0 is the η production threshold. The η-nucleus interaction is calculated by using (a) the chiral doublet model with $C = 0.2$ and (b) the chiral unitary model. The thick solid lines show the total spectra and the dashed lines represent dominant subcomponents as indicated in the figures. The neutron-hole states are indicated as $(n\ell_j)_n^{-1}$ and the η states as ℓ_η. The solid arrow indicates the peak due to the bound state in each model. $T_\pi = 650$ MeV case is discussed in Ref. 18 in detail.

existence of the bound state in the unitary model is predicted in Ref. 15. As seen in the figure, however, it is impossible to observe the signal of the bound state separately, because there is a larger contribution from the $(0p_{2/3})_n{}^{1} \otimes p_\eta$ subcomponent in the same energy and it masks the bound state peak in the $(0s_{1/2})_n^{-1} \otimes s_\eta$ component. The peak appearing in $(0p_{2/3})_n^{-1} \otimes p_\eta$ at the η production threshold has nothing to do with the bound state of the η meson and is produced by the threshold effect. This result also means that we should keep in mind that an experimentally observed peak does not always corresponds to a bound state and it should be compared to the theoretical calculation in order to be identified.

In the case of the chiral doublet model case, as reported in Ref. 17, there are some bound states obtained as solutions found in the complex-energy-plane of the Klein-Gordon equation with the optical potential (5) with $C = 0.2$. For η with $\ell_\eta = 0$, the bound states were found with the eigenenergies (B.E., Γ)=(91.3, 26.3) MeV for $0s$ and (75.1, 33.0) MeV for $1s$.[17] In Fig. 4 we can see that the $0s$ bound state appears in the spectrum as a small bump around $E_{\rm ex} - E_0 \sim -70 - -80$ MeV. These deep bound states correspond to the lower mode mentioned in Sec. 2 in the chiral doublet model ($C = 0.2$). However, the strength of the bump in the spectrum is too small to be observed in real experiments. The η bound states with $\ell_\eta = 1$ were also obtained at (79.3, 31.1) MeV for $0p$ and (72.1, 34.2) MeV for $1p$. The corresponding peak structure to the $0p$ bound state is not seen in the spectrum again due to its small strength. In the recoilless kinematics, the $1s$ and $1p$ bound states should give much less contributions in the spectrum, since there are no $1s$- and $1p$-hole states in the daughter nucleus for the carbon target case.

Much more prominent structure is seen in the quasi-free region $E_{\rm ex} - E_0 > 0$ in the case of the chiral doublet model. We see a considerably large bump structure around $E_{\rm ex} - E_0 \sim 60$ MeV as shown in Fig. 4(a). This peak comes from the N^*-*hole* mode coupled to the η meson in the medium, namely the upper mode shown in Fig. 1(b). As discussed in Sec. 2, the upper mode in the chiral doublet model is enhanced as a consequence of the level crossing associated with the reduction of the N-N^* mass gap in nuclear matter stemming from the partial restoration of chiral symmetry. On the other hand, in the case of the chiral unitary approach, we see only a smooth slope in the quasi-free region. Hence, the enhancement in the quasi-free region could be a signal for the reduction of the N-N^* mass gap in the nuclear medium, which supports the scenario of the partial restoration of chiral symmetry in the nuclear medium.

5. Conclusion

We studied the η-mesic nuclei formation of the (π, N) reactions with nuclear targets to discuss the experimental feasibility in forthcoming experiments. Emphasizing in-medium properties of the $N^*(1535)$ baryon resonance in the context of chiral symmetry, we discussed the structure of the η-mesic nuclei formation spectra. Especially, the reduction of the N-N^* mass gap in the nuclear medium were investigated by two chiral models.

We showed the calculated spectra of the η-mesic nuclei by the (π, N) reactions. We confirmed that the magnitude of the cross section is large enough to be observed in experiments. We found that the (π, N) reactions were also appropriate to observe the interesting behaviors like deep bound states and the bump structure at the quasi-free region which can be understood by the concept of the level crossing phenomena caused by the partial restoration of chiral symmetry.[17] We conclude that we can get new information on the in-medium N^* properties through the formation of the η-mesic nuclei by the (π, N) reactions.

We believe that the present theoretical results are important to stimulate both theoretical and, especially, experimental activities to study the hadron properties in medium and to obtain new information on the partial restoration of chiral symmetry in the nuclear medium.

References

1. For example; C.J. Batty, E. Friedman and A. Gal, Phys. Rept. **287**, 385 (1997); E. Friedman, A. Gal, Phys. Rept. **452**, 89 (2007).
2. H. Toki, S. Hirenzaki, T. Yamazaki, Nucl. Phys. **A530**, 679 (1991); S. Hirenzaki, H. Toki, T. Yamazaki, Phys. Rev. **C44**, 2472 (1991).
3. H. Gilg *et al.*, Phys. Rev. C **62**, 025201 (2000); K. Itahashi, *et al.*, *ibid.* **62**, 025202 (2000).
4. H. Geissel *et al.*, Phys. Rev. Lett. **88**, 122301 (2002).
5. P. Kienle, and T. Yamazaki, Phys. Lett. B **514**, 1 (2001); H. Geissel, *et al.*, *ibid.* **549**, 64 (2002); K. Suzuki, *et al.*, Phys. Rev. Lett. **92**, 072302 (2004).
6. R.S. Hayano, S. Hirenzaki, A. Gillitzer, Eur. Phys. J. A **6**, 99 (1999).
7. K. Tsushima, D.H. Lu, A.W. Thomas, K. Saito, Phys. Lett. **B443**, 26 (1998); K. Tsushima, D.H. Lu, A.W. Thomas, Phys. Rev. **C59**,1203 (1999).
8. D. Jido, H. Nagahiro and S. Hirenzaki, Phys. Rev. C**66**, 045202 (2002); Nucl. Phys. **A721**, 665c(2003).
9. H. Nagahiro, D. Jido and S. Hirenzaki, Phys. Rev. C**68**, 035205 (2003).
10. Q. Haider and L. C. Liu, Phys. Lett. B **172** (1986) 257, L. C. Liu and Q. Haider, Phys. Rev. C **34** (1986) 1845.
11. M. Kohno and H. Tanabe, Phys. Lett. **B** 231(1989) 219-223, M. Kohno and H. Tanabe, Nucl. Phys. A **519** (1990) 755.
12. H. C. Chiang, E. Oset and L. C. Liu, Phys. Rev. C **44** (1991) 738.
13. T. Waas and W. Weise, Nucl. Phys. A **625** (1997) 287.
14. T. Inoue and E. Oset, Nucl. Phys. A **710** (2002) 354.
15. C. Garcia-Recio, J. Nieves, T. Inoue and E. Oset, Phys. Lett. **B550**, 47 (2002).
16. H. Nagahiro, D. Jido and S. Hirenzaki, Nucl. Phys. A **761**,92 (2005).
17. D. Jido, E. E. Kolomeitsev, H. Nagahiro and S. Hirenzaki, Nucl. Phys. A **811**, 158 (2008).
18. H. Nagahiro, D. Jido and S. Hirenzaki, Phys. Rev. C **80** (2009) 025205.
19. R. E. Chrien *et al.*, Phys. Rev. Lett. **60** (1988) 2595.
20. C. DeTar and T. Kunihiro, Phys. Rev. D **39** (1989) 2805.
21. D. Jido, M. Oka and A. Hosaka, Prog. Theor. Phys. **106** (2001) 873.
22. D. Jido, Y. Nemoto, M. Oka and A. Hosaka, Nucl. Phys. A **671** (2000) 471; Y. Nemoto, D. Jido, M. Oka and A. Hosaka, Phys. Rev. D **57** (1998) 4124.
23. N. Kaiser, P. B. Siegel and W. Weise, Phys. Lett. B **362**, 23 (1995).
24. T. Inoue, E. Oset and M. J. Vicente Vacas, Phys. Rev. C **65**, 035204 (2002).

BARYON RESONANCE PRODUCTION IN THE $\pi + d$ REACTION AND SEARCH FOR η-MESIC NUCLEI AT J-PARC

H. FUJIOKA

Department of Physics, Kyoto University,
Kyoto, 606-8502, Japan
E-mail: fujioka@scphys.kyoto-u.ac.jp

K. ITAHASHI

RIKEN Nishina Center for Accelerator-Based Science, RIKEN,
Saitama, 351-0198, Japan
E-mail: itahashi@riken.jp

A double-scattering reaction $\pi^+ + d \to p + p + \eta$, where an η meson is rescattered, may provide new information on the $\eta N \to \eta N$ scattering. An idea to measure this reaction is proposed. Moreover, experimental search for η-mesic nuclei by the same (π, N) reaction is discussed.

Keywords: baryon resonance; $N^*(1535)$; ηN interaction; η-mesic nuclei; η-nucleus interaction

1. Introduction

The possible existence of meson-nucleus bound systems is now one of the hot topics in hadron physics. Especially, an antikaon is expected to be bound in a finite nucleus, thanks to a strong attraction between an $I = 0$ $\overline{K}N$ pair,[1] and recent experimental findings of the signature of such a kaon-nucleus bound state[2,3] have brought a heated debate on whether it really exists with a decay width narrow enough to be observed experimentally.

In contrast, the interaction between an η meson and a nucleon is also known to be also attractive, while its strength is not well-determined. The ηN scattering length has a large ambiguity, between 0.270 fm and 1.050 fm for its real part and between 0.190 fm and 0.399 fm for its imaginary part.[4] It is not possible to carry out neither an X-ray measurement of mesic atom nor a scattering experiment, in order to derive information on ηN interaction, different from the case for an antikaon. Instead, experimental data of hadronic production and photoproduction of η mesons are mainly used in theoretical analyses. It should be noted that the $S_{11}(1535)$ resonance plays a dominant role in near-threshold η production, since it lies just about 50 MeV above the ηN threshold, and strongly couples with the ηN channel, as seen in its large decay branching ratio into $\eta + N$ (45–60%[5]).

If the existence of η-mesic nuclei is confirmed experimentally, experimental observables such as the binding energy and decay width will impose a constraint on theoretical models about (in-medium) ηN interaction and the $S_{11}(1535)$ resonance. Haider and Liu predicted their existence for the first time,[6] by using the scattering length of $(0.28+0.29i)$ fm or $(0.27+0.22i)$ fm, and they found a bound state with the mass number $A \geq 12$ may exist. If the scattering length is larger than they applied, there may be a possibility for lighter η-mesic nuclei to exist.

In this paper, we discuss experimental ideas to explore ηN interaction and η-mesic nuclei by using the (π, N) reactions on different nuclear targets which will be feasible at the J-PARC facility.

2. (π, N) reaction at J-PARC

Secondary particles, such as pions and kaons, are produced by the bombardment of intense proton beam accelerated by the J-PARC 50 GeV Proton Synchrotron onto the production target. Then, they are extracted and transported into experimental areas.

At present, there are two beamlines, K1.8 (available momentum up to $\sim$ 2 GeV/c) and K1.8BR (up to 1.1 GeV/c), and further beamlines are planned. For simplicity, we discuss experimental plans at the K1.8BR area, where most of the detectors for the E15 experiment[7] may be reused after a small modification.

The E15 experiment aims to search for a deeply-bound kaonic nuclear state K^-pp by the ^{3}He(K^-, n) reaction. The layout of the experimental area together with the detectors to be installed is shown in Fig. 1. A K^- is transported through the K1.8BR beamline, and momentum-analyzed. Scattered neutrons by the (K^-, n) reaction on the helium-3 target are detected by the Neutron Counter, located 15 m downstream from the target. The decay particles of $K^-pp \to \Lambda + p \to p + \pi^- + p$ are detected by the Cylindrical Detector System which surrounds the target. Their trajectories can be reconstructed by the Cylindrical Drift Chamber installed in a solenoid magnet which operates at 0.5 T parallel to the beam axis. The Cylindrical Detector Hodoscopes, which locates outside of the CDC, are used for the

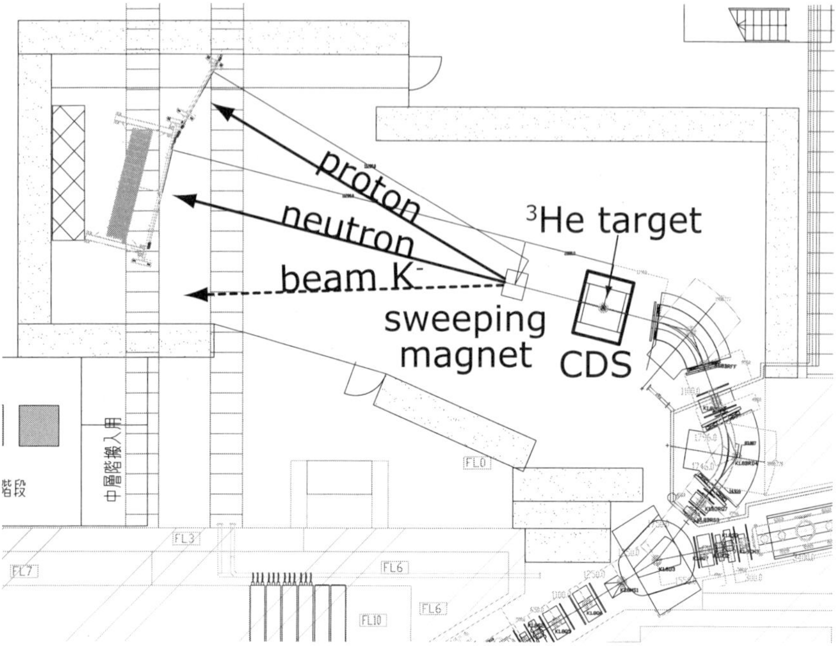

Fig. 1. Setup for the J-PARC E15 experiment.

trigger purpose and the particle identification.

Recently, a new measurement of the ^{3}He(K^-, p) reaction in the E15 experiment, by adding a detector system for scattered protons, was proposed (J-PARC P28[8]), and the proposal has been approved as a part of the E15 experiment. Because a beam sweeping magnet is installed downstream of the target, scattered protons are also bent away from the beam axis (see Fig. 1). Hence, an array of TOF counters for scattered protons will be located next to the charge veto counters for neutrons. In order to enable an inclusive measurement, a set of drift chambers will be installed downstream of the target, so that the trajectories of the protons can be reconstructed.

The present experimental setup is, of course, optimized for the E15 experiment. However, the ($\pi^{\pm}$, n) and/or ($\pi^{\pm}$, p) reactions on a different nuclear target can be investigated with almost the same setup. As discussed below, it is vitally important to detect rescattered particles in the $\pi^+ + d$ reaction and decay particles from η-mesic nuclei. Therefore, the usage of the Cylindrical Detector System will be desirable, too.

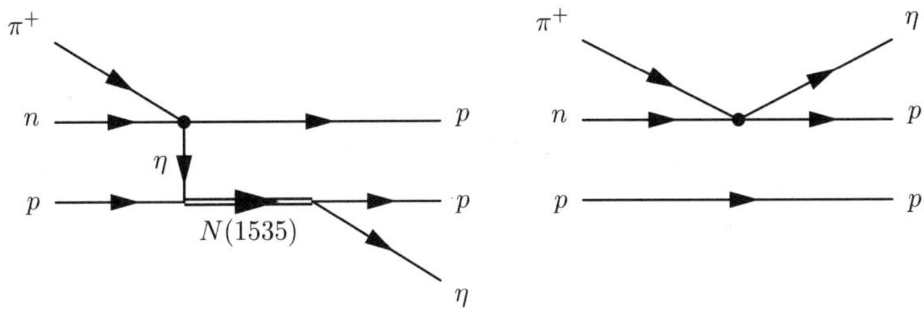

(a) double-scattering reaction (b) quasi-free reaction

Fig. 2. Diagrams for the $\pi^+ d \to pp\eta$ reaction.

3. Baryon resonance production in the $\pi + d$ reaction

In this section, we consider a *double-scattering reaction*:

$$\pi^+ + d \to p + p + \eta, \tag{1}$$

where both the protons in the final state are involved (Fig. 2(a)). This reaction must be distinguished from a *quasi-free reaction*:

$$\pi^+ + n \to p + \eta, \tag{2}$$

where the proton in the deuteron is a spectator. (Fig. 2(b)).

The double-scattering reaction is a two-step process, namely the quasi-free reaction (2) followed by the rescattering:

$$\eta + p \to \eta + p. \tag{3}$$

The experimental goal is to obtain the missing-mass spectrum for the $d(\pi^+,\ p)X$ reaction with tagging X as the $p\eta$ final state. It is expected that a bump corresponding to the $S_{11}(1535)$ resonance due to the rescattering (3) will be observed in the spectrum.

Since we are interested in ηN interaction at low energy, the beam momentum should not be far from the recoilless condition for the $n(\pi^+,\ p)\eta$ reaction. As shown in Fig. 3, the optimal beam momentum will be around $0.8 - 1.1\,\text{GeV}/c$. The momentum of the $S_{11}(1535)$ (p^*) resonance will be almost at rest (Fig. 3), as expected.

Then, one possibility to separate the double-scattering reaction from the quasi-free one is to detect a proton from the S_{11} decay ($p^* \to p + \eta$). While a proton from the decay of a S_{11} with its mass $1.535\,\text{GeV}/c^2$ has its momentum of $\sim 186\,\text{MeV}/c$, the Fermi momentum of a nucleon in the deuteron hardly exceed $200\,\text{MeV}/c$. Moreover by detecting a decay proton,

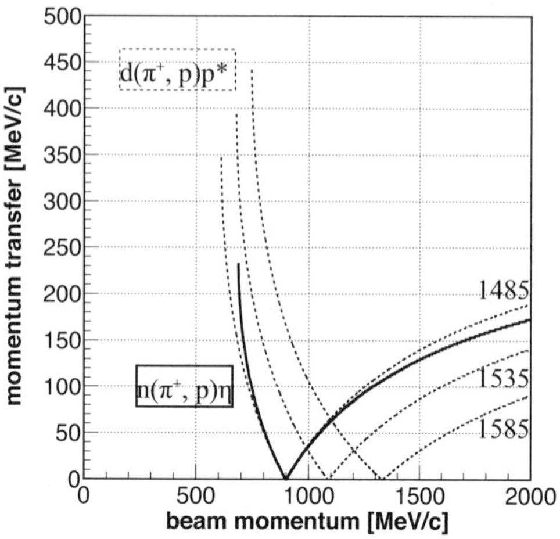

Fig. 3. Momentum transfer for the $n(\pi^+, p)\eta$ reaction (solid line) and the $d(\pi^+, p)p^*$ reaction (dashed lines) for three different p^* masses (1485, 1535, 1585 MeV/c^2).

a missing η can be identified by use of the missing-mass of the $d(\pi^+, pp)X$ reaction.

From an experimental point of view, the present CDC for the E15 experiment does not allow for the detection of such a slow proton because of the existence of its inner support. For instance, the range of a 200 MeV/c proton is only 5 mm in a plastic scintillator. Accordingly, a target has to be thin enough, less than a few mm[a], and we have to equip a new detector system between the CDC and the target. For example, it is considered to install two barrels of plastic scintillators with different distances from the center; the counter in the inner barrel measures the total energy of a proton, and that in the outer barrel serves as a veto counter to confirm the proton is stopped in the inner counter.

By the way, the $\pi^- + p \to \eta + n$ reaction has been investigated by several groups.[9,10] The total cross section increases very rapidly from the threshold (center-of-mass energy $W \sim 1.487$ GeV), and reaches more than 2 mb in the $S_{11}(1535)$ resonance region. While the bump below $W \sim 1.6$ MeV can be explained by the predominant contribution of the $S_{11}(1535)$ resonance, the cross section above 1.6 GeV is strongly model-dependent.[11] It is because more resonances, such as $S_{11}(1650)$, $P_{11}(1710)$, $D_{13}(1520)$, $D_{13}(1700)$, $P_{13}(1720)$, and so on, contribute to the reaction.

[a]Therefore, a solid CD_2 will be used, instead of liquid D_2.

A similar behavior should hold for the $\pi^+ + n \to \eta + p$ reaction, due to the isospin symmetry. Therefore, an incident pion momentum around $0.8\,\mathrm{GeV}/c$, which corresponds to $W \sim 1.56\,\mathrm{GeV}$, may be suitable in order to avoid difficulties in model calculations. A measurement with a higher beam momentum may provide information on different resonances, including the second S_{11} resonance.

It should be noted that not only η but also π and vector mesons (ρ, ω) can propagate as an intermediate meson, even if the center-of-mass energy of the pion and the neutron is adjusted to the $S_{11}(1535)$ region. This is also the case for the $p + p \to p + p + \eta$ reaction near the threshold, which has been experimentally studied by COSY-11 and COSY-TOF (see, for example, Ref. 12). Theoretical calculations for the $p + p$ reaction[13] take into account these intermediate mesons between two nucleons.

Therefore, a new calculation for the $\pi^+ + d$ reaction, based on a model for the $\pi^+ + p$ and $p + p$ reactions, is awaited. This novel reaction might reveal unique features, i.e., a possibility to make a constraint to parameters such as the a MNN^* coupling constant.

It is worth noting that an analogous reaction has been already discussed with regard to the $\Lambda(1405)$ resonance. In the chiral unitary model, it is shown that a $\Lambda(1405)$, which is dynamically generated in the meson-baryon scattering, is a superposition of two poles.[14] The lower pole ($\sim 1390\,\mathrm{MeV}$) and higher pole ($\sim 1420\,\mathrm{MeV}$) couple strongly with the $\pi\Sigma$ and $\overline{K}N$ channel, respectively. In order to confirm that the resonance in the $\overline{K}N$ channel locates around 1420 MeV instead of 1405 MeV, the scattering of $\overline{K}N \to \pi\Sigma$ has to be examined. However, it is impossible to produce the $\Lambda(1405)$ resonance in a direct $K^- + p$ scattering, because it lies below the $\overline{K}N$ threshold. Jido *et al.*[15] discussed a two-step process of the $K^- + d \to \Lambda(1405) + n$ reaction, where the resonance can be produced by the scattering of an nucleon in the deuteron and an intermediate antikaon, and compared their calculation with an old experiment by Braun *et al.*[16bc] In J-PARC, an experimental proposal to investigate the $d(K^-,\ n)$ reaction, together with the identification of the decay mode of $\Lambda(1405)$ into $\pi^0\Sigma^0$ as well as $\pi^\pm\Sigma^\mp$ has been submitted.[17]

Turning back to the double-scattering reaction (1), the situation is different in that the $S_{11}(1535)$ resonance is above the ηN threshold. Neverthe-

---

[b] Since they used a deuteron bubble chamber, the scattered neutron was not detected. Instead, they obtained the $\pi^+\Sigma^-$ invariant-mass spectrum.

[c] Not only $\Lambda(1405) + n$, but also $\Sigma^- + p$, $\Sigma^-(1385) + p$, and $\Lambda(1520) + n$ were identified as final states.

less, there is a possibility that one can access information on the $\eta N \to \eta N$ scattering, caused by an intermediate η meson, which cannot be investigated by a direct reaction with "η meson beam".

4. Search for η-mesic nuclei

After Haider and Liu predicted the existence of η-mesic nuclei in 1986,[6] η-nucleon and η-nucleus interaction have been of particular interest both theoretically and experimentally. Observation of η-mesic nuclei, as well as investigation of near-threshold η production, will provide much better understanding of the interaction.

The TAPS/MAMI experiment investigated the photoproduction of η-mesic ^{3}He by the $\gamma + {}^3\text{He} \to \eta + X$ and $\pi^0 + p + X$ reaction.[18] The total cross section of the coherent η production exhibits a peak-like structure at the threshold. An enhancement of correlated $\pi^0 p$ pairs with relative angles near 180° is also observed near the η production threshold, which may be attributed to the decay of η-mesic nuclei. By a simultaneous fit of the cross sections for the two channels, the binding energy -4.4 ± 4.2 MeV and the width 25.6 ± 6.1 MeV are obtained. However, Hanhart claimed that it is not possible to conclude whether the observed enhancement corresponds to a bound state or a virtual state, due to the lack of statistics.[19]

Recently, the COSY-GEM experiment investigated the ^{27}Al(p, ^{3}He) reaction at the recoilless kinematics.[20] They observed back-to-back π^--p pairs, whose energy corresponds to $S_{11}(1535)$ at rest. The missing-mass spectrum, with requiring back-to-back π^--p pairs in coincidence, shows an enhancement below the $^{25}\text{Mg} + \eta$ threshold. They obtained the binding energy (-13.13 ± 1.64 MeV) and the width (10.22 ± 2.98 MeV) of the $^{25}\text{Mg} \otimes \eta$ system.

Both the experiments tagged a pair of a pion and a nucleon as the decay product of η-mesic nuclei. Since an η meson and a nucleon strongly couples with a $S_{11}(1535)$ resonance (N^*), a main decay channel is considered to be $N^* \to N\pi$. The $N^* \to N\eta$ decay, which has a large branching ratio in free space, is suppressed because of the Pauli blocking, even if the state is above the η emission threshold. Other decay modes, namely $N^*N \to NN$ and $NN\pi$, also contributes to the total decay width. The width of NN decay is estimated to be negligibly small.[21,22] Experimentally, the $N^* \to N\pi$ is easiest to identify, as its back-to-back topology is a clean signal.

Historically, the first experiment to search for η-mesic nuclei was done at BNL by using the (π^+, p) reaction on lithium, carbon, oxygen, and aluminum targets,[23] motivated by a theoretical calculation by Liu and

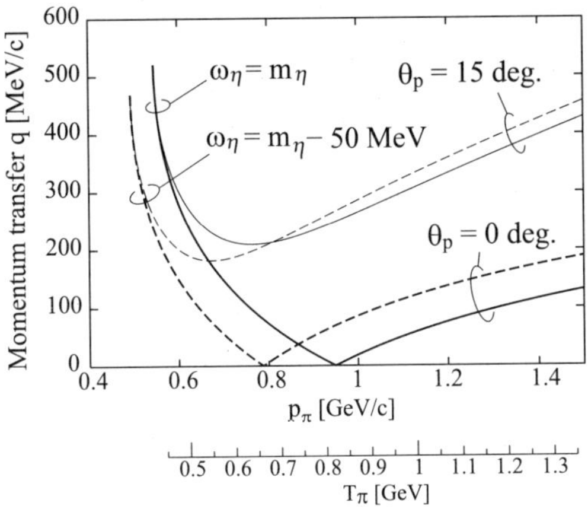

Fig. 4. Momentum transfer for the $A(\pi, N)$ reaction (heavy mass limit) as a function of incident pion momentum (p_π) or kinetic energy (T_π), for different scattering angle (θ_p) of 0° and 15°. The solid (dashed) line corresponds to the η energy at $\omega_\eta = m_\eta$ ($m_\eta - 50\,\text{MeV}$). This figure is taken from Ref. 26.

Haider.[24] A peak structure corresponding to η-mesic nuclei was not observed for any target.

We consider that it is worthwhile to reconsider this reaction for two reasons, and plan to carry out a new experiment at J-PARC.[25]

First, the proton spectrometer was set to 15° against the beam in the BNL experiment, and the momentum transfer is as large as $200\,\text{MeV}/c$ for the incident beam momentum $800\,\text{MeV}/c$ (see Fig. 4). A theoretical calculation by Nagahiro *et al.* revealed that many subcomponents of different configurations of a nucleon-hole and an η meson will contribute with substantial fractions, and that they will mask a possible peak structure of η-mesic nuclei.[26]

On the other hand, the recoilless condition can be satisfied by measuring a proton scattered at almost zero degree. As shown in Fig. 4, the magic momentum, for the η energy $\omega_\eta = m_\eta - 50\,\text{MeV}$ and m_η, where m_η is the mass of an η meson, is $777\,\text{MeV}/c$ and $950\,\text{MeV}/c$, respectively. In this condition only substitutional states are enhanced, since the angular momentum of the target nucleus is conserved after the reaction.

Secondly, the BNL experiment investigated only the inclusive reaction. A signal of η-mesic nuclei, if exists, may be overwhelmed by a large background irrelevant to η production. Such a background will be reduced by detecting $N\pi$ pairs in back-to-back simultaneously.

Taking these points into consideration, we proposed to study the (π^-, n) reactions on ^{7}Li target at J-PARC[25] (see also Ref. 26). We have chosen ^{7}Li, in addition to ^{12}C as a candidate of the target, because of less contribution of a proton in the $p_{3/2}$-shell. It is replaced by an η meson with its orbital angular momentum 1 after the reaction (p-wave component), and it is foreseen that the structure from the s-wave component from a proton in the $s_{1/2}$-shell is much more interesting than that from the p-wave component.

Furthermore, the missing-mass spectrum will be affected by the in-medium property of the $N^*(1535)$ resonance. Nagahiro *et al.* derived the η-nucleus optical potentials, based on two kinds of chiral models — *chiral doublet model* and *chiral unitary model* — and calculated the spectrum for each potential.[26]

In the chiral doublet model, the N^* is regarded as the chiral partner of the nucleon, and the mass gap between N^* and N decreases in a finite density because of partial restoration of chiral symmetry. Thereby, a level crossing between the N^*-N^{-1} and η modes may take place in a nucleus, which results in a deeply-bound η state and a bump structure in the unbound region.

On the other hand, the mass gap will not change largely in nuclear medium in case of the chiral unitary model. A shallow bound state, which is similar to the result in Ref. 27, will appear in the missing-mass spectrum.

Therefore, we expect to obtain information on the in-medium N^* property in relation with the level crossing, via a missing-mass spectroscopy covering the unbound region as well as the bound region.

5. Conclusion

We discussed experimental ideas to utilize the (π, N) reactions for two purposes: investigation of the properties of the $S_{11}(1535)$ resonance, and search for η-mesic nuclei. It seems that both studies will shed light on the η-nucleon and η-nucleus interaction.

Acknowledgments

We acknowledge stimulating comments from Dr. M. Döring. We would like to thank Dr. H. Nagahiro, Dr. D. Jido, and Prof. S. Hirenzaki for valuable discussions on η-mesic nuclei.

References

1. See, for example, A. Dote, in these proceedings (2010).

2. M. Angello *et al.* (FINUDA Collaboration), Phys. Rev. Lett. **94**, 212303 (2005).
3. M. Maggiora *et al.* (DISTO Collaboration), arXiv:0912.5116 [hep-ex].
4. Q. Haider and L. C. Liu, Phys. Rev. C **66**, 045208 (2002).
5. C. Amsler *et al.* (Particle Data Group), Phys. Lett. **B667**, 1 (2008).
6. Q. Haider and L. C. Liu, Phys. Lett. **B172**, 257 (1986); *ibid.* **174**, 465(E) (1986).
7. M. Iwasaki, T. Nagae *et al.*, J-PARC E15 proposal (2006). (http://j-parc.jp/NuclPart/pac_0606/pdf/p15-Iwasaki.pdf)
8. H. Fujioka *et al.*, J-PARC P28 proposal (2009). (http://j-parc.jp/NuclPart/pac_0907/pdf/Fujioka.pdf)
9. T. W. Morrison, Ph.D. thesis, George Washington University (1999); R. M. Brown *et al.*, Nucl. Phys. **B153**, 89 (1979); N. C. Debenham *et al.*, Phys. Rev. D **12**, 2545 (1975); J. Feltesse *et al.*, Nucl. Phys. **B93**, 242 (1975); B. W. Richards *et al.*, Phys. Rev. D **1**, 10 (1970); F. Bulos *et al.*, Phys. Rev. **187**, 1827 (1969); W. Deinet *et al.*, Nucl. Phys. **B11**, 495 (1969); F. Bulos *et al.*, Phys. Rev. Lett. **13**, 486 (1964).
10. S. Prakhov *et al.*, Phys. Rev. C **72**, 015203 (2005).
11. See, for example, J. He and B. Saghai, Phys. Rev. C **80**, 015207 (2009); J. Durand, B. Juliá-Díaz, T.-S. H. Lee, B. Saghai, and T. Sato, Phys. Rev. C **78**, 025204 (2008); K. Nakayama, Y. Oh, and H. Haberzettl, arXiv:0803.3169v3 [hep-ex]; X.-H. Zhong, Q. Zhao, J. He, and B. Saghai, Phys. Rev. C **76**, 065205 (2007).
12. P. Moskal *et al.*, arXiv:0912.1592v1 [nucl-ex].
13. See, for example, K. Nakayama, Y. Oh, and H. Haberzettl, arXiv:0803.3169v3 [hep-ex], and references therein.
14. D. Jido, J. A. Oller, E. Oset, A. Ramos, U.-G. Meißner, Nucl. Phys. **A725**, 181 (2003).
15. D. Jido, E. Oset, and T. Sekihara, Eur. Phys. J. **A42**, 257 (2009).
16. O. Braun *et al.*, Nucl. Phys. **B129**, 1 (1977).
17. H. Noumi *et al.*, J-PARC P31 proposal (2009). (http://j-parc.jp/NuclPart/pac_0907/pdf/Noumi.pdf)
18. M. Pfeiffer *et al.*, Phys. Rev. Lett. **92**, 252001 (2004).
19. C. Hanhart, Phys. Rev. Lett. **94**, 049101 (2005).
20. A. Budzanowski *et al.* (COSY-GEM Collaboration), Phys. ReV. C **79**, 012201(R) (2009).
21. H. C. Chiang, E. Oset, L. C. Liu, Phys. Rev. C **44**, 738 (1991).
22. H. Nagahiro, D. Jido, and S. Hirenzaki, Phys. Rev. C **68**, 035205 (2003).
23. R. E. Chrien *et al.*, Phys. Rev. Lett. **60**, 2595 (1988).
24. L. C. Liu and Q. Haider, Phys. Rev. C **34**, 1845 (1986).
25. K. Itahashi, H. Fujioka, S. Hirenzaki, D. Jido, and H. Nagahiro, Letter of Intent for J-PARC (2007). (http://j-parc.jp/NuclPart/pac_0707/pdf/LoI-itahashi.pdf)
26. H. Nagahiro, D. Jido, and S. Hirenzaki, Phys. Rev. C **80**, 025205 (2009); H. Nagahiro, in these proceedings (2010).
27. C. García-Recio, T. Inoue, J. Nieves, E. Oset, Phys. Lett. **B550**, 47 (2002).

K^-pp STUDIED WITH COUPLED-CHANNEL COMPLEX SCALING METHOD

A. DOTÉ

High Energy Accelerator Research Organization (IPNS/KEK)
1-1 Oho, Tsukuba, Ibaraki 305-0801, Japan
E-mail: dote@post.kek.jp

T. INOUE

Graduate School of Pure and Applied Sciences, Univ. of Tsukuba
1-1-1 Tenoudai Tsukuba, Ibaraki 305-8571, Japan
E-mail: takash-i@sophia.ac.jp

In the study of K^-pp that is the most essential $\bar{K}$-nuclear system, the coupling dynamics of $\bar{K}NN$ and $\pi\Sigma N$ is important. In case that the K^-pp is considered as a $\bar{K}NN$-$\pi\Sigma N$ coupled channel system, we need to treat it as a resonant state, because the $\pi\Sigma N$ channel is an open channel. We propose the application of the Complex Scaling Method (CSM), which is a powerful tool for the treatment of resonant states, to the study of K^-pp. In this article, we report the result of our test calculation for the two-body case, $I = 0$ $\bar{K}N$-$\pi\Sigma$ system corresponding to $\Lambda(1405)$ studied with a phenomenological $\bar{K}N$ potential.

Keywords: $\bar{K}$-nuclear system; Resonant state; Few-body system.

1. Introduction

Nuclear system with strangeness has been expected to have interesting properties. One of hot topics in strange nuclear physics is "Kaonic nuclei" (nuclei involving anti-kaon, finite nuclear system with $\bar{K}$ mesons), because of strong $\bar{K}N$ attraction. Many researchers in nuclear and hadron physics have been interested in kaonic nuclei, especially since possible existence of deeply bound kaonic nuclei was proposed. According to studies with a phenomenological $\bar{K}N$ potential reproducing low-energy $\bar{K}N$ data, a K^- meson is found to be deeply bound in light nuclei with more than 100 MeV binding energy, and such $\bar{K}$-nuclear systems can exist as a quasi-bound state with narrow decay width.[1] The strong $\bar{K}N$ attraction makes a nucleus shrunk to be highly dense (up to 4 times normal density in average)

Table 1. Summary of recent results of theoretical calculations of K^-pp. "DHW" etc is a label indicating each calculation. The binding energy and decay width ($K^-pp \to \pi Y$) obtained in each calculation are shown as "B.E." and "Width (πY)" in unit of MeV, respectively. "Method" indicates the method of each calculation. The type of employed $\bar{K}N$ potential is given in the line of "$\bar{K}N$ int.": "Chiral" means a potential based on chiral SU(3) theory, while "Pheno." means a phenomenological potential. The actual form of the potential is also mentioned as "Gaussian in r-space" etc. "Channels" indicates channels dealt with explicitly in each calculation.

	DHW[6]	AY[7]	IS[8]	SG[9]
B.E.	20 ± 3	48	60-95	50-70
Width (πY)	40 - 70	61	45-80	90-110
Method	Variational	Variational	Faddeev	Faddeev
$\bar{K}N$ int.	Chiral	Pheno.	Chiral	Pheno.
	Gaussian, r-space		Separable, p-space	
Channels	Single ($\bar{K}N$)		Coupled ($\bar{K}N$, πY)	

involving so strange structures that we have never seen in the study of ordinary nuclei.[2] However, such exotic features of kaonic nuclei have not been established yet and situation of the $\bar{K}$-nuclear study has been still controversial.[3]

These years, researchers have focused on a prototype of kaonic nuclei "K^-pp",[a] before going to more complicated systems. We expect that essential properties of kaonic nuclei are clarified by investigating such a simple system composed of two protons and a K^- meson. Indeed, due to the simplicity, the K^-pp has been investigated in various ways.[4,5] In Table 1 we summarize theoretical studies of K^-pp structure with s-wave $\bar{K}N$ potential which is constrained by experimental data on $\bar{K}N$ system; low energy scattering data, kaonic hydrogen atom data and the observed $\Lambda(1405)$.[b] All calculations listed in this table show that the K^-pp system can be bound. However, the predicted binding energy and decay width are strongly dependent on employed methods and used interactions, in spite of the constraint by the existing experimental data on $\bar{K}N$ system.

Difference between the results of two variational calculations, DHW[6] and AY,[7] is attributed to different interpretation of the observed $\Lambda(1405)$; $\Lambda(1405)$ in a phenomenological model and $\Lambda(1420)$ in a chiral-based model. The AY potential forms a $I = 0$ $\bar{K}N$ resonance at 1405 MeV, just corre-

[a] We consider $\bar{K}NN$-πYN three-body system with $J^\pi = 1/2^-$, $T = 1/2$. Such a coupled system is symbolically represented as K^-pp in this article.

[b] A study including p-wave $\bar{K}N$ potential is reported in Ref.[10]

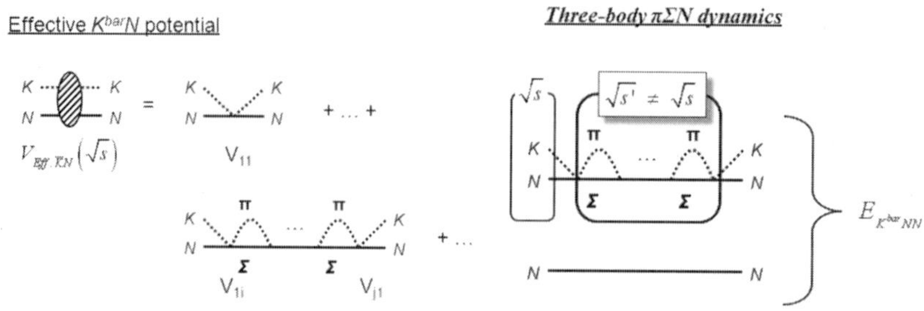

Fig. 1. Diagrammatic explanation of the effective $\bar{K}N$ potential used in DHW (left panel) and three-body $\pi\Sigma N$ dynamics (right panel).

sponding to the nominal position of the observed $\Lambda(1405)$. On the other hand, it appears at 1420 MeV in the chiral-based potential used in DHW calculation. In the chiral model, the observed $\Lambda(1405)$ is described with two poles. (*Double-pole nature of* $\Lambda(1405)$[11,12]) One pole around 1420 MeV is strongly coupled to $\bar{K}N$ channel and the other one involving large width is strongly coupled to $\pi\Sigma$ channel. As a result of different description of $\Lambda(1405)$, the $I = 0$ $\bar{K}N$ attraction in the DHW calculation is about a half of that in the AY calculation and then the energy of the three-body system of K^-pp is also different in these calculations.

As the origin of large discrepancy between DHW and IS[8] results, both of which are constrained with the chiral SU(3) theory, several reasons are considered such as finite-range potential in r-space or separable potential in p-space, and non-relativistic or semi-relativistic treatment. An interesting origin, "three-body $\pi\Sigma N$ dynamics", has been suggested in Ref. 13.

In the DHW study, the variational calculation is performed in a single $\bar{K}N$ channel with an effective $\bar{K}N$ potential which is constructed so as to reproduce the $\bar{K}N$ scattering amplitude predicted by the coupled-channel chiral dynamics.[12] The $\pi\Sigma$ channel is eliminated and incorporated into the effective $\bar{K}N$ potential, as explained diagrammatically in the left panel of Fig. 1. This procedure can be performed exactly in case of two-body system. Applying the effective $\bar{K}N$ potential to a three-body system, we have a problem occurred: Using this effective $\bar{K}N$ potential, the intermediate $\pi\Sigma$ energy is fixed to the energy of the initial $\bar{K}N$ state, it should be variable in the three-body $\bar{K}NN$-$\pi\Sigma N$ system. In the three-body system, the total energy $E_{\bar{K}NN}$ is conserved but the energy of the two-body subsystem is not so. In other words, the energy of the intermediate $\pi\Sigma$ state can differ from the initial $\bar{K}N$ energy, due to the existence of the third nucleon. (See the right panel of Fig. 1) Such three-body $\pi\Sigma N$ dynamics is obviously taken into account in the Faddeev calculation, because the $\pi\Sigma$ degree of freedom is explicitly treated, differently from the variational calculation. This dynamics is expected to bring extra attraction to the K^-pp system.

2. Complex scaling method

We wish to investigate detailed properties of K^-pp, including the above-mentioned effect of the $\pi\Sigma N$ dynamics, based on the variational scheme with novel NN and $\bar{K}N$ potentials. To realize our attempt, explicit treatment of the $\pi\Sigma$ channel is needed in the variational calculation as performed in Faddeev calculations. However, such a coupled-channel calculation seems impossible in a naive variational scheme used for bound-states calculation.[c] From recent studies, the K^-pp state is expected to exist between $\bar{K}NN$ and $\pi\Sigma N$ thresholds. If this is the case, the K^-pp state is surely a bound state for the $\bar{K}NN$ channel, but it is a resonant state for the $\pi\Sigma N$ channel. The variational calculation can not be applied to this case, because the variational principle is useless to search for resonant states. Here, we employ Complex Scaling Method (CSM)[14] to deal with resonant states, because this method is considered to be an extension of variational method as explained later.

In the CSM, Hamiltonian H and a wave function $|\Phi\rangle$ are transformed by complex scaling for a coordinate $\boldsymbol{r}$:

$$U(\theta) : \boldsymbol{r} \to \boldsymbol{r}e^{i\theta}, \ \boldsymbol{k} \to \boldsymbol{k}e^{-i\theta} \tag{1}$$

$$\Longrightarrow \ H_\theta \equiv U(\theta)HU^{-1}(\theta), \ |\Phi_\theta\rangle \equiv U(\theta)|\Phi\rangle \tag{2}$$

where θ is a scaling angle. The energy is calculated as $E_\theta = \langle\tilde{\Phi}_\theta|H_\theta|\Phi_\theta\rangle$ with a bi-orthonormal state $|\tilde{\Phi}_\theta\rangle$. Energies of bound and resonant states are independent of the scaling angle θ while those of continuum states, which are complex value due to the complex scaling, change with θ. (*ABC theorem*[14]) Investigating the θ-dependence of the energy of each state, we can find resonant states. It is noted that continuum states appear on the line of $-\mathrm{Im}\,E\,/\,\mathrm{Re}\,E = \tan 2\theta$ (2θ *line*) in the complex energy plane when the scaling angle is θ, because the kinetic energy is proportional to $\boldsymbol{k}^2e^{-2i\theta}$.

An advantage of the CSM is that resonant states can be calculated quite in the same way as bound states. Here, let us consider a case of two-body

[c] With the aid of multiple scattering theory, such a coupled-channel calculation in the variational scheme is realized in Ref.[10]

system. A wave function of a resonant state is expressed as

$$\Phi_R(r) \sim e^{ik_R r} = e^{i(\kappa - i\gamma)r}, \tag{3}$$

where κ and γ are real numbers and γ is positive. Applying the complex scaling to this wave function, then it is transformed as

$$\begin{aligned} U(\theta)\Phi_R(r) &\sim e^{ik_R r e^{i\theta}} \\ &= \exp[(\gamma\cos\theta - \kappa\sin\theta)r] \cdot \exp[i(\gamma\sin\theta + \kappa\cos\theta)r]. \end{aligned} \tag{4}$$

When a condition $\theta > \tan^{-1}(\gamma/\kappa)$ is satisfied, the first exponential term becomes a dumping function. In other words, the boundary condition for resonant states is modified to the same one for bound states. Therefore, we can obtain resonant states by diagonalizing the complex-scaled Hamiltonian H_θ with Gaussian base as done in usual studies of bound states.

3. Application of CSM to $I = 0$ $\bar{K}N$-$\pi\Sigma$ system

Here, as the first step of our attempt, we consider the simplest case of a two-body system with two channels; an s-wave $I = 0$ $\bar{K}N$-$\pi\Sigma$ system (corresponding to $\Lambda(1405)$), which is studied with an energy-independent $\bar{K}N$ potential.

3.1. *Hamiltonian and wave function*

Hamiltonian used in the present study is

$$\hat{H} = \sum_{\alpha=\bar{K}N,\pi\Sigma} \left(\Delta M_\alpha + \frac{\boldsymbol{p}^2}{2\mu_\alpha}\right) |\alpha\rangle\langle\alpha| + \hat{V}_{MB}, \tag{5}$$

where a mass difference ΔM_α is $(M_N + m_K) - (M_\Sigma + m_\pi)$ for the $\bar{K}N$ channel and zero for the $\pi\Sigma$ channel. μ_α is a reduced mass for the channel α. We employ a phenomenological $\bar{K}N$ potential[1] as $\hat{V}_{MB}$:

$$\hat{V}_{MB} = \textstyle\sum_{\alpha,\beta} V_{\alpha,\beta} \exp[-(r/0.66\,\mathrm{fm})^2]\ |\alpha\rangle\langle\beta|, \tag{6}$$

$$V_{\bar{K}N,\bar{K}N} = -436\,\mathrm{MeV}, \quad V_{\pi\Sigma,\pi\Sigma} = 0\,\mathrm{MeV}, \tag{7}$$

$$V_{\bar{K}N,\pi\Sigma} = V_{\pi\Sigma,\bar{K}N} = -412\,\mathrm{MeV}. \tag{8}$$

The wave function contains the $\bar{K}N$ and $\pi\Sigma$ components. The spatial parts of each component, $\phi_{\bar{K}N}(r)$ and $\phi_{\pi\Sigma}(r)$, are expanded with Gaussian base $G_n^\alpha(r)$ in which width parameters $\{b_n^\alpha\}$ are geometrically generated from minimum and maximum width parameters, b_{min} and b_{max}. In the present study, b_{min} is chosen to be 0.1 fm, while various values of b_{max} are

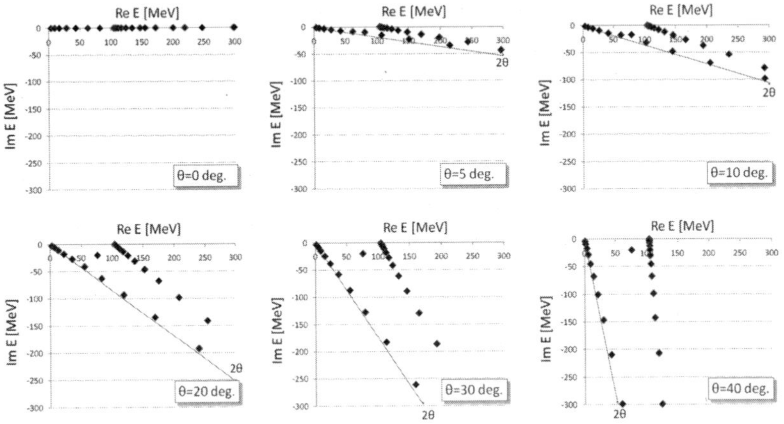

Fig. 2. θ trajectory of s-wave $I = 0$ $\bar{K}N$-$\pi\Sigma$ system in the complex energy plane. ($N = 30$, $b_{max} = 10$ fm) Symbols of filled diamond are complex energy eigenvalues.

tested. (so-called b trajectory) Coefficients $\{C_n^\alpha\}$ are unknown parameters to be determined.

$$|\Phi^{I=0}_{\bar{K}N-\pi\Sigma}\rangle = \phi_{\bar{K}N}(r)|\bar{K}N\rangle \;+\; \phi_{\pi\Sigma}(r)|\pi\Sigma\rangle \tag{9}$$

$$\phi_\alpha(r) = \sum_n C_n^\alpha G_n^\alpha(r) = \sum_n C_n^\alpha \exp[-(r/b_n^\alpha)^2] \tag{10}$$

The Hamiltonian $\hat{H}$ given in Eq. (5) is complex-scaled with the scaling angle θ. Diagonalizing the complex-scaled Hamiltonian $\hat{H}_\theta$ with the base $G_n^\alpha(r)|\alpha\rangle$, we obtain complex-energy eigenvalues. Checking the θ dependence of each eigenvalue, we can distinguish resonant states from lots of eigenstates.

3.2. *Result*

Since this study is our first attempt, we show the intermediate procedure of our calculation as well as the final result of s-wave $I = 0$ $\bar{K}N$-$\pi\Sigma$ resonance obtained with the complex scaling method.

Figure 2 shows the distribution of the complex-energy eigenvalues of the complex-scaled Hamiltonian at various scaling angle θ. These panels show so-called θ trajectory. Here, the number of Gaussian base N and the maximum value of the width parameter of Gaussian b_{max} are fixed as (N, b_{max})=(30, 10 fm). Most of eigenvalues are varied with the change of θ. However, we can find a stable point in the figure. In the panels for $\theta > 10$

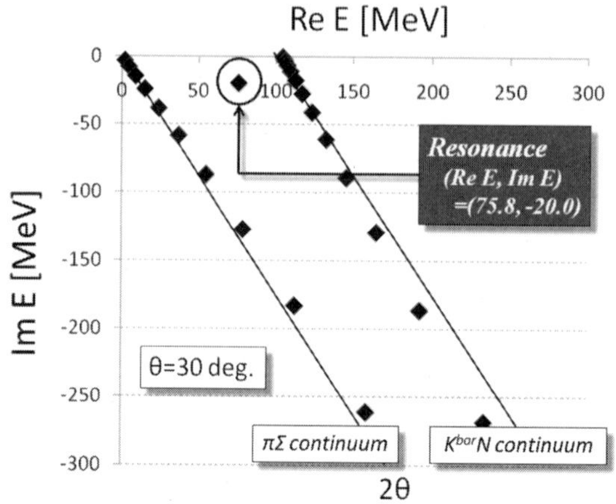

Fig. 3. A resonant state found in complex energy plane by complex scaling method.

degree, it is found that one eigenvalue, (Re E, Im E)=(75.8, −20.0) MeV, is independent of θ and it keeps staying at the same position in the complex energy plan. The state indicated with this eigenvalue is a resonant state.

In the present calculation, the energy is measured from the $\pi\Sigma$ threshold. The eigenvalues on the 2θ line starting from the origin indicates $\pi\Sigma$ continuum states. There are also the eigenvalues on the other 2θ line starting from the energy (Re E, Im E)=(104, 0) MeV. Since this energy is the energy of $\bar{K}N$ threshold measured from the $\pi\Sigma$ threshold, these states on this 2θ line are $\bar{K}N$ continuum states. (Figure 3)

By measuring the energy from the $\bar{K}N$ threshold, we can define the $\bar{K}N$ binding energy as B.E. $(\bar{K}N) \equiv (M_N + m_K) - \mathrm{Re}\,E$. We remember also the fact of $\mathrm{Im}\,E = -\Gamma/2$ where Γ means the decay width. Then, (B.E. $(\bar{K}N)$, Γ) = (28.2, 40.0) MeV for the above mentioned resonant state. This resonant state is found to be $\Lambda(1405)$, because these values are consistent with the experimental data of $\Lambda(1405)$; (B.E. $(\bar{K}N)$, Γ) = (27, 50) MeV.

We have checked the stability of the obtained resonant state for other parameters, b_{max} and N. In other words, we have investigated so-called b trajectory and n trajectory which are given in Table 2. As shown in the Table 2, it is confirmed that the resonant state is very stable for the variation of b_{max}. This resonant state is stable also for the change of N as given in the last column of the table.

4. Summary and future plans

We propose the application of the complex scaling method (CSM) to the study of the most essential $\bar{K}$-nuclear system, K^-pp. The CSM enables us to treat resonant states quite in the same way as bound states; we can obtain the resonant states by the diagonalization method with Gaussian base. With the aid of CSM, the K^-pp system will be treated as a $\bar{K}NN$-πYN coupled-channel system, advantages of past variational calculations being succeeded. By the explicit treatment of the πYN degree, we will be able to consider the binding mechanism of the K^-pp system in more detail, such as the importance of $\pi\Sigma N$ dynamics that was pointed out by the Faddeev calculation.[13]

As the first step of our attempt, we have investigated a two-body system with two channels using an energy-independent $\bar{K}N$ potential. Applying the CSM to an s-wave $I = 0$ $\bar{K}N$-$\pi\Sigma$ system and using a phenomenological $\bar{K}N$ potential,[1] we obtained a resonant state corresponding to $\Lambda(1405)$, as mentioned in the original paper of this $\bar{K}N$ potential.

We have two comments for this phenomenological $\bar{K}N$ potential. First, certainly also in our calculation, only a single resonant state has been obtained. Since this potential has no interaction between π and Σ as shown in Eq. (7), $V_{\pi\Sigma,\pi\Sigma} = 0$, and has no energy dependence, there is no resonant state in the $\pi\Sigma$ channel and then only $\pi\Sigma$ continuum state is there. Second, if we use only the $\bar{K}N$-$\bar{K}N$ direct potential ($V_{\bar{K}N,\bar{K}N}$) of this potential, the $\bar{K}N$ system is bound with very small binding energy of 3.2 MeV. Therefore, as a result that this $\bar{K}N$ bound state is coupled with the $\pi\Sigma$ continuum state through the coupling potential $V_{\bar{K}N,\pi\Sigma}$, the $\bar{K}N$ binding energy increases to be 28 MeV and simultaneously the decay width of 40 MeV is generated. This is the mechanism of formation of $\Lambda(1405)$ in this phenomenological $\bar{K}N$ potential.

The study reported in this article is only the repetition of the original study proposing the phenomenological $\bar{K}N$ potential, because the com-

Table 2. b trajectory at $N = 30$ and $\theta = 30$ degree. A part of n trajectory is also given in the last column, where N is changed to be 60. "B.E. ($\bar{K}N$)" is the resonant energy measured from the $\bar{K}N$ threshold. "Γ" is the decay width. Both are in units of MeV.

b_{max} N	5 30	10 30	15 30	20 30	30 30	40 30	50 30	50 60
B.E. ($\bar{K}N$)	28.167	28.170	28.170	28.169	28.170	28.169	28.170	28.199
Γ	40.061	40.058	40.058	40.058	40.058	40.059	40.060	40.047

plex scaling method was employed when the authors constructed their phenomenological potential.[1] Future plans of our study using the CSM are as follows:

- The $\bar{K}N$-πY system with an energy-dependent $\bar{K}N$ potential
  A $\bar{K}N$ potential derived from a chiral SU(3) theory has an energy dependence and somewhat strong attraction in $\pi\Sigma$-$\pi\Sigma$ channel. As mentioned in the Introduction, there are two poles in such a chiral-based potential since another resonant state in $\pi\Sigma$ channel is formed due to both the effects. We will see how the $I = 0$ $\bar{K}N$-πY system is obtained in the CSM scheme with the chiral-based potential.

- Study of three-body system $\bar{K}NN$-πYN
  As explained in the Introduction, we will check the effect of the three-body $\pi\Sigma N$ dynamics, comparing to the previous study with variational method.[6,7] Furthermore, we have interested in the case that the chiral-based potential is employed; How will the double-pole effect appear in the three-body system?

- Analysis of the obtained resonant state
  In the complex scaling method, we have a wave function. Since various physical quantities can be calculated with the CSM wave function,[15] we will investigate detailed property of obtained resonant states.

For a couple of years, new experimental results related to the K^-pp system have been reported.[16,17] DISTO group has reported the observation of a clear peak structure in the Λp invariant spectrum.[17] (B. E. = 115 MeV and Γ = 100 MeV if this peak is a bound state of the K^-pp system) Since the observed peak structure is just on the $\pi\Sigma N$ threshold, we expect that such a state is dominated by the degree of $\pi\Sigma N$ rather than that of $\bar{K}NN$. Our study will give some hints to understand the state reported by DISTO group. In J-PARC (Japan Proton Accelerator Research Complex), two experiments related to $\bar{K}$-nuclear physics are planned as Day-1 experiment: E15 "A search for deeply bound kaonic nuclear states by ^{3}He(in-flight K^-, n) reaction" and E17 "Precise spectroscopy of kaonic ^{3}He atom $3d \to 2p$ X-rays" . (E17 are now going on.) We hope that these experiments will provide us useful information on $\bar{K}$-nuclear systems.

Acknowledgement

One of authors (A. D.) is thankful to Dr. Hyodo for fruitful discussion and appreciates Dr. Myo for his kind advice about the complex scaling method. This work is supported by the Grant for Scientific Research (No. 19740163) from the Ministry of Education, Culture, Sports, Science and Technology (MEXT) of Japan.

References

1. Y. Akaishi and T. Yamazaki, Phys. Rev. C **65**, 044005 (2002).
2. A. Doté, H. Horiuchi, Y. Akaishi and T. Yamazaki, Phys. Lett. **B590**, 51 (2004); Phys. Rev. C **70**, 044313 (2004).
3. M. Agnello *et al.*, (FINUDA collaboration), Phys. Rev. Lett **94**, 212303 (2005); V. K. Magas, E. Oset, A. Ramos, and H. Toki, Phys. Rev. C **74**, 025206 (2006); T. Kishimoto *et al.*, Prog. Theor. Phys. **118**, 181 (2007); M. Sato *et al.*, Phys. Lett. **B659**, 107 (2008).
4. A. Arai, M. Oka and S. Yasui, Prog. Theor. Phys. **119**, 103 (2008); T. Nishikawa and Y. Kondo, Phys. Rev. **C77**, 055202 (2008).
5. T. Koike and T. Harada, Phys. Lett. **B652**, 262 (2007); Phys. Rev. C **80**, 055208 (2009).
6. A. Doté, T. Hyodo and W. Weise, Phys. Rev. C **79**, 014003 (2009).
7. T. Yamazaki and Y. Akaishi, Phys. Rev. C **76**, 045201 (2007).
8. Y. Ikeda and T. Sato, Phys. Rev. C **76**, 035203 (2007).
9. N. V. Shevchenko, A. Gal, J. Mareš, and J. Révai, Phys. Rev. C **76**, 044004 (2007).
10. S. Wycech and A. M. Green, Phys. Rev. C **79**, 014001 (2009).
11. D. Jido, J. A. Oller, E. Oset, A. Ramos, U. -G. Meißner, Nucl. Phys. **A725**, 181 (2003).
12. T. Hyodo and W. Weise, Phys. Rev. C **77**, 035204 (2008).
13. Y. Ikeda and T. Sato, Phys. Rev. C **79**, 035201 (2009).
14. S. Aoyama, T. Myo, K. Kato, and K. Ikeda, Prog. Theor. Phys. **116**, 1 (2006).
15. T. Myo, R. Ando and K. Kato, Phys. Rev. C **80**, 014315 (2009).
16. M. Agnello *et al.*, Nucl. Phys. **A775**, 35 (2006); T. Suzuki *et al.*, arXiv:0711.4943 [nucl-ex]; Mod. Phys. Lett. **A23**, 2520 (2008).
17. T. Yamazaki *et al.*, arXiv:0810.5182 [nucl-ex], in EXA'08 proceedings.

CHIRAL PROPERTIES OF BARYONS

V. DMITRAŠINOVIĆ*

Vinča Institute of Nuclear Sciences, Physics lab 010
P.O.Box 522, 11001 Beograd, Serbia
**E-mail: dmitra@vinca.rs*

We consider baryon chiral multiplet mixing as an explanation of the nucleon's isovector $g_A^{(3)}$ (expt. = 1.267 ± 0.004) and flavor-singlet $g_A^{(0)}$ (expt. 0.33 ± 0.08) axial coupling constants. Three-quark nucleon interpolating fields in QCD have well-defined $U_A(1)$ and $SU_L(3) \times SU_R(3)$ chiral transformation properties, which means that the axial currents made from these fields have well defined flavor singlet $g_A^{(0)}$ and flavor octet $g_A^{(3)} = F + D$, $g_A^{(8)} = \frac{1}{\sqrt{3}}(3F - D)$ couplings. There are two independent local baryon fields, the $[(3, \overline{3}) \oplus (\overline{3}, 3)]$ (the "Ioffe current") and the $[(8, 1) \oplus (1, 8)]$. These two fields together with their (non-local) chiral "mirrors", $[(\overline{3}, 3) \oplus (3, \overline{3})]$ and $[(1, 8) \oplus (8, 1)]$ allow four possible scenarios of mixing with the non-local $[(6, 3) \oplus (3, 6)]$ interpolating field. Chiral multiplet mixing then predicts the nucleon flavor-singlet axial coupling $g_A^{(0)}$, with a value that depends on the chiral multiplet that is admixed to the $[(6, 3) \oplus (3, 6)]$ one. In this talk we explore all two- and three-state mixing scenarios and show that one can fit the observed data reasonably well, though not perfectly.

Keywords: Chiral multiplet mixing; Baryons; Axial couplings.

1. Introduction

Chiral symmetry, as one of the basic symmetries of QCD, is a key property of the strong interactions. When the chiral $SU_L(3) \times SU_R(3)$ symmetry is spontaneously broken to $SU_V(3)$, by the QCD ground state, it is widely believed that no left-overs of the Wigner-Weyl symmetry realization/phase, such as chiral symmetry multiplets, remain. Yet, Harari,[1] Gerstein and Lee,[2] and somewhat later Weinberg[3] considered mixing of chiral multiplets in the spontaneously broken symmetry phase a.k.a. Nambu-Goldstone phase and found some interesting/encouraging results, which were then left without adequate follow-up for more than 40 years.

In general, such chiral representation mixing tends to be most powerful when only a few multiplets are involved. That is perhaps best visible in the case of heavy-light flavored mesons (e.g. D and D_s mesons) in which case powerful mass sum rules have been derived[4] from the premise of a single chiral multiplet. But as the number of admixed multiplets grows this method becomes increasingly complicated and thus loses its predictive power. The number of chiral multiplets grows as the number of light valence quarks increases, and in the case of three light (valence) quarks in the baryons, the number of chiral multiplets is three: $[(3,\overline{3}) \oplus (\overline{3},3)]$, $[(8,1) \oplus (1,8)]$ and $[(6,3) \oplus (3,6)]$ (or six, if one counts the so-called mirror multiplets, when one has to add the $[(\overline{3},3) \oplus (3,\overline{3})]$ and $[(1,8) \oplus (8,1)]$ and $[(3,6) \oplus (6,3)]$).

In one of the early applications of this idea, Harari,[1] Gerstein and Lee,[2] and somewhat later Weinberg,[3] used the mixing of the $[(\frac{1}{2},0) \oplus (0,\frac{1}{2})]$ and the $[(1,\frac{1}{2}) \oplus (\frac{1}{2},1)]$ $SU_L(2) \times SU_R(2)$ chiral multiplets[a] to fit the nucleon's isovector axial coupling constant $g_A^{(1)}$ value at 1.267 and thus phenomenologically explain its being different from unity, as was seemingly demanded by the Gell-Mann–Levy model.[5]

The nucleon also has a flavor singlet axial coupling $g_A^{(0)}$, that has been estimated from spin-polarized lepton-nucleon DIS data as $g_A^{(0)} = 0.28 \pm 0.16$,[6] or more recently as $0.33 \pm 0.03 \pm 0.05$,[7] which value the non-relativistic quark model predicts to be unity. The basic is question now if the same (Weinberg) chiral mixing angle can simultaneously explain the anomalously low value of this coupling? The answer manifestly depends on the $U_A(1)$ chiral transformation properties of the two admixed nucleon fields.

At this stage one must note that, at about the same time, another alternative explanation of the iso-vector axial coupling constant was offered by Bjorken and Nauenberg[8] that depends on the presence of (arbitrary/unconstrained) derivative couplings. Bjorken and Nauenberg's interaction cannot be used to predict the flavor singlet nucleon axial coupling, but a new interaction of the same type can be written to fit it.

One may also note that the original idea predates QCD and does not even invoke quarks, whose very existence Weinberg disputed at that stage (see p.4 of Ref. 9). This idea may nevertheless be viable in QCD: indeed, in the early 1990's Weinberg revived this idea himself,[10] when he re-named it "mended symmetry".

In this talk I address this question using the $U_A(1)$ and $SU_L(3) \times SU_R(3)$ chiral transformation properties of nucleon fields as derived from the three-quark nucleon interpolating fields in QCD in Refs. 11–13. If the answer

[a]these are the two-flavor chiral symmetry analogons of the $[(3,\overline{3}) \oplus (\overline{3},3)]$, $[(8,1) \oplus (1,8)]$ and $[(6,3) \oplus (3,6)]$ two-flavor chiral multiplets.

to our question ("can one explain the anomalously low value of the flavor singlet axial coupling with the chiral mixing angle that fits the isovector axial coupling?") turns out in the positive, then we may speak about the chiral mixing idea being viable in QCD.

2. Three-quark nucleon interpolating fields

We start by summarizing the transformation properties of various quark trilinear forms with quantum numbers of the nucleon as shown in Refs. 11,12. The nucleon interpolating fields appear to be rather formal objects with little intuition attached to them. So, in order to gain some physics intuition, we shall look at the non-relativistic version of three-quark nucleon interpolating fields first.

2.1. *Non-relativistic three-quark nucleon interpolating fields*

There is nothing in the interpolating field formalism that would limit it to relativistic (Dirac) quark fields: indeed a straightforward, extension of the interpolating field formalism to three non-relativistic second-quantized Schrödinger quark fields leads to two apparently independent nucleon interpolating fields

$$\mathcal{N}_1 = \epsilon_{abc}(q_a^T q_b) q_c \tag{1}$$

$$\mathcal{N}_2 = \epsilon_{abc}(q_a^T \tau^i \sigma^j q q_b) \tau^i \sigma^j q_c, \tag{2}$$

where ϵ_{abc} is the completely antisymmetric tensor, the indices $a, b, c = 1, 2, 3$ correspond to the color degree of freedom and σ and τ are the spin- and isospin Pauli matrices, respectively. These two field operators are equivalent, however. One can see that in two different ways: 1) Formally we can prove this equivalence starting from the Fierz identities

$$\mathcal{F}\left[\begin{pmatrix}\mathcal{N}_1 \\ \mathcal{N}_2\end{pmatrix}\right] = -\frac{1}{4}\begin{pmatrix}1 & 1 \\ 9 & 1\end{pmatrix}\begin{pmatrix}\mathcal{N}_1 \\ \mathcal{N}_2\end{pmatrix}, \tag{3}$$

which lead to the identity:

$$\mathcal{N}_2 = 3\mathcal{N}_1. \tag{4}$$

Therefore, only one of the two nucleon operators is independent. 2) Intuitively, one expects that each field operator corresponds to the two spin-isospin wave functions/combinations shown on the right hand-side of Eqs. (5,6).

$$\mathcal{N}_1 \longleftrightarrow \chi^{\rho_{12}}\phi^{\rho_{12}} + \chi^{\rho_{23}}\phi^{\rho_{23}} + \chi^{\rho_{31}}\phi^{\rho_{31}}; \tag{5}$$

$$\mathcal{N}_2 \longleftrightarrow \chi^{\lambda_{12}}\phi^{\lambda_{12}} + \chi^{\lambda_{23}}\phi^{\lambda_{23}} + \chi^{\lambda_{31}}\phi^{\lambda_{31}} \tag{6}$$

The $\mathcal{N}_{1,2}$ are operators, however, not wave functions, that are also anti-symmetric with respect to the color degree of freedom. This means that after complete anti-symmetrization one obtains both the ρ and the λ components in the w.f.:

$$\chi^{\rho_{12}}\phi^{\rho_{12}} + \chi^{\rho_{23}}\phi^{\rho_{23}} + \chi^{\rho_{31}}\phi^{\rho_{31}} = \chi^{\rho_{12}}\phi^{\rho_{12}} + \chi^{\lambda_{12}}\phi^{\lambda_{12}}; \tag{7}$$

$$\chi^{\lambda_{12}}\phi^{\lambda_{12}} + \chi^{\lambda_{23}}\phi^{\lambda_{23}} + \chi^{\lambda_{31}}\phi^{\lambda_{31}} = 3\left(\chi^{\rho_{12}}\phi^{\rho_{12}} + \chi^{\lambda_{12}}\phi^{\lambda_{12}}\right). \tag{8}$$

Note the factor 3 in the wave functions in Eq. (8) that is directly related to the factor 3 in the Fierz identity Eq. (4).

In other words, these two interpolating field operators are equivalent. This should have been expected: we knew all along that there is only one viable candidate for the non-relativistic 3-quark ground state. This non-relativistic example is instructive because it shows how the interpolating field operators correspond to the spin-flavor wave functions and how the Pauli principle reduces their number. The same is true for relativistic interpolating fields, except in that case there is more than one independent field.

The $\mathrm{SU_{FS}}(6)$ flavor-spin symmetry leads to (partially) conserved Noether currents defined by way of the SU(6) transformation properties of these fields:

$$\delta_{\vec{b}}^{\vec{a}}\left(\mathcal{N}_1 + \frac{1}{3}\mathcal{N}_2\right) = i\left[\frac{5}{3}(\boldsymbol{\tau}\cdot\boldsymbol{a})(\boldsymbol{\sigma}\cdot\boldsymbol{b})\left(\mathcal{N}_1 + \frac{1}{3}\mathcal{N}_2\right) + 4a^a b^i \Delta^{a,j}\right]. \tag{9}$$

The factor $\frac{5}{3}$ is the iso-vector axial coupling of the nucleon in the SU(6) symmetric quark model. Therefore we may say that the two fields, $(\mathcal{N}_1 + \frac{1}{3}\mathcal{N}_2)$ and $\Delta^{a,j}$, form the **56** representation of the $SU_{\mathrm{FS}}(6)$ symmetry. In the next subsection we shall use this method of constructing Noether currents, only with respect to the chiral $SU_L(3) \times SU_R(3)$ transformations in that case.

2.2. *Relativistic three-quark nucleon interpolating fields*

In Refs. 11,12,14 the local (non-derivative) spin $\frac{1}{2}$ baryon operators

$$N_1 = \epsilon_{abc}(\tilde{q}_a q_b)q_c, \tag{10}$$

$$N_2 = \epsilon_{abc}(\tilde{q}_a \gamma^5 q_b)\gamma^5 q_c, \tag{11}$$

were classified according to their Lorentz, chiral $SU_L(2)\times SU_R(2)$ (so-called "naive" chiral multiplet, whose axial charge is positive) and $U_A(1)$ group representations. Here we have introduced the "tilde-transposed" quark field $\tilde{q}$ as $\tilde{q} = q^T C\gamma_5(i\tau_2)$, where $C = i\gamma_2\gamma_0$ is the Dirac field charge conjugation operator, τ_2 is the second isospin Pauli matrix. Once one allows for the presence of one derivative in the field, such as in

$$N_1' = \epsilon_{abc}(\tilde{q}_a q_b) i\partial_\mu \gamma^\mu q_c, \tag{12}$$

$$N_2' = \epsilon_{abc}(\tilde{q}_a \gamma^5 q_b) i\partial_\mu \gamma^\mu \gamma^5 q_c. \tag{13}$$

whose axial charge is negative, Refs.15–17, we see that the so-called "mirror" fields $(0, \frac{1}{2}) \oplus (\frac{1}{2}, 0)$ become available. And the $(1, \frac{1}{2}) \oplus (\frac{1}{2}, 1)$ nucleon chiral representation

$$N_3' = i\partial_\mu(\tilde{q}\gamma_\nu q)\Gamma^{\mu\nu}_{3/2}\gamma_5 q, \tag{14}$$

$$N_4' = i\partial_\mu(\tilde{q}\gamma_\nu\gamma_5\tau^i q)\Gamma^{\mu\nu}_{3/2}\tau^i q, \tag{15}$$

are also Pauli allowed, see Table 1. Here $\Gamma^{\mu\nu}_{3/2} = g^{\mu\nu} - \frac{1}{4}\gamma^\mu\gamma^\nu$.

In Table 1 we show the Abelian and non-Abelian chiral properties of the nucleon interpolating fields in QCD, Refs. 11–13. It turns out that the relativistic nucleon, i.e., spin-1/2 (Lorentz group representation $(\frac{1}{2}, 0) \oplus (0, \frac{1}{2})$) and isospin 1/2, field comes in two varieties: one with "mirror" and another with "triple-naive" Abelian chiral properties. These two fields have the same non-Abelian $SU_L(2) \times SU_R(2)$ transformation properties,[11,12] though different $SU_L(3) \times SU_R(3)$ transformation properties: $N_1 - N_2$ belongs to $[(3, \overline{3}) \oplus (\overline{3}, 3)]$ multiplet, $N_1 + N_2$ to $[(8, 1) \oplus (1, 8)]$ and $(N_3' + \frac{1}{3}N_4')$ to $[(6, 3) \oplus (3, 6)]$, see Ref. 13.

Table 1. The Abelian and the non-Abelian axial charges (+ sign indicates "naive", - sign "mirror" transformation properties) and the non-Abelian chiral mutiplets of $J^P = \frac{1}{2}$, Lorentz representation $(\frac{1}{2}, 0)$ nucleon fields. The field denoted by 0 belongs to the $(1, \frac{1}{2}) \oplus (\frac{1}{2}, 1)$ chiral multiplet and is the basic nucleon field that is mixed with various $(\frac{1}{2}, 0)$ nucleon fields in Eq. (16).

case	field	$g_A^{(0)}$	$g_A^{(1)}$	$SU_L(2) \times SU_R(2)$	F	D	$SU_L(3) \times SU_R(3)$
I	$N_1 - N_2$	-1	$+1$	$(\frac{1}{2}, 0) \oplus (0, \frac{1}{2})$	0	$+1$	$(3, \overline{3}) \oplus (\overline{3}, 3)$
II	$N_1 + N_2$	$+3$	$+1$	$(\frac{1}{2}, 0) \oplus (0, \frac{1}{2})$	$+1$	0	$(8, 1) \oplus (1, 8)$
III	$N_1' - N_2'$	$+1$	-1	$(0, \frac{1}{2}) \oplus (\frac{1}{2}, 0)$	0	-1	$(\overline{3}, 3) \oplus (3, \overline{3})$
IV	$N_1' + N_2'$	-3	-1	$(0, \frac{1}{2}) \oplus (\frac{1}{2}, 0)$	-1	0	$(1, 8) \oplus (8, 1)$
0	$\partial_\mu(N_3^\mu + \frac{1}{3}N_4^\mu)$	$+1$	$+\frac{5}{3}$	$(1, \frac{1}{2}) \oplus (\frac{1}{2}, 1)$	$+\frac{2}{3}$	$+1$	$(6, 3) \oplus (3, 6)$

Thus we found that indeed, as Gell-Mann and Levy[5] had postulated, the lowest-twist (non-derivative) J= $\frac{1}{2}$ nucleon field(s) form a $(\frac{1}{2}, 0)$ chiral multiplet, albeit there are two such independent fields. There is only one set of J= $\frac{1}{2}$ Pauli-allowed sub-leading-twist (one-derivative) interpolating fields that form a $(1, \frac{1}{2})$ chiral multiplet, however.

Moreover, we note that the pair of sub-leading-twist (one-derivative) interpolating fields Eqs. (12),(13) that form the "mirror" $(0, \frac{1}{2})$ chiral multiplets, is not the only possible one: it is possible to construct non-Abelian $SU_L(2) \times SU_R(2)$ symmetry "mirror" interpolating fields from four quark and one anti-quark field, see Ref.15, and perhaps, also from higher Fock-space configurations. Thus the source of the "mirror" components in the nucleon flavor-spin "wave function" is not unique, consequently their interpretation in terms of "meson cloud", or "$q\bar{q}$" components remains open.

We shall use these results as the (theoretical) input into our subsequent calculations. This constitutes a minimal assumption, as one has no other guide to the chiral representations of the nucleon.

3. Mixing of two chiral representations

Next we consider the mixing of one of the fundamental chiral representations, as shown in Table 1 and the "higher" representation $(1, \frac{1}{2})$ for the nucleon,

$$\begin{aligned} g_{A\ \mathrm{mix.}}^{(1)} &= g_{A,\ \alpha}^{(1)} \cos^2\theta + g_{A\ (1,\frac{1}{2})}^{(1)} \sin^2\theta, \\ &= g_{A,\ \alpha}^{(1)} \cos^2\theta + \frac{5}{3} \sin^2\theta = 1.267. \end{aligned} \tag{16}$$

Here the suffix α corresponds to one of I-IV and the corresponding values of $g_{A,\ \alpha}^{(1)}$ are given in Table 1. We have also used the fact that $g_{A\ (1,\frac{1}{2})}^{(1)} = \frac{5}{3}$, see Refs. 1–3,12.

This provides a possible solution to the nucleon's axial coupling problem in QCD. Three-quark nucleon interpolating fields in QCD have well-defined two-fold $U_A(1)$ chiral transformation properties, see Table 1, that can be used to (naively) predict the isoscalar axial coupling $g_{A\ \mathrm{mix.}}^{(0)}$ as follows

$$g_{A\ \mathrm{mix.}}^{(0)} = g_{A,\ \alpha}^{(0)} \cos^2\theta + g_{A\ (1,\frac{1}{2})}^{(0)} \sin^2\theta, \tag{17}$$

together with the mixing angle θ extracted from Eq. (16). Note, however, that due to the different (bare) non-Abelian $g_A^{(1)}$ and Abelian $g_A^{(0)}$ axial couplings, see Table 1, the mixing formulae Eq. (17) give substantially different predictions from one case to another, see Table 2. We can see in Table 2 that the two candidates are cases I and IV, with $g_A^{(0)} = -0.2$ and $g_A^{(0)} = 0.4$, respectively, the latter being within 1-σ of the measured value $g_A^{(0)} = 0.33 \pm 0.08$. The nucleon field in case I is the well-known "Ioffe current", which reproduces the nucleon's properties in QCD lattice and sum rules calculations. The nucleon field in case IV is the "mirror" opposite of

Table 2. The values of the baryon isoscalar axial coupling constant predicted from the naive mixing and $g_{A\ \mathrm{expt.}}^{(1)} = 1.267$; compare with $g_{A\ \mathrm{expt.}}^{(0)} = 0.33 \pm 0.03 \pm 0.05$.

	$(g_A^{(1)}, g_A^{(0)})$	$g_{A\ \mathrm{mix.}}^{(1)}$	θ	$g_{A\ \mathrm{mix.}}^{(0)}$	$g_{A\ \mathrm{mix.}}^{(0)}$
I	$(+1,-1)$	$\frac{1}{3}(4-\cos 2\theta)$	$\pm 39.3^o$	$-\cos 2\theta$	-0.20
II	$(+1,+3)$	$\frac{1}{3}(4-\cos 2\theta)$	$\pm 39.3^o$	$2+\cos 2\theta$	2.20
III	$(-1,+1)$	$\frac{1}{3}(1-4\cos 2\theta)$	$\pm 67.2^o$	1	1.00
IV	$(-1,-3)$	$\frac{1}{3}(1-4\cos 2\theta)$	$\pm 67.2^o$	$-(1+2\cos 2\theta)$	0.40

the orthogonal complement to the Ioffe current, an interpolating field that, to our knowledge, has not been used in QCD thus far.

4. Three-field mixing

A linear superposition of yet another field (except for the mixture of cases II and III above) ought to give a perfect fit to both experimental values. Such an admixture introduces new free parameters (besides the two already introduced mixing angles, e.g. θ_1 and θ_4, we have the relative/mutual mixing angle θ_{14}, as the two nucleon fields I and IV may also mix). One may subsume/redefine the sum and the difference of the two angles θ_1 and θ_4 into the new angle θ, whereas one may define $\theta_{14} \doteq \varphi$ (this relationship depends on the precise definition of the mixing angles θ_1, θ_4 and θ_{14}); thus we find two equations with two unknowns of the general form:

$$\frac{5}{3}\sin^2\theta + \cos^2\theta \left(g_A^{(1)} \cos^2\varphi + g_A^{(1)\prime} \sin^2\varphi \right) = 1.267 \tag{18}$$

$$\sin^2\theta + \cos^2\theta \left(g_A^{(0)} \cos^2\varphi + g_A^{(0)\prime} \sin^2\varphi \right) = 0.33 \tag{19}$$

The values of the mixing angles obtained from this simple fit to the two baryon axial coupling constants are shown in Table 3. This, however, is not just a mere fit: when extending to the $SU_L(3) \times SU_R(3)$ symmetry, chiral transformation properties of the two nucleon fields differ, see Table 1. From these chiral $SU_L(3) \times SU_R(3)$ symmetry assignment we can also predict the F and D couplings (un-corrected for the explicit $\mathrm{SU}_F(3)$ symmetry breaking) in Table 3, which can be compared with the experimental numbers. We have not calculated the $\mathrm{SU}_F(3)$ symmetry breaking corrections, as yet, so we have not taken into account the "error bars" on the mixing angle(s), which remains a task for the future. At any rate, it should be clear that the predicted values are "in the right ball park". Note that all three admissible scenarios (i.e. choices of pairs of fields admixed to the (6,3) one that

Table 3. The values of the mixing angles obtained from the fit to the baryon axial coupling constants and the predicted values of axial F and D couplings. Experimental values have evolved from F=0.459±0.008 and D=0.798±0.008 in Ref. 18 to F=0.477±0.001 and D=0.835±0.001 in Ref. 19. Note that the new values are more than 2-σ away from the old ones, and that the new F,D add up to F+D $= 1.312 \neq 1.269 \pm 0.002$.

case	$g_{A\ \mathrm{expt.}}^{(0)}$	$g_{A\ \mathrm{expt.}}^{(1)}$	θ	φ	F	D
I-II	1.267	0.33	39.3^o	26.6^o	0.399	0.868
I-III	1.267	0.33	50.7^o	23.9^o	0.399	0.868
I-IV	1.267	0.33	63.2^o	53.9^o	0.399	0.868

lead to real mixing angles) yield the same values of F and D! This is due to the fact that all three-quark baryon fields satisfy the following relation $g_A^{(0)} = 3F - D = \sqrt{3}g_A^{(8)}$, Ref. 20. The first relation $g_A^{(0)} - 3F - D$ was not expected, as the flavor-singlet properties, such as $g_A^{(0)}$ are generally expected to be independent of the flavor-octet ones, such as F, D. Yet, it is not unnatural, either, as it indicates the absence of polarized $s\bar{s}$ pairs in these SU(3) symmetric, three-quark nucleon interpolators. The $g_A^{(0)} = 3F - D$ relation appears to be particular for the three-quark interpolating fields and may well change when one considers e.g. five-quark interpolating fields. In that sense a deviation of the measured values from this relation may be seen as a sign of of higher-order configurations' contribution.

5. Summary and discussion

We have shown that one can reproduce, within 1-σ uncertainty, the (unexpectedly small) flavor singlet axial coupling of the nucleon by mixing (only) two (out of five independent) nucleon interpolating fields [b] by fitting the isovector- axial coupling. This solution to the nucleon spin problem does not necessarily invoke exotica such as a) hidden strangeness; or b) polarized gluon components in the nucleon wave function, in agreement with recent results of the COMPASS experiment,[21],[22] but does not exclude them either. This scenario is quantitatively reproduced in a simple dynamical model[17] which then predicts the existence of the S_{31} resonance at 2160 MeV, in agreement with the PDG tables.[23]

By mixing three nucleon interpolating field chiral multiplets one may simultaneously fit both the isovector and the flavor singlet axial couplings

[b] or equivalently relativistic component in the nucleon's Bethe-Salpeter wave function

and predict the SU(3) F and D couplings, which have the correct order of magnitude, within the expected $\mathcal{O}(20\%)$ SU(3) symmetry breaking corrections. Special cases of such three-field chiral mixing have been considered by Harari,[1] and by Gerstein and Lee[2] in the context of the ("collinear") $U(3) \times U(3)$ current algebra at infinite momentum, long time ago. We emphasize here that our results are based on the $SU_L(3) \times SU_R(3)$ chiral algebra of space-integrated time components of currents, without any assumptions about saturation of this algebra by one-particle states, or dependence on any one particular frame of reference. Indeed, our nucleon interpolating fields transform as the $(\frac{1}{2}, 0) + (0, \frac{1}{2})$ representation of the Lorentz group, the currents are (fully) Lorentz covariant and therefore our results hold in any frame.

Thus, the chiral multiplet mixing is a viable theoretical scenario for the explanation of the nucleon isoscalar axial couplings in QCD.

Acknowledgments

The work reported in this talk was done jointly with Hua-Xing Chen, Keitaro Nagata and Atsushi Hosaka.[11–14,16,17,20] The author wishes to thank these gentlemen for valuable collaborations, as well as Profs. Daisuke Jido, Akira Ohnishi and Makoto Oka for valuable conversations. Last, but not least the author acknowledges the kind hospitality and financial support extended to him by RCNP, Osaka University and Yukawa Institute for Theoretical Physics, Kyoto.

References

1. H. Harari, *Phys. Rev. Lett.* **17**, 56 (1966).
2. I. S. Gerstein and B. W. Lee, *Phys. Rev. Lett.* **16**, 1060 (1966).
3. S. Weinberg, *Phys. Rev.* **177**, 2604 (1969).
4. W. A. Bardeen, E. J. Eichten and C. T. Hill, *Phys. Rev.* **D68**, p. 054024 (2003).
5. M. Gell-Mann and M. Levy, *Nuovo Cim.* **16**, p. 705 (1960).
6. B. W. Filippone and X.-D. Ji, *Adv. Nucl. Phys.* **26**, p. 1 (2001).
7. S. D. Bass, *The Spin structure of the proton* (2007), World Scientific. (ISBN 978-981-270-946-2 and ISBN 978-981-270-947-9). 212 p.
8. J. D. Bjorken and M. Nauenberg, *Ann. Rev. Nucl. Part. Sci.* **18**, 229 (1968).
9. D. J. Gross, *Proc. Nat. Acad. Sci.* **102**, 9099 (2005).
10. S. Weinberg, *Phys. Rev. Lett.* **65**, 1177 (1990).
11. V. Dmitrašinović, A. Hosaka and K. Nagata, *Mod. Phys. Lett.* **A23**, 2381 (2008).
12. K. Nagata, A. Hosaka and V. Dmitrašinović, *Eur. Phys. J.* **C57**, 557 (2008).

13. H.-X. Chen, V. Dmitrašinović, A. Hosaka, K. Nagata and S.-L. Zhu, *Phys. Rev.* **D78**, p. 054021 (2008).
14. V. Dmitrašinović, K. Nagata and A. Hosaka, *"Bled Workshops in Physics"* **Vol. 8**, 17 (2007).
15. D. Jido, M. Oka and A. Hosaka, *Prog. Theor. Phys.* **106**, 873 (2001).
16. V. Dmitrašinović, A. Hosaka and K. Nagata, *Int. J. Mod. Phys.* **E 19**, p. 91 (2010).
17. V. Dmitrašinović, A. Hosaka and K. Nagata, *Mod. Phys. Lett.* **A 25**, p. 233 (2010).
18. P. G. Ratcliffe, *Phys. Lett.* **B242**, 271 (1990).
19. T. Yamanishi, *Phys. Rev.* **D76**, p. 014006 (2007).
20. H.-X. Chen, V. Dmitrašinović and A. Hosaka, *Phys. Rev.* **D81**, p. 054002 (2010).
21. E. S. Ageev *et al.*, *Phys. Lett.* **B647**, 330 (2007).
22. M. Alekseev *et al.*, **hep-ex**, 0802.3023.
23. W. M. Yao *et al.*, *J. Phys.* **G33**, 1 (2006).

MEASUREMENT OF ϕp SCATTERING LENGTH FROM THE ϕ PHOTOPRODUCTION AT SPring-8/LEPS

S.Y. RYU

RCNP, Osaka University, Ibaraki, Osaka 567-0047, Japan
and RIKEN, Wako, Saitama 351-0198, Japan
E-mail: syryu@rcnp.osaka-u.ac.jp

FOR THE LEPS COLLABORATION

We present preliminary results on the measurement of ϕp scattering length from the ϕ photoproduction near its threshold energy at LEPS/SPring-8.

Keywords: ϕ photoproduction; scattering length; SPring-8/LEPS.

1. Scientific Motivation

Recent experimental results on in-medium modification of ϕ-meson properties have been reported in 12-GeV pA interactions.[1,2] The in-medium modification can be interpretated as due to a partial restoration of the chiral symmetry at a normal nuclear density. However, the ϕ-meson can rescatter off a spectator nucleon in the finite nucleus, so that it is important to understand nucear many-body effects such as collisional broadening and Fermi motion.

Photoproduction of the ϕ-mesons from the proton near threshold is suited for exploring the ϕp scattering cross section. The quark model predicts the ϕN cross section ($\sigma_{\phi N}$) to be 13.0 ± 1.5mb.[3] Another estimate from a QCD van der Waals potential calculation results in $\sigma_{\phi N} = 56$ mb at $q_\phi = 0$.[4] While the ϕN cross section was estimated to be $7.7 - 8.7$ mb from the measurement of $d\sigma/dt$ for the ϕ photoproduction at $E_\gamma = 4.6\text{–}6.7$GeV,[5] the LEPS collaboration[6] reported the $\sigma_{\phi N}$ to be 35^{+17}_{-11}mb from the A-dependence measurement of ϕ photoproduction.

Photoproduction of ϕ mesons are consistent with diffractive production via Pomeron-exchange as a dominant mechanism. The ϕp cross section can be evaluated in the Vector Dominance Model from the $\gamma p \to \phi p$ reaction. Within VDM the hadron-like photon is a superposition of all possible vector meson states. Forward differential cross section can then be written as

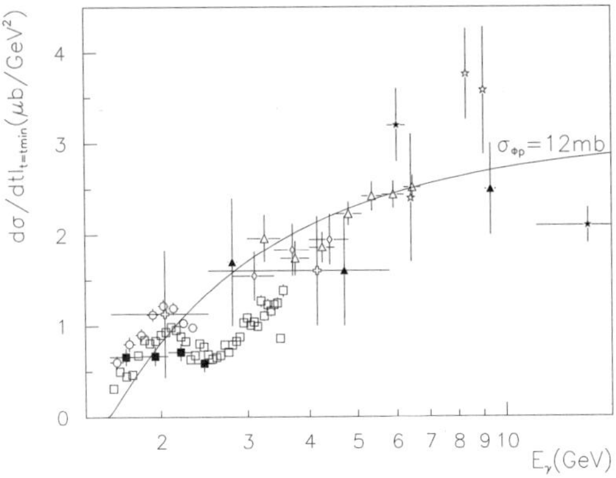

Fig. 1. Forward differential cross sections for ϕ photoproduction near threshold. Data points quoted here are taken from Refs. 5,8,9.

$$\frac{d\sigma}{dt}\bigg|_{t=t_{\min}}^{\gamma p\to\phi p} = \frac{\alpha}{16\gamma_\phi^2}\frac{q_\phi^2}{q_\gamma^2}(1+\eta_\phi^2)\sigma_{\phi p}^2 \underbrace{\longrightarrow}_{\eta_\phi=0} \text{const}\cdot\frac{q_\phi^2}{q_\gamma^2}\cdot\sigma_{\phi p}^2$$

where $\eta_\phi = \mathrm{Re}T_{\gamma p\to\phi p}/\ \mathrm{Im}T_{\gamma p\to\phi p}$ and $q_{\gamma,\phi}$ is γ/ϕ momentum of the center-of-mass system. The ϕp scattering length ($a_{\phi p}$) is given by $\sigma_{\phi p} = 4\pi a_{\phi p}^2$.[7]

2. Near-threshold ϕ Photoproduction Experiment at SPring-8/LEPS

We measured ϕ photoproduction from threshold to 2.4 GeV with linearly polarized photons at SPring-8. The LEPS spectrometer detects charged particles from the ϕ photoproduction at forward angles. The reaction of interest is $\gamma p \to \phi p$ close to its threshold energy (1.5732 GeV). In this analysis we first used K^-p detection mode from the 2002/2003 LH_2 data set. The other K^+p and K^-K^+ detection modes were analyzed for a consistency check. The K^+p detection mode has the same kinematical acceptance as the K^-p detection mode, but it contains large background from K^+ photoproduction associated with hyperons. Photon energy E_γ was calculated from the energy-momentum balance in $\gamma p \to K^-K^+p$ reaction.

Figure 2 shows an K^-K^+ invariant-mass spectrum (left) and a missing-mass spectrum for $p(\gamma, K^-p)X$ reaction by detecting a K^-p pair in the LEPS spectrometer. The ϕ signal is indicated by red hatched area spanning from 1.009 to 1.029 GeV/c^2.

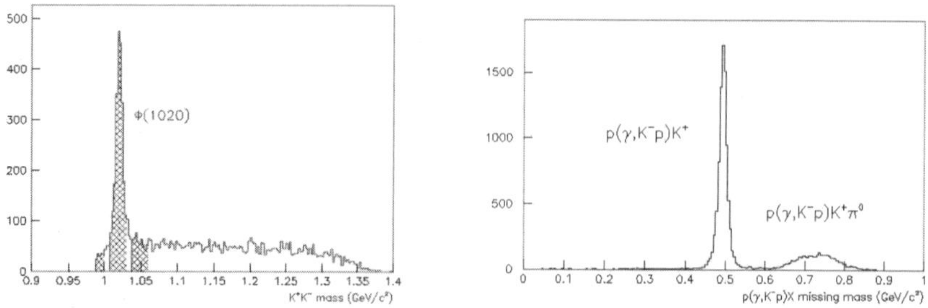

Fig. 2. Invariant mass spectrum for K^-K^+ systems and a missing-mass spectrum for $p(\gamma, K^-p)X$ by detecting a K^-p pair.

Possible competing reactions are therefore: (1) $\gamma p \to (K^+K^-)_{\text{S-wave}}p \to K^+K^-p$ ($E_\gamma^{th} = 1.5069\text{GeV}$) (2) $\gamma p \to \Lambda(1520)K^+ \to K^-pK^+$ ($E_\gamma^{th} = 1.6906\text{GeV}$) Close to the ϕ production threshold energy only the background reaction (1) contributes to the ϕ signal region. The $\Lambda(1520)$ production reaction appears beyond the ϕ threshold. We can neglect a contribution from the $\Lambda(1520)$ close to the ϕ production threshold, as indicated in Figure 2, although it contributes significantly to the background in the high energy region.

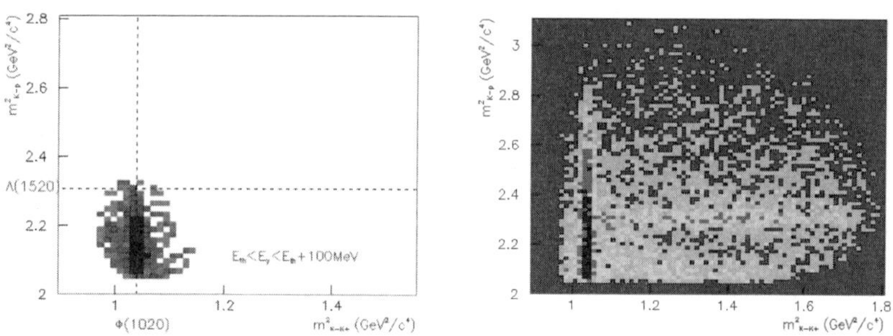

Fig. 3. Dalitz plots for the K^-p and K^+K^- systems in the energy region from the ϕ production threshold to 100 MeV (left) and in the whole energy region. Overlaid lines indicate the mass-squared values for $\Lambda(1520)$ and ϕ, respectively.

Near threshold the underlying background could largely be due to S-wave non-resonant K^+K^- pair production. Its different decay angular distribution could yield a different acceptance for the K^-p detection mode.

3. Preliminary Results and Discussion

About 750 $p(\gamma, K^-p)X$ events in the energy range from the threshold to 2.4 GeV have been selected falling in K^+K^- mass bins $1.009 \leq m_{KK} \leq 1.029$ GeV and in the missing-mass region $0.9 \leq M_X \leq 1.0$ GeV.

The acceptance-corrected t-distributions are fitted with the function $\frac{d\sigma}{dt} = N_0 \cdot \exp(-b|t - t_{\min}|)$, where $N_0 = d\sigma/dt|_{t=t_{\min}}$ is the intercept and b represents the slope parameter. The extrapolated cross section intercepts (optical points) are displayed in Fig. 4. The solid square points denote the current preliminary results compared with the previous results from the LEPS experiment[9] represented by the solid circle points. The ϕp scattering length will be estimated if further systematic studies are made with more recent data sets included.

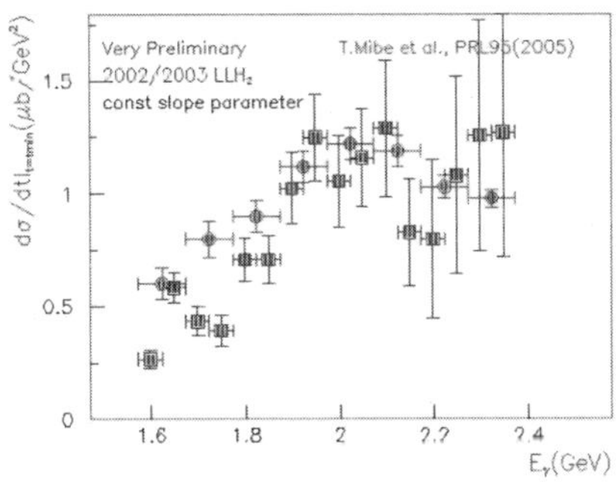

Fig. 4. Forward differential cross sections $d\sigma/dt|_{t=t_{\min}}$ for the ϕ photoproduction as a function of E_γ from the threshold to 2.4 GeV.

References

1. R. Muto *et al.*, Phys. Rev. Lett. **98** 042501 (2007).
2. T. Tabaru *et al.*, Phys. Rev. **C74** 025201 (2006).
3. M. Effenberger and A. Sibirtsev, Nucl. Phys. **A632** 99 (1998).
4. H. Gao, T.-S. Lee and V. Marinov, Phys. Rev. **C63**, 022201 (2001).; S. Liska, H. Gao, W. Chen, and X. Qian, *arXiv:0707.1009v1* (2007).
5. H.-J. Behrend *et al.*, Phys. Lett. **B56** 408 (1975).
6. T. Ishikawa *et al.*, Phys. Lett. **B608** 215 (2005).
7. A. Sibirtsev *et al.*, Eur. Phys. Jour. **A29** 209 (2006).
8. R. Erbe *et al.*, Phys. Rev. **175** 1669 (1968).; R.L. Anderson *et al.*, Phys. Rev. **D1** 27 (1970).; G. McClellan *et al.*, Phys. Rev. Lett. **26** 1593 (1971).; J. Ballam *et al.*, Phys. Rev. **D7** 3150 (1973).
9. T. Mibe *et al.*, Phys. Rev. Lett. **95** 182001 (2005).

ϕ-PHOTOPRODUCTION NEAR THE THRETHOLD WITHIN AN EFFECTIVE LAGRANGIAN APPROACH

HUI-YOUNG RYU* and A. HOSAKA

Research Center for Nuclear Physics (RCNP), Osaka University, Ibaraki, Osaka,567-0047, Japan
**E-mail: hyryu@rcnp.osaka-u.ac.jp*

HYUN-CHUL KIM

Department of Physics, Inha University, Incheon, 402-751, Korea
E-mail: hchkim@inha.ac.kr

We present in this talk a recent investigation on the $\gamma p \to \phi p$ reaction within the effective Lagrangian approach. We introduce the exchange of a particle with exotic quantum number $J^{PC} = 1^{-+}$ and loop corrections with $K^{\pm}$. We compare the results with the experimental data and discuss them.

Keywords: ϕ photoproduction, $J^{PC} = 1^{-+}$ particle exchange, loop corrections

1. Introduction

It is of great importance to understand the ϕ photoproduction, since it sheds light on the excitation of the hidden strangeness in the reactions. Moreover, the Okubo-Zweig-Iizuka (OZI) rule suppresses Born diagrams, which leads to introduce the soft pomeron exchange. This pomeron exchange describes phenomenologically well the increasing behavior of the differential cross setion as energy is getting large. Recently, the LEPS experiment has reported a bump-like structure in $d\sigma/dt$ near the threshold energy. It has been soon realized that the pomeron exchanges are not enough to reproduce such a narrow bump structure. Moreover, while the pomerons describe generally the data, we still do not understand clearly their physical meaning.

In this work, we propose a hadronic description of the ϕ photoproduction, based on an effective Lagrangian approach. We introduce a vector meson with allowed quantum number $J^{PC} = 1^{-+}$, since the vector meson exchange makes the differential cross section increases as a function of the energy. We also pay a special attention to the gauge invariant set of diagrams(s, t and u) and the momentum dependence of the amplitude, which may help to understand the underlying reaction mechanism. In addition, we include loop diagrams with $K^{\pm}$. The results are discussed in the context of the recent LEPS data.

2. Effective Lagrangian approach

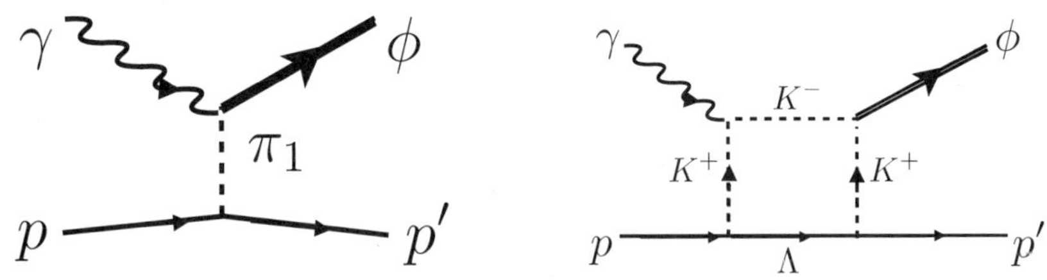

Fig. 1. The t channel diagram of π_1-exchange (left panel) and $K^{\pm}$ box diagram (right panel).

We start with the effective Lagrangians relevant to the diagrams in Fig. 1

$$\begin{aligned}\mathcal{L}_{\gamma\pi_1\phi} &= eg_{\gamma\pi_1\phi}(\partial_\mu A_\nu - \partial_\nu A_\mu)\phi^\mu\pi_1^\nu,\\ \mathcal{L}_{\pi_1 NN} &= eg_{\pi_1 NN}\bar{N}\gamma_\mu N\pi_1^\mu,\\ \mathcal{L}_{\gamma KK} &= ie(K^+\partial_\mu K^- - K^-\partial_\mu K^+)A^\mu, \end{aligned} \quad (1)$$

$$\begin{aligned}\mathcal{L}_{KN\Lambda} &= ig_{KN\Lambda}\bar{\Lambda}\gamma_5 K^+ p,\\ \mathcal{L}_{KK\phi} &= ig_{KK\phi}(K^+\partial_\mu K^- - K^-\partial_\mu K^+)\phi^\mu,\\ \mathcal{L}_{KN\Lambda} &= ig_{KN\Lambda}\bar{p}\gamma_5 K^-\Lambda, \end{aligned} \quad (2)$$

where A_μ, ϕ, N, $K^{\pm}$, and Λ represent the photon, the ϕ meson, the nucleon, the kaon, and the Λ, respectively. The π_1 denotes a vector particle with $J^{PC} = 1^{-+}$. The coupling constants are determined either by experiments or by a simple model. In addition, we introduce form factors in a gauge-invariant manner at each vertex.

3. Results and discussion

Figure 2 shows the result for the differential cross section $d\sigma/dt$ as a function of t. We have fitted the coupling of the $\pi_1(J^P = 1^{-+})$ in such a way that it reproduces the data. However, we still cannot describe the bump structure in the vicinity of $E_\gamma = 2.5$ GeV.

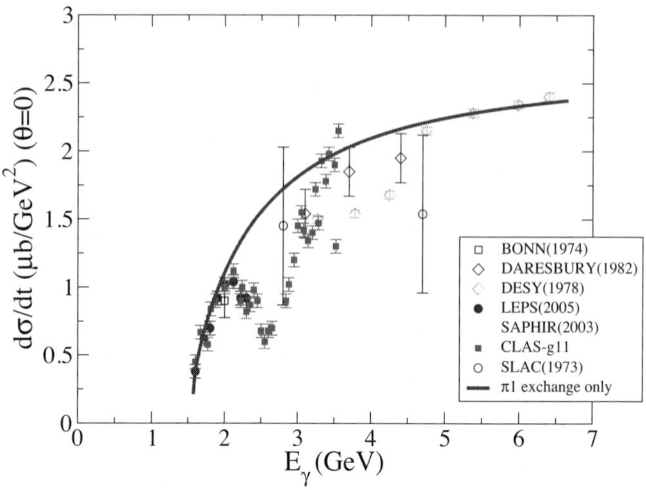

Fig. 2. The differential cross section of the reaction $\gamma p \to \phi p$ with the $J^{PC} = 1^{-+}$ particle exchange introduced.

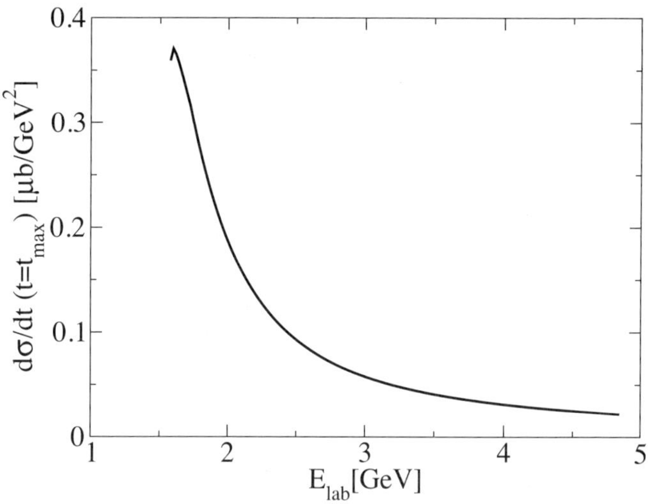

Fig. 3. The contribution of the imaginary part of the loop diagrams to the differential cross section.

In Fig. 3, we draw the result of the differential cross section with the kaon loop diagrams considered. As shown in Fig. 3, the contribution of the loop diagrams decreases drastically as the E_γ increases, so that it is not enough to explain the experimental data.

4. Summary and outlook

In the present talk, we have briefly reported a recent investigation on the ϕ photoproduction within an effective Lagrangian approach. We have introduced the vector particle with allowed quantum number $J^{PC} = 1^{-+}$. The result of the differential cross section has been generally well described. However, the present method cannot describe the bump structure around $E_\gamma = 2.5$ GeV.

While vector-meson exchanges tend to enhance the cross section of photoproductions in general, it is not possible to include them at the tree level because of the charge conjugation. However, we can introduce them in the loop diagrams. The corresponding investigation is under way.

References

1. A. I. Titov, T.-S. H. Lee, H. Toki. and O. Streltsova, Phys. Rev. **C 60**, 035205 (1999).
2. A. I. Titov and B. Kämpfer, Phys. Rev. **C 76**, 035202 (2007).
3. T. Mibe *et al.*, Phys. Rev. Lett. **95**, 182001 (2005).
4. Bengt Friman, Madeleine Soyeur, Nucl. Phys. **A600** (1996)
5. J.-M. Laget, Phys. Lett. **B 489** (2000).
6. A. I. Titov *et al.*, Phys. Rev. **C 58** (1998).
7. J. Ballam *et al.*, Phys. Rev. D 7 (1973).
8. H. J. Besch *et al.*, Nucl. Phys. B 70 (1974).

REGGE APPROACH IN CHARGED K^* PHOTOPRODUCTION

S. OZAKI* and A. HOSAKA
Research Center for Nuclear Physics (RCNP), Osaka University, Ibaraki, Osaka, 567-0047, Japan,
**E-mail: sho@rcnp.osaka-u.ac.jp*

H. NAGAHIRO
Department of physics, Nara Women's University, Nara 630-8506, Japan

We investigate $\gamma p \to K^{*+}\Lambda$ reaction within a Regge approach. For the gauge invariance of the scattering amplitude, we reggeize the s-channel and contact term amplitudes as well as the t-channel amplitude. We obtain the decreasing behavior of the total cross section as the CLAS's preliminary data show. We also calculate spin density matrices, and find clear differences between our Regge model and the previous Feynman (isobar) model.

Keywords: meson photoprodution, Regge model; Proceedings; World Scientific Publishing.

1. Introduction

Recently strangeness photoproduciton has been of interest in hadron physics. Several photon facilities such as CLAS, LEPS, SAPHIR and CBELSA/TAPS provide accurate data of strangeness photoproduction. In these data there are very interesting phenomena, for instance, pentaquark production,[1,2] peak structures near the threshold in ϕ photoproduction[3] and also in Λ resonance photoproduction.[4,5] To understand those phenomena, we have to understand a fundamental mechanism of open strangeness photoproduction.

From a theoretical point of view, Regge phenomenology is very successful in strangeness photoproduction above 2 GeV. The model reproduces the energy- and t-dependence of cross sections and spin observables of several strangeness photoproduction in the forward angle region. Especially what we pay attention to is the energy-dependence of cross sections predicted by this model. The Regge phenomenology successfully reproduces the decreasing behavior of cross sections of the kaon photoproduction in terms of kaon and K^* trajectories.[6–9] On the other hand, in ϕ photoproduction, the experimental data show a monotonically increasing behavior of the cross section. This behavior is explained by the Pomeron exchange.[10,11]

In this paper, we study the charged K^* photoproduction by using the Regge approach. The feature of the charged K^* photoproduction is that K^* (vector meson) exchange is allowed, whereas it is forbidden in the neutral K^* photoproduction. Since K^* exchange leads to an amplitude increasing with the energy, it is difficult to reproduce monotonically decreasing behavior of the total cross section as the CLAS's preliminary data of $\gamma p \to K^{*+}\Lambda$[12] show. However we can naively expect that Regge amplitude can naturally reproduce the decreasing behavior of the cross section due to kaon, K^* and κ trajectories.

In Sec. II we simply review the Regge amplitude for the charged K^* photoproduction. We show and discuss our results in Sec III. More detail for this study is described in Ref. 13.

2. Regge amplitude

The Regge amplitude is obtained by replacing the t-channel propagator in the Feynman amplitudes[14] by the Regge propagator as[7–9]

$$\frac{1}{t-m_K^2} \to \mathcal{P}^K_{regge} = \left(\frac{s}{s_0}\right)^{\alpha_K(t)} \frac{1}{\sin(\pi\alpha_K(t))} \frac{\pi\alpha_K'}{\Gamma(1+\alpha_K(t))}, \tag{1}$$

$$\frac{1}{t-m_\kappa^2} \to \mathcal{P}^\kappa_{regge} = \left(\frac{s}{s_0}\right)^{\alpha_\kappa(t)} \frac{1}{\sin(\pi\alpha_\kappa(t))} \frac{\pi\alpha_\kappa'}{\Gamma(1+\alpha_\kappa(t))}, \tag{2}$$

$$\frac{1}{t-m_{K^*}^2} \to \mathcal{P}^{K^*}_{regge} = \left(\frac{s}{s_0}\right)^{\alpha_{K^*}(t)-1} \frac{1}{\sin(\pi\alpha_{K^*}(t))} \frac{\pi\alpha_{K^*}'}{\Gamma(\alpha_{K^*}(t))}. \tag{3}$$

Formally one can reggeize the t-channel amplitude by multiplying the t-channel Feynman amplitude by $(t-m_{K^*}^2)\mathcal{P}_{regge}$. We choose constant phases for all trajectories and set $s_0 = 1$ GeV. Meson trajectories are given by

$$\alpha_K(t) = 0.70 \text{ GeV}^{-2}(t-m_K^2), \tag{4}$$

$$\alpha_\kappa(t) = 0.70(t-m_\kappa^2), \tag{5}$$

$$\alpha_{K^*}(t) = 1 + 0.85 \text{ GeV}^{-2}(t-m_{K^*}^2), \tag{6}$$

To maintain the gauge invariance, we need to reggeize the set of the gauge invariant terms of the electric s-channel, contact, and K^* exchange terms simultaneously. This means that in accordance with Eq.(6), we perform the following replacement

$$(\mathcal{M}_s^{elec} + \mathcal{M}_c) \to (\mathcal{M}_s^{elec} + \mathcal{M}_c) \times (t - m_{K^*}^2)\mathcal{P}_{regge}^{K^*}. \tag{7}$$

This procedure has been shown to be important to reproduce the very forward angle behavior ($\theta \sim 0$) in the charged kaon photoproduction.[6,7,9]

3. Results and discussion

Let us now discuss the result of our approach. As shown in Fig.1-(a), we find that the Regge model can reproduce the experimental data very well and describe the decreasing behavior of the total cross section. Figure 1-(b) shows each contribution of our model. In this model K^* trajectory and the reggeized contact term contributions are both decreasing as the energy is increased. Here the gauge invariant prescription Eq.(7) plays a crucial role in explaining the decreasing behavior.

To see the difference between the previous Feynman model[14] and our Regge model for charged K^* photoproduction, we calculate spin observables in the reaction $\gamma p \to K^{*+}\Lambda$. Spin observables are very important to understand the reaction mechanism; for instance, Refs. 15,16 pointed out an importance of chiral anomaly in kaon photoproduction by using the photon beam asymmetry. To focus on the t-channel contributions, we study spin density matrices at a forward angle, $\theta = 20$ degrees. Figure 2 shows ρ_{00}^0 which represents spin one flip process in the Gottfried-Jackson(GJ) system. In this system, a scalar or pseudoscalar exchange makes this matrix element exactly zero, but a vector meson exchange makes a finite contribution to this matrix element. Since in the previous Feynman model the dominant contribution is the kaon (psudoscalar meson) exchange, ρ_{00}^0 is almost zero. On the other hand, in our Regge model, ρ_{00}^0 is finite due to the sizable contribution of K^* trajectory. The significantly different behaviors in the two models in the spin density matrices are expected to be observed in future experiments which give us important information on the K^* photoproduction mechanism.

References

1. T. Nakano *et al.* [LEPS Collaboration], Phys. Rev. Lett. **91**, 012002 (2003).
2. T. Nakano *et al.* [LEPS Collaboration], Phys. Rev. C **79**, 025210 (2009).
3. T. Mibe *et al.* [LEPS Collaboration], Phys. Rev. Lett. **95**, 182001 (2005).
4. M. Niiyama *et al.*, Phys. Rev. C **78**, 035202 (2008).
5. H. Kohri *et al.* [LEPS Collaboration], arXiv:0906.0197 [hep-ex].

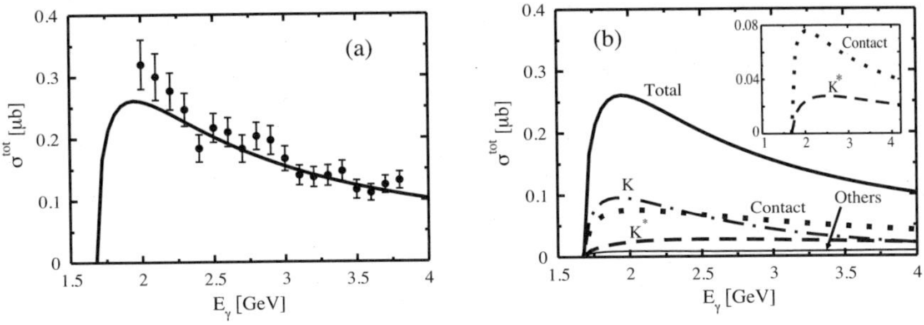

Fig. 1. (a) Total cross section of the reaction $\gamma p \to K^{*+}\Lambda$ in the Regge model. The data are taken from Ref. 12. (b) Various contributions to the total cross section. The solid line is total cross section, the dashed line is K^* exchange, the dotted line is kaon exchange, the dot-dash line is the contact term and the thin solid line is other channel contributions including κ exchange.

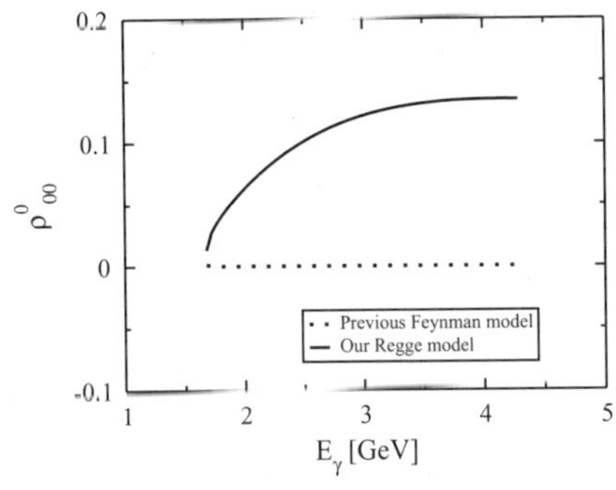

Fig. 2. Spin density matrices of $\gamma p \to K^{*+}\Lambda$ reaction, ρ^0_{00}. In the calculation, the K^* angle is fixed at $\theta = 20$ degrees. The solid line is our Regge model result, and the dotted line is the previous Feynman model result.

6. T. Mart and C. Bennhold, arXiv:nucl-th/0412097.
7. T. Corthals, J. Ryckebusch and T. Van Cauteren, Phys. Rev. C **73**, 045207 (2006).
8. T. Corthals, D. G. Ireland, T. Van Cauteren and J. Ryckebusch, Phys. Rev. C **75**, 045204 (2007).
9. M. Guidal, J. M. Laget and M. Vanderhaeghen, Nucl. Phys. A **627**, 645 (1997).
10. A. I. Titov and B. Kampfer, Phys. Rev. C **76**, 035202 (2007).
11. S. Ozaki, A. Hosaka, H. Nagahiro and O. Scholten, Phys. Rev. C **80**, 035201 (2009).
12. L. Guo and D. P. Weygand [CLAS Collaboration], arXiv:hep-ex/0601010.
13. S. Ozaki, H. Nagahiro and A. Hosaka, arXiv:0910.0384 [hep-ph].
14. Y. Oh and H. Kim, Phys. Rev. C **73**, 065202 (2006).
15. S. Ozaki, H. Nagahiro and A. Hosaka, Phys. Lett. B **665**, 178 (2008).

16. S. Ozaki, H. Nagahiro and A. Hosaka, Mod. Phys. Lett. A **23**, 2325 (2008).

BEAM ASYMMETRY OF $K^*\Sigma^+$ PHOTOPRODUCTION FROM THRESHOLD TO 3 GeV

S.H. HWANG[1] and J.K. AHN[2]

Department of Physics, Pusan National University, Busan, 609-735, Korea
[1]*E-mail: shhwang@pusan.ac.kr and* [2]*ahnjk@pusan.ac.kr*

K. HICKS[3]

Department of Physics and Astronomy, Ohio University, Athens, OH 45701, USA
[3]*E-mail: hicks@ohio.edu*

FOR THE LEPS COLLABORATION

Recently, the CLAS collaboration at Thomas Jefferson Nation Acceleration Facility reported the total cross section data for K* photoproduction, and found that the t-channel K exchange dominates the production amplitudes at small scattering angles. The measurement of forward-angle beam spin asymmetries for this reaction, where the theoretical calculations show an unambiguous signal, would establish the role of the $\kappa(800)$. We have measured the photon beam asymmetry for the $\gamma p \to K^{*0}\Sigma^+$ reaction on a Liquid Hydrogen target in the energy from threshold to 3.0 GeV by using a linearly polarized photon beam at SPring-8.

Keywords: Beam asymmetry; $K^*(892)$ photoproduction; SPring-8/LEPS.

1. Physics Motivation

The nature of the scalar mesons is yet to be clarified and there are many models on the structure of scalar meson nonet.[1] The Particle Data Group,[2] in their note on scalar mesons, states: "In contrast to the vector and tensor mesons, the identification of the scalar mesons is a long-standing puzzle". In particular, the κ-meson with a resonance pole at about 800 MeV is seen in many phenomenological analysis[3] and also in analyses of D-meson decay,[4] yet its existence is still contraversial.

Differential cross section data for K^{*0} photoproduction was reported by the CLAS collaboration for the reaction $\gamma p \to K^{*0}\Sigma^+$.[6] The angular distributions are forward peaked as shown in Figure 1, because the t-channel exchange makes a strong contribution this reaction. The available exchange particles are the kaon(K) and the kappa(κ). The scalar κ has unnatural parity and the pseudoscalar kaon has unnatural parity. A recent theoretical paper has shown that the scalar meson know as the $\kappa(800)$ contributes to K^{*0} photoproduction through t-channel exchange.[5] Figure 1 shows total cross section data CLAS and theoretical calculation of the total cross section and beam asymmetry in the reaction $\gamma p \to K^{*0}\Sigma^{+}$. The solid lines are the results with κ-exchange and the dashed lines are for kaon-exchange only. The measurement of forward-angle beam spin asymmetries for this reaction, where the theoretical calculations show an unambiguous signal, would establish the role of the $\kappa(800)$.

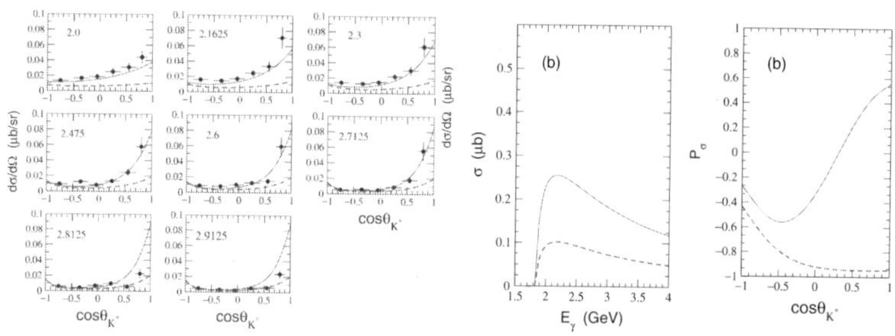

Fig. 1. CLAS data overlaid with Oh's theoretical estimate.

2. K^* Photoproduction Experiment at SPring-8/LEPS

Our experiment has been carried out at the Laser-Electron-Photon beam line (LEPS) at SPring-8. A photon beam was produced by the backward Compton scattering of the deep UV laser(257nm) from 8-GeV electrons in the storage ring. The energy range of the tagged photon beam was reached 1.4 – 2.97 GeV as shown in Figure 2(c). The linearly polarized laser light produced linearly polarized GeV energy photon beam. The photon energy was determined by the measuring energy of the recoiling electron which lost energy due to the collision. The liquid hydrogen (LH_2) target was exposed in the photon beam. The LEPS spectrometer consists of a plastic scintillator after a target, and silicon-strip vertex detector, three drift chambers with dipole magnet, and Time-of-Flight(ToF) wall as shown in Figure 2(a). For the K^{*0} photoproduction experiment, we removed the Aerogel Cherenkov count to get high momentum kaon and pion events, and a new plastic

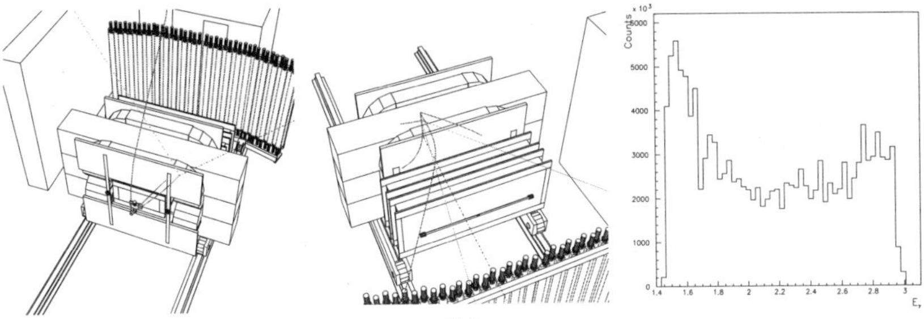

(a) (b) (c)

Fig. 2. Experimental setup((a),(b)) and tagged photon energy spectrum(c)

scintillator veto counter was added down stream of the drift chambers as shown in Figure 2(b). In order to measure the momentum of the charged particles, their trajectories are measured with silicon-strip vertex detectors and drift chambers. The particle velocity was measured by the ToF wall.

3. Preliminary Results and Discussion

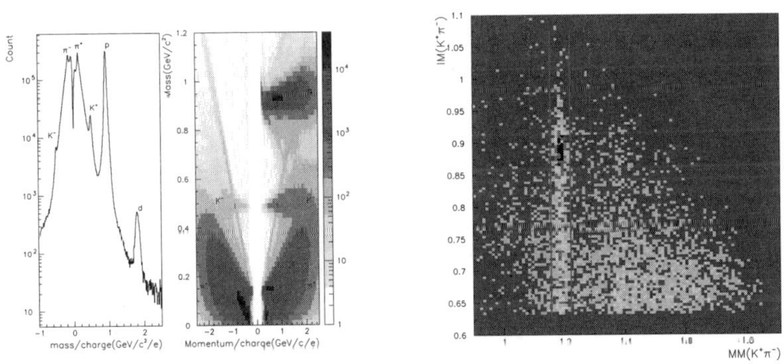

(a) (b)

Fig. 3. Reconstructed mass distribution of charged particles and a two-dimensional plot of a missing-mass for $p(\gamma, K^+\pi^-)X$ versus an invariant mass for the $K^+\pi^-$ system.

Figure 3(a) shows the mass spectrum and the two-dimensional plot of the momentum/charge versus mass. The peaks corresponding to pions, kaons, protons and deutrons. To select events from the $\gamma p \to K^{*0}\Sigma^+$ photoproduction, we required the following cut conditions: select K^+, π^- particles, remove decay-in-flight events, select reaction vertex in the target, reject the accidental event electron and positron in the e^+e^- veto count, select a recoil electron by compton scattering and reject shower and accidential hits in the tagging counter. The high momentum region of the K^+ events is contaminated by high momentum pions. We select a 4-σ band for π^- and a 2-σ band for K^+ to reduce the contaminated pion events. Also, these contamination events will not form peaks in the mass spectra shown next.

A two-dimensional plot of the invariant mass versus missing mass of $K^+\pi^-$ system is shown in the right side of the Figure 3. The spot is for $K^{*0}\Sigma^+$ production. Figure 4 shows the invariant mass distribution for horizontal beam polarization and vertical beam polarization, which corresponds to a 3-σ cut on the Σ^+ missing mass. The horizontal beam polarization collected more events compare with vertical beam polarization. The background can be determined from side band cuts, shown with the dashed vertical lines in figure 3. Side band events shows no K^* peak.(hatched area in figure 4). The background is from contributions due to hyperon production and 3-body phase space. The measured mean value of the K^* is 885.63 $\pm$ 23.7 MeV. The world average of the K^{*0} mass is 896.00 MeV $\pm$ 0.9 MeV. The total number of collected events is about 500.

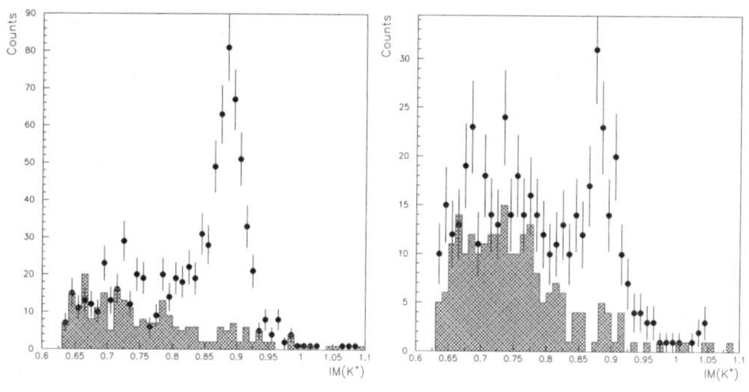

Fig. 4. $K^+\pi^-$ invariant mass spectra for horizontal beam polarization (left) and and vertical beam polarization (right).

4. Summary and Prospects

We have measured the $\gamma p \rightarrow K^{*0}\Sigma^+$ reaction with beam energies from threshold to 3.0 GeV by using a linearly polarized photon beam at SPring-8. We collected about 500 K^{*0} events. We plan to correct the acceptance by simulation and extract the beam asymmetry. Additional data using a 3 GeV photon beam was taken in 2009. If included, the statistics will be doubled. We will measure the beam asymmetry which gives information on t-channel exchange processes for K^* photoproduction.

References

1. Y. Nambu and J.J. Sakurai. Phys. Rev. Lett. **11**, 42 (1963); R. L. Jaffe. Phys. Rev. D **15**, 267 (1977); J. Weinstein and N. Isgur, *ibid.* **41**, 2236 (1990); F. E. Close *etal.*, Phys. Lett. **B319**, 291 (1993); D. Black, A. H. Fariborz, F. Sannino, and J. Schechter, Phys. Rev. D **59**, 074026 (1999); C. M. Shakin and H. Wang, *ibid.* **63**, 014019 (2001); J. R. Peláez, Phys. Rev. Lett. **92**, 102001 (2004).
2. Particle Data Group, Phys. Lett. **B592**, 1 (2004).
3. D.V. Bugg, Phys. Latt.**B572**, 1 (2003).
4. E.M. Aitala *etal.*,Phys. Rev. D **15**, 267 (1977).
5. Y. Oh and H. Kim, Phys. Rev. C **74**, 015208 (2006).
6. CLAS Collaboration(I. Hleiqawi *etal.*), Phys. Rev.C **75**, 042201 (2007).

HOLOGRAPHIC DUAL MODEL OF QUANTUM HADRODYNAMICS

KYUNG-IL KIM* and SU HOUNG LEE

Institute of Physics and Applied Physics, Yonsei University,
Seoul 120-749, Korea
**E-mail: hellmare@yonsei.ac.kr*

YOUNGMAN KIM

Asia Pacific Center for Theoretical Physics and Department of Physics, Pohang University of Science and Technology, Pohang, Gyeongbuk 790-784, Korea

We study baryons in dense matter using a holographic QCD (hQCD) model. Baryons are introduced as the bulk Dirac field in five dimensional anti-de Sitter (AdS) space. We turn on the baryon number density ρ_B by a standard AdS/CFT dictionary and the scalar density ρ_s as a bi-fermion condensate of the nucleon field in hQCD. We calculate the in-medium mass of the nucleon and observe that the mass decreases with increasing baryon number density.

Keywords: AdS/QCD, Quantum Hadrodynamics(QHD), Nuclear Matter

1. Introduction

AdS/CFT correspondence[1] opens a new and promising route in understanding the non-perturbative features of quantum chromodynamics (QCD) under the name of holographic QCD model or AdS/QCD. A simple bottom-up hQCD model, called hard wall model,[2,3] has been constructed in a slice of the AdS_5 space,

$$ds^2 = \frac{1}{z^2}\left(-dz^2 + dx^\mu dx^\nu \eta_{\mu\nu}\right), \tag{1}$$

with $0 \leq z \leq z_m$. Here z_m is the IR cut-off corresponding to the confinement scale. The model contains a bulk scalar X, which is holographically dual to the quark bilinear $\bar{\psi}\psi$, and bulk gauge fields A_L and A_R, which are dual to the left and right quark number currents. Though this much simple, the model goes well with the phenomenology of QCD at low energy.

2. Baryons in holographic model with finite baryon density

The action for bulk baryons in the hQCD model[4,5] is given by,

$$S = \int d^5x\sqrt{G_5}\{i\bar{N}_1 e^M_A \Gamma^A D_M N_1 + i\bar{N}_2 e^M_A \Gamma^A D_M N_2$$
$$- \frac{5}{2}\bar{N}_1 N_1 + \frac{5}{2}\bar{N}_2 N_2 - g\bar{N}_1 X N_2 - g\bar{N}_2 X^\dagger N_1\}, \quad (2)$$

where N_1 and N_2 are 5D Dirac spinors. The covariant derivative acting on N_1 is defined by $D_M = \partial_M + \frac{i}{4}\omega_M^{AB}\Gamma_{AB} - i(A_L)_M$, where $\Gamma^{MN} = \frac{1}{2i}\left[\Gamma^M, \Gamma^N\right]$ with $\Gamma^M = \{\gamma^\mu, -i\gamma^5\}$.

Now we introduce the baryon (or quark) number density following an AdS/CFT dictionary. In AdS/CFT, the baryon chemical potential and the baryon number density are encoded in the time component of bulk U(1) gauge field. More precisely, the classical solution to the equation of motion for the time component of the bulk $U(1)$ vector gauge field V_μ is given by

$$V_0(z) = \mu_b + c_b z^2\,, \quad (3)$$

where μ_b (c_b) is proportional to baryon chemical potential (baryon number density). Similarly, the current quark mass and the chiral condensate are included in the vacuum expectation value of scalar field $\langle X\rangle \equiv X_0$. The classical solution to the equation of motion for X_0 has the form, $X_0 = \frac{1}{2}m_q z + \frac{1}{2}\sigma z^3$, where we take $\sigma = (0.311\text{GeV})^3$ with $m_q = 0$. To realize the fact that the nucleon mass is mostly from the chiral symmetry breaking, *i.e.* non-zero X_0, we introduce a Yukawa (or mass) coupling between the bulk scalar X and the bulk spinors, N_1 and N_2,

$$\mathcal{L}_{int} = -g\bar{N}_1 X N_2 - g N_2 X^\dagger N_1. \quad (4)$$

The coupling g (=8.7) is fitted to reproduce the nucleon mass $m_N = 0.882$ GeV (*in the chiral limit*) as the lowest mass eigenvalue of the Kaluza-Klein modes.

With $N(x,z) = f_L(z)B_L(x) + f_R(z)B_R(x)$, we have the eigenvalue equation of motion for baryons at finite baryon number density,

$$\begin{pmatrix} \partial_5 - \frac{\Delta}{z} & -g\frac{X_0}{z} \\ -g\frac{X_0}{z} & \partial_5 - \frac{4-\Delta}{z} \end{pmatrix}\begin{pmatrix} f_{1L} \\ f_{2L}\end{pmatrix} = \left(-m_N - c_b z^2\right)\begin{pmatrix} f_{1R} \\ f_{2R}\end{pmatrix},$$
$$\begin{pmatrix} \partial_5 - \frac{4-\Delta}{z} & g\frac{X_0}{z} \\ g\frac{X_0}{z} & \partial_5 - \frac{\Delta}{z} \end{pmatrix}\begin{pmatrix} f_{1R} \\ f_{2R}\end{pmatrix} = \left(m_N + c_b z^2\right)\begin{pmatrix} f_{1L} \\ f_{2L}\end{pmatrix}, \quad (5)$$

where $\Delta = \frac{9}{2}$. By solving the coupled equations with suitable boundary conditions,[4,5] we obtain the in-medium nucleon mass. We plot the $\tilde{c}_b$ dependence of the ratio between nucleon masses at finite density and in free space in Fig.1. Here $\tilde{c}_b$ is a dimensionless parameter defined as $\tilde{c}_b = c_b z_m^3$ with $z_m = 1/0.33\ \text{GeV}^{-1}$.

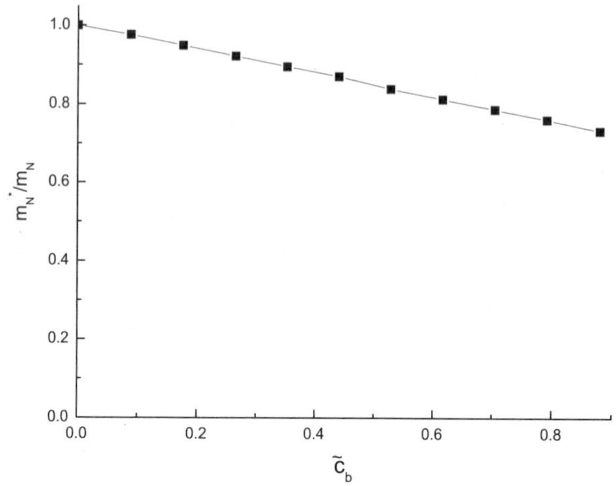

Fig. 1. $\tilde{c}_b$ dependence of nucleon masses

Now we turn on the scalar density following the mean field approach :[5]

$$\langle \bar{N}_2 N_1 + \bar{N}_1 N_2 \rangle = (f_{2R}^2 - f_{1R}^2) \langle \bar{\psi}(x)\psi(x) \rangle \Longrightarrow (f_{2R}^2 - f_{1R}^2)\rho_s \,. \quad (6)$$

Then the equation of motion for X_0 is modified by the Yukawa term in Eq. (4), as

$$\left(\partial_z^2 - \frac{3}{z}\partial_z + \frac{3}{z^2} \right) X_0 - \frac{g}{4z^2}(f_{2R}^2 - f_{1R}^2)\rho_s = 0. \quad (7)$$

Since $\rho_s \sim \rho_B$ at low density, we perform a numerical calculation by varying ρ_B. The solution X_0 of the above equation will be different from (3) due to non-zero density, and therefore we need to substitute it to the Eq.(5) to obtain nucleon mass and 5D nucleon wave functions, f_{1R}, f_{2R}. Then we plug the nucleon wave functions into the Eq.(7) again until the value of nucleon mass is stable. We confirm that the value of nucleon mass is stable against the iteration.

3. Summary

We studied baryons in dense matter using a holographic hQCD model with baryons. We introduced the baryon number density ρ_B through the time component of bulk U(1) vector field and the scalar density ρ_s as a bi-fermion condensate of the nucleon field. Note that X_0 depends on ρ_s through its equation of motion. In the sense that classical solution of the bulk U(1) vector and X_0 correspond to the expectation value of the omega field and that of scalar field, respectively, in quantum hadrodynamics (QHD), our approach is dual to the QHD. We then calculated the in-medium mass of the nucleon and found that the mass decreases with increasing baryon number density.

Coupling constants such as the vector meson-nucleon coupling $g_{\rho NN}$ and also the binding energy of nuclear matter are being studied in this model, which will be reported elsewhere.

Acknowledgments

Y.K. acknowledges the Max Planck Society(MPG), the Korea Ministry of Education, Science, Technology(MEST), Gyeongsangbuk-Do and Pohang City for the support of the Independent Junior Research Group at the Asia Pacific Center for Theoretical Physics(APCTP).

The work of S.H.L. and K.I.K. was supported by the Korea Research Foundation KRF-2006-C00011.

References

1. J. Maldacena, Adv. Theor. Math. Phys. **2** (1998) 231; S.S. Gubser, I.R. Klebanov and A.M. Polyakov, Phys. Lett. **B428** (1998) 105; E. Witten, Adv. Theor. Math, Phys. **2** (1998) 253; J. Polchinski and M. J. Strassler, Phys. Rev. Lett. **88** (2002) 031601; A. Karch and E. Katz, JHEP **0206**, 043 (2002).
2. J. Erlich, E. Katz, D. T. Son and M. A. Stephanov, Phys. Rev. Lett. **95**, 261602 (2005).
3. L. Dà Rold and A. Pomarol, Nucl. Phys. B **721**, 79 (2005).
4. D. K. Hong, T. Inami and H. -U. Yee, Phys. Lett. B **646**, 165 (2007).
5. Y. Kim, C. -H. Lee and H. -U. Yee, Phys. Rev. D **77**, 085030 (2008).
6. S. K. Domokos and J. A. Harvey, Phys. Rev. Lett. **99**, 141602 (2007).

R-PARITY VIOLATING SUPERSYMMETRIC CONTRIBUTION TO THE NUCLEAR BETA DECAY

N. YAMANAKA*, T. SATO and T. KUBOTA

Department of physics, Osaka University,
Toyonaka Osaka 560-0043, Japan
**E-mail: yamanaka@kern.phys.sci.osaka-u.ac.jp*
www.osaka-u.ac.jp

The contribution of the R-parity violating (RPV) minimal supersymmetric standard model (MSSM) to the neutron beta decay is discussed. The RPV effects contribute to the scalar interaction at the tree level. From it, we construct the effective interaction by using the nucleon mass difference and give its contribution to the angular correlations of the neutron beta decay within the recent phenomenological limits, i.e. the experimental result of the R correlation of the neutron beta decay given by PSI and the analyses of the Superallowed Fermi transitions.

Keywords: Supersymmetry; Neutron; Beta decay; R-parity violation.

1. Introduction

In recent years, the search for new physics beyond the Standard model (SM) is very actively performed or planned, and many candidates exist. Among them, the MSSM is the leading candidate. The supersymmetric extension of the SM allows lepton number or baryon number violating interactions, so we usually assume the conservation of R-parity to forbid them. But this assumption is *ad hoc*, and we must investigate its violation within phenomenological constraints.

Herczeg pointed that RPV interaction generates scalar/pseudoscalar interaction of beta decay and that the pion decay at the tree level[1,2] or the atomic electric dipole moment at the loop level is very sensitive to them.[3]

Recently, the R correlation of the neutron beta decay was measured by the group of Kozela at the PSI.[4] On the other hand, Hardy and Towner made an analysis of superallowed Fermi transitions and built a constraint on the Fierz interference term.[5] These two observables are sensitive to the scalar interaction of the neutron beta decay.

It seems then to be timely to investigate the neutron beta decay within RPV MSSM.

Our discussion is organized as follows. We first briefly review in the next section the R-parity violation and show how it contributes to the neutron beta decay. Section 3 is devoted to the discussion of the phenomenological bound obtained from the two recently published results.

2. RPV contribution to the neutron beta decay

The R-parity is defined by using the baryon number B, lepton number L and spin s, as $R = (-1)^{3B-L-2s}$. It is required for the conservation of the baryon or lepton number, since the Supersymmetry does not forbid such types of interaction. The RPV interactions needed in the beta decay are the following:

$$\begin{aligned}\mathcal{L} = &-\tfrac{1}{2}\textstyle\sum_{ijk}\lambda_{ijk}\left\{\tilde{e}^{\dagger}_{Rk}\bar{\nu}^{c}_{i}P_{L}e_{j} + \tilde{e}_{Lj}\bar{e}_{k}P_{L}\nu_{i} + \tilde{\nu}_{i}\bar{e}_{k}P_{L}e_{j} + \cdots\right\}\\ &-\textstyle\sum_{ijk}\lambda'_{ijk}\left\{-\tilde{e}_{Li}\bar{d}_{k}P_{L}u_{j} + \cdots\right\}\\ &+ \text{h.c.}\ .\end{aligned} \qquad (1)$$

They are of the Yukawa-type and lepton number violating. These RPV interactions are constrained phenomenologically.

By using the RPV interactions, we obtain the tree level diagrams as shown in Figure 1. The first diagram (Fig 1 (a)) is the SM contribution, with V-A interaction. The second diagram (Fig 1 (b)) is the RPV contribution, of scalar and pseudoscalar types with scalar-electron exchanged. The third one (Fig 1 (c)) is another RPV contribution. We can show that after Fierz transformation, it is of the V-A form, thus absorbed in the SM term.

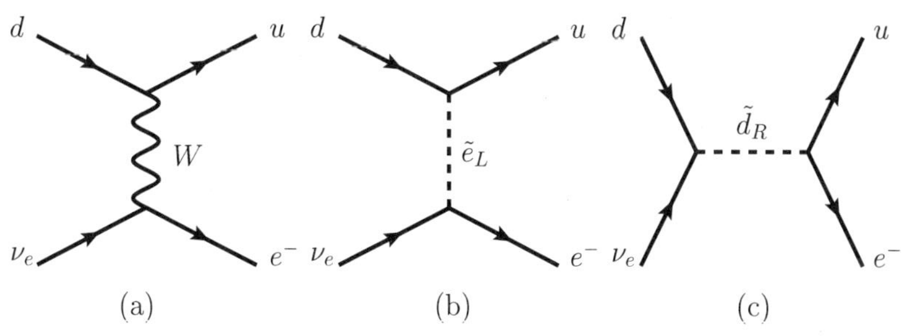

Fig. 1. Tree level amplitude of the quark beta decay within RPVMSSM.

By using the preceding quark level contributions, we can write the effective interaction of the neutron beta decay by using nucleon matrix elements, as follows:

$$H_\beta = \bar{p}\gamma^\mu(C_V - C_A\gamma_5)n\,\bar{e}\gamma_\mu(1-\gamma_5)\nu_e + C_S\bar{p}n\,\bar{e}(1-\gamma_5)\nu_e\ . \tag{2}$$

The first term (with C_V and C_A) is the SM contribution, the second term (with C_S) is the RPV scalar interaction. The pseudoscalar interaction was neglected since it vanishes in the non-relativistic limit. The scalar coupling C_S is expressed as

$$C_S = g_S \sum_{i=2,3} \frac{\lambda_{1i1}\lambda'^*_{i11}}{4m^2_{\tilde{e}_{Li}}}\ , \tag{3}$$

with λ and λ' the RPV couplings used in Figure 1 (b) and with the mass of the intermediate scalar-electron $m^2_{\tilde{e}_{Li}}$. g_S was calculated by using the relation of the nucleon matrix element $\langle p|\bar{u}d|n\rangle = \langle p|\bar{u}u - \bar{d}d|p\rangle$. According to Gasser and Leutwyler,[6] the isospin symmetric part of the nucleon form factor can be written as

$$\delta = \frac{m_d - m_u}{2M_N}\langle p|\bar{u}u - \bar{d}d|p\rangle\ , \tag{4}$$

which is just the mass difference of the proton and neutron $M_n - M_p$ to the lowest order in quark mass expansion. From these, we obtain $g_S = 0.49 \pm 0.17$.[a]

The observables of the neutron beta decay are the angular correlations which can be written as follows[8,9] :

$$\begin{aligned}\omega(E_e,\Omega_e,\Omega_\nu) \propto 1 + a\tfrac{\vec{p}_e\cdot\vec{p}_\nu}{E_eE_\nu} + b\tfrac{m_e}{E_e}&\\
+\,\vec{\sigma}_n\cdot&\left\{A\tfrac{\vec{p}_e}{E_e} + B\tfrac{\vec{p}_\nu}{E_\nu} + D\tfrac{\vec{p}_e\times\vec{p}_\nu}{E_eE_\nu}\right\}\\
+\,\vec{\sigma}_e\cdot&\left\{G\tfrac{\vec{p}_e}{E_e} + H\tfrac{\vec{p}_\nu}{E_\nu} + K\tfrac{\vec{p}_e}{E_e+m_e}\tfrac{\vec{p}_e\cdot\vec{p}_\nu}{E_eE_\nu} + L\tfrac{\vec{p}_e\times\vec{p}_\nu}{E_eE_\nu}\right\}\\
+\,\vec{\sigma}_e\cdot&\left\{N\vec{\sigma}_n + Q\tfrac{\vec{p}_e}{E_e+m}\tfrac{\vec{\sigma}_n\cdot\vec{p}_e}{E_e} + R\tfrac{\vec{\sigma}_n\times\vec{p}_e}{E_e}\right\}\\
+\,\vec{\sigma}_e\cdot&\left\{S\vec{\sigma}_n\tfrac{\vec{p}_e\cdot\vec{p}_\nu}{E_eE_\nu} + T\vec{p}_e\tfrac{\vec{\sigma}_n\cdot\vec{p}_\nu}{E_eE_\nu} + U\vec{p}_\nu\tfrac{\vec{\sigma}_n\cdot\vec{p}_e}{E_eE_\nu} + V\tfrac{\vec{p}_\nu\times\vec{\sigma}_n}{E_\nu}\right.\\
&\left.\qquad + W\tfrac{\vec{p}_e}{E_e+m_e}\tfrac{\vec{\sigma}_n\cdot\vec{p}_e\times\vec{p}_\nu}{E_eE_\nu}\right\}.\end{aligned} \tag{5}$$

The Fierz interference term ($b\frac{m_e}{E_e}$) and the R correlation are observables sensitive to the scalar interaction of the neutron beta decay. The Fierz interference term is the spectrum shape correction, sensitive to the real part

[a] We have used $M_n - M_p = (2.05 \pm 0.30)$ MeV, $m_d = 9.3 \pm 0.9$ MeV, $m_u/m_d = 0.553 \pm 0.043$.[7]

of the scalar interaction. The R coefficient is the CP-odd angular correlation between initial neutron polarization ($\vec{\sigma}_n$), emitted electron polarization ($\vec{\sigma}_e$) and electron momentum ($\vec{p}_e$), sensitive to the imaginary part. Both observables vanish in the SM, but receives contribution from RPV MSSM.

3. Discussion

We show the limit brought by Hardy & Towner to the Fierz term, and experimental value of the PSI of the R coefficient as follows:

$$\begin{aligned} b/2 &= 0.0011 \pm 0.0013 && \text{(Hardy \& Towner)} , \\ R &= 0.008 \pm 0.011 \pm 0.005 && \text{(PSI)} . \end{aligned} \tag{6}$$

By considering these two values, it is now possible to make new limits on the RPV couplings via scalar interaction, which are shown as follows:

$$\begin{aligned} Rc \textstyle\sum_{i=2,3} \lambda_{1i1} \lambda'^{*}_{i11} &= (7.2 \pm 8.5) \times 10^{-4} [m_{\tilde{e}_L}]^2 , \\ Im \textstyle\sum_{i=2,3} \lambda_{1i1} \lambda'^{*}_{i11} &= (0.012 \pm 0.017 \pm 0.008) [m_{\tilde{e}_L}]^2 . \end{aligned} \tag{7}$$

Our limit was derived from the scalar interaction and has the mass dependence on scalar-electron. By comparing these values with limits given by Herczeg, we see that the limit on the real part of the same combination of RPV couplings was derived via pseudoscalar interaction, and is at the same order.[2] The limit on the imaginary part was given by atom EDM, via one-loop level analysis, and give stronger constraints, but with other mass dependence.[3]

Our analysis was made at the tree level, and further evaluation at the loop level may be interesting. We found also another interesting angular correlations sensitive to the scalar interaction such as L, S, U, V and W in eq. (5).

References

1. P. Herczeg, Prog. Part. Nucl. Phys., **46**, 413 (2001).
2. P. Herczeg, Phys. Rev. **D52**, 3949 (1995).
3. P. Herczeg, J. Res. Natl. Inst. Tech., **110**, 453 (2005).
4. A. Kozela *et al.*, Phys. Rev. Lett. **102**, 172301 (2009).
5. J.C. Hardy and I.S. Towner, Phys. Rev. **C79**, 055502 (2009).
6. J. Gasser and H. Leutwyler, Phys. Rept., **87**, 77 (1982).
7. H. I. Leutwyler, Phys. Lett. **B378**, 313 (1996).
8. J. Jackson, S. Treiman and H. Wyld, Phys. Rev. **106**, 517 (1957).
9. M. E. Ebel and G. Feldman, Nucl. Phys. **4**, 213, (1957).

DEEPLY BOUND PIONIC STATES IN Sn AT RIKEN RI BEAM FACTORY

N. IKENO*, H. NAGAHIRO and S. HIRENZAKI

Department of Physics, Nara Women's University,
Nara, 630-8506, Japan
**E-mail: jan_ikeno@cc.nara-wu.ac.jp*

J. YAMAGATA-SEKIHARA

Departamento de Física Teórica and IFIC,
Centro Mixto Universidad de Valencia-CSIC, Valencia, 46071, Spain

D. JIDO

Yukawa Institute for Theoretical Physics, Kyoto University,
Kyoto, 606-8502, Japan

We study the structure of pionic atoms in ^{121}Sn, for which a new experiment will be performed at RIKEN RI Beam Factory. We evaluate numerically the effective nuclear densities probed by pion and potential parameter dependence of the binding energies and widths.

Keywords: Deeply Bound Pionic Atoms, (d,^{3}He) reactions, effective nuclear density.

1. Introduction

The study of the s-wave part of the pion-nucleus interaction is extremely interesting,[1] since the isovector s-wave potential strength provides information on the pion decay constant f_π, which can be related to the order parameter of chiral symmetry breaking of QCD, the quark condensate $\langle\bar{q}q\rangle$.[2,3] For this reason, it is highly required to observe the pionic s states precisely because the s state properties are strongly dependent on the s-wave potential[4] and, thus the $1s$ states in heavy nuclei (N>Z) provide key information on the isovector part of the s-wave potential.[1]

As one of the characteristic features of the pionic atoms, it is found that the shallow pionic states observed in the X-ray experiments only probed the certain value of the nucleon density ($\sim \frac{1}{2}\rho_0$), nearly independent on the nuclide and the π^- quantum number. As the consequence of this feature, Seki and Masutani found the presence of a strong correlation between potential parameters.[5] Toki *et al.* found that the same correlation also exists for the deeply bound $1s$ and $2p$ states in ^{208}Pb.[6,7]

Thus, it is quite interesting to study the effective nuclear densities probed by various deeply bound pionic states to obtain the information on pion-nucleus potential strength at different nuclear densities. In this report, we consider the pionic states in ^{121}Sn since a new experiment with extremely high energy resolutions will be performed soon at RIKEN RI Beam Factory. In this experiment, we can expect to observe the non-yrast pionic $2s$ state together with the deepest $1s$ state in the $(d,{}^3\mathrm{He})$ reactions.[8,9]

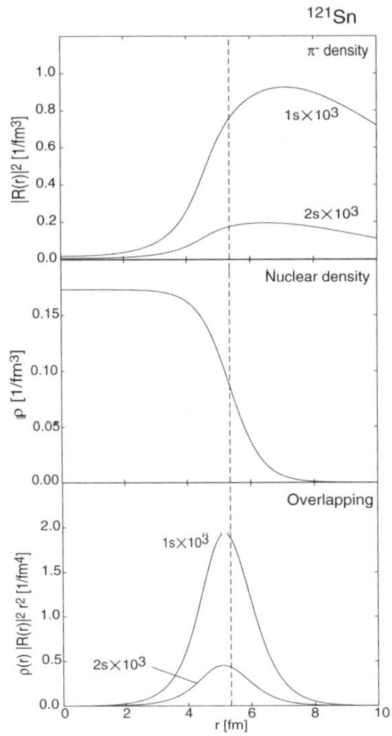

Fig. 1. Overlapping probabilities (lower frame) of the π^- densities (upper frame) with the nuclear density (middle frame) in pionic bound $1s$ and $2s$ states in ^{121}Sn. The vertical broken line shows the half-density radius of the nuclear density of ^{121}Sn.

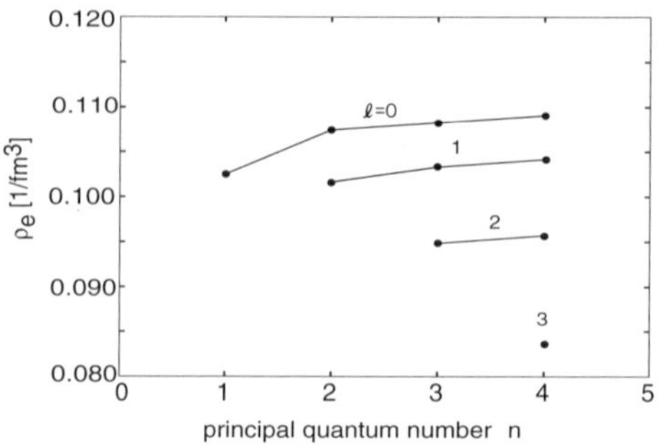

Fig. 2. The effective nuclear density ρ_e probed by $(n\ell)$ pionic states in ^{121}Sn.

2. Pion-Nucleus optical potential

We use a standard parameterization of the pion-nucleus optical potential[5,10] together with the Coulomb interaction to study the pionic states,

$$2\mu V(r) = -4\pi[\varepsilon_1\{b_0\rho(r)+b_1[\rho_n(r)-\rho_p(r)]\}+\varepsilon_2 B_0\rho^2(r)]+(p-\text{wave terms}), \tag{1}$$

where ε_1 and ε_2 are defined as $\varepsilon_1 = 1+\mu/M$ and $\varepsilon_2 = 1+\mu/2M$ with pion-nucleus reduced mass μ and nucleon mass M. The b_0, b_1 and B_0 parameters are to be determined by experimental data phenomenologically[5] and/or by theoretical calculations based on chiral effective theory. We solve the Klein-Gordon equation to study pion bound states.[6] The b_1 parameter expresses the isovector potential strength and is related to the pion decay constant f_π.[1–3] As for the correlation between b_0 and B_0, Seki and Masutani reported that the shallow pionic atom data are reproduced with a set of parameters which satisfy the relation,[5]

$$b_0 + \frac{\varepsilon_2}{\varepsilon_1}\rho_e B_0 = \beta_s = (-0.03+0.01i)\mathrm{m}_\pi^{-1}, \tag{2}$$

which can be interpreted that the properties of the pionic atoms are determined by the potential strength at an effective density $\rho_e \simeq \frac{1}{2}\rho_0$. The slope of contour plots of the observables in the b_0-B_0 plane is related to the nuclear density probed by the pionic states.

3. Results and discussions

We show in Fig. 1 calculated π^- densities of the $1s$ and $2s$ states ($|R_{n\ell}(r)|^2$), the nucleon density of ^{121}Sn ($\rho(r)$), and the overlapping densities[7] defined as,

$$S(r) = \rho(r)\,|R_{n\ell}(r)|^2\,r^2. \tag{3}$$

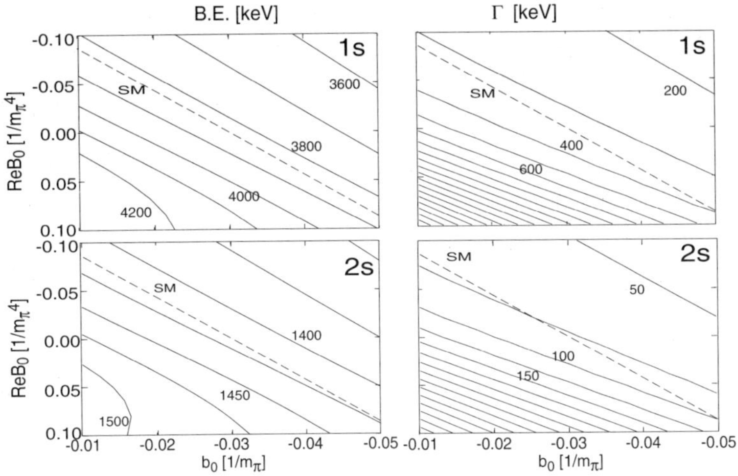

Fig. 3. Contour plots of the binding energies (left panels) and widths (right panels) for the $1s$ and $2s$ states in ^{121}Sn in the b_0-Real(B_0) plane. The dashed lines depicted by SM are the parameter sets of Seki and Masutani.[5]

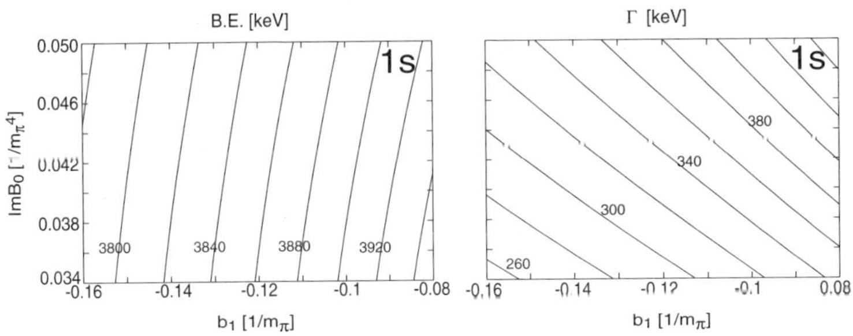

Fig. 4. Contour plots of the binding energy (left panel) and width (right panel) for the $1s$ state in ^{121}Sn in the b_1-Im(B_0) plane.

The overlapping densities have the peaks at nuclear surface as indicated by Seki-Masutani[5] for $1s$ and $2s$ pionic states. We calculate the pion densities for $1s$-$4f$ atomic states and show the effective nuclear densities ρ_e at the peak of the overlapping densities in Fig. 2. We find that the ρ_e increases for lower ℓ states as expected. As for the higher n states with a fixed ℓ value, the ρ_e changed slightly. We show that contour plots of the binding energies and widths for the $1s$ and $2s$ pionic states in b_0-Re(B_0) plane in Fig. 3. The contour line of the binding energy of the $1s$ state is almost parallel to the SM line shown in the figure, while that of the $2s$ state has slightly different slope from the SM line reflecting the difference of the effective density ρ_e. As for the contour lines of the widths, we found the different slope from the SM line both for $1s$ and $2s$ states. This feature is considered to be from the ρ^2 dependence of the imaginary potential term. Thus, the accurate data of the binding energies and widths of the $1s$ and $2s$ states may help to deduce the information on pion-nucleus interaction at different densities. We also show the contours in the b_1-Im(B_0) plane in Fig. 4. This contour plot was important in the analysis in Ref. 1 and indicated the accuracy of the b_1 parameter determination from the data with finite errors.

These numerical results indicate that the effective nuclear densities probed by pion depend on the atomic states, though the differences are small. The level widths also indicate probing the different densities due to the ρ^2 dependence of the imaginary potential.

References

1. K. Suzuki *et al.*, Phys. Rev. Lett. 92 (2004) 072302.
2. E. E. Kolomeitsev, N. Kaiser and W. Weise, Phys. Rev. Lett. 90 (2003) 092501.
3. D. Jido, T. Hatsuda and T. Kunihiro, Phys. Lett. B670 (2008) 109.
4. Y. Umemoto, S. Hirenzaki, K. Kume and H. Toki, Phys. Rev. C62 (2000) 024606.
5. R. Seki and K. Masutani, Phys. Rev. C27 (1983) 2799.
6. H. Toki, S. Hirenzaki, T. Yamazaki and R. S. Hayano, Nucl. Phys. A501 (1989) 653.
7. T. Yamazaki and S. Hirenzaki, Phys. Lett. B557 (2003) 20.
8. K. Itahashi *et al.*, "Precision Spectroscopy of Pionic Atoms in $(d,^3\mathrm{He})$ Nuclear Reactions" ; Exp. proporsal NP0702-RIBF-027 for RIBF, Dec. (2006).
9. K. Itahashi *et al.*, "Spectroscopy of Pionic Atom in $^{122}\mathrm{Sn}(d,^3\mathrm{He})$ Nuclear Reaction" ; Exp. proporsal NP0802-RIBF-054 for RIBF, Jan. (2008).
10. M. Ericson and T. E. O Ericson, Ann. Phys. (NY), 36 (1966) 496.

TENSOR OPTIMIZED SHELL MODEL USING BARE INTERACTION FOR LIGHT NUCLEI

TAKAYUKI MYO

Osaka Institute of Technology, Osaka, Osaka 535-8585, Japan
Research Center for Nuclear Physics (RCNP), Osaka University, Ibaraki, Osaka 567-0047, Japan
E-mail: myo@ge.oit.ac.jp

HIROSHI TOKI

Research Center for Nuclear Physics (RCNP), Osaka University, Ibaraki, Osaka 567-0047, Japan
E-mail: toki@rcnp.osaka-u.ac.jp

KIYOMI IKEDA

RIKEN Nishina Center, Wako, Saitama 351-0198, Japan
E-mail: k-ikeda@postman.riken.go.jp

We propose a new theoretical approach to calculate nuclear structure using bare nuclear interaction, in which the tensor and short-range correlations are described by using the tensor optimized shell model (TOSM) and the unitary correlation operator method (UCOM), respectively. We compare the obtained results using TOSM+UCOM for ^{4}He with the rigorous calculation.

Keywords: bare interaction, tensor correlation, short-range correlation.

1. Introduction

It is important to develop a theoretical framework to calculate nuclear structure using bare nucleon-nucleon interaction. Recently it becomes possible to calculate nuclei up to mass around $A \sim 12$ by using the Green's Function Monte-Carlo method.[1,2] The binding energies of nuclei were successfully reproduced by including three body interaction. One big surprise is a large contribution of the one-pion exchange interaction, which is about 70~80 % of the whole bare interaction. It is however extremely time consuming to apply this method to heavier nuclei. It is desired to develop a new method to calculate nuclei with large nucleon numbers by using the bare interaction.

The nucleon-nucleon interaction has distinctive features, namely there exist strong tensor interaction at intermediate distance caused by pion exchange and strong short-range repulsive interaction at a short distance caused by quark dynamics. Although these two interactions have totally different characteristics, it is customary to adopt the theory, such as the Brueckner Hartree-Fock, to integrate out the high-momentum components emerging from the tensor interaction and short-range repulsion on the same footing. In this way, one loses direct information on the tensor correlation and a short-range correlation in the wave function. Hence, we search for a powerful method of treating treat explicitly both the tensor interaction and the short-range interaction to study of not only light nuclei but also medium and heavy nuclei.

There are two important developments for this purpose. One is to find out that the tensor interaction is of intermediate range. We are able to express the tensor correlation in a reasonable shell model space.[3–5] We name this method as Tensor Optimized Shell Model (TOSM), where the wave function is constructed in terms of the shell model basis states with full optimization of the two particle-two hole ($2p2h$) states. We have shown that the tensor interaction could be treated by taking a reasonable amount of multipoles ($l \leq 5$) for the particle states.[3]

The other is the Unitary Correlation Operator Method (UCOM) for the treatment of the short-range correlation.[6,7] We shall combine two methods, TOSM and UCOM (TOSM + UCOM), to describe nuclei using bare interaction. We hope to apply a newly developed method to many nuclei.[5] We use TOSM for the strong tensor interaction utilizing the intermediate nature caused by finite angular momentum of the relative wave function, and use UCOM for the strong short-range interaction utilizing the short-range nature. Using TOSM+UCOM, we aim to understand the roles of the tensor and short-range correlations in nuclei using bare interaction. As a good start, we apply TOSM+UCOM to ^{4}He and compare the obtained results with rigorous calculations[8] and see how this method works by using bare interaction and how far we can describe each correlations and to find what we are supposed to do for further improvement.

2. Tensor Optimized Shell Model (TOSM)

We explain the framework of TOSM. We shall begin with many-body Hamiltonian,

$$H = \sum_i T_i - T_{\text{c.m.}} + \sum_{i<j} V_{ij}, \tag{1}$$

$$V_{ij} = v_{ij}^C + v_{ij}^T + v_{ij}^{LS} + v_{ij}^{Clmb} \,. \tag{2}$$

Here, T_i is the kinetic energy of each nucleons with $T_{\text{c.m.}}$ being the center of mass term. We take the bare interaction for V_{ij} such as AV8′[2] consisting of central (v_{ij}^C), tensor (v_{ij}^T) and spin-orbit (v_{ij}^{LS}) terms. v_{ij}^{Clmb} is the Coulomb term. We obtain the many-body wave function Ψ with the equation, $H\Psi = E\Psi$.

We begin with the TOSM wave function. We write the case of ^{4}He explicitly as an example including the variational coefficients $\{A_k\}$.

$$\Psi = A_0|0\rangle + \sum_{k=1} A_k|2p2h\rangle_k, \tag{3}$$

$$|2p2h\rangle_k = \left|[[\psi_{\alpha_1}^{n_1}(\vec{x}_1), \psi_{\alpha_2}^{n_2}(\vec{x}_2)]^J \otimes [\tilde{\psi}_{\alpha_3}^{n_3}(\vec{x}_1), \tilde{\psi}_{\alpha_4}^{n_4}(\vec{x}_2)]^J]^0\right\rangle \tag{4}$$

Here, $|0\rangle$ is $|(0s)^4\rangle$ for ^{4}He. $|2p2h\rangle$ are the $2p2h$ states with various radial components. The index k denotes representable quantum number of $2p2h$ states using the single particle (hole) wave functions ψ_α^n ($\tilde{\psi}_\alpha^n$) with the quantum numbers n and α; n is an index to distinguish the different radial wave functions. α are the sets of the spin-isospin quantum numbers to distinguish the single particle orbits. We describe the the various radial wave functions of single particle orbits by using the Gaussian expansion method.[3,11]

We explain the Gaussian expansion technique for single particle orbits. Each Gaussian basis function has the form of a nodeless harmonic oscillator wave functions (HOWF), except for $1s$ orbit. When we superpose a sufficient number of Gaussian bases with various length parameters, we can fully optimize the radial component of every orbit of every configuration in Eq. (3) of TOSM with respect to the total Hamiltonian in Eq. (1). We construct the following ortho-normalized single particle wave function ψ_α^n with a linear combination of Gaussian bases $\{\phi_\alpha\}$ having length parameters $b_{\alpha,\nu}$.

$$\psi_\alpha^n(\boldsymbol{r}) = \sum_{\nu=1}^{N_\alpha} d_{\alpha,m}^n \, \phi_\alpha(\boldsymbol{r}, b_{\alpha,\nu}) \qquad \text{for } n = 1, \cdots, N_\alpha, \tag{5}$$

where N_α is a number of basis functions for the orbit α, and ν is an index that distinguishes the bases with different values of $b_{\alpha,\nu}$. The explicit form of the Gaussian basis function is expressed as

$$\phi_\alpha(\boldsymbol{r}, b_{\alpha,\nu}) = \mathcal{N}\ r^l\ e^{-(r/b_{\alpha,\nu})^2/2}\ [Y_l(\hat{\boldsymbol{r}}), \chi^\sigma_{1/2}]_j \chi_{t_z}, \tag{6}$$

where l and j are the orbital and total angular momenta of the single particle states, respectively, and t_z is the projection of the nucleon isospin. The factor $\mathcal{N}$ is for the normalization of the basis function. The coefficients $\{d^n_{\alpha,\nu}\}$ are determined by solving the eigenvalue problem for the norm matrix of the non-orthogonal Gaussian basis set in Eq. (6) with the dimension N_α. Following this procedure, we obtain new single particle wave functions $\{\psi^n_\alpha\}$ in Eq. (5) for TOSM.

We choose the Gaussian bases for the particle states to be orthogonal to the occupied single particle states (hole states), which are the $0s$ state in the ^{4}He case. For $0s$ state, we employ one Gaussian basis function, namely, HOWF with length $b_{0s,\nu=1} = b_{0s}$. For $1s$ states, we introduce an extended $1s$ basis function, which is orthogonal to the $0s$ state and possessing a length parameter $b_{1s,\nu}$ that differs from b_{0s}.[3] In the extended $1s$ basis functions, we change the polynomial part from the usual $1s$ basis states to satisfy the conditions of the normalization and the orthogonality to the $0s$ state.

Two-body matrix elements in the Hamiltonian are analytically calculated using the Gaussian bases,[3,11] whose explicit forms are given in Appendix of Ref. 5 for central, LS and tensor interactions, respectively. In the numerical calculation, we prepare 9 Gaussian functions at most with various range parameters to get a convergence of the energy.

We have to furthermore take care of the center-of-mass (c.m.) excitations. We impose a constraint on the total wave function, in which c.m. motion should be the $0s$ state. For this aim, we take the well-tested method of introducing a c.m. term in the many body Hamiltonian to eliminate the c.m. excitations.[12] In the numerical calculation, the excitation of the spurious center-of-mass motion is suppressed to be less than 10 keV.

The variation of the energy expectation value with respect to the total wave function Ψ is given by

$$\delta \frac{\langle\Psi|H|\Psi\rangle}{\langle\Psi|\Psi\rangle} = 0\ , \tag{7}$$

which leads to the following equations:

$$\frac{\partial\langle\Psi|H-E|\Psi\rangle}{\partial b_{\alpha,\nu}} = 0\ , \quad \frac{\partial\langle\Psi|H-E|\Psi\rangle}{\partial A_k} = 0\ . \tag{8}$$

Here, E is a Lagrange multiplier corresponding to the total energy. The parameters $\{b_{\alpha,\nu}\}$ for the Gaussian bases appear in non-linear forms in the energy expectation value. We solve two kinds of variational equations in the following steps. First, fixing all the length parameters $b_{\alpha,\nu}$, we solve the linear equation for $\{A_k\}$ as an eigenvalue problem for H with partial waves up to $L_{\rm max}$. We thereby obtain the eigenvalue E, which is a function of $\{b_{\alpha,\nu}\}$. Next, we try to search various sets of the length parameters $\{b_{\alpha,\nu}\}$ in order to find the solution which minimizes the total energy. In this TOSM wave function, we can describe the spatial shrinkage with an appropriate radial form, which is important to describe the tensor correlation.[3]

3. Unitary Correlation Operator Method (UCOM)

We present UCOM for the short-range correlation,[6,7] in which the following unitary operator is introduced

$$C = \exp(-i\sum_{i<j} g_{ij}) = \prod_{i<j} c_{ij}, \qquad c_{ij} = \exp(-i\, g_{ij}). \tag{9}$$

We express the correlated wave function Ψ in terms of less sophisticated wave function Φ as $\Psi = C\Phi$. Hence, the Schrödinger equation becomes $\hat{H}\Phi = E\Phi$ where $\hat{H} = C^\dagger HC$. Since C is expressed with a two-body operator in the exponential, it is a many-body operator. In the case of the short-range correlation, we are able to truncate modified operators at the level of two-body operators.[6]

We define the two-body Hermite operator g in Eq. (9) as

$$g = \frac{1}{2}\left\{p_r s(r) + s(r) p_r\right\}, \tag{10}$$

$$\frac{dR_+(r)}{dr} = \frac{s\left(R_+(r)\right)}{s(r)}, \tag{11}$$

where p_r is the radial component of the relative momentum, which is conjugate to the relative coordinate r. $s(r)$ is the amount of the shift of the relative wave function and depending on r. In UCOM, we use $R_+(r)$ instead of $s(r)$. The function $R_+(r)$ corresponds to the transformed relative coordinate from r, and represents the correlation function to reduce the amplitude of the short-range part of the relative wave functions. The explicit transformation of the operator for the relative motion is given as $c^\dagger rc = R_+(r)$ for example.[6] We use TOSM as Φ to describe the tensor correlation.

In the calculation using UCOM, we parametrize $R_+(\mathrm{r})$ in the same manner as proposed by Neff-Feldmeier and Roth et al.[6,7,13] We assume the following forms for even and odd channels, respectively.

$$R_{+}^{\mathrm{even}}(r) = r + \alpha \left(\frac{r}{\beta}\right)^{\gamma} \exp[-\exp(r/\beta)], \tag{12}$$

$$R_{+}^{\mathrm{odd}}(r) = r + \alpha\,(1 - \exp(-r/\gamma))\exp[-\exp(r/\beta)]. \tag{13}$$

Here, α, β, γ are the variational parameters to optimize the function $R_{+}(r)$ and minimize the energy of the system. They are independently determined for four channels of the spin-isospin pair. In the actual procedure of the variation, once we fix the parameters included in $R_{+}(r)$, we solve the eigenvalue problem of the Hamiltonian using Eq. (8) and determine the configuration mixing of the shell model-type bases of TOSM in Eq. (3). Next, we try to search various sets of the $R_{+}(r)$ parameters to minimize the energy.

3.1. *Extension of UCOM –S-wave UCOM–*

It is generally possible to introduce the partial wave dependence in UCOM and then $R_{+}(r)$ functions are determined in each relative partial wave in the two-body matrix elements. Here, we consider the specific case of this extension of UCOM by taking care of the characteristics of the short range correlation. One of the simplest case of this extension is UCOM for only s-wave relative motion. Since all the other partial waves l except for s-wave ($l = 0$) have r^l behavior near at the origin, this r^l behavior cuts down the effect of the short range hard core largely. Only the s-wave function is finite at the origin and the behavior in the origin is determined by the hard core dynamics.

When we incorporate S-wave UCOM (S-UCOM, hereafter) into TOSM, we extract the relative s-wave component in every two-body matrix elements using the V type basis expanded by the Gaussian functions. For numerical calculation to carry out it, we prepare the completeness relation consisting of the T type basis functions $|T\rangle$ as

$$1 = \sum_i |T_i\rangle\langle T_i|, \qquad |T_i\rangle \;=\; |[[\psi_l(\boldsymbol{r}), \psi_L(\boldsymbol{R})]_{L'}, \chi_S^{\sigma}]_J\; \chi_T\rangle, \tag{14}$$

where the T type basis is expanded by the two coordinates of the relative part $\boldsymbol{r}$ and the center of mass part $\boldsymbol{R}$ of two nucleons, which are the set of Jacobi coordinate. The orbital angular momenta of each coordinate are l and L, respectively. It is easy to prepare the s-wave relative part by putting l to be zero in the T type basis. We construct the completeness relation of the T type basis states by diagonalizing the norm matrix expanded by the finite number of the Gaussian basis functions. We insert the prepared T

type completeness relation with $l = 0$ in Eq. (14) into the matrix elements of the arbitrary two-body operators.[5]

4. Variational Calculation of ^{4}He in TOSM+UCOM

First, we determine the UCOM functions $R_+(r)$ for the calculation of TOSM+UCOM. We optimize $R_+(r)$ function by changing the three parameters of α, β and γ to search for the energy minimum in TOSM+UCOM. The adopted $R_+(r)$ in UCOM for four channels are shown in Fig. 1.

We show the calculated results for the energy of ^{4}He in Fig. 2, where $L_{\rm max}$ is the maximum orbital angular momentum of the particle states. In Fig. 2, we get a good convergence of the energy with respect to $L_{\rm max}$. We show two results using UCOM and S-UCOM. The latter S-UCOM is the extended UCOM, in which UCOM is used only for the relative s-wave component of the even channel, while the original UCOM does not consider any partial wave dependence. In UCOM, it is found that the binding energy of ^{4}He is underestimated. The reason are understood as due to too large removal of the short-range part of the relative wave functions, in particular, in the d-wave part of the sd coupling of the tensor matrix element, where the tensor interaction possesses some amount of strength. We have also calculated the contributions beyond the $2p2h$ configurations in TOSM such as $3p3h$ and $4p4h$ configurations. When we include the $4p4h$ configurations within the p-shell, their contribution to the binding energy is approximately 50 keV. This fact denotes that these more complicated configurations contribute very little in the ^{4}He system.

In order to overcome this feature of UCOM, we have decided to restrict the use of UCOM to the relative s-wave only for the even channel (S-

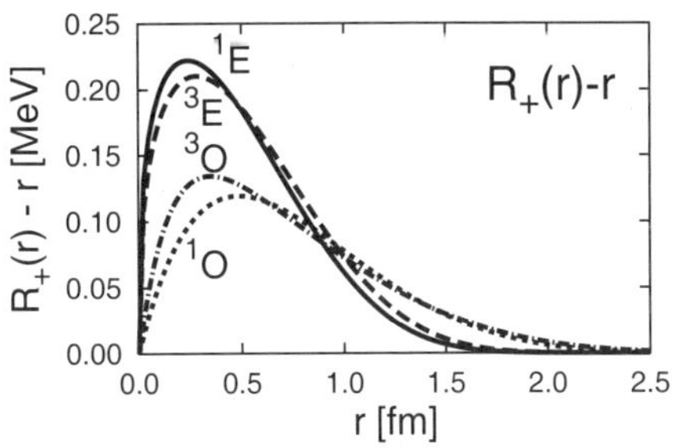

Fig. 1. The UCOM functions $R_+(r)$ in even ($^{1,3}E$) and odd ($^{1,3}O$) channels.

Table 1. Mixing probabilities in the ^{4}He ground state in %, where two subscripts of 00 and 10 are the spin-isospin quantum numbers.

$(0s)^4_{00}$	83.01
$(0s)^{-2}_{10}(0p_{1/2})^2_{10}$	2.45
$(0s)^{-2}_{10}[(1s_{1/2})(0d_{3/2})]_{10}$	2.31
$(0s)^{-2}_{10}[(0p_{3/2})(0f_{5/2})]_{10}$	1.90
$(0s)^{-2}_{10}[(0p_{1/2})(0p_{3/2})]_{10}$	1.45
$(0s)^{-2}_{10}[(0d_{5/2})(0g_{7/2})]_{10}$	0.80
$(0s)^{-2}_{10}(0d_{3/2})^2_{10}$	0.48
remaining part	7.60

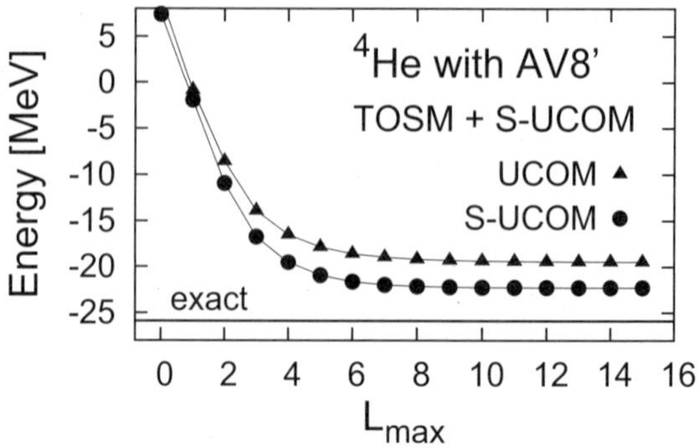

Fig. 2. Energy of ^{4}He in TOSM + UCOM / S-UCOM. Exact calculation is taken from Ref. 8.

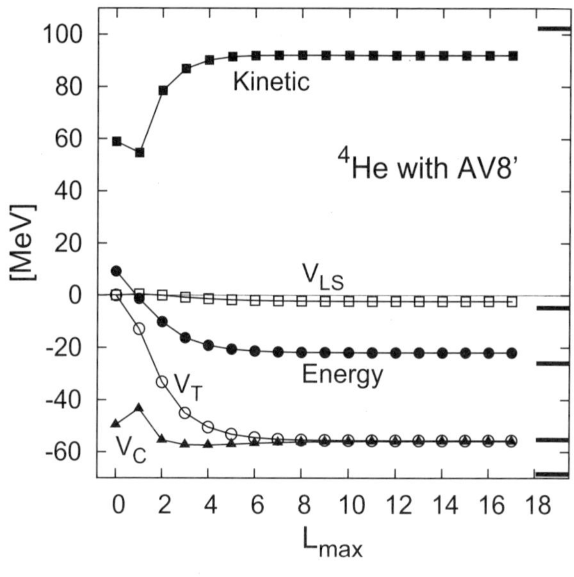

Fig. 3. Matrix elements of the central interaction (V_C), tensor interaction (V_T) and the spin-orbit interaction (V_{LS}) together with the kinetic energy (Kinetic) and total energy (Energy) in the Hamiltonian using AV8$'$ for ^{4}He as function of $L_{\max}$. These values are compared with the benchmark results of Ref. 8, which are indicated by the thick short solid lines on the right-hand side of the figure.

UCOM), where the treatment of the short-range repulsion is absolutely necessary. In other partial waves, we have the centrifugal potential that cuts out the short-range part from the wave functions of the higher partial waves. In this case, we can use the transformed interaction and the kinetic energy only for the relative s-wave component in the even channels. Since the use of UCOM for the odd partial wave is slightly better, we use UCOM for all odd partial waves. we have introduced S-UCOM and found that the situation was largely improved to give about 23 MeV of the binding energy. The tensor force contribution also gains by about 5 MeV in comparison with the UCOM case. This gain mainly comes from the sd coupling matrix elements of the tensor interaction.

We show all the components of the energy for ^{4}He in Fig. 3. All the energy components show the saturation behavior as function of $L_{\max}$. In the tensor component, the saturation is obtained at around $L_{\max}$ being 8. For the other components, their saturation points are seen at the similar $L_{\max}$. A very interesting feature is the kinetic energy, which goes up to a large value as the tensor interaction matrix element becomes large. As for the comparison with the rigorous calculation, we see that V_c satisfies the rigorous value, which is approximately -55 MeV. On the other hand, the tensor interaction matrix element, V_T converges to -55 MeV, while the rigorous one is -68 MeV. The kinetic energy is approximately 90 MeV, while the rigorous one is 102 MeV. The spin orbit matrix element is also a little bit smaller than the rigorous value. As the net value, the total energy, E, is -23 MeV and the rigorous value is -26 MeV. One of the possibilities for the lack of the energy in TOSM + (S)-UCOM is due to the separate treatment of the short-range and tensor correlations. Although the dominant part of the tensor interaction is of intermediate and long range, there may remain some small strength in the short-range part of the tensor interaction, which can couple with the short-range correlations. This effect can be included by extending the truncation of the UCOM transformation of the Hamiltonian to more than the two-body level. Three-body term of the UCOM transformation is one of the possibilities to overcome the lack of energy in TOSM+UCOM.[5,6] This term can contribute to increase the tensor matrix element and increase the high momentum component in the wave function.

To see the properties of the obtained ^{4}He wave function, the dominant configurations are listed in Table 1 with their probabilities. It is found that the several $2p2h$ states such as $(0s)^{-2}_{10}(0p_{1/2})^2_{10}$ are largely mixed and those configurations are essential for the tensor correlation in ^{4}He.[3,14] It is also found that the $0p0h$-$2p2h$ coupling exhausts 94% of the tensor matrix element.

5. Summary

We have developed a method to calculate nuclear ground state by using the bare interaction in the tensor optimized shell model. We have treated the tensor interaction in terms of TOSM, in which the $2p2h$ states are fully optimized to describe the excitations due to the tensor interaction. As for the short-range repulsive interaction, we have introduced the unitary correlation operator method (UCOM). We have furthermore considered the extension of UCOM by introducing the partial wave dependence on the correlation function, named as S-UCOM, which largely improves the results.

Acknowledgements

This work was supported by a Grant-in-Aid for Young Scientists from the Japan Society for the Promotion of Science (JSPS, No. 21740194) and also by the JSPS Core-to-Core Program.

References

1. S.C. Pieper and R. B. Wiringa, Annu. Rev. Nucl. Part. Sci. **51**, 53 (2001).
2. B. S. Pudliner, V. R. Pandharipande, J. Carlson, S. C. Pieper and R. B. Wiringa, Phys. Rev. **C56**, 1720 (1997).
3. T. Myo, S. Sugimoto, K. Katō, H. Toki, K. Ikeda, Prog. Theor. Phys. **117**, 257 (2007).
4. T. Myo, K. Katō, H. Toki and K. Ikeda, Phys. Rev. **C76**, 024305 (2007).
5. T. Myo, H. Toki, K. Ikeda, Prog. Theor. Phys. **121**, 511 (2009).
6. H. Feldmeier, T. Neff, R. Roth and J. Schnack, Nucl. Phys. **A632**, 61 (1998).
7. T. Neff and H. Feldmeier, Nucl. Phys. **A713**, 227 (2003).
8. H. Kamada *et al.*, Phys. Rev. **C64**, 044001 (2001).
9. T. Myo, K. Katō, K. Ikeda, Prog. Theor. Phys. **113**, 763 (2005).
10. T. Myo, K. Katō, H. Toki, K. Ikeda, J. of Phys. **G31**, S1681 (2005).
11. E. Hiyama, Y. Kino and M. Kamimura, Prog. Part. Nucl. Phys. **51**, 223 (2003).
12. D.H. Gloeckner, R.D. Lawson, Phys. Lett. **B53**, 313 (1974).
13. R. Roth, P. Papakonstantinou, N. Paar, H. Hergert, T. Neff and H. Feldmeier, Phys. Rev. **C73**, 044312 (2006).
14. H. Toki, S. Sugimoto and K. Ikeda, Prog. Theor. Phys. **108**, 903 (2002).

EXOTIC CLUSTER STRUCTURES EXCITED BY TWO NEUTRON TRANSFERS

MAKOTO ITO

Pure and Applied Physics, Kansai University,
Yamate-cho 3-3-35, Suita 564-8680, Japan
E-mail: itomk@ipcku.kansai-u.ac.jp

DAISUKE SUZUKI

National Superconductiong Cyclotron laboratory, Michigan State University,
1 Cyclotorn, East Lansing, MI 48824-1321, USA
E-mail: suzuki@nscl.msu.edu

The α+^{8}He low-energy reactions and the exotic structures of ^{12}Be are studied by the application of the generalized two-center cluster (α+α+4N) model. In the two-neutron transfer reactions, $\alpha+^8\text{He}_{g.s.} \rightarrow {}^6\text{He}_{g.s.}+{}^6\text{He}_{g.s.}$, a resonant peak with J^π=0$^+$ appears around $E \sim 1.3$ MeV above the $^6\text{He}_{g.s.}+^6\text{He}_{g.s.}$ threshold as the result of the formation of the covalent superdeformation, which has a hybrid structure of covalent and ionic configurations for the valence neutrons. The covalent superdeformed state becomes an isomeric state with a sharp width of $\Gamma \sim 400$ keV. Experimental conditions to identify the isomeric state are also discussed.

Keywords: Cluster structures, molecular resonances, neutron transfers

1. Introduction

In the last two decades, developments of experiments with secondary RI beam have extensively proceeded the studies on light neutron-rich nuclei. In particular, much efforts have been devoted to the investigation of molecular structure in Be isotopes. These isotopes are typical examples of two-center superdeformed systems which build on an α+α rotor of ^{8}Be. The low-lying states of these systems can be described by the formation of molecular orbitals (MOs) such as π^- and σ^+ orbitals associated with the covalent electrons of atomic molecules.[1] Experimentally, the molecular structures were mainly investigated by the breakup reactions[2] or inelastic α scatterings[3] by using intermediate-energy RI beams.

Furthermore, Low-energy reaction studies have also been conducted with an intense helium isotope beam provided by the ISOL technique.[2,4,5] In the recent experiment, ^{6}He resonant scattering by an α target was performed, and some characteristic enhancements are observed in angular distribution[6] and in the excitation function.[5] A similar measurement is also possible for the neutron-drip-line nucleus, ^{8}He, and the scattering of α+^{8}He has already been measured at GANIL.[7] Since ^{8}He has the α+$4N$ structure with four weakly bound neutrons and an α cluster, which is quite stable and inert, the valence four neutrons can easily be exchanged between two α particles during the collision. Due to these conditions, α+^{8}He can be considered transparent, and the resonances could be prominently excited by the neutron-transfer process.

In order to investigate the resonance formation in the slow α+^{8}He$_{g.s.}$ scattering, the reaction process should be treated consistently with the low-lying bound states of the compound system of ^{12}Be, because the resonances correspond to excited states, which are embedded in the continuum above the particle decay threshold. In Be isotopes including ^{12}Be, the MO model can successfully describe the low-lying states.[1] In the study of α+^{8}He$_{g.s.}$ scattering, therefore, the covalent MO configurations and the scattering process above the α+^{8}He$_{g.s.}$ threshold must be treated in a unified manner. For this purpose, we apply the generalized two-center cluster model (GTCM)[8–11] to the α+^{8}He$_{g.s.}$ resonant scattering. The GTCM can cover the low-lying MO configurations[1] as well as the ionic ones excited as the ^{X}He+^{Y}He molecular resonances (MRs, X,Y=4∼8)[12,13] in the continuum; hence, treating resonant phenomena observed in slow scattering is possible. In this paper, we study the excitation and the decay scheme of the resonances through the neutron-transfer reaction induced by the α+^{8}He$_{g.s.}$ scattering.

2. Framework

The unified treatment of the formation of the intrinsic states (MRs and MOs) and their coupling to the scattering continuum can be achieved by the combined method of GTCM[9,10] and Kamimura's method,[14,15] which has been developed in Refs. 10,11. In this method, the total wave function of ^{12}Be is given by the following linear combination

$$\Psi^{J^\pi(+)} = \sum_\beta \varphi_\beta\, \chi_\beta^{(+)} + \sum_\nu b_\nu \hat{\Psi}_\nu^{J^\pi} \,. \tag{1}$$

The first term stands for the "open channels" labeled by β, on which the scattering boundary condition is explicitly imposed.[9,15] Here, φ_β and $\chi_\beta^{(+)}$ denote the internal and relative wave-functions of the two scattering nuclei, and the whole nucleons included in the nuclei are completely anti-symmetrized. We consider three rearrangement channels, α+$^8\mathrm{He}_{g.s.}$, $^6\mathrm{He}_{g.s.}$+$^6\mathrm{He}_{g.s.}$ and $^5\mathrm{He}_{g.s.}$+$^7\mathrm{He}_{g.s.}$ as open channels.

The second term stands for the "intrinsic states" confined within the interaction region, which is a linear combination of the solutions obtained by diagonalizing the total Hamiltonian. The basis of the intrinsic states, $\hat{\Psi}_\nu^{J^\pi}$ labeled by an eigenvalue number ν, is damped in the asymptotic region. Therefore, the second term describes the compound states formed before decaying into the binary open channels. The $\hat{\Psi}_\nu^{J^\pi}$ is calculated by the GTCM, and its explicit form is written as

$$\hat{\Psi}_\nu^{J^\pi} = \int dS \sum_{\mathbf{m}} C_{\mathbf{m}}^\nu(S)\, \Phi_{\mathbf{m}}^{J^\pi}(S) \tag{2}$$

with the basis function given by

$$\Phi_{\mathbf{m}}^{J^\pi}(S) = \hat{P}_{K=0}^{J^\pi}\, \mathcal{A}\left\{\psi_L(\alpha)\psi_R(\alpha)\prod_{j=1}^{4}\varphi_j(m_j)\right\}_S . \tag{3}$$

The detailed explanation of Eq. (3) is given in Ref. 10, and we briefly explain these expressions. $\psi_n(\alpha)$ (n=L, R) denotes the α-cluster centered at the left (L) or right (R) side with relative distance S.[16] The individual α core is expressed by the $(0\mathrm{s})^4$ configuration of the harmonic oscillator (HO). $\varphi_j(m_j)$ represents the atomic orbit (AO) state with the $0p$ orbital for the valence neutrons localized around one of the α clusters. Here, the index m is a set of quantum numbers needed to specify the AO state: the direction of the 0p orbital, the neutron's spin and its center (L or R). In Eqs. (2) and (3), $\mathbf{m}$ represents a set of AOs for the four neutrons, $\mathbf{m}$=(m_1, m_2, m_3, m_4). The basis functions in Eq. (3) are fully anti-symmetrized by $\mathcal{A}$ and projected to the eigenstate of the total spin-parity J^π and its intrinsic angular projection K, which is restricted to the axial symmetric case (K=0) by the projection operator $\hat{P}_K^{J^\pi}$. The restriction of the axial symmetric configurations is a good approximation in the slow scattering as demonstrated in Ref. 17. The $\hat{\Psi}_\nu^{J^\pi}$ describing the intrinsic state is finally given by taking the superposition of Eq. (3) over S and $\mathbf{m}$ as shown in Eq. (2).

As for the AO states of the four valence neutrons, $\mathbf{m}$ in Eq. (2), we include all the possible configurations, and hence, the model space of MO, where each valence neutron rotates around the two centers simultaneously, is also covered.[8] We use the Volkov No.2 and the G3RS for the central and spin-orbit parts of the nucleon-nucleon interaction, respectively. The parameters in the interactions and the size of the HO are the same as those applied in Refs. 9–11, which successfully describe the structural properties of ^{10,12}Be. The adopted parameter set reasonably reproduces the threshold energies of $\alpha+{}^8\mathrm{He}_{g.s.}$, ${}^6\mathrm{He}_{g.s.}+{}^6\mathrm{He}_{g.s.}$ and ${}^5\mathrm{He}_{g.s.}+{}^7\mathrm{He}_{g.s.}$. This is essential in the treatment of scattering phenomena.

3. Results

3.1. *Structural features*

The energy spectra of bound states are obtained by solving the eigenvalue problem with the total wave function in Eq. (2). In addition, we calculated the scattering matrices (S matrices) of the $\alpha+{}^8\mathrm{He}_{g.s.}$ reaction when the system is excited above the threshold energy of this channel. By imposing the appropriate boundary conditions, we obtain whole energy spectra from the bound states to the resonant ones as shown in Fig. 1 ($J^\pi=0^+$). In this figure, the S matrix for the two neutron transfer reactions, $\alpha+{}^8\mathrm{He}_{g.s.} \rightarrow {}^6\mathrm{He}_{g.s.}+{}^6\mathrm{He}_{g.s.}$, is also shown.

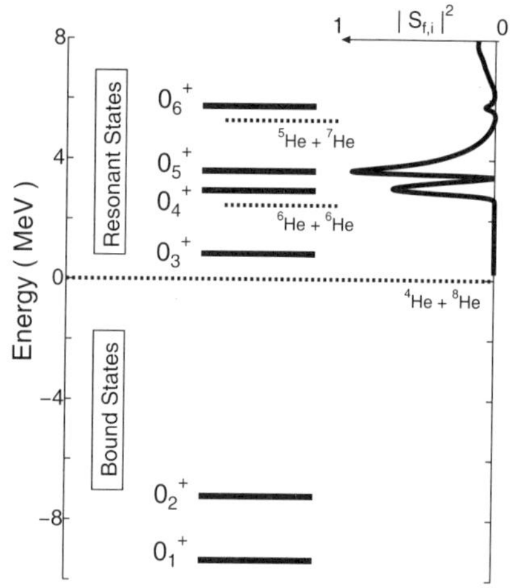

Fig. 1. Energy spectra for $J^\pi=0^+$. The solid curve at the right side shows the squared magnitude of the S matrix for $\alpha+{}^8\mathrm{He}_{g.s.} \rightarrow {}^6\mathrm{He}_{g.s.}+{}^6\mathrm{He}_{g.s.}$ with a respective scale at the top-right corner. The dotted lines represent the threshold energy of the open channels considered in the calculation.

The resonance energies are identified by checking the anti-clockwise behaviour of the S-matrix in the complex plane. The obtained 0^+ states are classified into three categories:

(i) MO states. Two bound states appear below the $\alpha+{}^8\mathrm{He}_{g.s.}$ threshold. The energy difference is about 2 MeV, and this difference agrees with a recent observation.[18] Here, the four neutrons are in MO motion around the

two α cores $((\pi^-_{3/2})^2(\sigma^+_{1/2})^2$ and $(\pi^-_{3/2})^2(\pi^-_{1/2})^2$ for the 0^+_1 and 0^+_2 states, respectively).

(ii) MR states. The resonance states, 0^+_3, 0^+_4, and 0^+_6 have the developed cluster configurations of α+^{8}He$_{g.s.}$, ^{6}He$_{g.s.}$+^{6}He$_{g.s.}$, and ^{5}He$_{g.s.}$+^{7}He$_{g.s.}$, respectively. Their energies are quite close to the corresponding cluster thresholds. The amplitude of the relative wave functions in these channels are enhanced around the ^{x}He$-^y$He barrier-top distance, although many kinds of other channels are strongly mixed in the inner region. Therefore, the intrinsic structure of these three resonances can be characterized as MR of the ^{x}He+^{y}He configurations.

(iii) Covalent SD state. The intrinsic structure of the 0^+_5 state is identified as a covalent superdeformed state (covalent SD), which has a hybrid configuration of covalent and ionic structures. In this configuration, two neutrons occupy the $0p$-wave AO, which is perpendicular to the $\alpha-\alpha$ axis (L_z=$\pm$1) around each α-cluster, like ^{5}He+^{5}He, while the remaining two neutrons form the $(\sigma^+_{1/2})^2$ bonding. Due to the formation of the σ^+-bonding, the $\alpha-\alpha$ distance is increased, which amounts to $\sim$ 5 fm. Since the radius of the α particle is taken to be $\sim$ 1.4 fm in the present model, this $\alpha-\alpha$ distance corresponds nearly to the hyperdeformation.

3.2. *Resonance formations in transfer reactions*

The detailed investigation of the S matrices are important to understand the reaction mechanism relevant to the resonance formation in the continuum energy region. The S matrices of the elastic scattering, one and two neutron transfer reactions are shown in Fig. 2. In this figure, we can clearly understand the important role of the intrinsic states in forming the sharp resonances. The dashed curves represent the S matrices calculated only from the first term in Eq. (1), the open channels. No resonant structures with sharp widths exist in the results of the open channels. In the full solutions shown by the solid curves, the sharp structures are generated by the coupling with the intrinsic states (double circles labeled by 0^+_3 $\sim$$0^+_6$). Each peak is generated by a linear combination of all the intrinsic states, but the dominant contribution in the individual peak is a single intrinsic state, which is shown by the arrows. Since the intrinsic states include the component of not only the open channels, but also of many kinds of closed channels, which have a higher excitation energy than the resonance energy, the coupling with the intrinsic states plays an important role in reducing the resonance width.

The present calculation predicts the sharp resonances corresponding to the four levels, $0_3^+ \sim 0_6^+$ (Fig. 1) in all the exit channels. The ratio of S matrices for various exit channels gives important informations on the intrinsic structures of the compound system, ^{12}Be. In particular, we focus on the ratio of the S matrices between the elastic scattering and the two neutron transfer reaction, because these two channels are particle bound systems and hence, their coincident measurement is available. The resonant state around $E_{c.m.} \sim 4$ MeV ($E \sim 1.3$ MeV with respect to the $^6He_{g.s}+^6He_{g.s.}$ threshold) corresponding to 0_5^+ has a large magnitude of S matrix of $^6He_{g.s.}+^6He_{g.s.}$ ($|S_{f,i}|^2 \sim 1$), while that of the elastic scattering is quite small. Therefore, the 0_5^+ state mainly decays into not $\alpha+^8He_{g.s.}$ but $^6He_{g.s.}+^6He_{g.s.}$.

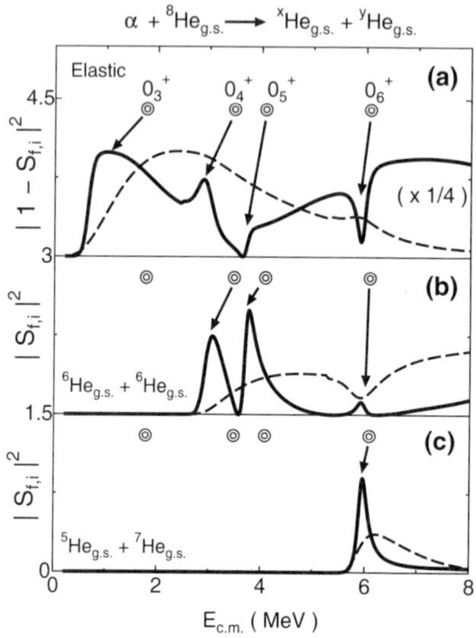

Fig. 2. S matrices for the central collision of $\alpha+^8He_{g.s.}$ ($J^\pi=0^+$). The panel (a) shows the S matrix of the elastic scattering, while (b) and (c) show those for the final channels of $^6He_{g.s.}+^6He_{g.s.}$ and $^5He_{g.s.}+^7He_{g.s.}$, respectively. The dashed and solid curves represent the results calculated with the three open channels and the full solutions of Eq. (1), respectively. The double circles represent the energy levels obtained by the diagonalization of the total Hamiltonian. The S matrices of the elastic scattering are divided by a factor of four.

Such the characteristic decay scheme is due to the intrinsic structure of 0_5^+. It has the covalent SD structure as explained in the previous subsection. Since the σ^+ orbital is constructed by $0p_z(L)-0p_z(R)$ where $0p_z(C)$ means $0p_z$-wave with the center C, the covalent SD is given by a linear combination of the (^{6}He+^{6}He)+(^{5}He+^{7}He) configurations. Here, ^{6}He+^{6}He and ^{5}He+^{7}He are constructed by a coherent mixing of the asymptotic channels of the respective configurations. This structure does not contain a component of $\alpha+^8He_{g.s.}$, which is an open channel at the resonance energy. Furthermore, the resonance energy is lower than the threshold of $^5He_{g.s.}+^7He_{g.s.}$, which is the lowest open channel in the ^{5}He+^{7}He configuration. Therefore, the decay channel of the covalent SD is restricted to $^6\mathrm{He}_{g.s.}$+$^6\mathrm{He}_{g.s.}$. This is the reason why the S matrix corresponding to the covalent SD reaches the unitary limit of $|S_{f,i}|^2 \sim 1$. Due to this limitation of the decay channel, the covalent SD becomes an isomeric state with a width of $\Gamma \sim$400 keV.

The enhancement of the decay into $^6\mathrm{He}_{g.s.}$+$^6\mathrm{He}_{g.s.}$ can be supported by a recent experiment on the nuclear breakup of ^{12}Be induced by α target.[3,19] In this measurement, the partial decay widths into the α+$^8\mathrm{He}_{g.s.}$ and $^6\mathrm{He}_{g.s.}$+$^6\mathrm{He}_{g.s.}$ channels together with their resonant spins are identified by performing multi-pole decomposition analysis (MDA). As a result of MDA, in the $^6\mathrm{He}_{g.s.}$+$^6\mathrm{He}_{g.s.}$ channel, the resonant peak is observed at $E_{exp} \sim$1.2 MeV above the $^6\mathrm{He}_{g.s.}$+$^6\mathrm{He}_{g.s.}$ threshold,[3] which nicely agrees with our result of $E_{cal} \sim$ 1.3 MeV. Furthermore, the observed resonance at $E_{exp} \sim$ 1.2 MeV does not have a non-zero partial decay width into the α+$^8\mathrm{He}_{g.s.}$ channel.[19] These observations are completely consistent with the covalent isomeric state obtained by the present calculation both in the decay scheme and in the energy position of the resonant peak.

3.3. *Gate of the scattering angle in two neutron transfers*

We have pointed out the characteristic decays in the two neutron transfers and, it is very interesting to measure the J^π=0^+ strength in realistic experiments, not the breakup of ^{12}Be but the slow scattering of α+^{8}He. However, in addition to the J^π=0^+ wave, the higher partial waves also contribute to the cross section measured in experiments. Therefore, the J^π=0^+ contribution should be extracted from the total cross section. Unfortunately, the magnitude of the partial cross section of J^π=0^+ is much smaller than that of J^π=2^+ contribution in the energy range of the resonances as shown in Fig. 3. The resonance behaviour in the J^π=0^+ contribution (bottom panel) is completely masked in the total cross section (top panel) due to the broad contribution of the J^π=2^+ strength (third panel from the top).

In this situation, we must introduce certain kinds of gate for the angular distribution of the scattered particles to extract the J^π=0^+ cross section. The general expression of the differential cross section for the spin-less reaction is simply given by

$$\frac{d\sigma}{d\Omega} = \frac{1}{4k^2}\left|\sum_{J=0,2}(2J+1)S^J P_J(\cos\theta)\right|^2 . \tag{4}$$

Here, S^J and $P_J(\cos\theta)$ denote the S matrix and the Legendre polynomial for the partial wave J, while k represents the wave number corresponding to the relative energy of the incident fragments. In this summation, only the $J = 0$ and 2 contributions is considered, because the partial wave of $J \geq 4$ is negligible in the range of the resonance energy (second panel in Fig. 3).

The contribution of $J^\pi=0^+$ is small due to kinematic weights of $2J+1$ in Eq. (4), but we can extract the $J^\pi=0^+$ contribution from the total summation by introducing the gate of the scattering angle θ. The Legendre polynomial of $J=2$ becomes exactly zero at the angles of $\theta_s \sim 55°$ and $125°$, while that of $J=0$ has an isotropic distribution in angles. Therefore, we can extract the resonance structure of $J^\pi=0^+$ by averaging the differential cross section around these specific angles θ_s.

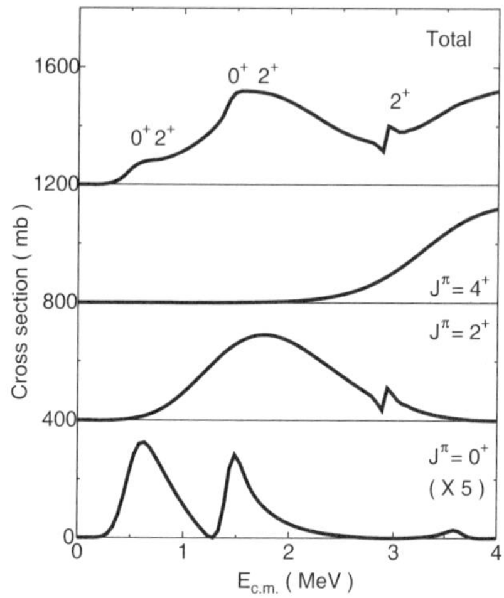

Fig. 3. Total and partial cross sections for the two neutron transfer reactions, $\alpha+^8$He $\rightarrow$ ^{6}He+^{6}He. The lower two panels represent the partial cross sections for $J^\pi=0^+$ and $J^\pi=2^+$, while the top panel and the second one from the top show the total cross section ($J^\pi=0^+$ + $J^\pi=2^+$) and the $J^\pi=4^+$ contribution, respectively. The magnitude of the $J^\pi=0^+$ cross section is multiplied by a factor of five.

We considered three kinds of gates for the angular distribution: (1) the extraction at specific angles of $\theta_s \sim 55°$ and $125°$. (2) the angular average with a range of $\theta_s \pm 5°$. (3) the same average but with $\theta_s \pm 15°$. The respective excitation functions are shown in Fig. 4. The gated excitation function at $\theta=55°$ and $125°$(the bottom panel) nicely reproduces the resonance behaviours appearing in the $J^\pi=0^+$ partial cross section of Fig. 3. The averaged function with the range of $\theta_s \pm 5°$ can also reproduce the similar energy dependence (middle panel). However, the shape of the gated cross section largely deviates from that of the $J^\pi=0^+$ cross sections, when the angular range is extended to about $\theta_s \pm 15°$ (top panel). In realistic experiments, therefore, the angular ranges of $\theta \sim 55° \pm 5$ and $125° \pm 5$ in the center-of-mass system are appropriate for the setting of the detections systems. The experiment has already been measured at GANIL.[7] Introducing of the angular gate is quite useful to extract the characteristic resonances in $J^{\pi}=0^{+}$ partial wave from the measured excitation function.

4. Summary

In summary, we have performed the unified studies of the structures of ^{12}Be and the reactions induced by the $\alpha+^{8}\text{He}_{g.s.}$ slow scattering. We have applied the GTCM and found that the characteristic structural changes appear as a function of the excitation energy. Above the particle decay threshold, the intrinsic structures and the decay scheme of the resonances are especially explored. The coupling effects between the open channels and the intrinsic states are essential for the formation of the sharp resonances.

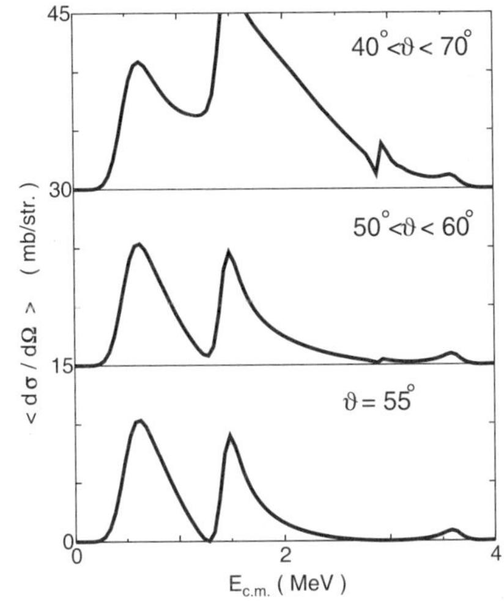

Fig. 4. Excitation functions with the angular gate. The bottom panel shows the result of the gate at specific angles, 55° (and 125°), where there is no contribution of J=2, while the middle and top panels shows the result of the angular average with the range of 55°±5° (and 125°±5°) and 55°±15° (and 125°±15°), respectively.

The present calculation predicts a characteristic decay scheme for the resonance corresponding to the covalent superdeformation. This resonance can decay only into the specific channel, $^{6}\text{He}_{g.s.}+^{6}\text{He}_{g.s.}$, and becomes an isomeric state due to this characteristic decay scheme, which is consistent with the recent observation of the nuclear breakup of ^{12}Be. In further studies, the coincident measurement of the elastic scattering and the two neutron transfers is strongly desired in the $\alpha+^{8}$He scattering experiment to investigate the intrinsic structure of excited resonances.

In the present study, we have pointed out the importance of the angular gate in the analysis of the excitation function. Recently, the resonant scattering of $\alpha+^{8}$He has already been measured at GANIL,[7] and the angular gate demonstrated in the present calculation must be considered in the analysis of the measured cross section.

This is the first study pointing out the formation of a covalent isomeric-state via a neutron-transfer reaction, which shall be generally observed in light neutron-rich nuclei, and systematic studies are now proceeding.

Acknowledgements

The authors would like to thank Prof. K. Ikeda and Prof. N. Itagaki for his valuable comments and encouragements. They would also like to thank Profs. T. Nakatsukasa, K. Kato, H. Sakurai and Dr. A. Saito for useful discussions and encouragements. This work has been supported by the Grant-in-Aid for scientific Research in Japan No. 21740211.

References

1. N. Itagaki and S. Okabe, Phys. Rev. C **61**, 044306 (2000); N. Itagaki, S. Okabe and K. Ikeda, Phys. Rev. C **62**, 034301 (2000), and references therein.
2. M. Freer et al., Phys. Rev. C **63**, 034301 (2001).
3. A. Saito et al., Suppl. Prog. Theor. Phys., **146**, 615 (2003); A. Saito et al., AIP conference proceedings Vol. 891, p205 (2006).
4. M. Milin *et al.,* Nucl. Phys. **A753**, 263 (2005), and references therein.
5. M. Freer et al., Phys. Rev. Lett. **96**, 042501 (2006).
6. R. Raabe et al., Phys. Rev. C **67**, 044602 (2003).
7. M. Freer, N. Orr and A. W. Ashwood, Private communication (2007).
8. M. Ito, K. Kato and K. Ikeda, Phys. Lett. B**588**, 43 (2004); *ibid.*, Mod. Phys. Lett. A **18**, 178 (2003).
9. Makoto Ito, Phys. Lett. B **636**, 293 (2006); *ibid*, Mod. Phys. Lett. A **21**, 2429 (2006).
10. M. Ito, N. Itagaki, H. Sakurai, and K. Ikeda, Phys. Rev. Lett. **100**, 182502 (2008); *ibid*, Jour. Phys. Con. Ser. **111**, 012010 (2008).
11. M. Ito and N. Itagaki, Phys. Rev. C **78**, 011602(R) (2008).
12. M. Ito and Y. Sakuragi, Phys. Rev. C **62**, 064310 (2000).
13. P. Descouvemont and D. Baye, Phys. Lett. B **505**, 71 (2001).
14. M. Kamimura, Prog. Theor. Phys. Suppl. **62**, 236 (1977).
15. M. Ito and K. Yabana, Prog. Theor. Phys. **113**, 1047 (2005).
16. H. Horiuchi et al., Suppl. Prog. Theor. Phys. **62**, 1 (1977).
17. M. Ito, K. Yabana, T. Nakatsukasa and M. Ueda, Phys. Lett. B **637**, 53 (2006).
18. S. Shimoura et al., Phys. Lett. B**654**, 87 (2007).
19. A. Saito and S. Shimoura, Private communication (2008).

NUCLEAR STRUCTURE AND ASTROPHYSICS r-PROCESS WITH COVARIANT DENSITY FUNCTIONAL THEORY

J. MENG[1,2,*], W. H. LONG[3], Z. M. NIU[2], B. SUN[1,4], and S. G. ZHOU[5]

[1]*School of Physics and Nuclear Energy Engineering, Beihang University, Beijing 100191, China*

[2]*State Key Laboratory of Nuclear Physics and Technology, School of Physics, Peking University, Beijing 100871, China*
[*]*E-mail: mengj@pku.edu.cn*

[3]*School of Nuclear Science and Technology, Lanzhou University, Lanzhou 730000, China*

[4]*Justus-Liebig-Universität Giessen, Heinrich-Buff-Ring 14, Giessen 35392, Germany*

[5]*Key Laboratory of Frontiers in Theoretical Physics, Institute of Theoretical Physics, Chinese Academy of Sciences, Beijing 100190, China*

The density functional theory (DFT) with a minimal number of parameters allows a very successful phenomenological description of ground state properties of nuclei all over the periodic table. Recent progresses on the application of the covariant density functional theory as well as its extensions by the group in Beijing for a series of interests and hot topics in nuclear astrophysics and nuclear structure are reviewed, including the rapid neutron-capture process, Th/U chronometer, halo and giant halo in density dependent relativistic Hartree-Fock-Bogoliubov, and neutron halo in deformed nuclei.

Keywords: rapid neutron-capture process; nuclear chronometer; halo phenomena.

1. Introduction

The density functional theory (DFT) with a minimal number of parameters have been successfully applied to the phenomenological description of ground state properties of nuclei all over the periodic table. The covariant version of DFT is based on a density functional which self-consistently takes into account Lorentz symmetry. During the past years, the covariant density functional theory has received wide attention due to its successful description of lots of nuclear phenomena.[1–4]

The covariant density functional theory can give more naturally the spin-orbit potential, the origin of the pseudo-spin symmetry[5,6] as a relativistic symmetry[7–9] and the spin symmetry in the anti-nucleon spectrum,[10] better reproduce the measurements of the isotopic shifts in the Pb region,[11] and be more reliable for nuclei far away from the line of β-stability, etc.. Furthermore the covariant density functional theory with density-dependent coupling constants can well reproduce the nuclear saturation properties in nuclear matter.[12,13] Obviously, covariant density functional theory is one of the best candidates for the description of exotic nuclei.

In this report, recent progresses on the application of the covariant density functional theory as well as its extensions by the group in Beijing for a series of interesting and hot topics in nuclear structure and nuclear astrophysics are reviewed, including

- the first application in the r-process nucleosynthesis study;
- the influence of nuclear masses and astrophysical conditions on thorium and uranium chronometer;
- halo and giant halo in density dependent relativistic Hartree-Fock-Bogoliubov (DDRHFB);
- neutron halo in deformed nuclei ^{44}Mg.

2. Results and Discussions

2.1. *Astrophysical r-process calculation based on the covariant density functional theory*

"How were the heavy elements from iron to uranium made?" is one of the 11 greatest unanswered questions of physics.[14] These nuclei were synthesized almost exclusively in the neutron-capture process: the slow process (s-process) and rapid process (r-process). Relative to the better known s-process, the r-process pathways are along extremely short-lived neutron-rich nuclei. Accurate nuclear physical properties for these nuclei, such as masses and half-lives, are essential for a basic understanding of the r-process nucleosynthesis in stars. So far, lots of efforts have been devoted to develop fast and extremely sensitive techniques with the radioactive beams over the world. One example that we were involved is the precise mass measurement of uranium fission fragments.[15] The masses of 35 neutron-rich nuclei in the element range from aluminum to barium have been directly determined with the novel large-scale Isochronous Mass Spectrometry at the FRS-ESR facility. These masses have further constrained the r-process simulation for light-element abundances.[16]

However, the majority of neutron-rich nuclei of relevance to the r-process are still out of the reach of present experimental capabilities, and therefore reliable predictions of theoretical predictions have to be used. It is of particular important to have a consistent, microscopic-rooted nuclear physics inputs. Recently, by using the state-dependent BCS method with a δ force, the first systematic study of the ground-state properties of over 7000 nuclei ranging from the proton drip line to the neutron drip line was performed in the relativistic mean field theory.[17] The corresponding rms deviation of neutron separation energy S_n with respect to the known experimental data is 0.7 MeV.

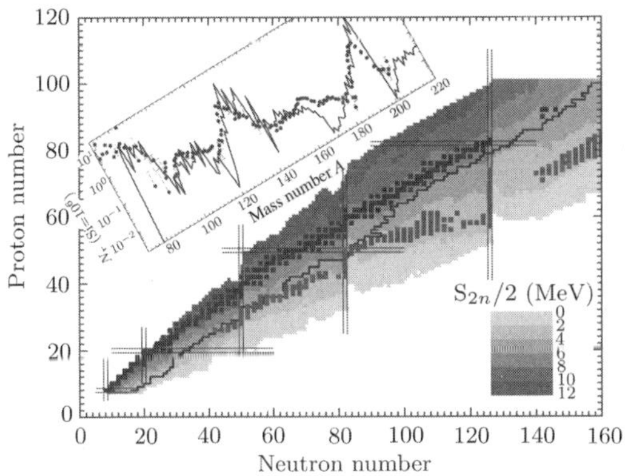

Fig. 1. Features of the r-process calculated using the new RMF mass table and the available experimental values. The solid line denotes the border of nuclei with known masses in the neutron-rich side. The dark grey squares show the r-process path. The observed and calculated solar r-process abundance curves are plotted versus the mass number A in the inset, whose x-axis is curved slightly to follow the r-process path. This figure is originally published in Ref. 19.

Initiated by the new development on nuclear masses, we preform the r-process calculations based on the classical r-process approach.[18,19] The best simulation using the relativistic mean field (RMF) mass table and the available experimental values[20] is presented in Fig. 1. The β-decay properties are taken from the finite-range droplet model plus quasiparticle random-phase approximation (FRDM+QRPA).[21] It is found that the solar r-process abundance distribution, in general, can be well reproduced. However, there are still large troughs before the second and third abundance peaks for the calculated r-process abundance distribution. These deficiencies in reproducing the solar r-process abundances can be mainly traced back to two aspects from the nuclear physics point of view: the shell effect and the location of nuclear shape transition. More detailed investigation on the nuclei with neutron numbers around magic numbers are demanded in microscopic nuclear physics models.

2.2. *The influence of nuclear masses and astrophysical conditions on thorium and uranium chronometer*

The age of the Universe is one of the most important physical quantities in cosmology. As the very metal-poor, i.e., $[\mathrm{Fe/H}] \equiv \log_{10}(\mathrm{Fe/H})_* - \log_{10}(\mathrm{Fe/H})_\odot < -2$, stars were formed at the early epoch of the Universe, their ages can set a lower limit on the age of the Universe. The ages of these stars can be determined by nuclear chronometers, which rely on the comparison of the present abundances of radioactive nuclei with the initial abundances at their productions. Recently, the Th and U lines were simultaneously detected in the metal-poor stars CS 31082-001,[22] BD +17°3248,[23] and HE 1523-0901,[24] so these allow us to estimate their ages from Th/U chronometers. A precise age determination via nuclear chronometers still requires reliable estimations of the initial r-process abundances of actinides, which can be derived from r-process calculations. However, the calculations rely on the astrophysical conditions of the r-process and the corresponding nuclear physics inputs, which are still not well understood.

In order to estimate the potential astrophysical uncertainties, we have performed three different fits to the scaled average element abundances of CS 31082-001 and CS 22892-052. In the first case, we fit abundances of elements in the range $38 \leqslant Z \leqslant 82$ (labeled as fit I). In addition, we also perform a fit to the abundances of elements in the range $56 \leqslant Z \leqslant 82$ (labeled as fit II), as well as a fit to the solar r-process isotopic abundances with mass number in the region $125 \leqslant A \leqslant 209$ (labeled as fit III). For nuclear physics inputs, available experimental data[15,20] are used, otherwise predictions of three newly developed mass models Hartree-Fock-Bogoliubov (HFB-17),[25] relativistic mean field (RMF),[26] and Koura-Tachibana-Uno-Yamada (KTUY)[27] or five commonly used mass models Duflo-Zuker DZ10[28] and DZ28,[29] extended Thomas-Fermi plus Strutinsky integral ETFSI-2[30] and ETFSI-Q,[31] finite-range droplet model (FRDM)[32] are employed. As for the β-decay rates, the predictions of the FRDM+QRPA method[33] are employed throughout the paper as a complementary to the experimental data.[20]

In Fig. 2, the calculated ages of the metal-poor stars HE 1523-0901 and

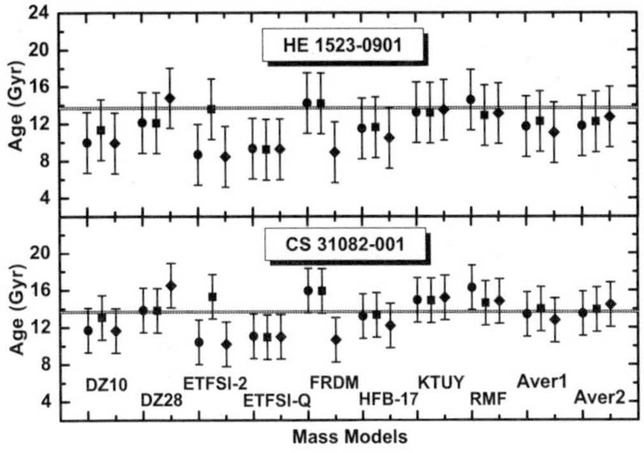

Fig. 2. Ages of the metal-poor stars HE 1523-0901 and CS 31082-001 determined using various mass models and fits. The circles (squares, diamonds) denote ages determined using the fit I (II, III). The average ages with all mass models and the three mass models DZ10, DZ28, KTUY are labeled as Aver1 and Aver2, respectively. The errors are attributed to uncertainties from observations of the corresponding metal-poor stars. The thick horizontal line corresponds to the age of the Universe derived from WMAP data.[35] This figure is originally published in Ref. 34.

CS 31082-001 are summarized. For a fixed fit, the deviation using different mass models is rather large. The corresponding difference can be as large as 5.9, 5.0, and 6.3 Gyr for fits I, II, and III, respectively. From the r-process calculations, it is found that the DZ10, DZ28, and KTUY simulations reproduce the stable element abundances better than the others. Therefore, these mass models might be more credible and are selected to estimate the age of the metal-poor star. To estimate the age of a star more reliably, one could adopt the average values for different simulations. The corresponding uncertainty is their root-mean-square deviation. In this way, the discrepancy between the average values of ages estimated using the three selected mass models for different fits decreases to 0.9 Gyr. For fit I, the age of the metal-poor star HE 1523-0901 is obtained as 11.8 ± 3.7 Gyr which agrees with the value of 13.2 Gyr in Ref. 24. The uncertainty of age determined here includes that of 1.6 Gyr from nuclear mass models and 3.3 Gyr from observation.[34] Similarly, ages of CS 31082-001 and BD $+17°3248$ are deduced as 13.5 ± 2.9 and 10.9 ± 2.9 Gyr, respectively.[34] These results can serve as the lower limit of the cosmic age and are consistent with results derived from the Wilkinson Microwave Anisotropy Probe (WMAP) data.[35]

2.3. *Halo and Giant halo in DDRHFB*

Halo phenomena in nuclei are driving forces for the development of the physics with radioactive ion beams. They are threshold effects[36] and have been first observed in the weakly bound system ^{11}Li.[37]

In this report, we take Cerium isotopes as the representative cases to study the halo phenomena within the DDRHFB theory.[38] For Cerium isotopes the proton number $Z = 58$ is tightly related to the pseudo-spin partner states $\pi 1\tilde{f}$ ($\pi 1g_{7/2}$ and $\pi 2d_{5/2}$). It is found that corresponding pseudo-spin symmetry (PSS) restoration is essential for the neutron shell effects around the neutron drip line, which further shows substantial effects on the stability of the halo structures.[39]

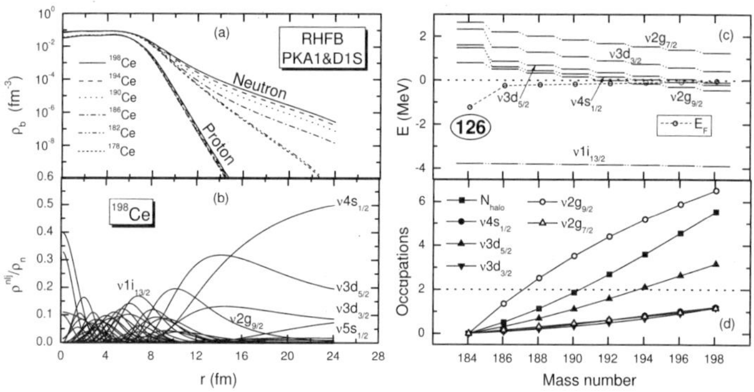

Fig. 3. (color online) (a) Neutron and proton densities, (b) the relative contributions of different orbits to the full neutron density in ^{198}Ce, (c) neutron canonical single-particle energies and Fermi energy E_F, and (d) occupations of neutron valence orbits beyond $N = 126$ shell for drip line Ce isotopes, where $N_{\rm halo}$ gives the total neutron number occupying in the halo orbits $4s_{1/2}$, $3d_{5/2}$, and $3d_{3/2}$. The results are calculated by RHFB with PKA1[40] plus the Gogny pairing force D1S.[42] The spherical box radius is adopted as $R_{\rm max} = 28$fm. This figure will publish in Ref. 39.

In Fig. 3 we show nuclear matter radial distributions (left panels) and neutron canonical single-particle configurations (right panels) for Cerium drip line isotopes. The neutron drip line calculated here with PKA1[40] is $N = 140$ whereas the calculations with PKO1[4] and DD-ME2[41] predict a shorter one as $N = 126$. From Fig. 3a, one can find distinct evidence of halo occurrence that the neutron densities become more and more diffuse after the isotopes ^{186}Ce ($N = 128$). In Fig. 3b it is shown that such extremely extensive matter distribution is mainly due to the low-l states, namely the halo orbits $\nu 4s_{1/2}$, $\nu 3d_{5/2}$ and $\nu 3d_{3/2}$. Additionally the occupations of the halo orbits N_{halo} in Fig. 3d indicate that the halos in ^{186}Ce, ^{188}Ce and ^{190}Ce are ordinary whereas ^{192}Ce, ^{194}Ce, ^{196}Ce and ^{198}Ce presumably have giant halos because more than two neutrons are occupying the halo orbits.

From Fig. 3c and Fig. 3d we find that the halo orbits ($\nu 4s_{1/2}$, $\nu 3d_{5/2}$ and $\nu 3d_{3/2}$) are located around the particle continuum threshold where they are gradually occupied. For the isotopes beyond ^{184}Ce ($N = 126$), the Fermi levels (in open circles) approach the continuum threshold rather closely such that the stability of these halo isotopes becomes sensitive to pairing effects. Nearby the low-l states, we find the high-l states $\nu 2g_{9/2}$ and $\nu 2g_{7/2}$. Because of the relatively large centrifugal barrier for g-orbits they do not contribute much to the diffuse neutron distributions. Nonetheless, the existence of the high-l states nearby halo orbits is still particulary significant, because it leads to a rather high level density around the Fermi surface, and evidently the pairing effects are enhanced to stabilize the halo isotopes.

RHFB calculations with the parameter set PKA1 (see Fig. 3) show clear evidence for the existence of halo structures in the Ce isotopes, while in the calculations of RHFB with PKO1[4] and RHB with DD-ME2,[41] the isotopic chain ends at $N = 126$, before the halo occurrence predicted by RHFB with PKA1. Further investigation indicates that the stability of neutron halo structures is tightly related to the PSS restoration on the proton side. The Fock terms, mainly the ρ-tensor couplings, present substantial contributions to the NN interaction in $T = 0$ channel, which accounts for the consistency between neutron halo structures and proton PSS restoration.[39]

2.4. *Neutron halo in deformed nuclei ^{44}Mg*

Considering that most open shell nuclei are deformed, the interplay between deformation and weak binding raises interesting questions, such as whether or not there exist halos in deformed nuclei and, if yes, what are their new features.

Halos in deformed nuclei were investigated in several mean field calculations in the past.[43–45] However, some other calculations[46,47] has raised doubt about the existence of deformed halos, even halos in deformed drip line nuclei. Therefore a model, which includes self-consistently the continuum, deformation effects, large spatial distributions, and the coupling among all these features, is essential to provide an adequate description of halos in deformed nuclei. The deformed relativistic Hartree Bogoliubov (DRHB) model using a spherical WS basis is especially suitable to study the halo phenomena in deformed exotic nuclei and it has shown that this model is fully equivalent to calculations in r space.[48]

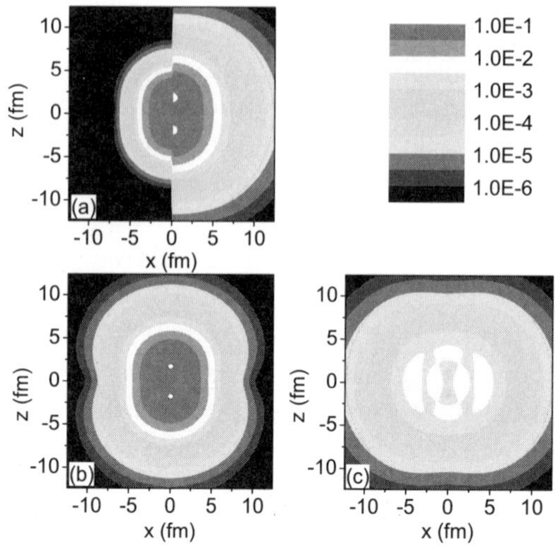

Fig. 4. Density distributions of ^{44}Mg with the z-axis as symmetry axis: (a) the proton density (for $x < 0$) and the neutron density (for $x > 0$), (b) the density of the neutron core, and (c) the density of the neutron halo. In each plot, a dotted circle is drawn for guiding the eyes. This figure is originally published in Ref. 49.

In magnesium isotopes, the last nucleus within the neutron drip-line is ^{46}Mg, an almost spherical nucleus. The neighboring nucleus ^{44}Mg is well deformed and it is therefore taken here as an example for a detailed investigation. The density distributions of all protons and all neutrons in ^{44}Mg are shown in Fig. 4a. Due to the large neutron excess, the neutron density not only extends much farther in space but it also shows a halo structure. The neutron density is decomposed into the contribution of the core in Fig. 4b and that of the halo in Fig. 4c. We find that the deformation of the halo is slightly oblate, although ^{44}Mg and its core are well deformed and prolate.[49] This indicates the decoupling between the deformations of core and halo. By studying the details of the neutron densities for core and halo, the single particle levels in the canonical basis, and the decomposition of the halo orbits in a spherical basis, it is found that the deformed neutron halo is formed by several orbits close to the threshold, either weakly bound or in the continuum.[49]

3. Summary

Recent progresses on the $r-$process, Th/U chronometer, halo and giant halo in DDRHFB, and neutron halo in deformed nuclei, are reviewed in this report. It is found that the solar r-process abundance distribution can be generally well reproduced by simulations using the RMF mass models. Together with the Th and U abundances from metal-poor stars HE 1523-0901, CS 31082-001, and BD +17°3248, their ages are determined as 11.8 ± 3.7, 13.5 ± 2.9, and 10.9 ± 2.9 Gyr, respectively. Moreover, by taking Cerium isotopes as the representative cases to study the halo phenomena within the DDRHFB theory, we find that corresponding PSS restoration is essential for the neutron shell effects around the neutron drip line, which further shows substantial effects on the stability of the halo structures. Furthermore, with recently developed DRHB model, we find that the deformation of the halo in ^{44}Mg is slightly oblate, although ^{44}Mg and its core are well deformed and prolate.

Acknowledgments

This work is partly supported by the Major State 973 Program 2007CB815000, NSFC (Nos. 10720003, 10775004, 10875157, 10947149, 10975008, and 10979066), and KIP of CAS (KJCX3-SYW-N02 and KJCX2-YW-N32).

References

1. J. D. Walecka, *Ann. Phys. (NY)* **83**, 491 (1974).
2. D. Vretenar, A. V. Afanasjev, G. A. Lalazissis, and P. Ring, *Phys. Rep.* **409**, 101 (2005).
3. J. Meng, H. Toki, S. G. Zhou, S. Q. Zhang, W. H. Long, and L. S. Geng, *Prog. Part. Nucl. Phys.* **57**, 470 (2006).
4. W. H. Long, N. V. Giai and J. Meng, *Phys. Lett.* **B 640**, 150 (2006).
5. A. Arima, M. Harvery, and K. Shimizu, *Phys. Lett. B* **30**, 517 (1969).
6. K. T. Hecht and A. Adler, *Nucl. Phys. A* **137**, 129 (1969).
7. J. N. Ginocchio, *Phys. Rev. Lett.* **78**, 436 (1997).
8. J. Meng, K. Sugawara-Tanabe, S. Yamaji, P. Ring, and A. Arima, *Phys. Rev. C* **58**, R628 (1998).
9. J. Meng, K. Sugawara-Tanabe, S. Yamaji, and A. Arima, *Phys. Rev. C* **59**, 154 (1999).
10. S. G. Zhou, J. Meng, and P. Ring, *Phys. Rev. Lett.* **91**, 262501 (2003).
11. M. M. Sharma, G. A. Lalazissis, and P. Ring, *Phys. Lett. B* **317**, 9 (1993).
12. R. Brockmann and R. Machleidt, *Phys. Rev. C* **42**, 1965 (1990).
13. R. Brockmann and H. Toki, *Phys. Rev. Lett.* **68**, 3408 (1992).
14. E. Haseltine, *Discover*, **23**(2), 37 (2002).
15. B. Sun *et al.*, *Nucl. Phys. A* **812**, 1 (2008).
16. B. Sun *et al.*, *Int. J. Mod. Phys. E*, **18**, 346(2008)
17. L. S. Geng, H. Toki, and J. Meng, *Prog. Theor. Phys.* **113**, 785 (2005).

18. B. Sun, F. Montes, L. S. Geng, H. Gessel, Yu. A. Litvinov, and J. Meng, *Phys. Rev. C* **78**, 025806 (2008).
19. B. Sun and J. Meng, *Chin. Phys. Lett.* **25**, 2429 (2008).
20. G. Audi, O. Bersillon, J. Blachot, and A. H. Wapstra, *Nucl. Phys. A* **729**, 3 (2003).
21. P. Möller, B. Pfeiffer, and K.-L. Kratz, *Phys. Rev. C* **67**, 055802 (2003).
22. R. Cayrel *et al.*, *Nature* **409**, 691 (2001).
23. J. J. Cowan *et al.*, *Astrophys. J.* **572**, 861 (2002).
24. A. Frebel, N. Christlieb, J. E. Norris, C. Thom, T. C. Beers, and J. Rhee, *Astrophys. J.* **660**, L117 (2007).
25. S. Goriely, N. Chamel, and J. M. Pearson, *Phys. Rev. Lett.* **102**, 152503 (2009).
26. L. S. Geng, H. Toki, and J. Meng, *Prog. Theor. Phys.* **113**, 785 (2005).
27. H. Koura, M. Uno, T. Tachibana, and M. Yamada, *Nucl. Phys. A* **674**, 47 (2000); H. Koura, T. Tachibana, M. Uno, and M. Yamada, *Prog. Theor. Phys.* **113**, 305 (2005).
28. J. Duflo and A. P. Zuker, `http://theory.gsi.de/~ksieja/ssp/DZ/du_zu_10.feb96`.
29. J. Duflo and A. P. Zuker, *Phys. Rev. C* **52**, R23 (1995).
30. S. Goriely, *AIP Conf. Proc.* **529**, 287 (2000).
31. J. M. Pearson, R. C. Nayak, and S. Goriely, *Phys. Lett. B* **387**, 455 (1996).
32. P. Möller, J. R. Nix, W. D. Myers, and W. J. Swiateck, *Atom. Data Nucl. Data Tables* **59**, 185 (1995).
33. P. Möller, B. Pfeiffer, and K.-L. Kratz, *Phys. Rev. C* **67**, 055802 (2003).
34. Z. Niu, B. Sun, and J. Meng *Phys. Rev. C* **80**, 065806 (2009).
35. E. Komatsu *et al.*, *Astrophys. J. Suppl. Ser.* **180**, 330 (2009).
36. A. S. Jensen *et al.*, *Rev. Mod. Phys.* **76**, 215 (2004).
37. I. Tanihata *et al.*, *Phys. Rev. Lett.* **55**, 2676 (1985).
38. W. H. Long, P. Ring, N. V. Giai, and J. Meng, *arXiv: 0812.1103v3 [nucl-th]*.
39. W. H. Long, P. Ring, J. Meng, N. V. Giai, and C. A. Bertulani, *arXiv: 1001.4976v1 [nucl-th]*.
40. W. H. Long, H. Sagawa, N. V. Giai, and J. Meng, *Phys. Rev. C* **76**, 034314 (2007).
41. G. A. Lalazissis, T. Nikšić, D. Vretenar, and P. Ring, *Phys. Rev. C* **71**, 024312 (2005).
42. J. F. Berger, M. Girod and D. Gogny, *Nucl. Phys. A* **428**, 23 (1984).
43. X. Li and P.-H. Heenen, *Phys. Rev. C* **54**, 1617 (1996).
44. J. Pei *et al.*, *Nucl. Phys. A* **765**, 29 (2006).
45. H. Nakada, *Nucl. Phys. A* **808**, 47 (2008).
46. I. Hamamoto, *Phys. Rev. C* **69**, 041306(R) (2004).
47. F. Nunes, *Nucl. Phys. A* **757**, 349 (2005).
48. S.-G. Zhou, J. Meng, and P. Ring, *Phys. Rev. C* **68**, 034323 (2003).
49. S.-G. Zhou, J. Meng, P. Ring, and E.-G. Zhao, *arXiv: 0909.1600v1 [nucl-th]*.

α-DECAY HALF-LIVES OF SUPERHEAVY NUCLEI AROUND THE $N = 152$ SHELL

DONGDONG NI[1,3] AND ZHONGZHOU REN[1,2,3,*]

[1]*Department of Physics, Nanjing University, Nanjing 210093, China*
[2]*Center of Theoretical Nuclear Physics, National Laboratory of Heavy-Ion Accelerator, Lanzhou 730000, China*
[3]*Kavli Institute for Theoretical Physics China, Beijing 100190, China*
* *E-mail: zren@nju.edu.cn*

The α-decay half-lives of superheavy nuclei around the $N = 152$ shell gap have been systematically investigated using theoretical Q values. These theoretical Q values are extracted from the local formula of binding energies for heavy and superheavy nuclei. Exact calculations are performed in two ways: the unified formula of half-lives for α decay and cluster radioactivity, and the generalized density-dependent cluster model. It is found that the available experimental α-decay half-lives are well reproduced in this two ways. Based on this fact, some useful predictions of α-decay half-lives are made for still unknown superheavy nuclei, which may guide future experiments.

Keywords: α decay; Superheavy nuclei; Half-lives; Theoretical Q values.

1. Introduction

One of hot topics in contemporary nuclear physics is the search of nuclei at the extremes. These extremes contain superheavy mass region, high isospins, large excitation energies, high spins, and so on. The very successful example is the synthesis of new superheavy nuclei in laboratories. Heavy elements up to 118, except 117, have been successfully synthesized, and the attempt to produce isotopes of the Z =117 and 120 has been made.[1–6] This not only extends our research field but also provides a stringent test of theoretical models. In view of the fact that the identification and knowledge of superheavy nuclei are mainly by the observation of α-decay chains, it is necessary and important to perform a detailed α-decay study of superheavy nuclei.[7–16]

Recently, besides the analytical formula of half-lives,[17] we have proposed the generalized density-dependent cluster model (GDDCM) for the evaluation of α-decay half-lives. Within this model, the calculated α-decay half-lives are in good agreement with the experimental data for a wide range of nuclei when we use the experimental Q values, and exotic α decays around the $N = 126$ magic shell have been investigated in detail.[18–20] As a further extension, the present study reports on a systematic calculation of α-decay half-lives for superheavy nuclei around the $N = 152$ shell, where theoretical Q values, obtained from a local formula of binding energies for heavy and superheavy nuclei,[21] are used. For comparison the analytical formula of half-lives is employed to evaluate α-decay half-lives as well.

2. Determination of theoretical Q values for α decay

As is well known, the Q value is a crucial quantity in evaluating α-decay half-lives, and various theoretical calculations always prefer the experimental Q values for a precise description of the half-life. In contrast, the theoretical Q values, Q_{th}, are used in the present calculations, which are extracted from the local formula of binding energies for heavy and superheavy nuclei in the following way,[21]

$$Q_{th}(Z, A) = B(Z-2, A-4) + B(^4He) - B(Z, A). \tag{1}$$

This binding-energy formula $B(Z, A)$ is obtained through detailed analysis of the available experimental data, and high precision is achieved in a straightforward manner.[21] For 117 nuclei with $Z \geq 90$, $N \geq 140$, the experimental binding energies are well reproduced with the standard deviation as low as 0.105 MeV. This gives us some guaranty for precise calculations and reliable predictions of α-decay half-lives with this local formula of binding energies.

3. Theoretical framework

α decay has traditionally been treated to proceed in two completely different steps. First an α cluster is formed at the nuclear surface and then penetrates through the potential barrier between the α cluster and the core nucleus. The decay constant is mainly determined by the latter step. In the previous studies, we have proposed both the analytical formula and the microscopic description for the interpretation of a rich amount of α-decay data. The analytical formula, based on the Wentzel-Kramers-Brillouin (WKB) approximation, is not only simple in form and easy to see the physical meanings but also gives a unified description of half-lives for α-decay and cluster radioactivity.[17] In the microscopic description, a new model of α decay is presented, that is, the GDDCM.[18–20] Within the GDDCM, the microscopic α-nucleus potential is numerically constructed in the well-established double folding model, and the exact solution of the Schrödinger equation with outgoing Coulomb wave boundary conditions is presented instead of the WKB barrier penetration probability. On the basis of the quantum theory, the decay width is ultimately computed from the overlap integral of the quasibound state wave function, the scattering state wave function, and the difference of potentials.

3.1. *Unified formula of half-lives for α decay and cluster radioactivity*

The unified formula of half-lives for α decay and cluster radioactivity is given by[17]

$$\log_{10} T_{1/2} = a\sqrt{\mu}\, Z_c Z_d\, Q^{-1/2} + b\sqrt{\mu}\, (Z_c Z_d)^{1/2} + c, \tag{2}$$

where the half-life $T_{1/2}$ is in seconds, the Q value is in MeV, Z_c and Z_d are respectively the atomic numbers of the cluster and the daughter nucleus, and μ is the reduced mass of the cluster-daughter system measured in unit of the nucleon mass, $\mu = A_c A_d/(A_c + A_d)$. Considering a single unpaired nucleon in odd-A nuclei and the proton-neutron coupling in odd-odd nuclei, there exist different hindrances for various kinds of α emitters. And it is pointed out that the hindrance, independent of the decay energy, has a direct effect on the cluster preformation probability.[22] Therefore, in the systematics of all kinds of α emitters, the parameters a and b remain the same for even-even, even-odd, odd-even, and odd-odd nuclei, while the parameter c has four different values.

Through a fitting procedure to the experimental data of 71 even-even, 68 even-odd, 52 odd-even, and 53 odd-odd nuclei (with $Z = 84 - 118$, $N = 128 - 176$), we find that the experimental data can be well reproduced by such a set of parameters:[17]

$$\begin{cases} a = \ 0.39961, \quad b = -1.31008; \\ c_{e-e} = -17.00698, \quad c_{e-o} = -16.26029, \\ c_{o-e} = -16.40484, \quad c_{o-o} = -15.85337. \end{cases}$$

The standard root-mean-square deviations are, respectively, 0.159, 0.692, 0.616, and 0.805.

3.2. *Generalized density-dependent cluster model for α decay*

In the GDDCM, the decaying system is regarded as a core nucleus plus a single α cluster moving in a microscopic potential. Such a potential is numerically constructed using the double-folding integral of the realistic nucleon-nucleon (NN) interaction with the density distributions of the α particle and the core nucleus, that is,[23–25]

$$V(\mathbf{r},\Omega_d) = \lambda \int d\mathbf{r_1}\ d\mathbf{r_2}\ \rho_1(\mathbf{r_1})\upsilon(\mathbf{s})\rho_2(\mathbf{r_2}), \tag{3}$$

where λ is the renormalized factor (λ=1 for the Coulomb potential), $\upsilon(\mathbf{s} = |\mathbf{r} + \mathbf{r_2} - \mathbf{r_1})$ is the effective NN interaction, and $\rho_1(r_1)$ and $\rho_2(r_2)$ are the density distributions of the α particle and the residual core nucleus, respectively. Based on the popular M3Y-Reid-type NN interaction and the standard Coulomb proton-proton interaction, this microscopic α-nucleus potential not only includes the zero-range potential for the single-nucleon exchange but also takes into full account the low density behavior in the surface region.[26,27] And thus it has been widely used to describe the potential barriers for fusion reactions, scattering, nuclear decays, and bound-state properties.

Next we numerically integrate the Schrödinger equation in the corresponding potential for the quasibound solution $u_{n\ell j}(r)$. The eigencharacteristic of the α-cluster quasibound state is described by the Wildermuth rule,[28] which relates the α cluster to the shell model and interprets the Pauli exclusion principle in a straightforward manner. In the case of a single α cluster, it is given by[28]

$$G \ = \ 2n + \ell \ = \ \sum_{i=1}^{4} g_i. \tag{4}$$

In this expression, n is the number of nodes in the radial wave function, ℓ is the angular momentum carried by the α particle, g_i are the oscillator quantum numbers of the nucleons composing the cluster, whose values are chosen to guarantee the α cluster entirely outside of the shell occupied by the core nucleus. For superheavy nuclei, following Refs. 7,10,12, the G number is taken as $G = 22$. The details on the evaluation of the double-folding potential and the quasibound wave function can be found in Refs. 18,19,23,24.

In conclusion, the decay-width formula has a clear expression as follows:[16,18,19,29–31]

$$\Gamma = \frac{4\mu}{\hbar^2 k}\left| \int_0^\infty F_\ell(r)[V_N(r) + V_C(r) - V_C^p(r)]u_{n\ell j}(r)dr \right|^2, \tag{5}$$

where the wave number k is determined from the decay Q value, $F_\ell(r)$ is the regular Coulomb wave function associated with the scattering wave function, $V_C^p(r)$ is the point-like Coulomb potential corresponding to the α particle far away from the daughter nucleus, and $u_{n\ell j}(r)$ is the radial quasibound state wave function of the α cluster.

As the last step toward α-decay half-lives, we need knowledge of the α-preformation factor P_α, which represents the degree of the overlap between the actual wave functions of the parent α emitter and of the residual daughter nucleus coupled to one α cluster. The microscopic calculation of Varga *et al* indicates that the probability of α clustering is about 0.3 for the nucleus ^{212}Po.[32] It is also known from available experimental cases that the preformation factor varies smoothly in the open shell region and has a value less than unity.[33] In the present study, as before, we take the same preformation factor for a certain kind of nuclei (even-even, odd-A, and odd-odd). This keeps the number of free parameters used in the model to a minimum. In this convenient way, α-decay half-lives can be fairly well reproduced when we use the experimental Q values. In combination with the previous systematic calculations, we take the preformation factors as follows: $P_\alpha = 0.56$ for even-even nuclei, $P_\alpha = 0.17$ for odd-A nuclei, and $P_\alpha = 0.10$ for odd-odd nuclei. These values remain the same as those in the previous investigation on heavy and superheavy nuclei with $N > 126$.[16] This means that we do not introduce any adjustable parameters into the present calculation.

4. Results and discussion

Using the formalism described above, we have performed a systematic study on α decays in superheavy nuclei ($Z = 101 - 109$) around the $N = 152$ shell gap. Since we have little knowledge of the level schemes in superheavy nuclei, for all these nuclei we evaluate their α-decay half-lives with the assumption of favored α transitions (i.e., $\ell = 0$ decays).

Based on the Q_{th} values obtained from Eq. (1),[21] calculations are, respectively, performed using the unified formula Eq. (2) and the GDDCM. In Figs. 1 and 2, the comparison of the available experimental α-decay half-lives with the theoretical ones is plotted versus the proton number of parent nuclei. One can see that the theoretical bars follow the experimental ones

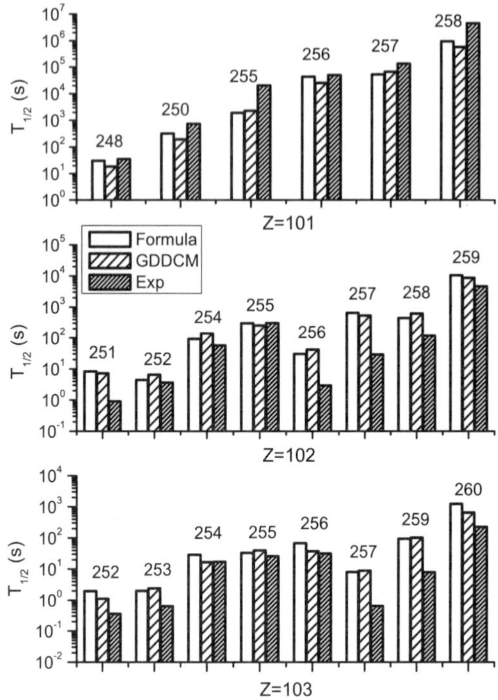

Fig. 1. Comparison of the available experimental α-decay half-lives with the theoretical ones for superheavy nuclei with $Z = 101 - 103$, where the mass number of parent nuclei is indicated on the top of each bar. The hollow bar stands for the results calculated using the unified formula Eq. (2), the bar with spares fill for the results calculated using the GDDCM, and the bar with dense fill denotes the experimental α-decay half-lives, taken from Ref. 34.

well, no matter it concerns the unified formula Eq. (2) or the GDDCM. Encouraged by this success, we have confidence in reliable predictions of α-decay half-lives for still unknown superheavy nuclei. Table 1 displays the detailed results calculated using the unified formula Eq. (2) and the GDDCM, together with the Q_{th} values involved in the calculations. These results may guide future experiments on superheavy nuclei. It will be very interesting to compare the present theoretical predictions with the future experimental measurements.

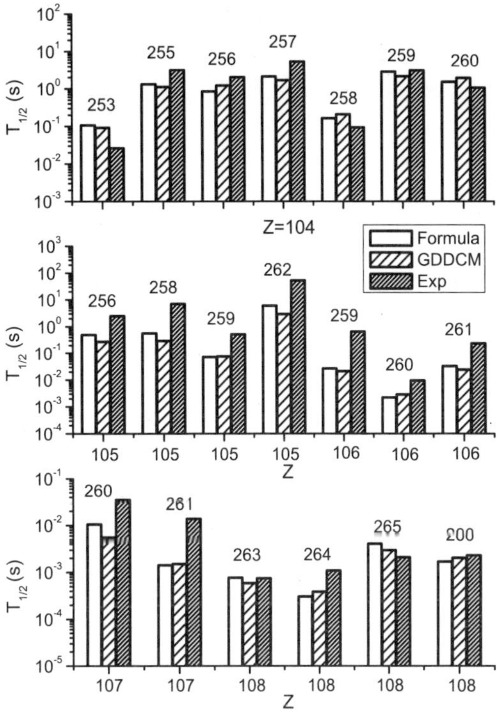

Fig. 2. Same as in Fig. 1 except for superheavy nuclei with $Z = 104 - 108$.

5. Summary

In summary, we have performed a systematic calculation on the α-decay half-lives of superheavy nuclei with $Z = 101 - 109$ around the $N = 152$ shell gap. The Q_{th} values, used in the present calculation, are extracted from the local formula of binding energies for heavy and superheavy nuclei. The α-decay half-lives are, respectively, evaluated using the unified formula of half-lives for α decay and cluster radioactivity, and the generalized density-dependent cluster model. The calculated α-decay half-lives are found to be in good agreement with the available experimental data. More importantly, we make some predictions of α-decay half-lives for still unknown superheavy nuclei, which may be useful for future experiments. In addition, we would like to point out that the local formula of binding energies is not satisfactory for heavier nuclei, because there is lack of information on experimental binding energies in heavier mass region. Therefore, at present we cannot give reliable predictions of half-lives for heavier α emitters. Nevertheless, we believe that this problem can be solved with the accumulation of experimental data in the future.

Acknowledgments

This work is supported by the National Natural Science Foundation of China (Grants 10535010, 10675090, 10775068, 10735010, and 10975072), by the 973 National Major State Basic Research and Development of China (Grant 2007CB815004), by CAS Knowledge Innovation Project No. KJCX2-SW-N02, and by the Research Fund of Doctoral Point (RFDP), No. 20070284016.

References

1. S. Hofmann and G. Münzenberg, *Rev. Mod. Phys.* **72**, 733 (2000).
2. T. N. Ginter *et al.*, *Phys. Rev. C* **67**, 064609 (2003).
3. A. Türler *et al.*, *Eur. Phys. J. A* **17**, 505 (2003).
4. Yu. Ts. Oganessian, *J. Phys. G* **34**, R165 (2007).
5. K. Morita *et al.*, *J. Phys. Soc. Jpn.* **76**, 045001 (2007).
6. Yu. Ts. Oganessian *et al.*, *Phys. Rev. C* **79**, 024603 (2009).
7. B. Buck, A. C. Merchant, and S. M. Perez, *Phys. Rev. C* **45**, 2247 (1992); B. Buck, A. C. Merchant, and S. M. Perez, *At. Data Nucl. Data Tables* **54**, 53 (1993).
8. D. S. Delion, S. Peltonen, and J. Suhonen, *Phys. Rev. C* **73**, 014315 (2006); S. Peltonen, D. S. Delion, and J. Suhonen, *ibid.* **78**, 034608 (2008).
9. V. Yu. Denisov and H. Ikezoe, *Phys. Rev. C* **72**, 064613 (2005); V. Yu. Denisov and H. Ikezoe, *ibid.* **80**, 034603 (2009).
10. P. Mohr, *Phys. Rev. C* **73**, 031301(R) (2006).
11. D. N. Poenaru *et al.*, *Phys. Rev. C* **74**, 014312 (2006).
12. J. C. Pei, F. R. Xu, Z. J. Lin, and E. G. Zhao, *Phys. Rev. C* **76**, 044326 (2007).
13. G. Royer, *J. Phys. G* **26**, 1149 (2000); G. Royer and H. F. Zhang, *Phys. Rev. C* **77**, 037602 (2008).
14. P. Roy. Chowdhury, C. Samanta, and D. N. Basu, *Phys. Rev. C* **77**, 044603 (2008); P. Roy. Chowdhury, C. Samanta, and D. N. Basu, *At. Data Nucl. Data Tables* **54**, 53 (1993).
15. Chang Xu, Zhongzhou Ren, and Yanqin Guo, *Phys. Rev. C* **78**, 044329 (2008).
16. Dongdong Ni and Zhongzhou Ren, *Nucl. Phys. A* **825**, 145 (2009).
17. Dongdong Ni, Zhongzhou Ren, Tiekuang Dong, and Chang Xu, *Phys. Rev. C* **78**, 044310 (2008).
18. Dongdong Ni and Zhongzhou Ren, *Nucl. Phys. A* **828**, 348 (2009).
19. Dongdong Ni and Zhongzhou Ren *Phys. Rev. C* **80**, 014314 (2009).

20. Dongdong Ni and Zhongzhou Ren *Phys. Rev. C* **80**, 051303(R) (2009).
21. Tiekuang Dong and Zhongzhou Ren *Phys. Rev. C* **77**, 064310 (2008).
22. D. Karlgren, R. J. Liotta, R. Wyss, M. Huyse, K. Van de Vel, and P. Van Duppen, *Phys. Rev. C* **73**, 064304 (2006).
23. M. J. Rhoades-Brown *et al.*, *Z. Phys. A* **310**, 287 (1983).
24. Chang Xu and Zhongzhou Ren, *Phys. Rev. C* **73**, 041301(R) (2006); *ibid.* **74**, 014304 (2006).
25. A. M. Kobos, B. A. Brown, P. E. Hodgson, G. R. Satchler, and A. Budzanowski, *Nucl. Phys. A* **384**, 65 (1982).
26. G. Bertsch, J. Borysowicz, H. Mcmanus, and W. G. Love, *Nucl. Phys. A* **284**, 399 (1977).
27. G. R. Satchler and W. G. Love, *Phys. Rep.* **55**, 183 (1979).
28. K. Wildermuth and Y. C. Tang, *A Unified Theory of the Nucleus*, (Academic Press, New York, 1997).
29. V. P. Bugrov, S. G. Kadmensky, V. I. Furman, and V. G. Khlebostroev, *Sov. J. Nucl. Phys.* **41**, 717 (1985).
30. S. Åberg, P. B. Semmes, and W. Nazarewicz, *Phys. Rev. C* **56**, 1762 (1997).
31. R. G. Lovas, R. J. Liotta, A. Insolia, K. Varga, and D. S. Delion, *Phys. Rep.* **294**, 265 (1998).
32. K. Varga, R. G. Lovas, and R. J. Liotta, *Phys. Rev. Lett.* **69**, 37 (1992).
33. P. E. Hodgson and E. Běták, *Phys. Rep.* **374**, 1 (2003).
34. G. Audi, O. Bersillon, J. Blachot, and A. H. Wapstra, *Nucl. Phys. A* **729**, 3 (2003).

Table 1. Calculated α-decay half-lives using the unified formula Eq. (2) and the GDDCM. The theoretical Q_{th} values involved are obtained from the local formula of binding energies for heavy and superheavy nuclei.[21]

Nucleus	Q (MeV)	$T_{1/2}^{Formula}$	$T_{1/2}^{GDDCM}$	Nucleus	Q (MeV)	$T_{1/2}^{Formula}$	$T_{1/2}^{GDDCM}$
^{245}Md	8.941	0.33 s	0.42 s	^{246}Md	8.785	3.41 s	2.08 s
^{247}Md	8.637	2.72 s	3.44 s	^{248}Md^{a}	8.478	30.5 s	18.4 s
^{249}Md	8.327	26.3 s	33.2 s	^{250}Md^{a}	8.166	5.34 min	3.24 min
^{251}Md	8.012	5.04 min	6.42 min	^{252}Md	7.850	66.6 min	41.0 min
^{253}Md	7.650	100 min	132 min	^{254}Md	7.716	3.40 h	2.03 h
^{255}Md^{a}	7.785	32.2 min	38.2 min	^{256}Md^{a}	7.569	12.0 h	7.0 h
^{257}Md^{a}	7.402	14.7 h	18.4 h	^{258}Md^{a}	7.226	10.9 d	6.7 d
^{259}Md	7.059	14.9 d	19.8 d	^{248}No	9.105	60.6 ms	91.4 ms
^{249}No	8.949	0.97 s	0.85 s	^{250}No	8.801	0.48 s	0.72 s
^{251}No^{a}	8.643	8.26 s	7.22 s	^{252}No^{a}	8.493	4.42 s	6.52 s
^{253}No	8.290	113 s	100 s	^{254}No^{a}	8.096	92.2 s	140 s
^{255}No^{a}	8.164	5.02 min	4.27 min	^{256}No^{a}	8.238	30.4 s	41.7 s
^{257}No^{a}	8.066	10.9 min	8.91 min	^{258}No^{a}	7.902	7.39 min	10.3 min
^{259}No^{a}	7.730	2.90 h	2.43 h	^{251}Lr	9.255	0.20 s	0.24 s
^{252}Lr^{a}	9.101	1.96 s	1.12 s	^{253}Lr^{a}	8.912	2.00 s	2.39 s
^{254}Lr^{a}	8.713	29.0 s	16.6 s	^{255}Lr^{a}	8.521	33.1 s	39.8 s
^{256}Lr^{a}	8.595	68.4 s	37.4 s	^{257}Lr^{a}	8.714	8.12 s	8.80 s
^{258}Lr	8.545	99.0 s	51.3 s	^{259}Lr^{a}	8.383	93.7 s	102 s
^{260}Lr^{a}	8.213	20.7 min	10.8 min	^{261}Lr	8.051	21.1 min	23.3 min
^{253}Rf^{a}	9.522	107 ms	91.2 ms	^{254}Rf	9.336	63.6 ms	91.3 ms
^{255}Rf^{a}	9.140	1.31 s	1.10 s	^{256}Rf^{a}	8.952	0.85 s	1.21 s
^{257}Rf^{a}	9.071	2.09 s	1.67 s	^{258}Rf^{a}	9.195	0.16 s	0.21 s
^{259}Rf^{a}	9.028	2.81 s	2.12 s	^{260}Rf^{a}	8.870	1.51 s	1.92 s
^{261}Rf	8.702	28.1 s	21.1 s	^{262}Rf	8.542	16.4 s	20.8 s
^{255}Db	9.743	41.2 ms	48.8 ms	^{256}Db^{a}	9.550	0.49 s	0.28 s
^{257}Db	9.407	0.35 s	0.40 s	^{258}Db^{a}	9.529	0.57 s	0.30 s
^{259}Db^{a}	9.655	71.6 ms	75.3 ms	^{260}Db	9.494	0.71 s	0.35 s
^{261}Db	9.337	0.55 s	0.57 s	^{262}Db^{a}	9.172	5.95 s	2.90 s
^{263}Db	9.015	4.91 s	4.98 s	^{258}Sg	9.867	10.2 ms	14.3 ms
^{259}Sg^{a}	9.991	26.9 ms	21.2 ms	^{260}Sg^{a}	10.121	2.22 ms	2.84 ms
^{261}Sg^{a}	9.961	32.3 ms	23.9 ms	^{262}Sg	9.810	14.6 ms	18.0 ms
^{263}Sg	9.648	0.22 s	0.16 s	^{264}Sg	9.493	0.11 s	0.13 s
^{258}Bh	10.446	9.94 ms	5.62 ms	^{259}Bh	10.309	6.13 ms	7.12 ms
^{260}Bh^{a}	10.437	10.5 ms	5.55 ms	^{261}Bh^{a}	10.568	1.41 ms	1.50 ms
^{262}Bh	10.412	12.1 ms	6.03 ms	^{263}Bh	10.263	8.04 ms	8.24 ms
^{264}Bh	10.106	72.4 ms	34.7 ms	^{265}Bh	9.953	51.3 ms	50.7 ms
^{262}Hs	11.021	60.5 μs	79.8 μs	^{263}Hs^{a}	10.867	0.77 ms	0.59 ms
^{264}Hs^{a}	10.721	0.31 ms	0.39 ms	^{265}Hs^{a}	10.566	4.06 ms	2.95 ms
^{266}Hs^{a}	10.418	1.69 ms	2.04 ms	^{264}Mt	11.306	0.39 ms	0.20 ms
^{265}Mt	11.162	0.23 ms	0.24 ms	^{266}Mt	11.010	1.83 ms	0.90 ms
^{267}Mt	10.864	1.13 ms	1.14 ms				

Note: ^{a} denotes the cases where the experimental α-decay half-lives are known.

ALPHA-PARTICLE CONDENSED STATES IN NUCLEI

YASURO FUNAKI[1], TAIICHI YAMADA[2], HISASHI HORIUCHI[3,4], GERD RÖPKE[5], PETER SCHUCK[6,7] AND AKIHIRO TOHSAKI[3]

[1]*Institute of Physics, University of Tsukuba, Tsukuba 305-8571, Japan*
[2]*Laboratory of Physics, Kanto Gakuin University, Yokohama 236-8501, Japan*
[3]*Research Center for Nuclear Physics (RCNP), Osaka University, Osaka 567-0047, Japan*
[4]*International Institute for Advanced Studies, Kizugawa 619-0225, Japan*
[5]*Institut für Physik, Universität Rostock, D-18051 Rostock, Germany*
[6]*Institut de Physique Nucléaire, CNRS, UMR 8608, Orsay, F-91406, France*
[7]*Université Paris-Sud, Orsay, F-91505, France*

Low density states near the 3α and 4α breakup threshold in ^{12}C and ^{16}O, respectively, are discussed in terms of the α-particle condensation. Calculations are performed in OCM (Orthogonality Condition Model) and THSR (Tohsaki-Horiuchi-Schuck-Röpke) approaches. The 0_2^+ state in ^{12}C and the 0_6^+ state in ^{16}O are shown to have dilute density structures and give strong enhancement of the occupation of the S-state c.o.m. orbital of the α-particles. The 0_6^+ state in ^{16}O has a large component of α+^{12}C(0_2^+) configuration, which is another reliable evidence of the state to be of 4α condensate nature. The possibility of the existence of α-particle condensed states in heavier $n\alpha$ nuclei is also discussed.

1. Introduction

It is well established that α-clustering plays a very important role for the structure of lighter nuclei.[1,2] The importance of α-cluster formation has also been discussed in infinite nuclear matter, where α-particle type condensation is expected at low density,[3] quite in analogy to the recently realised Bose-Einstein condensation of bosonic atoms in magneto-optical traps.[4] On the other hand, for trapped fermions, quartet condensation is also an emerging subject, discussed, so far, only theoretically.[5] In nuclei the bosonic constituents always are only very few in number, nevertheless possibly giving rise to clear condensation characteristics, as is well known from nuclear pairing.[6] Concerning α-particle condensation, the Hoyle state, i.e. the 0_2^+ state in ^{12}C has clearly been established. Several papers of the past[7–10] and also more recently[11–14] have by now established beyond any doubt that the Hoyle state, only having about one third of saturation density, can be described, to good approximation, as a product state of three α-particles, condensed, with their c.o.m. motion, into the lowest mean field $0S$-orbit.[15,16] This shall be the definition of a Bose-condensed state in finite nuclei, clearly reflecting the situation found in infinite matter in Ref. 3. Occasionally, we also shall call it a gas-like state. The establishment of this novel aspect of the Hoyle state naturally leads us to the speculation about 4α-particle condensation in ^{16}O.

In the present paper, two topics are addressed. A brief review of the status of work on the Hoyle state is given first. Secondly, very recent progress of work on the investigation of 4α-particle condensate state in ^{16}O is shown.

2. Description of the Hoyle state as the 3α condensate

The α clustering nature of the nucleus ^{12}C has been studied by many authors using various approaches.[2] Among these studies, solving the fully microscopic three-body problem of α clusters gives us the most important and reliable theoretical information of α clustering in ^{12}C. First solutions of the microscopic 3α problem where the antisymmetrization of nucleons is exactly treated, have been given by Uegaki et al.[9] and by Kamimura et al.[8] Their calculations reproduced reasonably well the observed binding energy and r.m.s. radius of the ground 0_1^+ state which is the state with normal density, while they both predicted a very large r.m.s. radius for the second 0_2^+ state which is larger than the r.m.s. radius of the ground 0_1^+ state by about 1 fm, i.e. by over 30%. The observed 0_2^+ state lies slightly above the 3α breakup threshold. The energies of the calculated 0_2^+ state reproduced reasonably well the observed value, together with the electron scattering form factors with respect to the 0_2^+ state.[8,9] The dilute character of the 0_2^+ state can be described by a gas-like structure of 3α-particles which interact weakly among one another, predominantly in relative S waves. This S-wave dominance in the 0_2^+ state had been already suggested by Horiuchi on the basis of the 3α OCM (Orthogonality Condition Model) calculation.[7]

Recently, based on the investigations of the possibility of α-particle condensation in low-density nuclear matter,[3] the last four authors of the present paper proposed a conjecture that near the $n\alpha$ threshold in self-conjugate $4n$ nuclei there exist excited states of dilute density which are composed of a weekly interacting gas of self-bound α particles and which can be considered as an $n\alpha$ condensed state.[11] This conjecture was backed by examining the structure of ^{12}C and ^{16}O using a new α-cluster wave function of the α-cluster condensate type, which has been denoted as THSR (Tohsaki-

Horiuchi-Schuck-Röpke) wave function, as follows:

$$\Phi_{n\alpha} \propto \mathcal{A}\left[\prod_{i=1}^{n} \exp\Big(-2\frac{(\boldsymbol{X}_i - \boldsymbol{X}_G)^2}{B^2}\Big)\phi(\alpha_i)\right]. \tag{1}$$

Here $\phi(\alpha_i)$, $\boldsymbol{X}_i$, $\boldsymbol{X}_G$, and B are the internal wave functions, c.o.m. coordinates, of the i-th alpha cluster, the total c.o.m. coordinate, and a variational parameter, respectively. It should be noted that (1) contains two limits exactly: the one of a shell-model relevant at higher densities and the one of an ideal α-particle condensate in the dilute limit.[11]

The detailed structure analyses of ^{12}C[12] showed that the 0_2^+ wave function of ^{12}C which was obtained in past by solving the full three-body problem of the microscopic 3α cluster model is almost completely equivalent to the wave function of the 3α THSR state. This result gives us strong support to our opinion that the 0_2^+ state of ^{12}C has a gas-like structure of 3α clusters with "Bose-condensation". The rms radius for this THSR state was calculated as $R(0_2^+)_{\text{THSR}} = 4.3$ fm which fits well with experimental data for the form factor of the Hoyle state, see Ref. 14,17. It confirms the assumption of low density as a prerequisite for the formation of an α-cluster structure for which the Bose-like enhancement of the occupation of the S orbit is possible.

A very interesting analysis of the applicability of the THSR wave function can be performed by comparing with stochastic variational calculations[15] and OCM calculations.[16] The α density matrix $\rho(\boldsymbol{r}, \boldsymbol{r}')$ defined by integrating out of the total density matrix all intrinsic α-particle coordinates, is diagonalized to study the single-α orbits and occupation probabilities in ^{12}C states. Fig. 1 shows the occupation probabilities of the L-orbits with S, D and G waves belonging to the k-th largest occupation number (denoted by L_k), for the ground and Hoyle state of ^{12}C obtained by diagonalizing the density matrix $\rho(\boldsymbol{r}, \boldsymbol{r}')$. We found that in the Hoyle state the α-particle S orbit with zero node is occupied to more than 70 % by the three α-particles (see also Ref. 15 and Fig. 1). Taking into account the finite size of the nucleus, a reduction of the condensate fraction from 100 % to about 70 % is not surprising, and the remaining fraction (about 30 %) is due to higher orbits originating from antisymmetrization among nucleons. This huge percentage means that an almost ideal α-particle condensate is realized in the Hoyle state. One should remember that superfluid ^{4}He has only 8 % of the particles in the condensate, what represents a macroscopic amount of particles nonetheless. On the other hand, in the ground state of ^{12}C, the α-particle occupations are equally shared among S_1, D_1 and

G_1 orbits, where they have two, one, and zero nodes, respectively, reflecting the SU(3)$(\lambda\mu)$ = (04) character of the ground state.[16] This fact thus invalidates a condensate picture for the ground state.

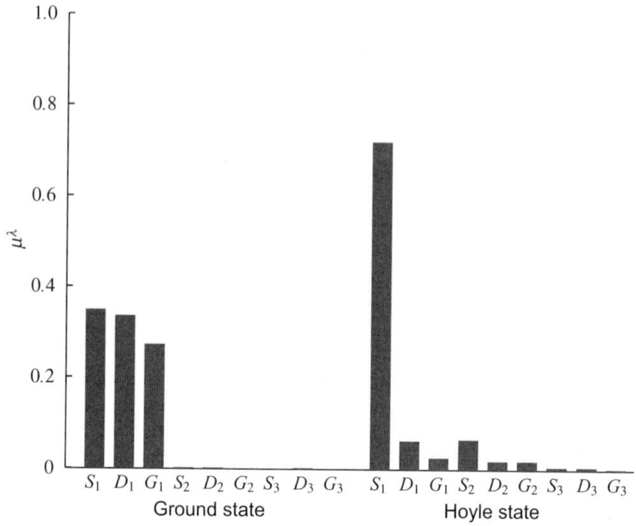

Fig. 1. (Color online) Occupation of the single-α orbitals of the ground state of ^{12}C compared with the Hoyle state.[16] For explanation see the text.

3. 4α condensation in ^{16}O

The situation in ^{16}O with respect to cluster states is already much more complicated than in ^{12}C. In ^{12}C, exciting one α-particle out of the ground state necessarily leads to a dilute α-gas state, because the remaining nucleus, ^{8}Be, is itself a loosely bound two α-object. On the other hand exciting an α-particle out of the ^{16}O ground state can lead to multiple ^{12}C-α configurations, be it only for the spectrum of 0^+ states. It is well documented[21] that the 0_2^+ state at 6.06 MeV is an α-particle orbiting in a $0S$-wave around the ground state of ^{12}C. However, there are many more possibilities. The α-particle can be in higher nodal S-wave, the α can orbit in a $0D$-wave around the 2_1^+ of ^{12}C and couple to a 0^+ state. It also can orbit in an odd parity wave around the 1^- and 3^- states in ^{12}C, etc., etc. The ^{12}C can, of course, also be in the Hoyle state which then leads us to the four α gas state in ^{16}O, which is the state of our interest. Experimentally, there are six 0^+ states in ^{16}O up to around the four α disintegration threshold at

14.4 MeV: the ground state 0_1^+, 0_2^+ at 6.06 MeV, 0_3^+ at 12.05 MeV, 0_4^+ at 13.6 MeV,[20] 0_5^+ at 14.01 MeV, and 0_6^+ at 15.1 MeV.

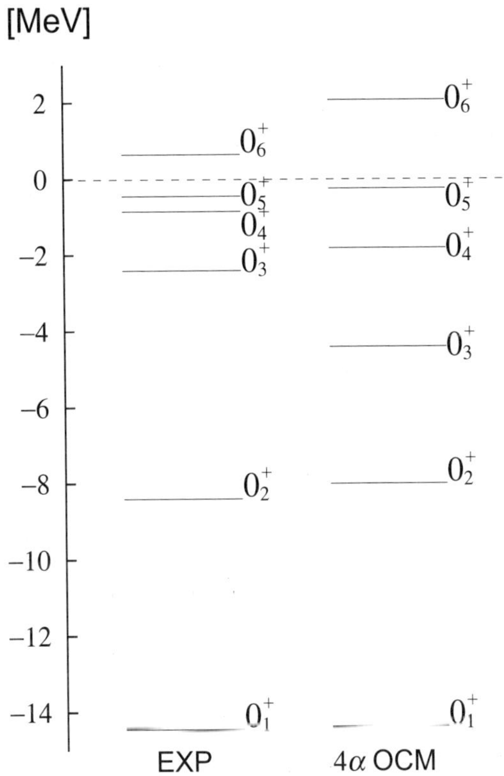

Fig. 2. Comparison of energy spectra between experiments[18] and 4α OCM calculation.[19] The 0_4^+ state in experiment is given in Ref. 20.

In analogy to the aforementioned OCM calculation for ^{12}C,[16] we recently performed a quite complete OCM calculation also for ^{16}O, including many of the cluster configurations.[19] We were able to reproduce the full spectrum of 0^+ states with 0_2^+ at 6.4 MeV, 0_3^+ at 9.4 MeV, 0_4^+ at 12.6 MeV, 0_5^+ at 14.1 MeV, and 0_6^+ at 16.5 MeV. As shown in Fig. 2, we tentatively make a one to one correspondence of those states with the six lowest 0^+ states of the experimental spectrum. In view of the complexity of the situation, the agreement can be considered as very satisfactory. Considering the properties of the various states in more detail, a certain number of rather big surprises came up. The analysis of the diagonalisation of the α-particle density matrix $\rho(\boldsymbol{r}, \boldsymbol{r}')$ (as was done in[15,16,22,23]) showed that the newly discovered 0^+ state at 13.6 MeV,[20] as well as the well known 0^+ state at 14.01 MeV, corresponding to our states at 12.6 MeV and 14.1 MeV, respectively, have, contrary to what we assumed previously,[24] very little condensate occupancy of the $0S$-orbit (about 20 %). On the other hand, the sixth 0^+ state at 16.5 MeV calculated energy, to be identified with the experimental state at 15.1 MeV, has 61 % of the α-particles being in the $0S$-orbit. The corresponding single-α $0S$ orbit is shown in Fig. 3. It has a strong spatially extended behaviour without any node ($0S$). This indicates that α particles are condensed into the very dilute $0S$ single-α orbit, see also Ref. 25. Thus, the 0_6^+ state clearly has 4α condensate character. We should note that the orbit is very similar to the single-α orbit of the Hoyle state.[15,16] We also show in Fig. 3 the single-α orbit for the ground state. It has maximum amplitude at around 3 fm and oscillations in the interior with two nodal ($2S$) behaviour, due to the Pauli principle and reflecting the shell-model configuration. The 0_6^+ state also has a very large radius of 5.6 fm, though this value may be somewhat over estimated because of the 10 % too high energy of the 0_6^+ state.

The condensate nature for the 0_6^+ state also can clearly be seen by the following analysis. We calculate an overlap amplitude, which is defined as follows:

$$\mathcal{Y}(r) = \Big\langle \Big[\frac{\delta(r' - r)}{r'^2} Y_L(\hat{\boldsymbol{r}}') \Phi_L(^{12}\mathrm{C}) \Big]_0 \Big| \Psi(0_6^+) \Big\rangle. \tag{2}$$

Here, $\Phi_L(^{12}\mathrm{C})$ is the wave function of ^{12}C, given by the 3α OCM calculation,[16] and r is the relative distance between the center-of-mass of ^{12}C and the α particle. From this quantity we can see how large is the component in a certain α+^{12}C channel which is contained in the 0_6^+ state. The amplitudes for the 0_6^+ state are shown in Fig. 4. It only has a large amplitude in the α+$^{12}\mathrm{C}(0_2^+)$ channel, whereas the amplitudes in other channels are much suppressed. The amplitude in the Hoyle-state channel has no oscillations and a long tail stretches out to $\sim$ 20 fm.

The reason why we previously tried to identify one of the two 0^+ states at 13.6 MeV or at 14.01 MeV with the analog of the Hoyle state, was that the calculation with our condensate wave function THSR[11] gave a third 0^+ state at 16.1 MeV as the highest 0^+ state of our calculation. A new analysis, however, shows that in Ref. 11 a fourth 0^+ state at 18.1 MeV with a very large radius of 6.0 fm was missed. We now identify this state

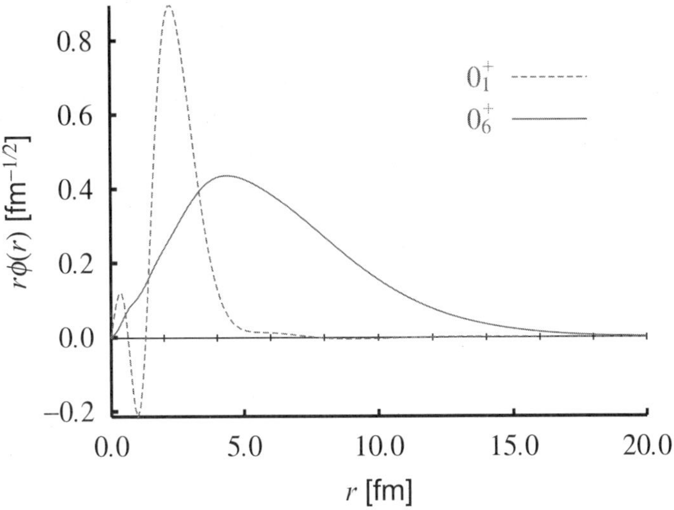

Fig. 3. (Colors online) Single-α orbits with $L = 0$ belonging to the largest occupation number, for the ground and 0_6^+ states. The radial part of the orbits are shown.

with the four α-particle condensate state, corresponding to the one at 16.5 MeV obtained with the OCM calculation. This was the second surprise. We also will publish details of this calculation in Ref. 26. Since our THSR wave function cannot describe ^{12}C-α configurations, the two intermediate states at 6.06 MeV and 13.6 MeV published already in Ref. 11 must be considered as trying to mock up such ^{12}C-α states in an insufficient, average way. Only the first 0^+ state and the newly discovered highest lying 0^+ state at 18.1 MeV, obtained with the THSR wave function have a clear physical interpretation as being the ground state and the analogue to the Hoyle state, respectively. We, thus, have produced two independent results, both confirming that the α-particle condensate state in ^{16}O lies, as in ^{12}C, above the α-particle disintegration threshold, giving us good confidence for our interpretation.

It, therefore, seems that the situation with respect to cluster states in ^{16}O has become more complex and at the same time more rich and exciting. We have the feeling to have opened a window for an understanding of the whole 0^+ spectrum up to around the α disintegration threshold, giving free the way for more exotic α cluster structures to be detected at even higher energies. It would now be of greatest importance that our results be

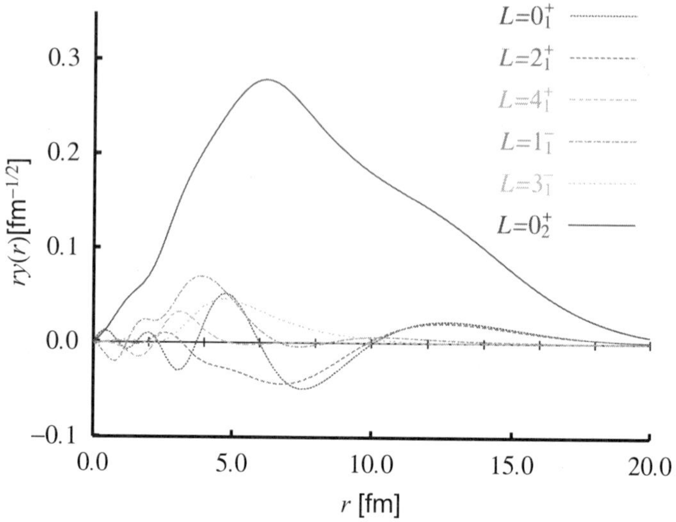

Fig. 4. (Colors online) $r\mathcal{Y}(r)$ for the 0_6^+ state is shown, where $\mathcal{Y}(r)$ is the α-particle overlap amplitude defined by Eq. (2). L denotes the orbital angular momentum of the remaining α particle coupling to ^{12}C.

confirmed by independent calculations. Also on the experimental side more information is very much needed. Of great interest would be, as in the past for ^{12}C, the measurement of inelastic form factors, as they give strong hints to the spatial extension of the various cluster states.

4. Conclusion and remarks

Multiple successful theoretical investigations, concerning the Hoyle state in ^{12}C, have established, beyond doubt, that it is a dilute gas-like state of three α-particles, held together only by the Coulomb barrier, and describable very well by a wave function of the THSR type, where the three bosons are condensed into the $0S$-orbital. There is no objective reason, why in ^{16}O,^{20}Ne,$\cdots$ there should not exist similar "Hoyle"-like states. At least the calculations with THSR and OCM approaches show this to be the case, systematically. In this work, we give preliminary results of a complete OCM calculation which reproduces the six first 0^+ states of ^{16}O to rather good accuracy . In view of the complexity of the situation, this is to be considered as quite an achievement. In that calculation the 0_6^+ state at 16.5 MeV, to be identified with the experimental 0^+ state at 15.1 MeV, shows the characteristics typical for the Hoyle state. This is also confirmed with an extended (with respect to Ref. 11) calculation of the THSR-type, where an additional very extended fourth 0^+ state has been put forward.

Further topics to be investigated in the future in the context of α-particle condensation are numerous. An interesting question is how many α's can maximally be in a self bound α-gas state. In this respect, a schematic investigation using an effective α-α interaction in an α-gas mean field calculation of the Gross-Pitaevskii type was performed.[27] Because of the increasing Coulomb repulsion, the Coulomb barrier fades away and our estimate yields a maximum of about eight α-particles that can be held together in a condensate. However, a few extra neutrons can have a strong additional binding effect (see ^{9}Be and ^{10}Be[28,29]) and may stabilize larger condensates. Another exciting possibility is to observe expanding α-particle condensate states. Imagine that one excites ^{40}Ca, via a heavy-ion collision, to about 60 MeV, i.e. to the total α disintegration threshold. The α condensate, being formed with a certain probability, will start expanding, since there no longer exists any Coulomb barrier to confine it. With multiparticle detectors such as INDRA or CHIMERA, all decaying α-particles could be detected in coincidence, and the coherent state could be identified by its very low energy in the c.o.m. system. This would then be analogous to an expanding atomic condensate after switching off the confining trap potential.[4] Experiments in this direction are being analysed at IPN-Orsay.[30] Another interesting idea concerning α-particle condensates was put forward by von Oertzen and collaborators.[31,32] α-particles outside a strongly bound core (e.g. ^{40}Ca) can form a condensate at the multi-α-particle threshold.[31] For the condensate with a fixed particle number, the emission of two α's and three α's must be enhanced. In fact the observation of the emission of ^{12}C in the state from the compound nucleus ^{52}Fe has been observed[33] and a very strong deviation from statistical model predictions is observed. Similar ideas have been advanced by Ogloblin,[34] who hypothesizes a three α-particle cluster state on top of ^{100}Sn, and earlier by Brenner et al.[35] who reports evidence of a gaseous α-particles in ^{28}Si and ^{32}S on top of an inert ^{16}O core. Also, very interesting recent experimental work on loosely bound α-structures in light nuclei has been performed by T. Kawabata et al.[36] and M. Freer et al.[37]

References

1. K. Wildermuth and Y. C. Tang, *A Unified Theory of the Nucleus* (Vieweg, Braunschweig, 1977).

2. K. Ikeda *et al.*, Prog. Theor. Phys. Suppl. No. 68, 1 (1980).
3. G. Röpke *et al.*, Phys. Rev. Lett. **80**, 3177 (1998).
4. F. Dalfovo *et al.*, Rev. Mod. Phys. **71**, 463 (1999).
5. A. S. Stepanenko *et al.*, arXiv: cond-mat/9901317; B. Doucot *et al.*, Phys. Rev. Lett. **88**, 227005 (2002); H. Kamei *et al.*, J. Phys. Soc. Jpn. **74**, 1911 (2005); S. Capponi *et al.*, Phys. Rev. A **77**, 013624 (2008).
6. P. Ring, and P. Schuck, *The Nuclear Many-Body Problem* (Springer-Verlag, Berlin, 1980).
7. H. Horiuchi, Prog. Theor. Phys. **51**, 1266 (1974); **53**, 447 (1975).
8. Y. Fukushima *et al.*, Suppl. of J. Phys. Soc. Japan, **44**, 225 (1978); M. Kamimura, Nucl. Phys. A **351**, 456 (1981).
9. E. Uegaki *et al.*, Prog. Theor. Phys. **57**, 1262 (1977); E. Uegaki *et al.*, Prog. Theor. Phys. **59**, 1031 (1978); **62**, 1621 (1979).
10. P. Descouvemont *et al.*, Phys. Rev. C **36**, 54 (1987).
11. A. Tohsaki *et al.*, Phys. Rev. Let. **87**, 192501 (2001).
12. Y. Funaki *et al.*, Phys. Rev. C **67**, 051306(R) (2003).
13. Y. Funaki *et al.*, Eur. Phys. J. A **24**, 321 (2005).
14. M. Chernykh *et al.*, Phys. Rev. Lett. **98**, 032501(2007).
15. H. Matsumura *et al.*, Nucl. Phys. A **739**, 238 (2004).
16. T. Yamada *et al.*, Euro. Phys. J. A **26**, 185 (2005).
17. Y. Funaki *et al.*, Eur. Phys. J. A **28**, 259 (2006).
18. F. Ajzenberg-Selove, Nucl. Phys. **A 460**, 1 (1986).
19. Y. Funaki *et al.*, Phys. Rev. Lett. **101**, 082502 (2008).
20. T. Wakasa *et al.*, Phys Lett B **653**, 173 (2007).
21. H. Horiuchi and K. Ikeda, Prog. Theor. Phys. **40**, 277 (1968); Y. Suzuki, Prog. Theor. Phys. **55**, 1751 (1976); **56**, 111 (1976).
22. Y. Suzuki *et al.*, Phys. Rev. C **65**, 064318, (2002).
23. T. Yamada *et al.*, Phys. Rev. A **78**, 035603 (2008).
24. Y. Funaki *et al.*, Nucl. Phys. News Lett., **17**(04), 11 (2007).
25. Y. Funaki *et al.*, Phys. Rev. C **77**, 064312 (2008).
26. Y. Funaki, *et al.*, in preparation.
27. T. Yamada, P. Schuck, Phys. Rev. C **69**, 024309 (2004).
28. S. Okabe and Y. Abe, Prog. Theor. Phys. **61**, 1049 (1979).
29. W. von Oertzen *et al.*, Phys Rep. **432**, 43 (2006).
30. B. Borderie, M. F. Rivet, private communication.
31. W. von Oertzen et al, Eur. Phys. J. A **29**, 133 (2006).
32. Tz. Kokalova *et al.*, Phys. Rev. Lett. **96**, 192502 (2006).
33. T. Kokalova *et al.*, Eur. Phys. J. A **23**, 19 (2005).
34. A. A. Ogloblin *et al*, Proceedings of the International Nuclear Physics Conference, Peterhof, Russia, June 28-July 2, 2005.
35. M. W. Brenner et al, Proceedings of the International Conference "Clustering Phenomena in Nuclear Physics", St. Petersburg, published in Physics of Atomic Nuclei (Yadernaya Fizika), 2000.
36. T. Kawabata et al., Phys. Lett. B **646**, 6 (2007).
37. M. Freer *et al.*, Phys. Rev. C **71**, 047305 (2005); **76**, 034320 (2007).

DENSE HADRONIC MATTER AND FINITE NUCLEI IN AN RMF MODEL WITH A CHIRAL POTENTIAL

K. TSUBAKIHARA*

Department of Physics, Hokkaido University,
Sapporo 060-0810, Japan
**E-mail: tsubaki@nucl.sci.hokudai.ac.jp*

A. OHNISHI

Yukawa Institute for Theoretical Physics, Kyoto University,
Kyoto 606-8502, Japan
E-mail: ohnishi@yukawa.kyoto-u.ac.jp

In constructing the dense matter equation of state (EOS), it is desired to respect both chiral symmetry and hypernuclear physics. In dense matter, strangeness is expected to play a decisive role and the partial restoration of chiral symmetry would modify the hadron properties. For chiral symmetry side, We have recently developed the chiral SU(2) symmetric RMF model with logarithmic sigma potential derived in the strong coupling limit (SCL) of the lattice QCD. In this SCL model, we can describe not only symmetric nuclear matter but also bulk properties of finite nuclei. We find that the EOS of symmetric matter is softened by the scalar meson with hidden strangeness, which couples with sigma meson through the determinant interaction. In this paper, we present this RMF model including chiral SU(3) potential and its results. We discuss the importance of hyperon in neutron star in this chiral SU(3) RMF model and show an effect to nuclear star maximum mass by introducing this potential. We also inquire the effect of meson-meson-baryon coupling which is corresponding to density dependent couplings.

Keywords: Neutron star, Hypernuclei

1. Introduction

The equation of state (EOS) of dense hadronic matter is one of the key ingredients in nuclear physics. In addition, astrophysical compact events, such as neutron stars and/or supernovae, are strongly affected by the properties of the dense matter EOS. It is, however, essentially difficult to observe the character of dense matter EOS directly. Thus, it is preferable to examine the properties of dense matter EOS through the theoretical models which can give a reasonable description about the experimental data of finite nuclei. In particular, hyperons are expected to appear at relatively small densities, $\sim (2-3)\rho_0$ and to make the hadronic star matter EOS softened. But YN and YY interactions can not be measured directly because of the short life times of hyperons for weak decay. From this point of view, it is also desired to investigate the property of hyperons through hypernuclei, hyperon bound states in nuclei.

Besides, the chiral symmetry is known as another important ingredient in dense matter. It is one of fundamental symmetries of QCD with massless quarks, and its spontaneous symmetry breaking generates constituent quark and thus hadron masses.[1] Therefore, it is preferable for hadronic many-body theories to include the chiral symmetry by considering its spontaneous breaking and partial restoration at finite baryon densities. We regard a relativistic mean field (RMF) model [2–7] as a candidate which satisfy such a theoretical need.

Recently, we have constructed the chiral SU(2) RMF model, the SCL2 RMF model,[6] with a logarithmic σ potential and we have applied its extension, the SCL3 RMF model[7] to the hypernuclear systems. They have succeeded in reproducing not only the properties of "normal" nuclear systems, for example, the saturation property of the symmetric nuclear matter and the BEs and charge rms radii of finite nuclei, but also the ones of Λ and Σ hypernuclear states. However the SCL3 RMF model failed to explain the observed maximum mass of neutron star if we adopt the parameter sets based on normal nuclear and hypernuclear experimental data. In this work, we examine the extension of the SCL logarithmic chiral potential where the finite temperature effect is taken into account. In addition, we also study $\bar{\Psi}_B \sigma\omega \Psi_B$ type NLO coupling regarded as density dependent type coupling since naively the σ meson field is proportional to the baryon density ρ_B.

This paper is organized as follows. In the Sec. 2, we present brief chiral RMF model description including a new chiral potential derived from the finite temperature calculation in the strong coupling limit of lattice QCD. We also introduce the density dependent type coupling for the meson-baryon coupling part of RMF Lagrangian. Next, we reveal the present results of these extended SCL3 RMF model abbreviated as SCL3ΛSH and SCL3DDC in Sec. 3. Finally, we summarize in Sec. 4.

2. Model description

In this section, we introduce the formalization of the extended SCL3 RMF model.

Lagrangian to be solved in an RMF model can be written as

$$\begin{aligned}
\mathcal{L} = & \sum_i \bar{\psi}_i \left[i\partial\!\!\!/ - M_i^* - \gamma_\mu U_i^\mu \right] \psi_i \\
& + \frac{1}{2}\partial_\mu \varphi_\sigma \partial^\mu \varphi_\sigma - \frac{1}{2} m_\sigma^2 \varphi_\sigma^2 + \frac{1}{2}\partial_\mu \varphi_\zeta \partial^\mu \varphi_\zeta - \frac{1}{2} m_\zeta^2 \varphi_\zeta^2 \\
& - \frac{1}{4}\omega_{\mu\nu}\omega^{\mu\nu} + \frac{m_\omega^2}{2}\omega_\mu \omega^\mu - \frac{1}{4} R_{\mu\nu} R^{\mu\nu} + \frac{m_\rho^2}{2} R_\mu R^\mu \\
& - \frac{1}{4}\phi_{\mu\nu}\phi^{\mu\nu} + \frac{m_\phi^2}{2}\phi_\mu \phi^\mu - \frac{1}{4} F_{\mu\nu} F^{\mu\nu} \\
& + \frac{c_\omega}{4}\left(\omega_\nu \omega^\nu\right)^2 - V_{\sigma\zeta}\left(\varphi_\sigma, \varphi_\zeta\right) ,
\end{aligned} \tag{1}$$

where the couplings between scalar mesons and baryons correspond to attractive parts of baryon interaction contrary to the ones between vector mesons and baryons corresponding to repulsive parts. For meson fields, we include effective potential of mesons as shown in Eq. (1) with $V_{\sigma\zeta}$ and ω^4 potential as same as Ref. 4 which succeeded in reproducing normal nuclear property. Then, it should be investigated what form the meson potential should have. In previous work,[6,7] we regard chiral symmetry between σ and π as a constraint and thus, we apply the chiral potential derived from the strong coupling limit of lattice QCD(SCL-LQCD). In those works, the $SU_f(2)$ and SU_3 chiral potentials are examined about normal nuclear and hypernuclear properties.

The $SU_f(2)$ and $SU_f(3)$ chiral potential has a logarithmic form derived from analytic calculation on SCL-LQCD and written as

$$\begin{aligned}
V_\chi = & -\frac{a_\sigma}{2} \log\left(\det MM^\dagger\right) + \frac{b_\sigma}{2}\,\mathrm{tr}(MM^\dagger) - c_\sigma \sigma \\
= & -a_\sigma \log(\sigma^2 + \boldsymbol{\pi}^2) + \frac{b_\sigma}{2}\left(\sigma^2 + \boldsymbol{\pi}^2\right) - c_\sigma \sigma \\
\simeq & -2a_\sigma\, f_{\mathrm{SCL}}\left(\frac{\varphi_\sigma}{f_\pi}\right) + \frac{1}{2} m_\sigma^2 \varphi_\sigma^2 + \frac{1}{2} m_\pi^2 \boldsymbol{\pi}^2 , \\
& f_{\mathrm{SCL}}(x) = \log(1-x) + x + \frac{x^2}{2} \quad a_\sigma = \frac{f_\pi^2\left(m_\sigma^2 - m_\pi^2\right)}{4} ,
\end{aligned} \tag{2}$$

where the explicit chiral symmetry breaking term $-c_\sigma \sigma$ is introduced. SCL3 RMF model which include $SU_f(3)$ logarithmic chiral potential, however, has failed in explaining the maximum mass of neutron star if we apply the parameter sets determined hypernuclear data such as S_Λ and/or $\Delta B_{\Lambda\Lambda}$. In Fig. 5 of Ref. 7, we show the scalar and vector potential of nucleon in symmetric nuclear matter as a function of ρ_B. Both scalar and vector potentials seem to be in good agreement with RBHF results[8] around ρ_0. Vector potential shows certain deviation from the RBHF result on the contrary to that scalar potential in the wide range of ρ_B. The strength of introduced ω^4 effective potential c_ω in SCL3 model($c_\omega \sim 300$) is 2.5 times larger than the one introduced in TM model and this potential is regarded as a suppressor of vector potential. Thus, we need some mechanism which stabilize nuclear matter and do not need too strong c_ω. In this work, we examine arcsinh type chiral potential and density dependent type coupling between meson and baryon.

First, we introduce arcsinh type chiral potential extended from the logarithmic chiral potential. If we calculate Lattice action in SCL-LQCD on finite temperature and then we take zero temperature limit, logarithmic chiral potential should be corrected as

$$
\begin{aligned}
&-\frac{a}{2}\log(y) + by^2 + c_\sigma y \quad y = \sqrt{\frac{\sigma^2}{+}\pi^2 f_\pi^2} \\
\rightarrow &-\frac{a'}{2}\mathrm{arcsinh}(\alpha y) + b'y^2 + c_\sigma y \\
= &-a'\left\{\mathrm{arcsinh}\left(\alpha\frac{f_\pi+\varphi}{f_\pi}\right) - \beta\frac{\varphi}{f_\pi} + \frac{\beta^3}{2}\left(\frac{\varphi}{f_\pi}\right)^2\right\} + \frac{m_\sigma^2}{2}\varphi + \frac{m_\pi^2}{2}\pi^2 \qquad (3)
\end{aligned}
$$

where $\beta = \alpha/\sqrt{1+\alpha^2}$. In the case of the arcsinh chiral potential, we have an additional parameter α compared to the logarithmic one as shown in Eq. (3). We adjust it so as to reproduce TM1 non-linear σ potential since TM1 does not need so much c_ω to reproduce normal nuclear property. This result is shown in Fig. 1. Logarithmic chiral potential has an infinite barrier at $\sigma = 0$ and thus, full chiral restoration does not occur with this potential in all range of temperature. On the other hand, in arcsinh chiral potential has only finite (negative) derivative but no divergent behavior at $\sigma = 0$. Therefore, in finite temperature, chiral symmetry may restore and this point may be important when we calculate the equation of state in finite temperature for examining supernovae process.

Second, we study density NLO meson-bayron coupling written as

$$
\mathcal{L}_{\mathrm{DDC}} = \bar{\Psi}_i g_{\sigma\omega i}\frac{\varphi_\sigma}{f_\pi}\not{\omega}\Psi_i \qquad (4)
$$

where this coupling regarded as density dependent type coupling (DDC) since σ field is naively proportional to scalar density ρ_s, $\varphi_\sigma \sim \sum_i \frac{g_{\sigma i}}{m_\sigma^2}\rho_{si}$ in RMF models. As seen in Fig. 5 of Ref. 7, σ field seems to saturate in high density phase. Thus, this coupling should be more effective around normal density than in high density. From this point, this coupling can be regarded

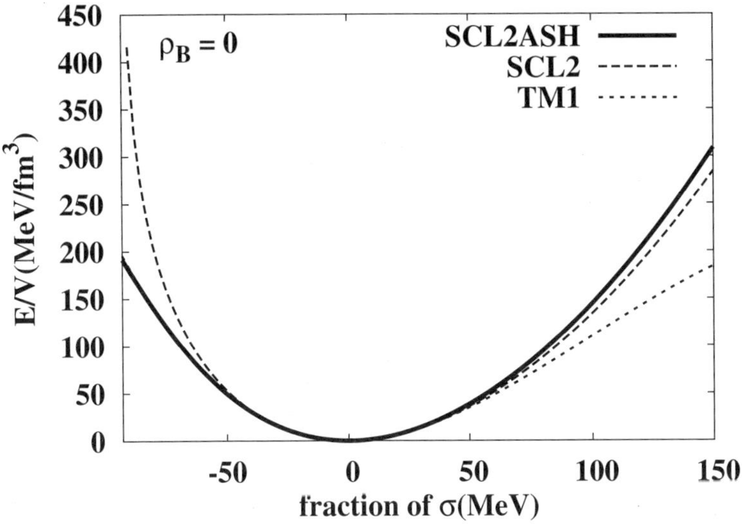

Fig. 1. Chiral potential of SCL3 and SCL2ASH RMF model. We also present effective non-linear potential of TM1.[4]

as a substitute of ω^4 potential around ρ_0 but should not suppress vector potential compared to ω^4 potential case.

There are several model parameters in two RMF models, SCLfASH RMF model and SCLfDDC RMF model (f:the number of flavors) which include arcsinh chiral potential and/or NLO density dependent type coupling, respectively. First, the parameters of chiral potential are constrained by the masses of mesons. As we mentioned, additional parameter α is determined by reproducing TM1 effective σ potential in the case of SCLfASH RMF model. The remained parameter m_σ and the strength of ω^4 potential, c_σ, and the coupling constants between mesons and baryons are determined so as to explaining the saturation property of symmetric nuclear matter, BE and charge rms radii of normal stable nuclei, S_Λ of single Λ hypernuclei, and $\Delta B_\Lambda\Lambda$ of ${}^{6}_{\Lambda\Lambda}$He. For the coupling constants between vector mesons and hyperons, we adopt $SU_f(3)$ relation briefly introduced in Ref. 7. In SCLfDDC case, DDC parameters should be determined so as to reduce c_ω compared to previous SCL3 RMF model.

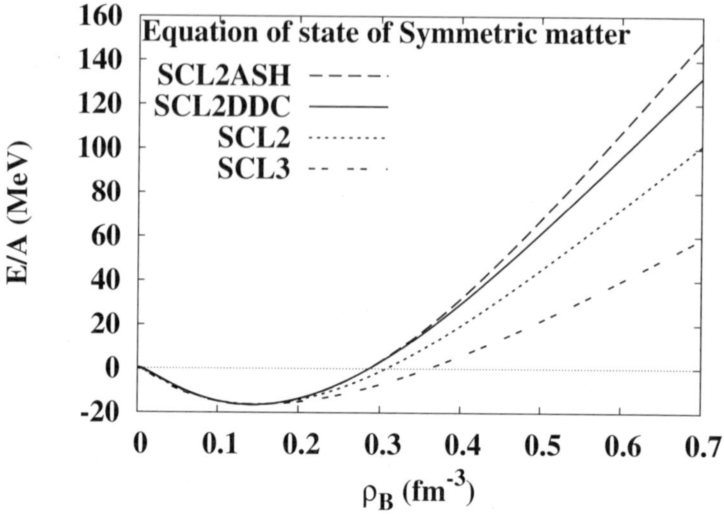

Fig. 2. Symmetric nuclear matter EOS of SCL2ASH and SCL2DDC RMF model.

3. Numerical results

In this section, we present brief results of the SCL3ASH RMF model and the SCL2DDC RMF model, respectively.

First, we show the symmetric nuclear matter in SCL2ASH RMF model and SCL2DDC RMF model with NL, TM, SCL2, and SCL3 RMF model. The parameters are determined by reproducing the saturation property and *BE* and charge rms radii of normal stable nuclei. Both calculated symmetric nuclear matters of SCL2ASH and SCL2DDC succeed in explaining the saturation property of symmetric nuclear matter as shown in Fig. 2. Their incompressibilities are also in the empirically acceptable range.

In addition, both two models succeed in reducing c_ω values in each their model, which are equal to 71.3 (66.5) in SCL2ASH (SCL2DDC) RMF model, respectively. In Fig. 3 and 4, we show calculated neutron star matter EOS and neutron star mass. Here, the results of SCL2ASH RMF model are almost similar to the ones of SCL2DDC RMF model. Therefore, we omit to plotting the results of SCL2ASH in these figures. As easily seen, calculated NS maximum masses of both two model show a enhancement compared to SCL2 RMF model result. From this result, these two extension is important for explaining the observed property of neutron star matter EOS.

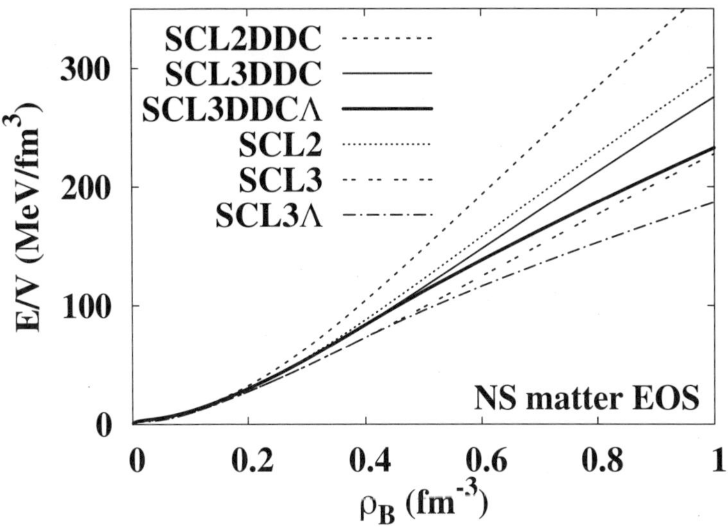

Fig. 3. Calculated neutron star matter EOS and neutron star mass.

Second, we introduce hyperon degree of freedom. Unfortunately, it is not well established to extend $SU_f(2)$ arcsinh chiral potential to $SU_f(3)$ one. Therefore, we concentrate on adopting SCL3DDC RMF model for hypernuclear physics. Calculated S_Λ in SCL3DDC RMF model are presented in Fig. 5. DDC parameter of Λ are determined by naive quark counting, $2g_{\sigma\omega N}/3$. Combining S_Λ and $\Delta B_{\Lambda\Lambda}$, we can fix two parameter for Λ, $g_{\sigma\Lambda}$ and $g_{\zeta\Lambda}$.

With these parameters, we can calculate neutron star matter EOS based on hypernuclear data. Back to Figs. 3 and 4, we show the neutron star matter EOS and neutron star mass on SCL3DDC RMF model where Λ hyperon is taken into account. Calculated maximum mass of neutron star is equal to 1.65 $M_\odot$ in SCL3DDC RMF model even if we include Λ hyperon effects (SCL3DDCΛ). From these results, hyperons make EOS softened certainly but calculated maximum mass of neutron star exceeds observed one.

4. Summary and discussion

In this paper, we examine two RMF models, SCLfASH RMF model and SCLfDDC RMF model which include arcsinh chiral potential and/or NLO

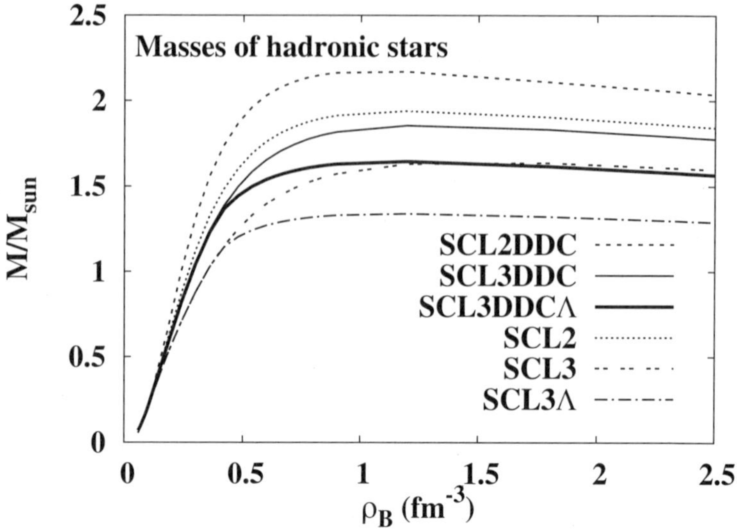

Fig. 4. Calculated neutron star matter EOS and neutron star mass.

density dependent type coupling, respectively. They are extended from SCL3 RMF model so as to explain observed maximum mass of neutron star with a chiral potential derived from SCL-LQCD.

In SCLfASH RMF model, the arcsinh chiral potential derived from finite temperature calculation is adopted in place of logarithmic chiral potential of SCL3 RMF model. Besides, we examine NLO density dependent type coupling between mesons and baryons in SCLfDDC RMF model. With these two RMF models, we can explain the saturation point of symmetric nuclear matter and *BE* of normal nuclei. Unfortunately, it is not well established to extend SCL-LQCD to $SU_f(3)$ case in finite temperature. Thus, we study hypernuclear systems with only SCL3DDC RMF model so far. Combining the $SU_f(3)$ symmetric relation for the coupling between vector mesons and baryons, the parameters of Λ hyperon, $g_{\sigma\Lambda}$ and $g_{\zeta\Lambda}$ are determined by reproducing S_Λ of single Λ hypernuclei and $\Delta B_{\Lambda\Lambda}$ of ${}^{6}_{\Lambda\Lambda}$He. With these parameters, SCL3DDC RMF model succeeds in explaining the maximum mass of neutron star even if it includes Λ hyperon. Thus, these results indicate that the chiral RMF model derived from SCL-LQCD can reproduce not only finite nuclear properties including hypernuclear data but also explain the observed neutron star maximum mass. We should pro-

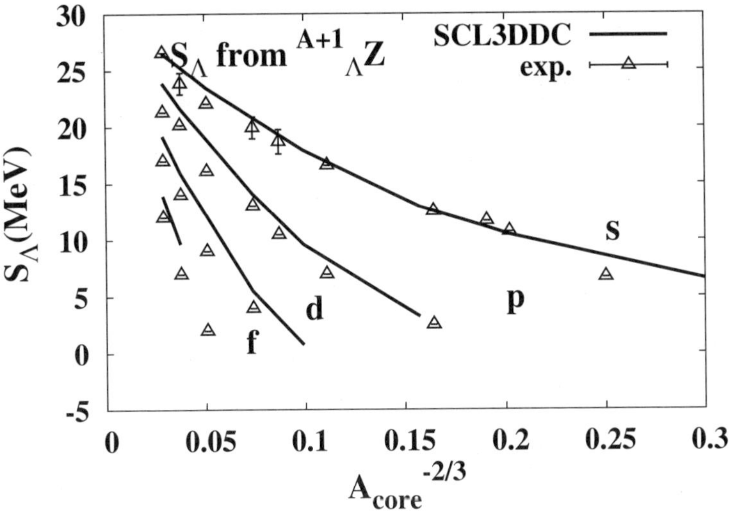

Fig. 5. Separation energy of Λ from single Λ hypernuclei.

ceed to further investigation about Σ and Ξ hyperons in order to discuss neutron star matter more properly.

Acknowledgments

This work was supported in part by KAKENHI from MEXT and JSPS under the grant numbers, 17070002, 19540252 and 20·4326, Global COE Program "The Next Generation of Physics, Spun from Universality and Emergence", and the Yukawa International Program for Quark-hadron Sciences (YIPQS).

References

1. Y. Nambu and G. Jona-Lasinio, Phys. Rev. **122**, 345 (1961); Phys. Rev. **124**, 246 (1961).
2. P. -G. Reinhard, M. Rufa, J. Maruhn, W. Greiner and J. Friedrich, Z. Phys. A **323**, 13 (1986); Suk-Joon Lee, J. Fink, A. B. Valantekin, M. R. Strayer, A. S. Umar, P. G. Reinhard, J. A. Maruhn and W. Greiner, Phys. Rev. Lett. **57**, 2916 (1986).
3. G.A. Lalazissis, J. König and P. Ring, Phys. Rev. C **55**, 540 (1997).
4. Y. Sugahara and H. Toki, Nucl. Phys. **A579**, 557 (1994).
5. B. Y. Sun, H. Toki and J. Meng, Phys. Lett. B **683**, 134 (2010).

6. K. Tsubakihara and A. Ohnishi, Prog. Theor. Phys. **117**, 903 (2007).
7. K. Tsubakihara, H. Maekawa, H. Matsumiya and A. Ohnishi, arXiv:0909.5058 [nucl-th].
8. R. Brockmann and R. Machleidt, Phys. Rev. C **42**, 1965 (1990); R. Brockmann and H. Toki, Phys. Rev. Lett. **68**, 3408 (1992).

THREE α CLUSTER STRUCTURE IN THE EXCITED STATES OF ^{13}C

N. FURUTACHI

Meme Media Laboratory, Hokkaido University 060-8628, Sapporo, Japan

M. KIMURA

Creative Research Initiative "Sousei", Hokkaido University, Sapporo 001-0021, Japan

Cluster structures in the excited states of ^{13}C have been investigated by using a microscopic 3α+n model. We have found two excited rotational bands that have developed α cluster structures. Unlike the dilute cluster-gas structure of 0_2^+ state in ^{12}C, three α clusters in the lower rotational band has bended linear-chain like configuration.

Keywords: Cluster structure.

1. Introduction

^{12}C is well known to have the developed three α cluster structures. 0_2^+ state of ^{12}C has dilute cluster-gas like structure, which is proposed to be α-condensed state.[1,2] Another 0^+ state that has dominant reduced α width amplitude in ^{8}Be(2^+)+α with D-wave relative motion channel has been proposed by some theoretical calculations.[3,4] In Ref. 4, it was shown that this state has banana-like intrinsic structure that is close to the linear-chain configuration of 3α clusters.

Recently, investigations of neutron-rich Be isotopes has revealed that the valence neutrons bring about the exotic cluster structures in these nuclei.[5,6] Therefore, it is interesting to investigate how valence neutrons change the cluster structures in ^{12}C. The investigation of cluster structures in neutron-rich C isotopes using molecular-orbit model has predicted the cluster structures with the intriguing geometrical configurations of 3α clusters.[7,8] In Ref. 7, cluster structures of excited states in ^{14}C were investigated, and it was found that the 3α cluster structure with a equilateral triangular configuration is stabilized by two valence neutrons. In Ref. 8, stabilization of

3α linear-chain structure was discussed, and it was found that the linear-chain configuration of 3α clusters is stabilized in ^{16}C by valence neutrons in molecular σ orbit. These intriguing structures predicted in these studies should be investigated more in detail with precise treatment of three α cluster configurations and valence neutrons.

In this paper, we have studied the cluster structures in the excited states of ^{13}C as the starting point to investigate the exotic cluster structures in C isotopes. As for ^{13}C, K^π=$3/2^\pm$ bands that has α cluster structure have been proposed on the basis of the systematic analysis of transfer reactions, inelastic excitations one other data.[9] We analyze the geometrical configuration of α clusters in ^{13}C, and discuss how the dilute cluster structure found in ^{12}C is changed by adding one neutron, and whether the linear-chain configuration of 3α clusters is stabilized or not.

To investigate cluster structures in ^{13}C, we have used a microscopic 3α+n model. We describe the 3α cluster structure using generator coordinate method (GCM), and the neutron wave function is optimized for each basis of GCM. We have obtained two rotational bands that have developed three α cluster structures. These structures are discussed focusing on the stabilization of three α linear-chain configuration by analyzing the energy surface with respect to the bending angle of α clusters and the α-α distance.

2. Theoretical framework

To investigate the cluster structures in the excited states of ^{13}C, the microscopic 3α+n cluster model was used.

The wave function that describes 3α+n system is given as

$$\Phi(\boldsymbol{S}) = \mathcal{A}[\varphi_{\alpha_1}(\boldsymbol{S}_1)\varphi_{\alpha_2}(\boldsymbol{S}_2)\varphi_{\alpha_3}(\boldsymbol{S}_3)\varphi_v]. \tag{1}$$

Here, the wave functions of α clusters and valence neutron are given as

$$\varphi_{\alpha_i}(\boldsymbol{S}_i) = \prod_{j=1}^{4} \phi_\alpha(\boldsymbol{S}_i)\chi_j\tau_j, \quad \phi_\alpha(\boldsymbol{S}_i) = \exp\left[-b\left(\boldsymbol{r} - \boldsymbol{S}_i/\sqrt{b}\right)^2\right], \tag{2}$$

$$\chi_j\tau_j = \{p\uparrow, p\downarrow, n\uparrow, n\downarrow\}, \tag{3}$$

$$\varphi_v = \sum_a C^a\phi_v^a(\boldsymbol{Z}^a)\chi_v^a\tau_v, \quad \phi_v^a(\boldsymbol{Z}^a) = \exp\left[-\nu^a\left(\boldsymbol{r} - \boldsymbol{Z}^a/\sqrt{\nu^a}\right)^2\right] \tag{4}$$

$$\chi_v^a = (1/2+\xi^a)\chi_\uparrow + (1/2-\xi^a)\chi_\downarrow, \quad \tau_v = neutron. \tag{5}$$

The width parameter $1/\sqrt{2b}$ of α cluster wave function is 1.46 fm.

The Hamilonian is given as $\hat{H} = \hat{T} + \hat{V}_n + \hat{V}_c - \hat{T}_g$, where $\hat{T}$ is the total kinetic energy and $\hat{T}_g$ is the energy of the center-of-mass motion. We used

Volkov No. 2 force[10] for the central part and G3RS force[11] for the spin-orbit part of the effective nuclear force $\hat{V}_n$. Majorana parameter of Volkov No. 2 force and strength parameter of G3RS force that reproduce the scattering phase shifts of α+n and α+α system[12] are 0.6 and 2000 MeV, respectively.

The valence neutron wave function φ_v is optimized by energy variation with variational parameters C^a, ν^a, $\boldsymbol{Z}^a$ and ξ^a. Parity projected wave function $\Phi^{\pm}=\hat{P}^{\pm}\Phi$ is employed as a variational function. The number of superposed Gaussian wave packet in the valence neutron wave function a is 6, which is enough for the energy convergence against the increase of a.

Finally, we perform GCM calculation using 3α configuration $\boldsymbol{S} = \{\boldsymbol{S}_1, \boldsymbol{S}_2, \boldsymbol{S}_3\}$ as generator coordinate. The wave function that describes certain J^π state of ^{13}C is given as

$$\Psi_n^{J\pm} = \sum_i c_i^n P_{MK}^{J\pm}\Phi(\boldsymbol{S}(i)). \tag{6}$$

Here, c_i^n is determined by Hill-Wheeler equation.[13]

As for GCM bases, 3α configurations used in GCM calculation of ^{12}C performed by Uegaki et al.[3] are adopted in this study. 3α configurations that represent three normal modes of vibration belonging to the equilateral triangle are parametrized by inter-α distance D_A and coordinates e_x and e_y. Definitions of D_A, e_x and e_y are given in Fig. 1. D_A is ranged from 1 fm to 6 fm, and the adequate e_x and e_y are chosen for each D_A: [D_A=1 fm, (e_y=e_x=0)], [D_A=2, (e_y=0,±1/2)], [D_A=3, (e_y=0,±1/2,1)], [D_A=4, (e_y=0,±1/3,±2/3,1), (e_x=1/2)], [D_A=5, (e_y=0,±1/3, 1/2, ±2/3), (e_x=1/3, 2/3)], [D_A=6, (e_y=0,±1/3, 1/2, ±2/3), (e_x=1/3, 2/3)]. Details for these 3α configurations are given in Ref. 3.

To analyze the geometrical configurations of 3α clusters in the excited states, we calculate the energy surface using the wave function orthogonalized to the ground state. The wave function that is orthogonal to the ground state is described as

$$\begin{aligned}\tilde{\psi}^{J\pm}(D_A,\theta) &= \psi^{J\pm}(D_A,\theta) - \langle\Psi_0^{J\pm}|\psi^{J\pm}(D_A,\theta)\rangle\Psi_0^{J\pm},\\ \psi^{J\pm}(D_A,\theta) &= \sum_i a_i^0 P_{MK_i}^{J\pm}\Phi(D_A,\theta).\end{aligned} \tag{7}$$

Here, $\psi^{J\pm}(D_A,\theta)$ is the superposition of the GCM bases that have different K quantum numbers, and $\Phi(D_A,\theta)$ is the same wave function with that given in Eq. 1. To exhibit the geometrical configurations of 3α clusters, $\boldsymbol{S}$ of GCM basis is translated to the inter-α distance D_A and the angle between α clusters θ that is defined in Fig. 1. We include wave functions that are not incorporated in the GCM wave function Eq. 2 to calculate smooth energy surface.

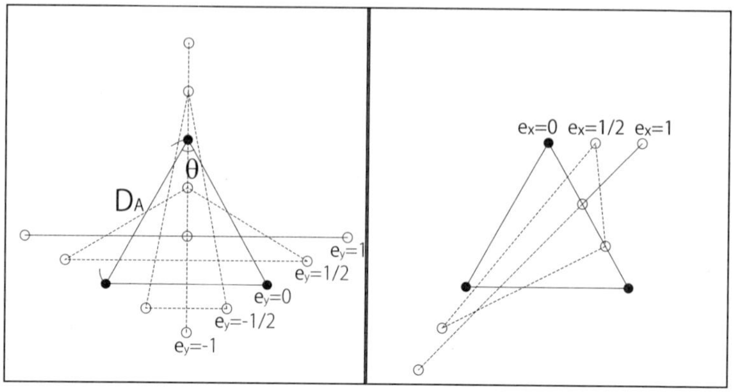

Fig. 1. 3α configurations for GCM bases and the definitions of D_A and θ. Details for 3α configurations are given in Ref. 3.

3. Results

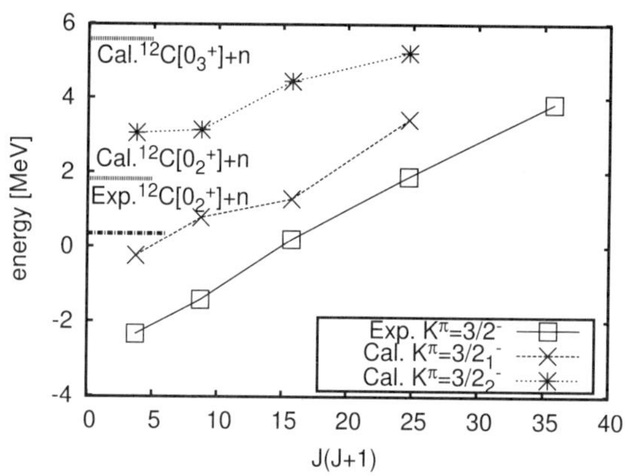

Fig. 2. Calculated rotational bands of ^{13}C.

We have investigated the α cluster structure of ^{13}C using a microscopic 3α+n cluster model. For comparison, ^{12}C is also calculated using 3α GCM with the same effective interaction parameter. Calculated and experimental ground state energies (α-α-α-n threshold energies) of ^{13}C are -94.5 (-82.8) and -97.1 (-84.9) MeV, respectively. Calculated ground state energy measured from α-α-α-n threshold energy well agrees with the experimental value (-11.7 and -12.2 MeV). We have found two rotational bands that have developed α cluster structures, which are labeled $K^\pi = 3/2_1^-$ and $K^\pi = 3/2_2^-$ bands, respectively. They are shown in Fig. 2 together with the experimental K^π=3/2$^-$ band that is proposed to have α cluster structure by M. Milin et al.[9] The proton radii of the members of $K^\pi = 3/2_1^-$ band are around 3.2 fm, which are much larger than the ground state proton radius of 2.38 fm because of the development of the 3α cluster structure. On the other hand, they are smaller than the proton radius of 0_2^+ state of ^{12}C that is calculated to be 3.53 fm. The smaller proton radius indicate that the cluster structure of $K^\pi = 3/2_1^-$ is not so dilute cluster-gas like one as in 0_2^+ state of ^{12}C. Moment of inertia of experimental and calculated $K^\pi = 3/2^-$ bands are similar to each other. Although the excitation energies of calculated $K^\pi = 3/2^-$ are higher than those of experimental one, calculated energy of $3/2^-$ state measured from ^{12}C(0_2^+)+n threshold energy (-2.2 MeV) is close to that of experimental one (-2.70 MeV). Therefore, the present $K^\pi = 3/2_1^-$ band is a good candidate of the experimentally proposed $K^\pi = 3/2^-$ α cluster band.

To analyze α cluster structure of K^π=$3/2_1^-$ band in detail focusing on the geometrical shape of α clusters, we have calculated two-dimensional energy surface with respect to inter-α distance D_A and the angle between α clusters θ that are defined in Fig. 1, and the overlap between GCM wave function and GCM basis $\langle\psi(D_A,\theta)|\Psi(3/2_2^-)\rangle$. To calculate the energy surface and the overlap, the wave functions that are not incorporated in Eq. 2 are also used.

We discuss the result of $3/2_2^-$ state that is the band head state of K^π=$3/2_1^-$ band. $3/2_2^-$ state has the energy minimum at (D_A,θ)$\sim$(142 deg, 3.5 fm), and this large D_A value means that $3/2_2^-$ state has the developed α cluster structure. The global behavior of the energy surface is similar to that of 0_2^+ state of ^{12}C, which indicate that $3/2_2^-$ state of ^{13}C is formed by adding one neutron to 0_2^+ state of ^{12}C. On the other hand, while the energy surface of 0_2^+ state shows remarkable softness against the increase of D_A and θ in D_A >4 fm region, the energy surface of $3/2_2^-$ state is steeper than that of 0_2^+ state around the energy minimum. This hardness against D_A and θ indicate that $3/2_2^-$ state is not so dilute as 0_2^+ state, but has a stabilized geometrical configuration of 3α clusters with large θ. A clear difference between 0_2^+ state and $3/2_2^-$ states is found in the overlaps shown in Fig. 4. While the wave functions that have large overlaps with 0_2^+ state are

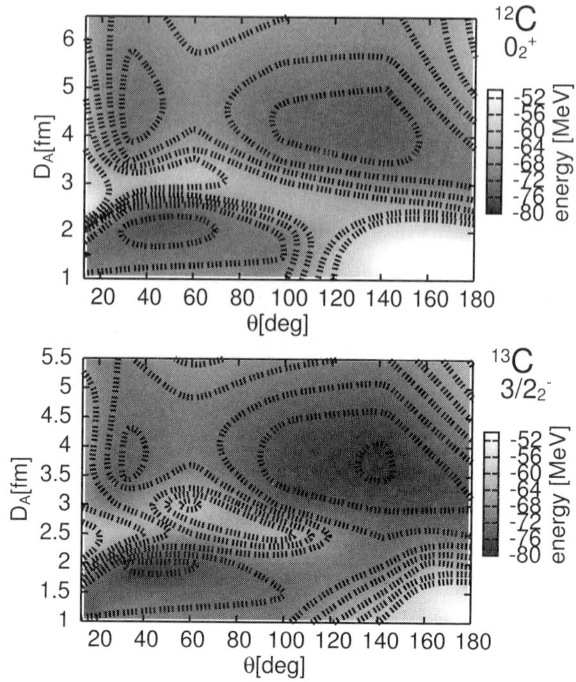

Fig. 3. Energy surface of $3/2_2^-$ state of ^{13}C. For comparison, that of 0_2^+ state of ^{12}C is also shown. Contour lines are plotted by intervals of 2.5 MeV.

distributed around $\theta \sim 60$ deg, $3/2_2^-$ state has the largest overlap with the wave function at $(D_A,\theta)\sim$(4 fm, 142 deg), and small overlaps with the wave functions in $\theta < 60$ region. This concentration of the dominant component in the large θ region also indicate the stabilization of 3α configuration. Fig. 5 shows the density distribution of the wave function that has the largest overlap with $3/2_2^-$ state of ^{13}C. Contour lines and gray plots exhibit the density distribution of 3α clusters and the valence neutron, respectively. As indicated by large θ, 3α clusters have bended linear-chain like configuration. The valence neutron occupy the orbit around 3α clusters, which is considered to be the molecular orbit constructed by the linear combination of p-orbits around α clusters. From these analysis, we found that the dilute cluster-gas like structure of ^{12}C is changed to bended linear-chain like configuration of 3α clusters in ^{13}C by the valence neutron that occupy the molecular orbit around α clusters.

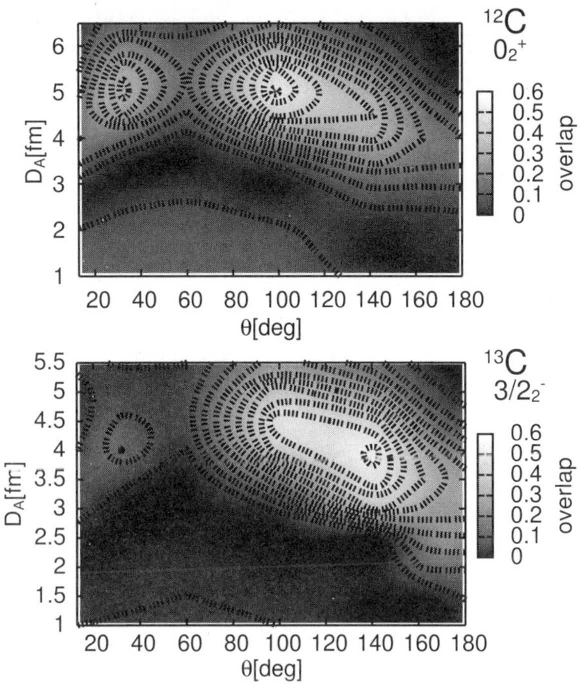

Fig. 4 Overlap between GCM wave function and GCM basis for $3/2_2^-$ state $\langle\psi(D_A,\theta)|\Psi^{3/2_2^-}\rangle$. For comparison, that for 0_2^+ state of ^{12}C $\langle\psi(D_A,\theta)|\Psi^{0_2^+}\rangle$ is also shown.

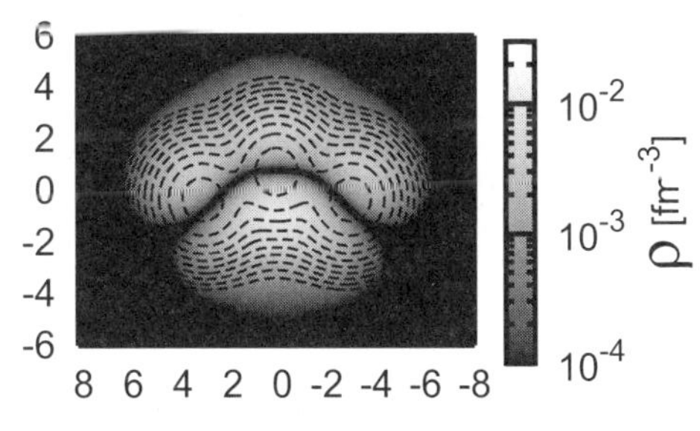

Fig. 5. Density distribution of the wave function that have largest overlap with $3/2_2^-$ state of ^{13}C. Contour lines and gray plots show the density distributions of 3α clusters and the valence neutron, respectively. Units on both axes are fm.

4. Summary

To investigate the structure adding one neutron to α cluster structure in ^{12}C, we have performed a microscopic 3α+n model calculation based on 3α

GCM in ^{13}C.

We have found K^π=$3/2^-_{1,2}$ rotational bands that have developed α cluster structures. K^π=$3/2^-_1$ band has similar moment of inertia with the experimentally proposed K^π=$3/2^-$ band,[9] although the calculated excitation energy is rather higher than the experimental one. Since the experimental K^π=$3/2^-$ band is proposed to be α cluster band, the calculated K^π=$3/2^-_1$ band is a good candidate of the experimental K^π=$3/2^-$ band.

The structure of $3/2^-_2$ state that is the band head state of K^π=$3/2^-_1$ band was discussed focusing on its geometrical shape by analyzing the energy surface and the overlap. We compared them with those of 0^+_2 state of ^{12}C, and found that $3/2^-_2$ state is not so dilute as 0^+_2 state of ^{12}C, but has increased stability with the bended linear-chain like 3α configuration by the valence neutron that occupy the molecular orbit around α clusters.

References

1. A. Tohsaki, H. Horiuchi, P. Schuck, and G. Röpke, Phys. Rev. Lett. **87**, 192501 (2001).
2. P. Schuck, Y. Funaki, H. Horiuchi, G. Röpke, A. Tohsaki, and T. Yamada, Nucl. Phys. **A738**, 94 (2004).
3. E. Uegaki, S. Okabe, Y. Abe and H. Tanaka, Prog. Theor. Phys. **57**, 1262 (1977). E. Uegaki, Y. Abe, S. Okabe and H. Tanaka, Prog. Theor. Phys. **59**, 1031 (1978); Prog. Theor. Phys. **62**, 1621 (1979).
4. Y. Kanada-En'yo, Prog. Theor. Phys. bf 117, 655 (2007).
5. W. von Oertzen, Z. Phys. A **354** (1996), 37 ; **357** (1997), 355.
6. M. Freer *etal.*, Phys. Rev. Lett. **96**, 042501 (2006).
7. N. Itagaki, T.Otsuka, K.Ikeda, and S.Okabe, Phys. Rev. Lett. **92**, 142501 (2004).
8. N. Itagaki, S. Okabe, K. Ikeda, and I. Tanihata, Phys. Rev. **C 64**, 014301 (2001).
9. M. Milin and W. von Oertzen, Eur. Phys. J. A **14**, 295 (2002).
10. A. B. Volkov, Nucl. Phys. **74**, 33 (1965).
11. N. Yamaguchi, T. Kasahara, S. Nagata and Y. Akaishi, Prog. Theor. Phys. **62**, 1018 (1979).
12. S. Okabe and Y. Abe, Prog. Theor. Phys. **61**, 1049 (1979).
13. D. L. Hill and J. A. Wheeler, Phys. Rev. **89**, 1102 (1953).

ALGEBRAIC $N\alpha$ MODEL: APPLICATION TO ^{12}C

T. YOSHIDA

Center for Nuclear Study, University of Tokyo

K. KATO

Division of Physics, Graduate School of Science, Hokkaido University

N. ITAGAKI

Department of Physics, University of Tokyo

Algebraic nuclear models have successfully applied for various mass regions. In ^{12}C, we couple this model space with treatment of the so-called complex scaling method for resonance states. We use the $Sp(2,R)_z$ algebra to create ladder states on SU(3) bases which have large harmonic oscillation quanta. One advantage of this method is that it is possible naturally to extend to $N\alpha$ systems. The wave function is semi-microscopic and satisfies the Pauli principle (orthogonal to the Pauli forbidden states). We study basic properties of this wave function.

1. Introduction

The existence of gas like and linear chain structures of ^{12}C has been suggested. Such a kind of the peculiar 3α structure appears around the $N\alpha$ threshold energy region. From the experimental side, the 4α cluster structures are also suggested in the excited states of ^{16}O. Therefore, it has been required to propose some reasonable models for such structures. The studies of the α cluster structure based on a semi-microscopic treatment of the α cluster model such as the orthogonal condition method (OCM) are expected to reveal the interesting physics expected to be seen at the $N\alpha$ threshold energy region.

In the usual procedure of OCM calculations, the pseudo potential method is used. An advantage of this method is that the continuum states are easily taken into account into the model space. This is important because the resonant states around the α-decay threshold are considered to have coupling with continuum states as shown in the α gas like state.

Because OCM is a semi-microscopic theory, we can easily reproduce the threshold energies of various decay-channel using appropriate inter-cluster potentials. In the present calculations, the Pauli allowed states are described by H.O. wave functions including large harmonic oscillator quanta so as to describe the coupling with the continuum states. Here, basis states for the large number of H.O. quanta are effectively prepared by using the $Sp(2,R)_z$ algebra on H.O. states of SU(3).

Moreover, there is a well known procedure of the complex scaling method (CSM)[1] which describes resonant states. In CSM, the boundary condition of the resonant states around the $N\alpha$ threshold are transformed to regular wave functions like the bound states. We couple OCM generated by the $Sp(2,R)_z$ truncation and CSM. As this is the first step of our studies, we focus on the two- and three-α nuclei and convince basic properties.

2. Formulation and results

There have been many studies based on the algebraic model, and our truncation of the 3α model space is one of such kinds of studies. The algebraic method has a general rule to specify the quantum number using the Young tableau. The total SU(3) quanta are given by a product of SU(3) quanta corresponding to the state of the Jacobi coordinate for each α-α relative motion:

$$SU(3) = SU(3) \otimes SU(3), \tag{1}$$

$$N = N_1 + N_2. \tag{2}$$

The obtained SU(3) label for the total wave function of the 3α system, therefore, becomes $N(\lambda,\ \mu)\rho$, where N is total H.O. quanta, $\lambda = N_z - N_y$, $\mu = N_y$ and ρ distinguishes the multiplicity of the states described by the linear combination of N_1 and N_2 pairs. The 3α Pauli allowed wave functions are expanded as follows:[2]

$$U_i(r,R) = \sum_{\rho,N_1,N_2} C_{\rho,\Lambda}^{N_1,N_2} A_{N_1,N_2}^{N(\lambda,\mu)\rho} V_{N_1,N_2}^{N(\lambda,\mu)_\rho,J}(r,R), \tag{3}$$

where the index i denotes an abbreviation of the J, Λ and $N(\lambda,\mu)$. The coefficients $A_{N_1,N_2}^{N(\lambda,\mu)\rho}$ are determined so as to be the orthogonal to all Pauli forbidden states of every α-α relative wave function. This requirement is also expressed by another relation,[4]

$$\hat{Q}|N(\lambda,\mu)k\rangle = q_k|N(\lambda,\mu)k\rangle. \tag{4}$$

The Pauli allowed states are obtained as the eigenstates of $q_k = 0$. The operator $\hat{Q}$ expresses the projection to the Pauli-forbidden states in the different Jacobi coordinate system. The index k denotes a set of the harmonic oscillator quanta of y and z directions; $\{n_{1z}, n_{1y}, n_{2z}$ and $n_{2y}\}$, for a given $N(\lambda, \mu)$ state.

On the other hand, the detailed expression of $V^{N(\lambda,\mu)_\rho,J}_{N_1,N_2}(r, R)$ in Eq. (3) is,

$$V^{N(\lambda,\mu)_\rho,J}_{N_1,N_2}(r, R) = \sum_{l_1,l_2} \langle N_1, l_1, N_2, l_2 | N(\lambda, \mu)_\rho, J\rangle [u_{N_1,l_1}(r) u_{N_2,l_2}(R)]_J. \quad (5)$$

Here, $\langle N_1, l_1, N_2, l_2 | N(\lambda, \mu)_\rho\rangle$ shows the CG coefficient for SU(3).[2] When we treat eigenstates related to the E0 or E2 transition from the ground states (which we have particular interests), it is expected that not all of these states are important. These states are generated by creation operators $a^+_{iz}a^+_{iz}$ which corresponds to $Sp(2, R)_z$. Therefore, we use the other index Λ which distinguishes the quanta of the $Sp(2, R)_z$ algebra. The z index shows that we take into account only the degree of freedom of the z direction. The generators are

$$L_+ = -(a^\dagger_{1z}a^\dagger_{1z} + a^\dagger_{2z}a^\dagger_{2z})/\sqrt{8}, \quad (6)$$
$$L_- = -(a_{1z}a_{1z} + a_{2z}a_{2z})/\sqrt{8}, \quad (7)$$
$$L_0 = -(a^\dagger_{1z}a_{1z} + a^\dagger_{2z}a_{2z})/4. \quad (8)$$

The coefficients C and the quantum number Λ are obtained as a solution and eigenvalue of the Casimir operator of this algebra.[3] We check the valid-

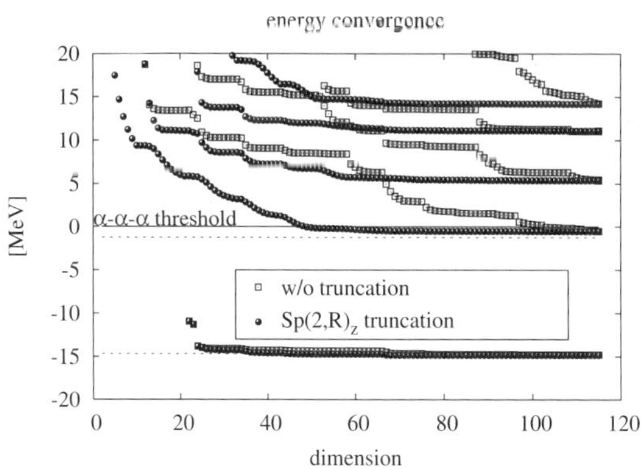

Fig. 1. Energy convergence of the 3α system as a function of the number of basis states with (black) or without (white) $Sp(2, R)_z$ truncation. The dotted lines are the energies obtained by using pseudo potential method.[5]

ity to use the $Sp(2,R)_z$ truncation method. In Fig. 1, we show the energy as a function of the number of basis states. When we apply the truncated bases (black dot), the energy convergence becomes much faster than the case without the $Sp(2,R)_z$ truncation (white dot) as shown in Fig. 1. As for the interaction, we use the α-α folding potential with the effective nucleon interaction of the so-called Schmidt-Wildermuth force. The Coulomb and three-body potentials are not taken into account for simplicity. When the same number of the basis states is prepared, however, a good energy convergence can be obtained even if we chose only lower Λ values. As a next step, we combined this algebraic formation with the complex scaling method to treat the resonance states.

Before continuing with the 3α calculation, we check the reliability in the 2α case. Expansion of the basis states in 2α is done by the following procedure which is the same as for 3α calculation: Instead of using the pseudo-potential to project out the Pauli forbidden states like as

$$(\hat{H} + \lambda V_{\text{pseudo}} - E\hat{N})P|\Phi\rangle = 0, \tag{9}$$

we solve the equation

$$(\hat{H} - E\hat{N})P_{\text{pas}}|\Phi\rangle = 0. \tag{10}$$

The complex scaling method is applied to the above equation, and we have

$$U_\theta(\hat{H} - E\hat{N})P_{\text{pas}}U_\theta^{-1}|\Phi\rangle = 0, \tag{11}$$

where U_θ is the complex scaling transformation. For the 2α case, P_{pas} is expressed as

$$P_{\text{pas}} = \sum_{n \in pas} |l,n;r,\nu_0\rangle\langle l,n;r,\nu_0|, \tag{12}$$

where ν_0 is the width parameter to reproduce the size of the α particle.

The complex energy of 2^+ states obtained by this method is shown in Fig. 2. The results of two calculations based on Eqs. (9) and (10) are compared. The width parameter b of Φ is set to be 1.4 fm and 1.8 fm. The θ value varies from 15° to 30°. Because of the fact that the complex scaling transformation does not conserve the norm, the numerical properties of the overlap matrix element $< l,m;r,\nu|U_\theta|l,n;r,\nu_0 >$ oscillates strongly for large n values when a large θ value is taken. This property is understood by using the analytic expression of the overlap matrix elements;

$$\langle l,m,\nu|U_\theta|l,n,\nu_0\rangle \sim {}_2F_1(-m,-n,l+3/2,-4\nu_\theta\nu_0/(\nu_0-\nu_\theta)^2), \tag{13}$$

where ${}_2F_1$ is the hyper geometrical function. For such a problem, it is solved by treating the numerical calculation carefully.[6] As a result, we obtained a good agreement with the pseudo potential method up to around θ =25°. In Fig. 2, we can see that the pseudo potential method (Eq. (9)) and the expansion using Pauli allowed states (Eq. (10)) are consistent each other.

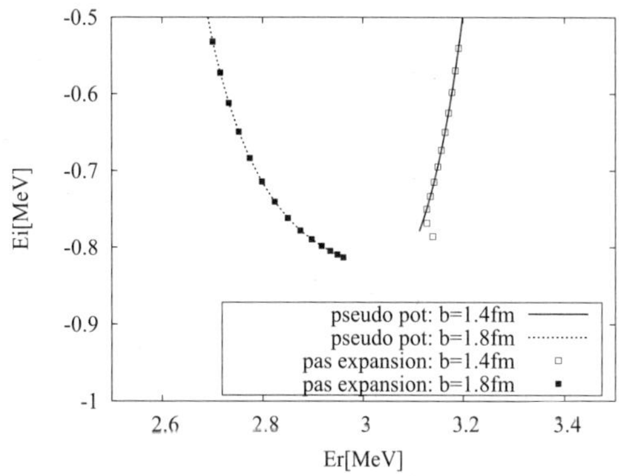

Fig. 2. The energy convergences of two different procedures based on Eqs. (9) and (10) on the complex energy plane of 2^+ states for the 2α system.

The similar calculation can be performed for the 3α cluster system. The complex scaled wave function is expanded by the SU(3) H.O. wave functions as

$$\Psi = \sum_{i \in total} c_n^\theta \psi_n(\theta). \tag{14}$$

The Pauli allowed Hamiltonian is given as

$$H_{\rm pas} = \hat{P}_{\rm pas} \hat{H} \hat{P}_{\rm pas}. \tag{15}$$

The matrix elelment therefore becomes

$$\langle \psi_m(\theta) | H_{\rm pas} | \psi_m(\theta) \rangle = \sum_{u,v \in {\rm pas}} < m|U_\theta|u >< u|\hat{H}|v >< v|U_\theta|n > . \tag{16}$$

$< u|\hat{H}|v >$ is exactly the same as the Hamiltonian used in Fig. 1 except that we add three-body and Coulomb interactions in Eq. (16). It is necessary to choose carefully the model space of Pauli allowed states (specified by u and v) and the complex scaled model space specified by m and n. For the m and n model spaces, q_k in Eq. (4) should not be restricted to zero. The size of m and n can be taken so large to obtain the good convergence. However, too large m and n values which excess the spatial size of the model space expressed by u and v are not necessary for the sake of completeness. In the present calculation, we check whether the ground state and some excited states are invariant under the complex scaling angle θ with the restricted model space. In Fig. 3, the complex energy of 0^+ states are shown for different width parameters b and rotation angles θ. We can find some points are almost invariant under the parameter change. These states correspond to the ground and the second 0^+ states.

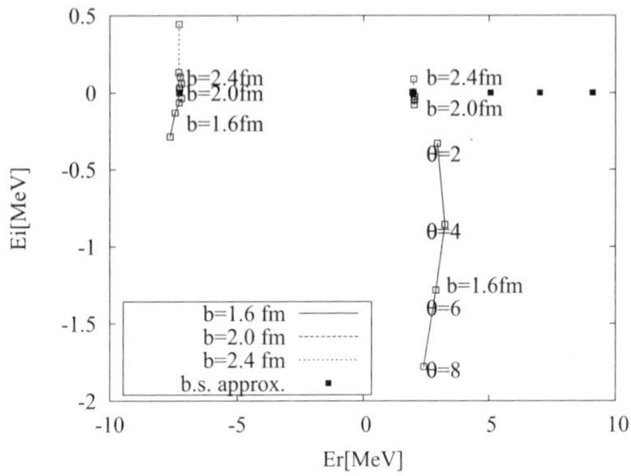

Fig. 3. Energy distribution on the complex energy plane for the 3α system.

3. Summary

We have proposed that the use of the $Sp(2,R)_z$ algebra to make the OCM bases is still effective even if we treat resonance states by using CSM. To ensure it, we calculated several states of two and three α systems. We used the truncation that the Pauli allowed states are directly taken into account by ladder states generated by $Sp(2,R)_z$. The result shows that we can use this procedure with a certain amount of precision. In future, we will sure to add a little device for the model space and enlarge them.

References

1. J. Agular and J. M. Combes, Commun. Math. Phys. **22**, 269 (1971).
   E. Balslev and J. M. Combes, Commun. Math. Phys. **22**, 280 (1971).
2. K. Katō, K. Fukatsu and H. Tanaka, Prog. Theo. Phys. **80**, 663 (1989).
3. K. Katō and H. Tanaka, Prog. Theo. Phys. **81**, 738 (1989).
4. H. Horiuchi, Prog. Theo. Phys. **58**, 204 (1989).
5. H. Horiuchi, Prog. Theo. Phys. **53**, 447 (1975).
6. R. E. SCRATON, BIT. **12**, 242-251 (1972).

ANTISYMMETRIZED MOLECULAR DYNAMICS WITH BARE NUCLEAR INTERACTIONS: BRUECKNER-AMD

T. TOGASHI* and K. KATŌ

Division of Physics, Graduate School of Science, Hokkaido University, Sapporo 060-0810, Japan
**E-mail: toga@nucl.sci.hokudai.ac.jp*

YUHEI YAMAMOTO

Department of Cosmosciences, Faculty of Science, Hokkaido University, Sapporo 060-0810, Japan
E-mail: yuhei@nucl.sci.hokudai.ac.jp

We develop the new antisymmetrized molecular dynamics (AMD) method which makes us perform the AMD calculations starting from the bare nuclear interactions. In this framework, the G-matrix and the single-particle energies of AMD are determined from the bare interactions self-consistently without any corrections. In this paper, we present the applicability of this method to describe not only the ground states but also the excited states for some light nuclei, especially the excited 0_2^+ state in ^{12}C which is not solved sufficiently by the present shell model approaches.

Keywords: AMD; G-matrix; *ab initio* calculation.

1. Introduction

As one of recent major subjects in theoretical nuclear physics, *ab initio* calculations[1–7] starting from the bare nuclear interactions are developed remarkablely. One hopes to understand the property of nuclei based on the bare nuclear interactions against the background of the progress of the fundamental understanding of nucler forces. As one of remarkable acheivements in *ab initio* approaches, the quantum Monte Carlo method[1] has presented the appearance of the $\alpha+\alpha$ structure in ^{8}Be starting from the bare nuclear interaction. That suggests that one has the ability to discuss cluster structures based on the bare nuclear interactions.

The existence of cluster states to be weakly bound systems of subunits appearing around thresholds is predicted by the Ikeda diagram.[8] The cluster states in several excited states of a light nucleus can be described as the states different from the shell-like structures. Cluster models are able to describe these states successfully. As one of the recent development of the cluster model, the antisymmetrized molecular dynamics (AMD) method,[9] where both the shell and cluster structures can be described without any assumption of configurations as the results of the energy variation, has succeeded in understanding and predicting the properties of many nuclei. However, the AMD calculations have been performed with phenomenological interactions so far. For the purpose of the fundamental understanding of nuclear structures, we have recently developed new AMD framework, "Brueckner-AMD",[10–12] which makes AMD available to us with bare nuclear interactions.

In the Brueckner-AMD, the Brueckner theory[13] to be the basic effective interaction theory is combined with the AMD+Hartree-Fock method[14] in AMD. In this framework, the single-particle orbits defined by the AMD+Hartree-Fock method are applied to the Bethe-Goldstone equation[15] with the bare interaction to determine the G-matrix and the single-particle energies self-consistently without any corrections. Furthermore, the correlation functions[11] generated from the solutions of the Bethe-Goldstion equaiton construct the central, spin-orbit, and tensor interactions in the G-matrix explicitly. In this work, we show the applicability of the Brueckner-AMD to describe not only the ground states but also some excited states for some light nuclei, especially the excited 0_2^+ state in ^{12}C which is indicated as the well-developed 3α cluster state[16–20] the present shell model approaches[21] are hardly able to describe.

In this paper, we explain the formulation of the Brueckner-AMD in Sect. 2. We report the applications of the Brueckner-AMD to some light nuclei and discuss their results in Sect. 3, and summarize this paper in Sect. 4.

2. Formulation

2.1. *AMD wave function*

The AMD wave function[9] $|\,\Phi\,\rangle$ is represented as an antisymmetrized product of single-particle wave functions for A nucleons (a single Slater determinant):

$$|\,\Phi\,\rangle = \frac{1}{\sqrt{A!}}\det|\,\phi_j\,\rangle \;\equiv\; \mathcal{A}\left\{\prod_j^A |\,\phi_j\,\rangle\right\} = \mathcal{A}\left\{\prod_j^A |\,\boldsymbol{Z}_j\,\rangle\cdot|\,\chi_j\,\rangle\right\}, \quad (1)$$

where $\mathcal{A}$ is the antisymmetrization operator, and the wave fuction of a nucleon, $|\,\phi_j\,\rangle$, is divided into a spatial part, $|\,\boldsymbol{Z}_j\,\rangle$, and a spin-isospin part, $|\,\chi_j\,\rangle$. The spatial wave function $|\,\boldsymbol{Z}_j\,\rangle$ has the form of a Gaussian wave packet:

$$\langle\,\boldsymbol{r}\,|\,\boldsymbol{Z}_j\,\rangle = \left(\frac{2\nu}{\pi}\right)^{3/4} \exp\left[-\nu\left(\boldsymbol{r} - \frac{\boldsymbol{Z}_j}{\sqrt{\nu}}\right)^2 + \frac{\boldsymbol{Z}_j^2}{2}\right]. \tag{2}$$

The spin-isospin wave function $|\chi_j\rangle$ is given by $\chi_j = p\uparrow, p\downarrow, n\uparrow$, or $n\downarrow$ for simplicity in this work. In the AMD method, the wave-packet positions $\boldsymbol{Z}_j$ in Eq. (2) are complex variational parameters in the A-body wave function. We use the frictional cooling method of molecular dynamics[9,22] to determine $\boldsymbol{Z}_j$. The brief expression of frictional cooling equation is given by

$$\frac{d\boldsymbol{Z}_j}{dt} = -\frac{\partial\mathcal{H}}{\partial\boldsymbol{Z}_j^*}, \quad \frac{d\boldsymbol{Z}_j^*}{dt} = -\frac{\partial\mathcal{H}}{\partial\boldsymbol{Z}_j}, \tag{3}$$

where $\mathcal{H}$ is the expectation value of Hamiltonian $\hat{H}$. In this calculation, the initial configurations of $\{\boldsymbol{Z}_j\}$ are chosen randomly, and the wave functions are determined as the solution of the energy variation with Eq. (3).

2.2. *G-matrix in the Brueckner-AMD*

A bare nuclear interaction has a strong repulsion at short-range distances. Therefore, it cannot be treated straightforward by the model wave functions such as the AMD wave function. In applying model wave functions, one needs to use effective interactions. The Brueckner theory[13] is a standard theory for deriving an effective interaction based on the bare nuclear interaction, G-matrix, between nucleons moving on single-particle orbits. In this theory, in order to calculate the G-matrix $\hat{G}$, we use the Bethe-Goldstone equation[15]

$$\hat{G} = \hat{v}_2 + \hat{v}_2 \frac{Q}{\varepsilon_\alpha + \varepsilon_\beta - (\,\hat{t}_1 + \hat{t}_2\,)} \hat{G}, \tag{4}$$

where $\hat{v}_2$, Q, ε_i and $\hat{t}_i$ are a bare two-body interaction, the Pauli projection operator, the single-particle energy, and the kinetic energy operator, respectively.

In the AMD method, the Brueckner theory can be applied straightforwards to single-particle orbits that are defined by the AMD+Hartree-Fock method.[14] The "Brueckner-AMD" is a framework for the application of the Brueckner theory to AMD. In the AMD+Hartree-Fock method, one can obtain the single-particle orbit $| \tilde{f}_\alpha \rangle$ by diagonalizing the Hartree-Fock Hamiltonian matrix:

$$h_{pq} \equiv \langle f_p \,|\, \hat{t} \,|\, f_q \rangle + \sum_r \langle f_p \, f_r \,|\, \hat{G} \,|\, \mathcal{A}\,\{ f_q \, f_r \} \,\rangle, \tag{5}$$

where $| \, f_p \, \rangle$ represents an orthonormal basis determined by diagonalizing the overlap matrix $B_{ij} \equiv \langle \, \phi_i \,|\, \phi_j \, \rangle$. The single-particle energy is obtained as

$$\varepsilon_\alpha \cdot \delta_{\alpha,\beta} \;=\; \langle \tilde{f}_\alpha \,|\, \hat{t} \,|\, \tilde{f}_\beta \,\rangle + \sum_\gamma \langle \tilde{f}_\alpha \, \tilde{f}_\gamma \,|\, \hat{G} \,|\, \mathcal{A}\,\{ \tilde{f}_\beta \, \tilde{f}_\gamma \} \,\rangle. \tag{6}$$

In this theory, both the single-particle energies ε_α and the G-matrix $\hat{G}$ are determined self-consistently. That means that the G-matrix within this framework can be determined theoretically without any corrections.

The Pauli projection operator $Q = 1 - P$ in the Bethe-Goldstone equation, Eq. (4), where P is the projection operator for occupied states, $P = \sum_{\alpha < \beta} | \, \tilde{f}_\alpha \, \tilde{f}_\beta \, \rangle \, \langle \, \tilde{f}_\alpha \, \tilde{f}_\beta \, |$, is introduced in order to take account of the Pauli principle. Following the prescription formulated by Bandō et al.,[23] which presents an appropriate means of treating Eq. (4) in the finite nuclear systems, we calculate the G-matrix in the following two steps: First, ignoring the Q-operator ($Q \to 1$), we solve the equation for $\hat{G}^0$:

$$\hat{G}^0 \;=\; \hat{v}_2 \;+\; \hat{v}_2 \, \frac{1}{\varepsilon_\alpha + \varepsilon_\beta \;-\; (\,\hat{t}_1 + \hat{t}_2\,)} \, \hat{G}^0. \tag{7}$$

Second, we solve the following equation to take into account the Q-operator:

$$\hat{G} \;=\; \hat{G}^0 \;+\; \hat{G}^0 \, \frac{Q \;-\; 1}{\varepsilon_\alpha + \varepsilon_\beta \;-\; (\,\hat{t}_1 + \hat{t}_2\,)} \, \hat{G}. \tag{8}$$

The explicit form of Eq. (8) becomes the algebraic equation for the G-matrix elements $\langle \, \tilde{f}_\gamma \, \tilde{f}_\delta | \, \hat{G} \, | \mathcal{A} \left\{ \tilde{f}_\alpha \, \tilde{f}_\beta \right\} \rangle$ and the G^0-matrix ones $\langle \, \tilde{f}_\gamma \, \tilde{f}_\delta | \, \hat{G}^0 \, | \mathcal{A} \left\{ \tilde{f}_\alpha \, \tilde{f}_\beta \right\} \rangle$ as described in Ref. 10. The way how to calculate the matrix elements for $\hat{G}^0$ to solve Eq. (8) is explained in the following.

Here, we present another expression of the Bethe-Goldstone equation as

$$\psi_{kl} = \varphi_{kl} + \frac{Q}{\varepsilon_\alpha + \varepsilon_\beta - (\hat{t}_1 + \hat{t}_2)} \, \hat{v}_2 \cdot \psi_{kl} \,, \tag{9}$$

where the following relationship is used: $\langle \varphi_{ij} | \hat{G} | \varphi_{kl} \rangle = \langle \varphi_{ij} | \hat{v}_2 | \psi_{kl} \rangle$. In the above equations, ψ_{kl} and φ_{kl} represent the solution of the Bethe-Goldstone equation and the two-body part of the model wave function, respectively.

Ignoring the Q-operator ($Q \to 1$) in Eq. (9), one can obtain the differential equation to solve the G^0-matrix as

$$\left[\hat{T}_r - \omega\right] \{\delta_{l' l_0} \cdot \varphi_{l_0 m}(r; \boldsymbol{Z}_{r_{kl}}) - \psi^{0\,JS}_{l' l_0 m}(r)\} = \sum_{l''} V^{JS}_{l' l''}(r)\, \psi^{0\,JS}_{l'' l_0 m}(r), \tag{10}$$

where $\hat{T}_r$, $\varphi_{l_0 m}(r; \boldsymbol{Z}_{r_{kl}})$, $\psi^{0\,JS}_{l' l_0 m}(r)$, and $V^{JS}_{l' l''}(r)$ represents the relative kinetic energy operator, the two-body AMD wave function of the relative part described in Ref. 11, the solution of this equation, and the potential for the partial wave in $\hat{v}_2$, repectively. In the above equation, the starting energy ω becomes $\omega = (\varepsilon_\alpha + \varepsilon_\beta) - T_C(ij : kl)$ where $T_C(ij : kl)$ is the expectation value of the two-body center-of-mass kinetic energy.

The G^0-matrix element is calculated schematically as $\langle \varphi_{ij} | \hat{G}^0 | \varphi_{kl} \rangle = \langle \varphi_{ij} | \hat{v}_2 | \psi^0_{kl} \rangle$ with the solution ψ^0_{kl} in Eq. (10). Here, we define the correlation function as $\hat{F}^0_{kl} = \psi^0_{kl}/\varphi_{kl}$. And then, one can find that $\langle \varphi_{ij} | \hat{G}^0 | \varphi_{kl} \rangle = \langle \varphi_{ij} | \hat{v}_2 \cdot \hat{F}^0_{kl} | \varphi_{kl} \rangle$. Explicitly, the correlation functions in this work are defined as

$$F^{0\,\{\alpha\beta\}}_{\;kl} \left[{}^{JS}_{l' l_0}\right](r) = \psi^{0\,JS}_{l' l_0\, m=0}(r) / \varphi_{l_0\, m=0}(r; \boldsymbol{Z}_{r_{kl}}), \tag{11}$$

where the denotation of $\{\alpha\beta\}$ means the dependences on the single-particle energies $\varepsilon_{\alpha,\beta}$ in ω of Eq. (10). The $\hat{G}^0$ for each J and S channel becomes

$$G^{0\,JS}_{\;l_0 l'}(r) = \sum_{l''} V^{JS}_{l' l''}(r) \cdot F^{0\,\{\alpha\beta\}}_{\;kl} \left[{}^{JS}_{l'' l_0}\right](r). \tag{12}$$

The $G^{0\,JS}_{\;l_0 l'}(r)$ does not include explicitly the spin-flip or time-reversal terms such as spin-orbit forces, and therefore, we determine the $\hat{G}^0$ with explicit non-central operators. For this aim, we utilize the following equation[24]

$$G^{0\,S}_{\;l_0 l', \lambda}(r) = \frac{(2\lambda + 1)}{\langle l_0 || \boldsymbol{R}_\lambda || l' \rangle \langle S || \boldsymbol{S}_\lambda || S \rangle} \sum_J (-1)^{J - l_0 - S} \times (2J + 1)\, W(l_0\, l' S S; \lambda J)\, G^{0\,JS}_{\;l_0 l'}(r), \tag{13}$$

where $W(l_0\, l' SS; \lambda J)$ is a Racah coefficient. The components of $\lambda = 0$, 1 and 2 correspond to the central, spin-orbit, tensor terms in a potential respectively. In this calculation, the angular momenta l_0 and l' in Eq. (13) are taken as the allowed lowest values for each two-body channel. As a result, the $\hat{G}^0$ becomes $G^0_{\lambda=0}(r) + G^0_{\lambda=1}(r)\, \boldsymbol{L} \cdot \boldsymbol{S} + G^0_{\lambda=2}(r)\, S_{12}$. One can obtain the G^0-matrix elements to solve G-matrix in Eq. (8) by using the potential with this form.

3. Applications

We apply the Brueckner-AMD method to various light nuclei and present the applicability of this method. In this work, we use the Argonne $v8$' (AV8')[25] as the bare nuclear interaction and switch off electromagnetic interactions.

3.1. *Energy-level scheme calculations with J^π projection*

In general, since the AMD wave functions are not good quantum states of the spin and parity, J^π, one cannot discuss the energy-level scheme by a single AMD wave function and must project out the AMD wave functions onto the spin and parity eigenstates, which means the superposition of space-reflected and rotated Slater determinants.[9] In these calculations, we determine the intrinsic wave functions $| \Phi \rangle$ by performing the energy variation for the parity-projected states $| \Phi^{\pm} \rangle$. Projecting out the solutions of the energy variation, $| \Phi^{\pm} \rangle$, onto each J state, we calculate the energy levels. In the following, we present the results of ^{4}He, ^{7}Li, ^{8}Be, and ^{9}Be using the above calculational procedure.

Fig. 1 shows the intrinsic densities obtained in this work. For ^{9}Be, both the lowest-energy negative-parity solution and the excited positive-parity one are shown. As seen in Fig. 1, the density of ^{4}He has a compact spherical shape. The density of ^{8}Be shows the well-known α+α structure. In this result, the density of ^{7}Li do not present the explicit cluster structure as seen in an almost symmetric distribution. In the densities of ^{9}Be, the structure of two α-clusters plus a valence neutron can be seen. In the negative-parity state of ^{9}Be, the density distribution of the valence neutron corresponds to a π-orbit; on the other hand, that of the positive-parity state corresponds to a σ-orbit.

The calculated energy-level schemes of ^{7}Li, ^{8}Be, and ^{9}Be are shown in Fig. 2. The experimental energy levels are compared with these results in the figure, where the groud states are normalized to zero energy. Each binding energy of the ground state is referred in the caption of the figure. The genuine three-body forces needed to reproducc the experimental binding energies are not considered in this framework. However, the obtained binding energy of the ground state 0^+ for ^{4}He is -24.6 MeV and its value is comparable to that obtained by few-body calculations,[7] -25.9 MeV. The result of ^{4}He indicates the quantitative reliability of this calculation. In the result of ^{8}Be, which is indicated to be the α+α structure, the rotational band is reproduced well as seen in Fig. 2. This framework can also provide

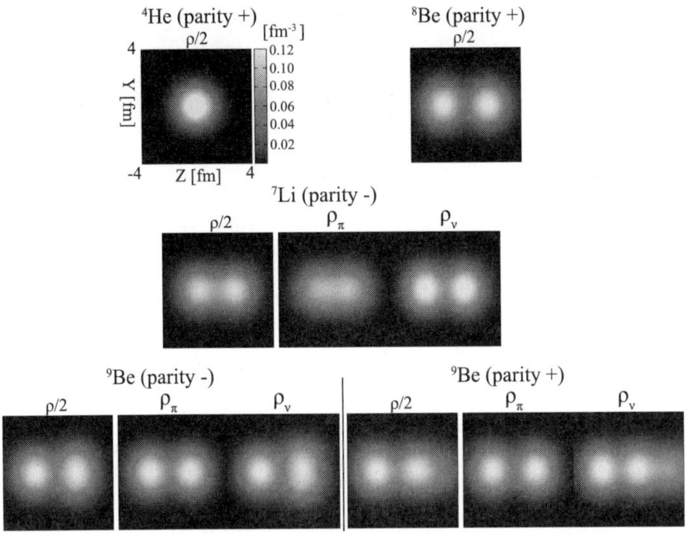

Fig. 1. Intrinsic density distributions of the lowest-energy parity solutions in ^{4}He, ^{8}Be, ^{7}Li and ^{9}Be. In the matter (ρ) density distributions, the densities are displayed as half values, and the proton (ρ_π) and neutron (ρ_ν) density distributions are shown in the odd nuclei. The size of these squares is shown in the panel of ^{4}He.

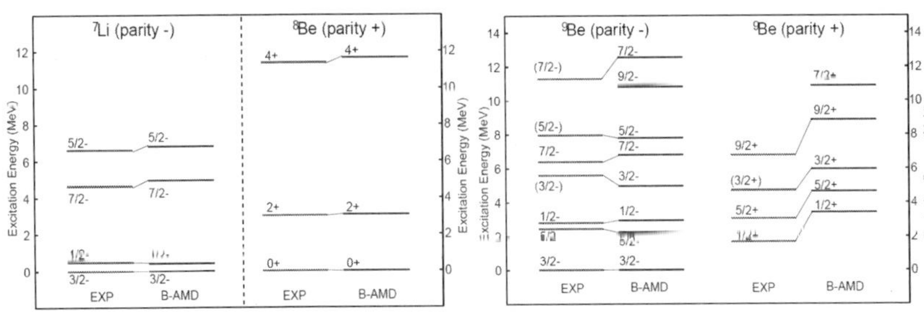

Fig. 2. Low-lying energy level schemes of the experiments and Brueckner-AMD calculations labeled "EXP" and "B-AMD", respectively. The results of ^{7}Li and ^{8}Be are in the left panel and that of ^{9}Be is in the right panel. The obtained ground-state binding energies of ^{7}Li, ^{8}Be, and ^{9}Be are -29.6 MeV, -44.0 MeV, and -41.9 MeV, respectively.

the reasonal results for odd nuclei. In the result of ^{7}Li, the energy levels are reproduced well. In the result of ^{9}Be, the energy levels including the first excited states, $3/2^-_2$, $5/2^-_2$, and $7/2^-_2$, which can be described by the K-mixing, are also reproduced well. Note that the results of the energy levels related to the spin-orbit splittig, for example, $3/2^-$ and $1/2^-$ in ^{9}Be are improved in comparison with our previous work[11] because the G-matrix includes the explicit spin-orbit term which generates the spin-orbit splittig in the energy level.

3.2. *Description of the 0_2^+ state in ^{12}C*

Here, we attempt to decribe not only the ground state 0^+ but also the excited 0_2^+ state in ^{12}C within the Brueckner-AMD framework. It is known that the present shell-model approaches[21] have difficulty in describing the 0_2^+ states in ^{12}C because huge shell-model space is needed. On the other hand, cluster models[16–20] have been succeeded in describing this state. To describe higher 0^+ states in ^{12}C, it is indicated that it is necessary to superpose a number of spatial configurations by the cluster models. In this work, we creat various states without any assumption for configurations. For this aim, we apply the cooling method under the constraint.[14,26] In this calculation, we take the root-mean-square mass radius R and the approximated deformation parameter β[14] as the quantities under the constraint. In addition, one should describe the configurations different from the ground state in viewpoint of the orthogonality of the excited state. Therefore, we perform the energy variation for the wave functions $|\,\Phi^{\,i}\,\rangle$ with the orthogonality to the ground state $|\Phi\{\boldsymbol{Z}^0\}\,\rangle$ as

$$|\,\Phi^{\,i}\,\rangle = |\,\Phi\,\{\boldsymbol{Z}^i\}\,\rangle - |\,\Phi\,\{\boldsymbol{Z}^0\}\,\rangle \cdot \frac{\langle\,\Phi\,\{\boldsymbol{Z}^0\}\,|\,\Phi\,\{\boldsymbol{Z}^i\}\,\rangle}{\langle\,\Phi\,\{\boldsymbol{Z}^0\}\,|\,\Phi\,\{\boldsymbol{Z}^0\}\,\rangle}, \tag{14}$$

where $\{\boldsymbol{Z}\}$ represent the spatial configuration of the AMD wave packets as $\{\boldsymbol{Z}\} \equiv \boldsymbol{Z}_1, \boldsymbol{Z}_2, \cdots, \boldsymbol{Z}_{12}$, and determine the excited configurations $\{\boldsymbol{Z}^i\}$. In this work, we determine the ground state $|\Phi\{\boldsymbol{Z}^0\}\,\rangle$ and the excited states $|\Phi\{\boldsymbol{Z}^i\}\,\rangle$ as not the J^π-projected states but the parity-projected ones for simplicity of calculations, therefore we impose it on the energy variation that the inertial axes of each state are fitted to prevent us from obtaining the solutions of the rotated ground state. Then we obtain the solutions $|\,\Phi_n^{J\,\pm}\,\rangle$ as the J^π-projected states,

$$|\,\Phi_n^{J\,\pm}\,\rangle = \sum_{i=0}\sum_{K} C_n^{J\,\pm}(i,K)\,|\,\Phi_{MK}^{J\,\pm}\,\{\boldsymbol{Z}^i\}\,\rangle, \tag{15}$$

where $|\,\Phi_{M\,K}^{J\,\pm}\,\{\boldsymbol{Z}^i\}\,\rangle$ represents the J^π-projected state of $|\Phi\{\boldsymbol{Z}^0\}\,\rangle$ or $|\Phi\{\boldsymbol{Z}^i\}\,\rangle$, and the weights of the states $C_n^{J\,\pm}(i,K)$ are determined by diagonalizing $\langle\,\Phi_{MK}^{J\,\pm}\,\{\boldsymbol{Z}^i\}\,|\,\hat{H}\,|\,\Phi_{MK'}^{J\,\pm}\,\{\boldsymbol{Z}^{i'}\}\rangle$ and $\langle\,\Phi_{MK}^{J\,\pm}\,\{\boldsymbol{Z}^i\}\,|\,\Phi_{MK'}^{J\,\pm}\,\{\boldsymbol{Z}^{i'}\}\rangle$ simultaneously.

In Fig. 3, the calculated energy-level scheme of the positive-partiy states in ^{12}C with the experimental data and the obtained intrinsic densities for

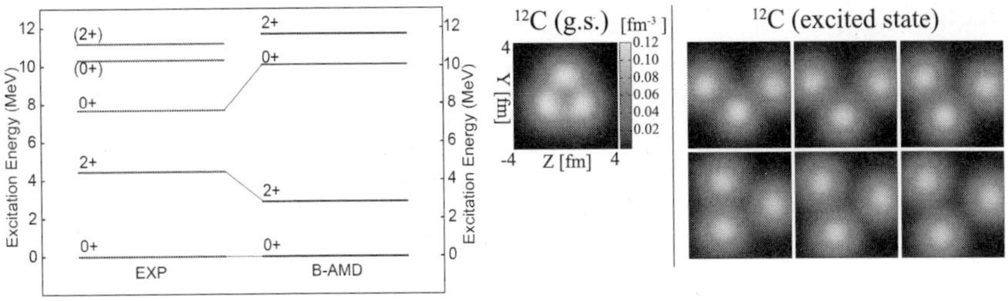

Fig. 3. The energy-level scheme of the positive-parity states with the excitation energy below 12 MeV in ^{12}C in the left panel. The obtained ground-state binding energy is -73.7 MeV. The density distribution of the ground-state solution and typical ones included in the excited 0_2^+ solution labeled "^{12}C(g.s.)" and "^{12}C(excited state)", respectively, in the right. The configurations of the 0_2^+ state have the root-mean-square mass radiuses $R = 2.8 - 3.4$ (fm) and the approximated deformation parameters $\beta = 0.49 - 0.76$.

the ground state and the excited 0_2^+ state are shown. We obtain the solution of the excited 0_2^+ state with the excitation energy $\sim$10 MeV. This excited 0_2^+ state contains of the explicit 3α configurations as seen in Fig. 3, which has no dominant 3α configuration because these states have almost equal weights in the 0_2^+ state. This feature seems to be indicated as that of the 3α condensed state[20] which is the present interpretation of the 0_2^+ state in ^{12}C. In the calculation with the condensed wave function,[20] it is indicated that this 0_2^+ state has very large mass radius with $\sim$4 fm, therefore, we may need to superpose more states with a larger spread to improve the description of 0_2^+ state as the following works.

4. Summary

In this work, we developed the framework of AMD with bare nuclear interactions based on the Brueckner theory, Brueckner-AMD. In the Brueckner-AMD, the G-matrix and the single-particle energies of AMD can be determined from the bare interactions self-consistently without any corrections. The correlation functions generated from the solutions of Bethe-Goldstone equation construct the G-matrix as the explicit effective interactions with non-central operators. In this paper, we decribed the reasonable cluster structures and reproduced the low-lying energy levels of some light nuclei starting from the Argonne $v8$'. Futhermore, we decribed the 0_2^+ state in ^{12}C which is not solved sufficiently by the shell model approaches.

References

1. R.B. Wiringa, S.C. Pieper, J. Carlson and V.R. Pandharipande, Phys. Rev. C **62**, 014001 (2000).
2. C. Forssén, E. Caurier and P. Navrátil, Phys. Rev. C **79**, 021303 (2009).
3. G. Hagen, T. Papenbrock, D.J. Dean, M. Hjorth-Jensen and B.V. Asokan, Phys. Rev. C **80**, 021306 (2009).
4. T. Neff and H. Feldmeier, Nucl. Phys. A **738**, 357 (2004).
5. S. Fujii, T. Mizusaki, T Otsuka, T Sebe and A. Arima, Phys. Lett. B **650**, 9 (2007).
6. T. Myo, H. Toki and K. Ikeda, Prog. Theor. Phys. **121**, 511 (2009).
7. H. Kamada *et al.*, Phys. Rev. C **64**, 044001 (2001).
8. K. Ikeda, T. Marumori, R. Tamagaki and H. Tanaka, Prog. Theor. Phys. Suppl. No. 52, 1 (1972).
9. Y. Kanada-En'yo and H. Horiuchi, Prog. Theor. Phys. Suppl. No.142, 205 (2001).
10. T. Togashi and K. Katō, Prog. Theor. Phys. **117**, 189 (2007).
11. T. Togashi, T. Murakami and K. Katō, Prog. Theor. Phys. **121**, 299 (2009).
12. T. Togashi, T. Murakami and K. Katō, Int. J. Mod. Phys. E **17**, 2081 (2008); J. Phys.: Conference Series **111**, 012026 (2008).
    T. Togashi, K. Katō and T. Murakami, Int. J. Mod. Phys. A **24**, 2142 (2009).
    T. Murakami, T. Togashi and K. Katō, Mod. Phys. Lett. A **24**, 1013 (2009).
13. K. A. Brueckner, C.A. Levinson and H.M. Mahmoud, Phy. Rev. **95**, 217 (1954).
14. A. Doté, Y. Kanada-En'yo and H. Horiuchi, Phys. Rev. C **56**, 1844 (1997).
15. H.A. Bethe, Phy. Rev. **103**, 1353 (1956).
    J. Goldstone, Proc. R. Soc. A **239**, 267 (1957).
16. E. Uegaki, Y. Abe, S. Okabe and H. Tanaka, Prog. Theor. Phys. **62**, 1621 (1979).
17. M. Kamimura, Nucl. Phys. A **351**, 456 (1981).
18. Y. Kanada-En'yo, Prog. Theor. Phys. **117**, 655 (2007).
19. M. Chernykh, H. Feldmeier, T. Neff, P.von Neumann-Cosel and A. Richter, Phys. Rev. Lett. **98**, 032501 (2007).
20. Y. Funaki, A. Tohsaki, H. Horiuchi, P. Schuck and G. Röpke, Eur. Phys. J. A **24**, 321 (2005).
21. P. Navrátil, J.P. Vary and B.R.Barrett, Phys. Rev. Lett. **84**, 5728 (2000).
22. A. Ono, H. Horiuchi, T. Maruyama and A. Ohnishi, Phys. Rev. Lett. **68**, 2898 (1992); Prog. Theor. Phys. **87**, 1185 (1992).
23. H. Bandō, S. Nagata and Y. Yamamoto, Prog. Theor. Phys. **44**, 646 (1970); Prog. Theor. Phys. **45**, 1515 (1971).
24. S. Nagata, H. Bandō and Y. Akaishi, Prog. Theor. Phys. Suppl. No. 65, 10 (1979).
25. R.B. Wiringa and S.C. Pieper, Phys. Rev. Lett. **89**, 182501 (2002).
26. Y. Taniguchi, M. Kimura and H. Horiuchi, Prog. Theor. Phys. **112**, 475 (2004).

COHERENT PIONS FROM NEUTRINO SCATTERING OFF NUCLEI

M. VALVERDE

Research Center for Nuclear Physics RCNP, Osaka University,
Ibaraki, 567-0047, Japan
E-mail: valverde@rcnp.osaka-u.ac.jp

J. E. AMARO

Dept. de Física Atómica, Molecular y Nuclear, Universidad de Granada
Granada, E-18071, Spain

E. HERNÁNDEZ

Grupo de Física Nuclear, Dept. de Física Fundamental e IUFFyM
Facultad de Ciencias, Univ. de Salamanca
Salamanca, E-37008, Spain.

J. NIEVES

Instituto de Física Corpuscular (IFIC), Centro Mixto CSIC-Universidad de Valencia,
Institutos de Investigación de Paterna
Valencia, Aptd. 22085, E-46071, Spain

M. J. VICENTE VACAS

Dpto. de Física Teórica and IFIC, Centro Mixto CSIC-Universidad de Valencia,
Institutos de Investigación de Paterna
Valencia, Aptd. 22085, E-46071, Spain

We describe a model for pion production off nucleons and coherent pions from nuclei induced by neutrinos in the 1 GeV energy regime. Besides the dominant Δ pole contribution, it takes into account the effect of background terms required by chiral symmetry. Moreover, the model uses a reduced nucleon-to-Δ resonance axial coupling, which leads to coherent pion production cross sections around a factor two smaller than most of the previous theoretical estimates. Nuclear effects like medium corrections on the Δ propagator and final pion distortion are included.

Keywords: Neutrino scattering; Coherent pion production; Axial coupling of the $\Delta(1232)$ resonance.

1. Introduction

Neutrinos have been in the forefront of research in particle and nuclear physics for a long time. One of these fields is the study of pion production off nuclei induced by neutrinos. A proper understanding of this process is necessary in the analysis of the present generation of precision neutrino oscillation experiments. For instance, the π^0 produced in neutral currents (NC) is the most important ν_μ-induced background to experiments like MiniBoone[1] that are trying to measure $\nu_\mu \to \nu_e$ oscillations in the neutrino energy range around 1 GeV. Also of importance is the background that appears from π^+ charged current (CC) production in $\nu_\mu \to \nu_x$ disappearance searches like T2K.[2] Moreover the pion is strongly coupled to the $\Delta(1232)$ resonance, and neutrino scattering is presently the best way to access to the axial nucleon-Δ transition couplings. The most complete information in this regard comes from the bubble chamber data of ANL[3,4] and BNL[5,6] where the target was cooled deuterium. However, in present oscillation experiments the target for neutrino interaction is finite nuclei, for instance ^{12}C (mineral oil in MiniBoone) or ^{16}O (water target in T2K). This introduces sizable many-body effects that are difficult to disentangle from the genuine single nucleon response to the neutrino probe.

In Sec. 2 of these paper we describe a phenomenological model[7] for pion production induced by neutrino scattering off free nucleons. This model takes into account non-resonant background processes that are usually neglected. These non-resonant processes are determined by chiral symmetry and thus do not introduce free parameters. We then perform a fit of the axial N-Δ parameters and discuss a possible violation of the non-diagonal Goldberger-Treiman relation. In Sec. 3 we extend the model to describe the CC coherent process in which the final nucleus is left in its ground state. We will further try to discuss how the coherent reaction can show some light on the values of the axial N-Δ couplings. We shall focus on the CC process, but the model can be easily extended to antineutrino reactions and NC processes. The interested reader could find further details in Refs. 7–11.

2. Single Nucleon Pion Production

Here we review the model for the free nucleon reaction

$$\nu_l(k) + N \to l^-(k') + N' + \pi^+(k_\pi) \tag{1}$$

as introduced in Ref. 7. This model considers the dominant Δ pole mechanism in which the neutrino excites a $\Delta(1232)$ resonance that subsequently decays into $N\pi$. In our model we have also included non-resonant background terms as required by chiral symmetry, see Fig. 1. Some previous works[12–14] also considered background terms, though they were not consistent with the chiral counting.

The vector part of the interaction N-Δ can be related to the electromagnetic current by imposing conservation of the vector current. For photon induced reactions extensive experimental data exist and in Ref. 15 they were employed to fit the vector current couplings. We shall use this fit in our work. Unfortunately the axial N-Δ current is not so well studied. The usual approach is to parameterize the interaction in terms of four form factors $C^A_{3,4,5,6}(q^2)$. One can assume partial conservation of the axial current (PCAC) and obtain the relation $C^A_6 = C^A_5 M^2/(m^2_\pi - q^2)$, with M the nucleon mass. Furthermore, one can deduce[16] from dispersion relations the following conditions: $C^A_3(q^2) = 0$ and $C^A_4 = -C^A_5/4$. Thus we are left with only a free form factor, the dominant one $C^A_5(q^2)$. A different number of parameterizations have been proposed for this form factor, nevertheless the experimental data are quite limited and thus a simple dipole form

$$C^A_5(q^2) = \frac{C^A_5(0)}{\left(1 - q^2/M^2_A\right)^2} \tag{2}$$

should be enough. In order to keep the axial transition radius in the range 0.7–0.8 fm one expect the axial mass to have a value of around $M_A \sim$ 0.85–1.0 GeV. Furthermore one can assume the well known Goldberger-Treiman relation (GTR) for πNN coupling to be also valid for the $\pi N\Delta$ coupling and thus obtain

$$C^A_5(0) = \sqrt{\frac{2}{3}} f_\pi \frac{f^*}{m_\pi} = 1.2\,, \tag{3}$$

where $f_\pi = 93\,\text{MeV}$ is the pion decay constant and $f^* = 2.2$, the $\pi N\Delta$ coupling. Unfortunately, there are no constraints from Chiral Perturbation Theory[17] and the lattice QCD calculations[18] are still inconclusive.

Most of the approaches in the literature assume Δ dominance, that is, only include the first two diagrams in Fig. 1. We improve this situation by including non-resonant contributions[7] required by chiral symmetry. In addition to the $\Delta(1232)$ pole (ΔP) (first row) mechanism the model includes background terms required by chiral symmetry: nucleon (second row) pole terms contact and pion pole contribution (third row) and pion-in-flight term. We calculate them by using the SU(2) non-linear σ model Lagrangian. The only parameters in the theory are the pion and nucleon

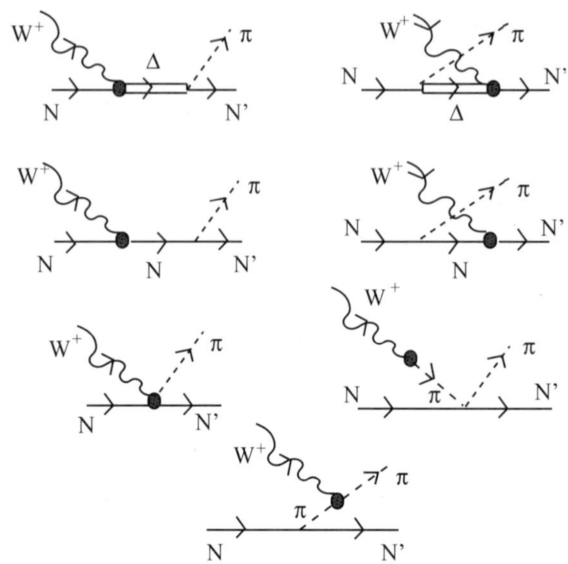

Fig. 1. Set of diagrams for the model for the $W^+N \to N'\pi$ reaction.

masses and the pion decay constant. All other couplings are completely fixed by the theory, so no new parameter is introduced.

We found (see Fig. 2) that these background terms produced significant effects in all channels, namely an enhancement of about 10% in the cross section that resulted in a disagreement with the ANL data. As a result we

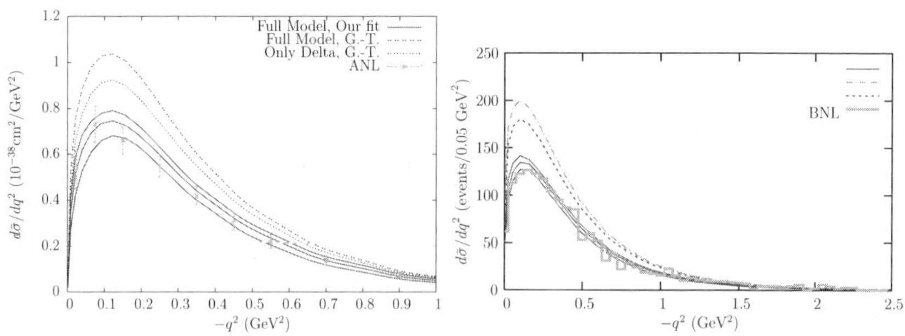

Fig. 2. Flux averaged q^2−differential $\nu_\mu p \to \mu^- p \pi^+$ cross section for the ANL (left) and BNL (right). Dashed lines stand for the contribution of the ΔP mechanism with the GTR assumption for C_5^A. We also plot results with the full model of Fig. 1, assuming GTR (dashed-dotted) and with our best fit parameters, Eq. (4).

had to readjust the strength of the dominant Δ pole contribution. The least known ingredients of the model are the axial nucleon-to-Δ transition form factors, of which C_5^A gives the largest contribution. This strongly suggested a refit of that form factor to the experimental data, which we did by fitting the flux-averaged $\nu_\mu p \to \mu^- p\pi^+$ ANL q^2-differential cross section for pion-nucleon invariant masses[a] $W < 1.4\,\mathrm{GeV}$. The obtained parameters were

$$C_5^A(0) = 0.87 \pm 0.08\,, \quad M_A = 0.985 \pm 0.082\,\mathrm{GeV} \tag{4}$$

with a $\chi^2/\mathrm{dof} = 0.4$ and a correlation coefficient $r = -0.85$, that amounts to a 30% reduction of the GTR prediction. Thus, our full model leads to an overall better description of the data for one-pion production reactions off the nucleon. This reduction of the $C_5^A(0)$ value is consistent with recent results in lattice QCD[18] and quark models.[19]

Recently other fits have been proposed. For instance, in Ref. 20 they keep the GTR but introduce a non-dipole form factor with additional parameters. As in the ANL data the relevant phase space is around $q^2 = 0.1\,\mathrm{GeV}^2$, they could keep the GTR at the cost of having a large dependence on q^2 for the form factor, that yields a large, somehow unphysical, axial transition radius of around 1.4 fm. Furthermore neither statistical errors nor correlation factors are given in that reference. Another analysis[21] raised new questions, namely the effect of deuterium wave function on the cross section and the flux uncertainties in the ANL and BNL data. The authors of this latter work took into account both effects, though we believe that their statistical analysis is not quite robust (see discussion in Ref. 11). Furthermore they only took into account the dominant Δ contribution. The inclusion of deuterium wave function reduces the cross section about an 8%, so this somehow compensates the neglect of the non-resonant background, and they obtained a best fit of $C_5^A(0) = 1.19 \pm 0.08$ in agreement with the GTR assumption.

Recently[11] we have improved our fit of Ref. 7, improving the lines suggested in Graczyk et al. In summary in this new fit

(i) all diagrams in Fig. 1 are included;
(ii) the fitted data are the full ANL data set and the BNL total cross sections at the three lowest neutrino energies (we neglect higher energies where the effect of higher resonances beyond the $\Delta(1232)$ must be

[a]This cut was introduced in order to avoid the effects of resonances higher than the $\Delta(1232)$.

addressed; the BNL q^2-differential cross sections were not taken into account as they are not normalized);

(iii) deuterium wave function effects were introduced following the prescription of Ref. 22;

(iv) the form factors C_3^A and C_4^A were tentatively included in one fit, though the data were found to be quite insensitive to their values so we decided to stick to the Adler's assumption; and

(v) the uncertainties in ANL and BNL flux normalization are introduced as fully correlated systematic errors.

In this way we obtain a best fit of $C_5^A(0) = 1.00 \pm 0.11$ and $M_A = 0.93 \pm 0.07\,\text{GeV}$ with a goodness of fit value of χ^2d.o.f. $= 0.42$. Thus we observe a violations of the off diagonal Goldberger-Treiman relation at the level of 2σ.

3. The Coherent Reaction

Here we describe our model[9] for the coherent reaction

$$\nu_l(k) + A_Z|_{gs}(p_A) \to l^-(k') + A_Z|_{gs}(p'_A) + \pi^+(k_\pi) \tag{5}$$

where the target nucleus A_Z is left in the ground state (gs). To calculate the amplitude of this process we sum over all individual nucleon wave functions, which are modelled by a Fermi gas in local density approximation. The individual nucleon amplitudes are modelled following the model of the previous section, using the fit of Eq. 4 for the C_5^A form factor. On top of that a number of many-body effects are introduced. In first place we take into account the in-medium modifications[23] of the $\Delta(1232)$ properties. This implies a shift in the pole mass towards lower energies and most importantly a net broadening of the width (the opening of new decay channels in the nuclear medium compensates the Pauli blocking of the πN decay channel). Also important is the distortion of the outgoing pion by strong interaction with the nucleus. Thus we consider the wave function of the pion to be the outgoing solution to the Klein-Gordon equation with a microscopic optical potential[24] whose imaginary part takes into account the inelastic interactions of the pion with the nucleus, that thus disappear from the coherent channel. We must emphasize here that solving the Klein-Gordon equation is the correct way of describing the distortion of the outgoing pion. Other approaches use either a Monte Carlo simulation[25] or include an attenuation factor fitted to the pion nuclei scattering cross section.[26] The first procedure, though physically sound, can be a bit misleading as it includes in its cross section processes (like quasi-elastic scattered pions) that do not leave the nucleus in it ground state, thus are not coherent. This kind of models are used in the analysis of MiniBoone experiment, thus making a bit messy the direct comparison between theoretical models and experimental results (see discussion in Ref. 9). The second approach is an oversimplification, as the pion-nucleus scattering is quite a different process from the neutrino induced pion production. As pion nucleus interaction is governed by the strong interaction, the incoming pion interacts strongly with the nuclear surface, thus the pion nucleus cross section is quite insensitive to the details of the nuclear core. On the other hand, neutrino scattering is a weak process, dominated by nuclear density, so pions are mostly produced in the deep, high density regions of the nucleus. The physics of pion interaction is thus quite different in pion scattering off nuclei and pion production by electroweak probes.

In left panel of Fig. 3 we show the pion momentum distribution for CC coherent pion production, in the peak energy region of the T2K experiment. Including Δ in-medium self-energy (long-dashed line) reduces the PWIA results (short-dashed line). Further inclusion of pion distortion (full model, solid line) reduces the cross section, and the peak is shifted towards lower energies, reflecting the strong absorption and the higher probability of a quasi-elastic collisions of the pion in the Δ kinematical region. The total cross section reduction is around 60%. Similar nuclear effects were already studied in Refs. 27,28. However, the authors of these references neglected the nucleon momenta in the Dirac spinors. The effect of this approximation (nucleons at rest, dotted line) results in a $\sim$ 15% decrease of the total cross section. In the right panel of Fig. 3 we show the pion angular distribu-

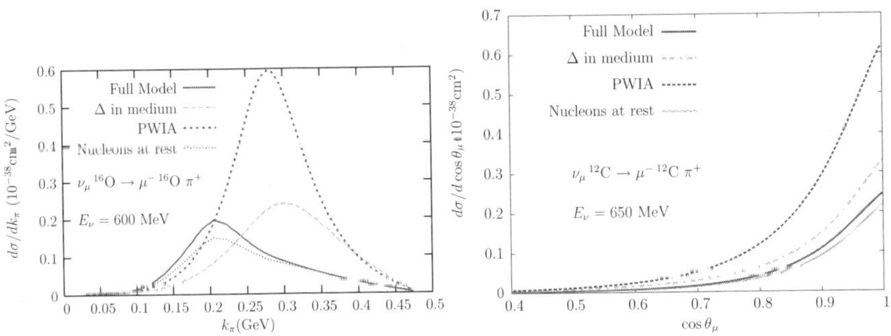

Fig. 3. Pion momentum (right) and angular (left) differential cross section.

tion with respect to the incoming neutrino direction. The reaction is very forward peaked, as expected due to the nucleus form factor. The angular distribution profile keeps its forward peaked behavior after introduction of nuclear medium effects. Furthermore we corrected some numerical errors in the mentioned papers. However one must be aware that this model does not take into account the non-localities[29] in the Δ propagation. We believe this effect is partially taken into account in an effective fashion by our treatment of the Δ in nuclear medium; nevertheless further studies would be interesting.

Table 1. Total cross sections for the coherent process. We neglect the highest 10% of the energy spectrum.

Reaction	Experiment	$\sigma(10^{-40}\ \mathrm{cm}^2)$	$\sigma(10^{-40}\ \mathrm{cm}^2)$ Exp
CC ν_μ+^{12}C	K2K		< 7.7
NC ν_μ+^{12}C	MiniBoone	3.33	$7.7 \pm 1.6 \pm 3.6$
CC ν_μ+^{12}C	MiniBoone	4.46	
CC ν_μ+^{16}O	T2K	4.19	
CC ν_μ+^{12}C	T2K	3.54	

In Table 1 we compare our model with present results of K2K experiment[30] and show our predicitions for MiniBoone and T2K. Our prediction, subject to sizable uncertainties, lies well below the K2K upper bound, due to the use of a low value for $C_5^A(0)$, while our prediction for the ν_μ NC MiniBoone cross section is notably smaller than that given in the PhD thesis of J. L. Raaf.[31] However, we believe (see discussion in Refs. 9,10 that the MiniBoone analysis might importantly overestimate this cross section, not only because some of the π^0s which undergo FSI collisions are accounted for instead of being removed, but also because a possible mis-match between the absolute normalisation of the background and coherent yields. Note that the K2K and MiniBoone results seems somehow incompatible with the approximate relation $\sigma_{\mathrm{CC}} \approx 2\sigma_{\mathrm{NC}}$, which would be expected from Δ dominance and neglecting finite muon mass effects.

Acknowledgments

M. V. acknowledges a Postdoctoral Fellowship from the Japanese Society for the Promotion of Science (JSPS). Research supported by DGI contracts FIS2008-01143, FIS2006-03438, FPA2007-65748 and CSD2007-00042, JCyL contracts SA016A07 and GR12, Generalitat Valenciana contract PROMETEO/2009/0090 and by EU HadronPhysics2 contract 227431.

References

1. A. A. Aguilar-Arevalo *et al.*, *Phys. Rev. Lett.* **98**, p. 231801 (2007).
2. K. Hiraide, *Nucl. Phys. Proc. Suppl.* **159**, 85 (2006).
3. S. J. Barish *et al.*, *Phys. Rev.* **D19**, p. 2521 (1979).
4. G. M. Radecky *et al.*, *Phys. Rev.* **D25**, 1161 (1982).
5. T. Kitagaki *et al.*, *Phys. Rev.* **D34**, 2554 (1986).
6. T. Kitagaki *et al.*, *Phys. Rev.* **D42**, 1331 (1990).
7. E. Hernandez, J. Nieves and M. Valverde, *Phys. Rev.* **D76**, p. 033005 (2007).
8. E. Hernandez, J. Nieves, S. K. Singh, M. Valverde and M. J. Vicente Vacas, *Phys. Rev.* **D77**, p. 053009 (2008).
9. J. E. Amaro, E. Hernandez, J. Nieves and M. Valverde, *Phys. Rev.* **D79**, p. 013002 (2009).
10. E. Hernandez, J. Nieves and M. J. Vicente-Vacas, *Phys. Rev.* **D80**, p. 013003 (2009).
11. E. Hernandez, J. Nieves, M. Valverde and M. J. Vicente-Vacas, arXiv:hep-ph/1001.4416, (2010).
12. G. L. Fogli and G. Nardulli, *Nucl. Phys.* **B160**, p. 116 (1979).
13. G. L. Fogli and G. Nardulli, *Nucl. Phys.* **B165**, p. 162 (1980).
14. T. Sato, D. Uno and T. S. H. Lee, *Phys. Rev.* **C67**, p. 065201 (2003).
15. O. Lalakulich, E. A. Paschos and G. Piranishvili, *Phys. Rev.* **D74**, p. 014009 (2006).
16. S. L. Adler, *Phys. Rev.* **135**, B963 (1964).
17. L. S. Geng, J. Martin Camalich, L. Alvarez-Ruso and M. J. Vicente Vacas, *Phys. Rev.* **D78**, p. 014011 (2008).
18. C. Alexandrou, T. Leontiou, J. W. Negele and A. Tsapalis, *Phys. Rev. Lett.* **98**, p. 052003 (2007).
19. D. Barquilla-Cano, A. J. Buchmann and E. Hernandez, *Phys. Rev.* **C75**, p. 065203 (2007).
20. T. Leitner, O. Buss, L. Alvarez-Ruso and U. Mosel, *Phys. Rev.* **C79**, p. 034601 (2009).
21. K. M. Graczyk, D. Kielczewska, P. Przewlocki and J. T. Sobczyk, *Phys. Rev.* **D80**, p. 093001 (2009).
22. L. Alvarez-Ruso, S. K. Singh and M. J. Vicente Vacas, *Phys. Rev.* **C59**, 3386 (1999).
23. E. Oset and L. L. Salcedo, *Nucl. Phys.* **A468**, 631 (1987).
24. J. Nieves, E. Oset and C. Garcia-Recio, *Nucl. Phys.* **A554**, 554 (1993).
25. A. A. Aguilar-Arevalo *et al.*, *Phys. Lett.* **B664**, 41 (2008).
26. S. X. Nakamura, T. Sato, T. S. H. Lee, B. Szczerbinska and K. Kubodera, arXiv:nucl-th:/0910.1057, (2009).
27. L. Alvarez-Ruso, L. S. Geng, S. Hirenzaki and M. J. Vicente Vacas, *Phys. Rev.* **C75**, p. 055501 (2007).
28. L. Alvarez-Ruso, L. S. Geng and M. J. Vicente Vacas, *Phys. Rev.* **C76**, p. 068501 (2007).
29. T. Leitner, U. Mosel and S. Winkelmann, *Phys. Rev.* **C79**, p. 057601 (2009).
30. M. Hasegawa *et al.*, *Phys. Rev. Lett.* **95**, p. 252301 (2005).
31. J. L. Raaf, Ph. D. Thesis. FERMILAB-THESIS-2005-20.

STRANGE AND NON-STRANGE BARYON-BARYON INTERACTIONS RECENT NIJMEGEN EXTENDED-SOFT-CORE MODELS

TH. A. RIJKEN*

IMAPP, University of Nijmegen, Nijmegen, The Netherlands, and RIKEN, Nishina Center for Accelerator-Based-Science, Wako 351-0198, Japan
**E-mail: t.rijken@science.ru.nl*

M.M. NAGELS

IMAPP, University of Nijmegen, Nijmegen, The Netherlands

Y. YAMAMOTO

Tsuru University, Tsuru, Yamanashi 402-8555, Japan

We report on recent results obtained with the extended-soft-core (ESC) interactions. The ESC-model gives a unified description of nucleon-nucleon (NN), hyperon-nucleon (YN), and hyperon-hyperon (YY), using (broken) flavor $SU_f(3)$-symmetry. This is achieved by fitting simultaneously NN-, YN-, and YY-data with single sets of meson parameters. The dynamics is derived from (i) one-boson-exchange (OBE), (ii) two-meson-exchanges (TME), and (iii) meson-pair-exchange (MPE). The changes with respect to the previous versions are explained, and the properties of the meson parameters are commented on. Emphasized are the relations with the quark-model (QM), QCD, chiral-symmetry, and the quark-structure of the baryons. The NN, YN, and YY results for this model are excellent. This is marked particularly by the NN-results, namely $\chi^2_{p.d.p.} = 1.182$ and 1.135 for respectively ESC08a and ESC08b. It is to be noted that the solutions suggest the existence of an S=-2 deuteron state in the $\Xi N(^3S_1, T = 1)$-channel. The predictions for the S=-3,-4-channels include the existence of a bound $^1S_0(\Xi\Xi, T = 1)$- state.

Keywords: Baryon-baryon, ESC-model, Quark-model, QCD, Hypernuclei, G-matrix.

1. Introduction

The reality of a Kaon-machine at J-PARC means great opportunities for studies in e.g. kaon-nucleon (KN) and baryon-baryon (BB) at low-, intermediate-, and high-energies. Here, we review briefly the most recent

ESC08-models for baryon-baryon systems with strangeness S=0,-1, and −2, using (broken) $SU_F(3)$-symmetry. The first results were presented and compared to those for ESC04[1–3] in Ref. 4. For a preliminary version of this new model, called ESC07, see Ref. 5. In the currently standard general physical framework, notably QCD, flavor SU(3), and chiral $SU(3)_L \otimes SU(3)_R$, we discuss the interactions between baryons in terms of meson-exchange from the perpectives of the quark-model (QM). In passing we emphasize that the ESC-models satisfy all criteria from chiral-symmetry in the non-linear realization.[6] Important topics on which the ESC-model gives information are: (i) properties of the Yukawian soft-core meson-exchange potentials: one-boson-exchange (OBE), two-meson-exchange (TME), meson-pair-exchange (MPE), (ii) the connection between the meson coupling constants and F/(F+D)-ratio's based on $SU_f(3)$ and the QM, (iii) multiple gluon-exchange, (iv) form-factors, (v) quark-core effects. The latter are based on the observation that the so-called Pauli-forbidden SU(6)-irrep [51] occurs with a weight around 1/2 for about all the baryon-baryon channels, except for the ${}^3S_1(\Sigma^+ p, T=3/2)$-, ${}^1S_0(\Sigma N, T=1/2)$-, and ${}^1S_0(\Xi N, T=1)$-channel. We describe this baryon-structure dependent feature by giving the pomeron-exchange an enhancement factor proportionally to the weight of the [51]-irrep. The treatment of the axial-vector mesons with $J^{PC}=1^{++}$ has been improved significantly. Furthermore, for these mesons we have included besides the $\gamma_\mu \gamma_5$- also the derivative coupling, As a further novelty, we introduce the Odderon-potentials. They represent the effects corresponding to color singlet n-gluon exchange, with odd $n = 3, 5, ...$ and which has a vector-exchange character.[7] These are complementary to the exchange of an even number of gluons (Pomeron-exchange). The details of these potentials will be published elsewhere.

With these ingredients a rather flexible dynamical framework is constructed. Namely, it appeared feasible to keep (a) the coupling constants $SU_f(3)$-symmetric, (b) the parameters of the model in reasonable accordance with the predictions of the ${}^3P_0$ quark-pair-creation model (QPC), although in ESC08a,b a certain admixture of ${}^3S_1$ quark-pair creation is present. This is the case for the meson- and meson-pair-baryon coupling constants and the $F/(F+D)$-ratio's as well. Compared to the ESC04-models[1,2] the main framework is unchanged and given by the ESC-potentials from OBE-, TME-, and MPE-exchanges. The guidance chosen are the couplings and $F/(F+D)$-ratio's as suggested by the QPC-model.[8] This, as it turns out, leads to a rather flexible dynamical system that allows very satisfactory simultaneous fits to the low-energy NN- and YN-data.

This with single sets of parameters that can be understood qualitatively in the QPC-model.

As is well known, at present the YN-data are scarce compared to the wealth of the precise NN-data for many types of observables. Nevertheless, the availables YN-scattering data and the requirement of the absence of bound states for the YN-channels is an important determining factor for the model and its parameters. For ESC08a and ESC08b we obtained for the 4300 selected NN-data $\chi^2_{p.d.p.} = 1.182$, respectively 1.135. The set of 38 YN-data was also fitted succesfully, while having no bound states. This set comprises the usual standard set of 35 data employed in the Nijmegen work on YN,[9] plus the recently measured Σ^+p X-sections of the KEK-PS E289 experiment.[10,11]
In Ref. 2 it is shown that, by exploiting the nuclear-density dependence of the vector-meson masses, the universal repulsion for the BB-interaction in a nuclear medium[12] can be described succesfully. Such a repulsion brings about three good things: (i) great improvement in nuclear saturation, (ii) stiffening of the equation of state (EOS), giving a realistic neutron-star mass, (iii) and improving the well-depth's U_Λ, U_Σ. This avoids a too soft EOS for ESC-models, involving only two-baryon forces. There remained, however, two problems with ESC04 to be solved. This is the Σ well-depth, and the spin-orbit splittings in Λ-hypernuclei. These problems will be addressed below in this paper.
The meson dynamical contents of the ESC08-model is best illustrated by looking at Table 1. The quantum numbers $^{2S+1}L_J$ refer to those of the $q\bar{q}$

Table 1. Meson and Meson-pair table ESC08-model. The SU(3)$_F$-irreps are denoted by $\{\mu\}$.

$^{2S+1}L_J$	J^{PC}	$T=1$	$T=0$	$T=1/2$	$\{\mu\}$
$^1S_0$	0^{-+}	$\pi(140)$	$\eta(495), \eta'(958)$	K(495)	$\{8\}\oplus\{1\}$
$^3S_1$	1^{--}	$\rho(760)$	$\omega(783), \phi(1019)$	$K^*(892)$	$\{8\}\oplus\{1\}$
$^3P_0$	0^{++}	$a_0(980)$	$\epsilon(760), S^*(993)$	$\kappa(900)$	$\{8\}\oplus\{1\}$
$^3P_1$	1^{++}	$a_1(1270)$	$D(1285), E(1420)$	$K_1(1400)$	$\{8\}\oplus\{1\}$
$^1P_1$	1^{+-}	$b_1(1235)$	$h_1(1170), h_1'(1380)$	$K_{1B}(1400)$	$\{8\}\oplus\{1\}$
$^3P_0$	0^{++}	—	$\pi\pi, \eta\eta, K\bar{K}$	—	$\{1\}$
$^3P_0$	0^{++}	$\pi\eta, K\bar{K}$	$\pi\pi, \eta\eta, K\bar{K}$	$\pi K, \eta K,$	$\{8_s\}$
$^3S_1$	1^{--}	$\pi\pi, K\bar{K}...$	—	$\pi K, \eta K$	$\{8_a\}$
$^3P_1$	1^{++}	$\pi\rho, K\bar{K}^*, ...$	$K\bar{K}^*, ..$	$K\rho, \pi K^*$	$\{8_a\}$
$^3P_1$	1^{++}	$\pi\sigma, K\bar{\kappa}, ...$	$K\bar{\kappa}, ..$	$K\sigma, \pi\kappa$	$\{8_a\}$
$^3P_1$	1^{+-}	$\pi\omega, K\bar{K}^*, ...$	$K\bar{K}^*, ..$	$K\rho, \pi K^*$	$\{8_s\}$

ground-states corresponding to the mesons. The lower part of this table shows the meson-pair exchange contents. Here, the $q\bar{q}$ quantum numbers refer to the dominant heavy-boson contributions to the pair-vertices. In ESC04 all these exchanges were also included, except for the axial-vector mesons of the second kind having $J^{PC} = 1^{+-}$. Besides the OBE- and MPE-potentials, also included are the uncorrelated two-pseudoscalar exchanges (TPE), and the diffractive pomeron/odderon potentials, whose quantum numbers are $0^{++}/1^{--}$.
In the text we refer to the models ESC08a,b as those that were also reviewed in Ref. 4. The version ESC08a', for which we give in this paper in particularly the results, is a slight update of model ESC08a having better $F/(F+D)$-parameters for the axial-vector couplings. Furthermore, for all these models we have now included also the quark-core effect in the $\Xi N(^1S_0, T=1)$-channel.

2. Short-range Phenomenology

After our long experience with the soft-core models, in order to generate sufficient repulsion for the well-depth U_Σ it is natural to look for some change in the short range potentials. From the work on the flavor(F) and spin(S) $SU(6)_{FS}$ quark-cluster model (QCM) of BB-interactions,[13–15] we found that the forbidden SU(6)-irrep [51] configuration $(0s)^6$ has an exceptional large weight in the $SU_F(3)$-irreps $\{10\}$ and $\{8_s\}$. So, in particularly for the $\Sigma^+p(^3S_1, T=3/2)$-, $\Sigma N(^1S_0, T=1/2)$- and $\Xi N(^3S_1, T=1)$-channels. For the other channels the weights are about 50%. Therefore, we represent the sort range repulsion due to two-gluon-exchange and the average effects of the forbidden [51] configuration by the pomeron, and effectuate the exceptional repulsion in specific BB-channels due to the $SU(6)_{FS}$-irrep [51] by giving here the pomeron an enhancement factor. This procedure solves the lack of the repulsion in these specific YN-channels.

In Table 2 for the six-quark-core $(0s)^6$-configurations we give the weights of the flavor-spin $SU(6)_{FS}$-irreps for the various BB-channels, see Ref. 14. As is well known these BB-channels can be expressed also in terms of the flavor SU(3)-irreps. Comparing this with Table 2 one easily finds the connections given in Table 3. So, we see that the [51]-irrep has a dominant large weight in the $\{10\}$- and $\{8_s\}$-irrep, which gives an argument for the presence of an exceptionally strong Pauli-repulsion in these $SU(3)_F$-irreps. In ESC08 we implement these features by adapting the Pomeron strength in the irreps $\{10\}$ and $\{8_s\}$, notably in the $\Sigma^+p(^3S_1, T=3/2)$ and $\Sigma N(^1S_0, T=1/2)$ channels. (A more refined treatment of the quark-core effects for all chan-

Table 2. Weights $SU(6)_{FS}$-irreps in various potentials (see Ref. 14, Table I)

	(S,I)	$V = aV_{[51]} + bV_{[33]}$
$NN \to NN$	$(0,1)$	$V_{NN} = \frac{4}{9}V_{[51]} + \frac{5}{9}V_{[33]}$
$NN \to NN$	$(1,0)$	$V_{NN} = \frac{4}{9}V_{[51]} + \frac{5}{9}V_{[33]}$
$\Lambda N \to \Lambda N$	$(0,1/2)$	$V_{\Lambda\Lambda} = \frac{1}{2}V_{[51]} + \frac{1}{2}V_{[33]}$
$\Lambda N \to \Lambda N$	$(1,1/2)$	$V_{\Lambda\Lambda} = \frac{1}{2}V_{[51]} + \frac{1}{2}V_{[33]}$
$\Sigma N \to \Sigma N$	$(0,1/2)$	$V_{\Sigma\Sigma} = \frac{17}{18}V_{[51]} + \frac{1}{18}V_{[33]}$
$\Sigma N \to \Sigma N$	$(1,1/2)$	$V_{\Sigma\Sigma} = \frac{1}{2}V_{[51]} + \frac{1}{2}V_{[33]}$
$\Sigma N \to \Sigma N$	$(0,3/2)$	$V_{\Sigma\Sigma} = \frac{4}{9}V_{[51]} + \frac{5}{9}V_{[33]}$
$\Sigma N \to \Sigma N$	$(1,3/2)$	$V_{\Sigma\Sigma} = \frac{8}{9}V_{[51]} + \frac{1}{9}V_{[33]}$
$\Lambda\Lambda \to \Lambda\Lambda$	$(0,0)$	$V_{\Lambda\Lambda} = \frac{1}{2}V_{[51]} + \frac{1}{2}V_{[33]}$
$\Xi N \to \Xi N$	$(0,0)$	$V_{\Xi N} = \frac{1}{3}V_{[51]} + \frac{2}{3}V_{[33]}$
$\Xi N \to \Xi N$	$(0,1)$	$V_{\Xi N} = \frac{7}{9}V_{[51]} + \frac{2}{9}V_{[33]}$

Table 3. Weights $SU(6)_{FS}$-irreps in the SU(3)$_F$-irreps.

$V_{\{27\}}$	=	$\frac{4}{9}V_{[51]} + \frac{5}{9}V_{[33]}$	,	$V_{\{1\}}$	=	$V_{[33]}$,
$V_{\{10^*\}}$	=	$\frac{4}{9}V_{[51]} + \frac{5}{9}V_{[33]}$	,	$V_{\{10\}}$	=	$\frac{8}{9}V_{[51]} + \frac{1}{9}V_{[33]}$,
$V_{\{8_a\}}$	=	$\frac{5}{9}V_{[51]} + \frac{4}{9}V_{[33]}$	,	$V_{\{8_s\}}$	=	$V_{[51]}$.

nels will be the topic of a next ESC-model.)

3. Mesons, Meson-pairs, Coupling- and Cut-off-parameters

In Table 4 we show the couplings obtained in a first NN⊕YN fit for the ESC08a' solution. The (rationalized) coupling constants and $F/(F+D)$-

Table 4. ESC08: Meson-couplings and $F/(F+D)$-ratio's.

J^{PC}	$T=0$		$T=1$		$F/(F+D)$		$\theta_{mix}(degr.)$
0^{-+}	$f_{\eta'} =$	0.166	$f_\pi =$	0.266	$\alpha_P =$	0.353	$\theta_P = -23.00$
1^{--}	$g_\omega =$	3.319	$g_\rho =$	0.767	$\alpha_V^e =$	1.00	$\theta_V = 37.50$
	$f_\omega =$	-0.582	$f_\rho =$	3.923	$\alpha_V^m =$	0.494	
0^{++}	$g_\epsilon =$	4.090	$g_\delta =$	1.174	$\alpha_S =$	0.762	$\theta_S = 37.50$
1^{++}	$g_{f_1'} =$	1.061	$g_{a_1} =$	1.309	$\alpha_A =$	0.019	$\theta_A = -42.70$
	$f_{f_1'} =$	0.071	$f_{a_1} =$	–0.189	$\alpha'_A =$	0.019	
1^{+-}	$g_{f_1'} =$	–0.076	$g_{b_1} =$	–0.041	$\alpha_B =$	0.353	$\theta_B = -23.00$

ratio's are given in Table 4. The f_η was not fitted but computed via meson-mixing angle θ_m and SU(3)-relations. The same for the couplings of the $\phi, S^*(993), D(1285)$, and $h(1170)$ meson. The gaussian cut-off parameters are $\Lambda_P(T=1) = 872.1$ MeV, $\Lambda_P(T=0) = 893.7$ MeV, $\Lambda_V(T=1) = 707.6$ MeV, $\Lambda_V(T=0) = 1116.5$ MeV, $\Lambda_S(T=1) = 1193.4$ MeV, $\Lambda_S(T=0) = 1246.0$ MeV.
The pomeron and odderon have T=0, being even/odd-gluon exchange. The mass of the Odderon is fitted as $m_O = 463.3$ MeV, and that of the Pomeron became $m_P = 267.7$ MeV. The Pomeron coupling is $g_P = 3.382$, and those for the Odderon are $g_O = 0.322$, $f_o = -1.894$. The enhancement factor for the pomeron-coupling in the waves $\Sigma^+ p(^3S_1, T=3/2)$, $\Sigma N(^1S_0, T=1/2)$, and $\Xi N(^1S_0, T=1)$ is $F_{[51]} = 1.5$.

In Table 5 the pair-couplings are displayed for ESC04d and ESC08a'. We compare with the heavy-boson-saturation (HBS)-predictions. In the last column we give the $F/(F+D)$-ratio's for ESC08a', which are very similar to those employed in ESC04d. These $F/(F+D)$-ratio's are reasonably close to those predicted by the QPC-model[8,16] for the corresponding heavy bosons. The very small value for $g_{(\pi\sigma)_1}$ can be explained by noting that it actually stands for the combined $(\pi\sigma)$- and (πP)-pair terms, which are expected to cancel each other largely. All this demonstrates that the employed pair-couplings are very reasonable. The values under HBS are computed using the predictions of the heavy boson couplings by QPC. From the pair couplings 6 were searched, giving in total 20 fit parameters, which are OBE- and Pair-couplings, and form-factor parameters. The meson mixing used are the standard ones for the pseudo-scalar- and vector-mesons.

Table 5. Heavy-boson-saturation (HBS) comparison Pair-couplings and $F/(F+D)$-ratio's.

J^{PC}	Coupling	HBS	HBS	ESC04d	ESC08a'	$F/(F+D)$
0^{++}	$g_{(\pi\pi)_0}$	–0.03	$f_0(760, 993), P$	0.00	0.00	—
	$g_{(\pi\eta)_1}$	–0.28	$a_0(980, 1450)$	–0.10	-0.02	1.00
1^{--}	$g_{(\pi\pi)_1}$	0.04	$\rho(760)$	0.03	0.00	1.00
	$f_{(\pi\pi)_1}$	0.16	$\rho(760)$	0.14	0.20	0.40
1^{++}	$g_{(\pi\rho)_1}$	0.42	$A_1(1270)$	0.83	0.61	0.07
	$g_{(\pi\sigma)_1}$	0.31	$A_1(1270)$	–0.04	–0.03	0.07
1^{+-}	$g_{(\pi\omega)_1}$	–0.16	$B_1(1235)$	–0.17	–0.09	0.35

4. Hyperon-Nucleon/Hyperon: ESC08'

The situation for the Λ well-depth U_Λ is changed essentially for ESC08 as compared with ESC04d, see Table 7. We have for ESC04d the value U_Λ =-44.1 MeV, whereas for ESC08a U_Λ = -35.6 MeV. This is still larger than the experimental value, however this difference can be changed by changing the k_F-parameter and/or can easily be attributed to e.g. medium effects.[2]

For the solution of the ΛN spin-orbit problem we include, the hitherto in the Nijmegen work neglected, potentials proportional to the CM-operator $P_8 = (\boldsymbol{\sigma}_A \cdot \mathbf{P})(\boldsymbol{\sigma}_B \cdot \mathbf{k}) - (\boldsymbol{\sigma}_A \cdot \mathbf{k})(\boldsymbol{\sigma}_B \cdot \mathbf{P})$, see[17] for the definition, giving anti-symmetric spin-orbit potentials. These terms were given by Brown, Downs, and Iddings,[18] and their possible importance was pointed out by Dover and Gal.[19] They show up for the $S \neq 0$-exchanges with K, K*, κ, etc. both in OBE and MPE. The inclusion of these novel ESC-potentials brings the spin-orbit much more in agreement with experiments.[20] The strength of the Λ spin-orbit splitting in ESC07 is, as seen from Table 6, much smaller than in ESC04d. This is due to the BDI-term[18] and the new $1/M$-corrections in the MPE-potentials for strangeness-exchange. As we will show below the present version the ESC-model can easily accomodate the small experimental spin-orbit splitting[20] in the ΛN-interaction.[21] The strengths of Λ spin-orbit potential-integrals is expressed by $K_\Lambda = K_{S,\Lambda} + K_{A,\Lambda}$ where $K_{S,\Lambda} = -\frac{\pi}{3} S_{SLS}$ and $K_{A,\Lambda} = -\frac{\pi}{3} S_{ALS}$ with the S-integrals

$$S_{SLS,ALS} = \frac{3}{q} \int_0^\infty r^3 j_1(qr) V_{SLS,ALS}(r) dr \ .$$

Then, for the Nijmegen models one obtains values for these K-integrals as given in Table 6. Now, as computed by Hiyama et al[22] for the NHC-D

Table 6. Strengths ΛN spin-orbit potentials.

	K_S	K_A	$K_\Lambda^{(0)}$	$K_\Lambda(BDI)$	$K_\Lambda(Pair)$	K_Λ
NHC-D	30.7	−5.9	24.8	(-3.4)	—	21.4
NHC-F	29.7	−6.7	23.0	(-3.8)	—	19.2
ESC04b	16.0	−8.7	7.3	(−2.4)	(−3.3)	1.6
ESC04d	22.3	−6.9	15.4	(−5.0)	(−6.9)	3.5
ESC07	20.9	−9.6	11.3	(4.7)	(−5.2)	1.4

these K_Λ give $\Delta E_{LS}({}^9_\Lambda Be) = 0.15$ MeV, whereas the experimental value is 0.043 ± 0.002 MeV.[20] Then, it is clear from Table 6 that ESC08 will be much closer to experiment. Similar results for ESC08a,b. Notice the importance of both the BDI- and pair-contributions. In the ESC08-model the extra short range repulsion due to the quark-core in the $\Sigma^+p(^3S_1)$-channel is realized through the $F_{[51]}$-parameter. For ESC08a the scattering length is $a(^3S_1, \Sigma^+p) = 0.76$ fm, and the phase shift at 900 MeV/c is $\delta(\Sigma^+p,^3S_1) = -65^o$. As has been pointed out by Dabrowski[23] the NHC-F model[24] is succesfull for the BNL $^9_\Sigma$Be-experiment. Now, the ESC08 $^1S_0$ phase is very similar to that in NHC-F model, and also the $^3S_1$-phase at $p_{Lab} = 900$ MeV/c is like that in NHC-F and QCM.[15]

Table 7. Partial wave contributions to $U_\Lambda(\rho_0)$

	$^1S_0$	$^3S_1$	$^1P_1$	$^3P_0$	$^3P_1$	$^3P_2$	D	sum
ESC08a'	–12.6	–23.6	3.1	0.1	1.4	–3.7	–1.6	–36.9
ESC08a	–12.7	–22.2	3.0	0.1	1.4	–3.6	–1.6	–35.6
ESC08b	–12.3	–19.7	2.7	–0.2	1.5	–4.2	–1.7	–34.0
ESC04a	–13.7	–20.5	0.6	0.2	0.5	–4.5	–1.0	–38.5
ESC04d	–13.6	–26.6	3.2	–0.2	0.9	–6.4	–1.4	–44.1
NSC97e	–12.7	–25.5	2.1	0.5	3.2	–1.2	–1.1	–34.7
NSC97f	–14.3	–22.4	2.4	0.5	4.0	–0.7	–1.2	–31.7

Table 8. Partial wave contributions to $U_\Sigma(\rho_0)$

		$^1S_0$	$^3S_1$	$^1P_1$	$^3P_0$	$^3P_1$	$^3P_2$	D	U_Σ
NSC97f	$T = 1/2$	14.9	–9.6	1.9	2.3	–4.0	0.4	–0.4	
	$T = 3/2$	–12.2	–4.2	–3.8	–1.8	5.5	–2.7	–0.2	–13.9
ESC04d	$T = 1/2$	6.5	–21.0	2.6	2.4	–6.7	–1.7	–0.9	
	$T = 3/2$	–10.1	14.0	–8.5	–2.6	5.9	–5.7	–0.2	–26.0
ESC08a'	$T = 1/2$	9.3	–22.5	2.1	1.7	–6.3	–1.4	–0.7	
	$T = 3/2$	–12.0	38.0	–7.8	-1.9	6.1	–1.0	–0.3	+3.6
ESC08a	$T = 1/2$	11.3	–23.9	2.3	1.7	–6.2	–1.4	–0.7	
	$T = 3/2$	–11.7	44.8	–7.2	-1.7	6.5	0.2	–0.2	+13.4
ESC08b	$T = 1/2$	10.3	–26.2	2.5	2.2	–7.9	–1.7	–0.8	
	$T = 3/2$	–10.6	52.7	–6.2	-2.0	7.4	0.8	–0.1	+20.3

Table 9. Partial wave contributions to $U_\Xi(\rho_0)$

		$^1S_0$	$^3S_1$	$^1P_1$	$^3P_0$	$^3P_1$	$^3P_2$	U_Ξ	Γ_Ξ
ESC08a'	$T = 0$	3.7	–1.8	–0.2	–3.4	1.4	–1.0		
	$T = 1$	17.1	–22.5	1.7	0.4	–3.8	–0.6	–9.0	5.9
ESC08a	$T = 0$	5.2	–0.9	–0.3	–2.8	1.4	–0.9		
	$T = 1$	21.9	–31.6	2.5	0.3	–3.7	–0.6	–9.6	7.8
ESC08b	$T = 0$	2.1	2.0	–0.6	–1.3	–0.1	–0.6		
	$T = 1$	26.2	–40.3	3.0	–0.6	–3.7	–1.3	–14.7	3.3

This is reflected in the well depths, see Table 8. We have now $U_\Sigma =$

+13.4 MeV in ESC08a, and $U_\Sigma = +20.3$ MeV in ESC08b. This illustrates the effect of the exceptional repulsion in the $\Sigma^+ p(^3S_1, T = 3/2)$-wave. In ESC08 we have not included the possible effect of the nuclear medium. This effect is illustrated by the difference between ESC04d and ESC04d*.[2]

The channels with strangeness S=-2 are for the $\Lambda\Lambda(^1S_0)$-channel similar to ESC04d, for which $\Delta B_{\Lambda\Lambda} \approx 0.80$ MeV, which is nicely in accordance with the Nagara event.[25] The $\Lambda\Lambda$ low-energy parameters in ESC08a are $a_{\Lambda\Lambda}(^1S_0) = -0.68$ fm and $r_{\Lambda\Lambda}(^1S_0) = 5.53$ fm. In contrast to ESC07,[5] the Ξ well-depths are in ESC08 attractive, in particularly for ESC08b. For ESC08b the well-depth is similar to that of ESC04d with $U_\Xi = -18.7$ MeV. However, this time the attraction is produced in the 'deuteron-like' S=-2-channel, which is more satisfying. In ESC08 U_Σ and U_Ξ are correlated. E.g. in ESC08a' for FI51= 0.5,0.7,0.9 we get $U_\Sigma = 3.6, 12.6, 20.2$ and $U_\Xi = -9.0, -6.2, -3.7$, respectively.

For the S=-4 the low-energy parameters in the case of ESC08a are: $a_{\Xi\Xi}(^1S_0) = -18.8$[fm], $r_{\Xi\Xi}(^1S_0) = 1.81$[fm], and $a_{\Xi\Xi}(^3S_1) = 0.73$[fm], $r_{\Xi\Xi}(^3S_1) = 0.16$[fm]. ESC08b has here a $^1S_0$-bound state. The latter possibility is not a surprise because of its close connention with NN($^1S_0$).

5. Conclusions and Outlook

The prospects for a completely satisfactory ESC-description of the BB-channels for S=0,-1,-2 are rather good. The extension from ESC04 to ESC08 solved the quest for more repulsion in the $\Sigma N(T = 3/2, ^3S_1)$ and the $\Sigma N(T = 1/2, ^1S_0)$ states, and attraction for the Ξ-nucleus interaction. We emphasize that the introduction of the quark-core effects will not influence strongly the BB-channels in the S=-2-sector. This, because here the SU(6)$_{FS}$ irrep [51] does not occur with large weights, except for the $\Xi N(^1S_0, T = 1)$-channel, which effect we included now and which decreases U_Ξ. The attraction in ESC08 in the $\Xi N(^3S_1, T = 1)$, which is $\{10^*\}$, is not destroyed by an exceptional Pauli-repulsion. The realization of solutions in fitting all the available BB-data, keeping the good features already realized and at the same time providing a sizeable Ξ-nuclear attraction has been achieved with the ESC08-models. This opens prospects for realistic predictions for the S=-3,-4 systems.

Finally, we would like to thank our colleagues, in particularly T. Motoba and E. Hiyama, for valuable and stimulating discussions.

References

1. Th.A. Rijken, Phys. Rev. C **73**, 044007 (2006).

2. Th.A. Rijken and Y. Yamamoto, Phys. Rev. C **73**, 044008 (2006).
3. Th.A. Rijken and Y. Yamamoto, arXiv:nucl-th/060874 (2006).
4. Th.A. Rijken, M.M. Nagels, and Y. Yamamoto, Proceedings X Int. Conf. on Hypernuclear and Strange Particle Physics 2009, Nucl. Phys. A (to be published).
5. Th.A. Rijken, M.M. Nagels, and Y. Yamamoto, Proceedings of the Sendai International Symposium *Strangeness in Nuclear and Hadronic Systems*, editors K. Maeda, S.N. Nakamura, and H. Tamura, World scientific, 2009.
6. See V. DeAlfaro, S. Fubini, G. Furlan, and C. Rosetti, *Currents in Hadron Physics*, chapter 5, North-Holland Pub. Comp., A'dam-London 1973.
7. E.R. Berger et al, Eur. Phys. J. C9(1999) 491; *ibid* C14 (2000) 673; C. Ewerz, *The Odderon in QCD*, arXiv:hep-ph/0306137 (2006).
8. L. Micu, Nucl. Phys. **B 10**, 521 (1969); R. Carlitz and M. Kislinger, Phys. Rev. **D 2**, 336 (1970); A. Le Yaouanc, L. Oliver, O. Péne, and J.-C. Raynal, Phys. Rev. D 8, 2223 (1973); ibid **D 11**, 1272 (1975).
9. M.M. Nagels, Th.A. Rijken, and J.J. de Swart, Phys. Rev. D**15**, 2547 (1977).
10. J.K. Ahn et al, Nucl. Phys. **A 761** (2005) 41.
11. H. Kanda, *Measurement of the cross sections of* $\Sigma^+ p$ *elastic scattering*, Ph. Thesis Kyoto University 2007.
12. S. Nishizaki, Y. Yamamoto, T. Takatsuka, Progr. Theor. Phys. **105** (2001) 607; ibid **108** (2002) 703.
13. S. Otsuki, R. Tamagaki, and M. Yasuno, Prog. Theor. Phys. Suppl. Extra number (1965), 578.
14. M. Oka, K. Shimizu, and K. Yazaki, Progr. Theor. Phys. Suppl. **137**, 1 (2000).
15. Y. Fujiwara, Y. Suzuki, and C. Nakamoto, Progr. in Part. and Nuclear Physics, **58** (2007) 439.
16. This pair-creation mechanism has been shown to be dominant in lattice QCD. See: N. Isgur and J. Paton, Phys. Rev. **D31**, 2910 (1985); R. Kokoski and N. Isgur, Phys. Rev. **D35**, 907 (1987).
17. J.J. de Swart, M.M. Nagels, Th.A. Rijken, and P.A. Verhoeven, Springer Tracts in Modern Physics, **60** 138-203 (1971).
18. J.T. Brown, B.W. Downs, and C.K. Iddings, Ann. Phys. (N.Y.) **60**, 148 (1970).
19. C. Dover and A. Gal, Progr. in Part. and Nuclear Physics, **12** (1984) 171
20. H. Tamura, Nucl.Phys. A691 (2001) 86c-92c.
21. E.Hiyama, Talk at the International Nuclear Physics Conference (INPC07), Tokyo, June 2007.
22. E. Hiyama et al, Phys. Rev. Lett. **85** (2000) 270.
23. J. Dabrowski, Nucl. Phys. **A 691** (2001) 58c-63c.
24. M.M. Nagels, Th.A. Rijken, and J.J. de Swart, Phys. Rev. D**20**, 1633 (1979).
25. H. Takahashi et al., Phys. Rev. Lett. **87** (2001) 212502.

SHELL-MODEL CALCULATIONS FOR TYPICAL *sd*-SHELL HYPERNUCLEI

A. UMEYA*

Nishina Center for Accelerator-Based Science, RIKEN,
Wako, Saitama 351-0198, Japan
**E-mail: aumeya@riken.jp*

T. MOTOBA and T. HARADA

Research Center for Physics and Mathematics, Osaka Electro-Communication University, Neyagawa, Osaka, 572-8530, Japan

As one of the typical *sd*-shell hypernuclei, the positive- and negative-parity energy levels of $^{20}_{\Lambda}$Ne have been calculated to investigate structures of low-lying states within the standard shell model. In addition to the conventional NN interactions, the ΛN G-matrix interactions derived from the Nijmegen Model-D and NSC97f potentials are employed.

Keywords: Hypernuclei; Shell model.

1. Introduction

Detailed hypernuclear studies have been mainly focused on structures of p-shell systems. In addition to the (π^+, K^+) reaction experiments done for wide mass number region,[1] the γ-ray measurements[2] from p-shell hypernuclei provide us with remarkable possibility of high-precision spectroscopic studies on theoretical side.[3] Recently, also the $(e, e'K^+)$ reaction experiments done at Hall-A and Hall-C of JLab[4,5] have been proved to be very fruitful in disclosing hypernuclear structure details within the energy resolution of few hundreds of keV.

These aspects encourage us to perform theoretical study of *sd*-shell hypernuclear structures, because even the Λ single-particle energies and its interplay with nuclear core excitations are not well known in these medium-mass systems. As a theoretical example of photoproduction study, one may refer to Ref. 6. In general, there should be more bound states of Λ in *sd*-shell hypernuclei so that we can expect novel aspect of hyperon coupling with nuclear rotational excitations. We anticipate innovated (K^-, π^-) reaction experiments to be done at J-PARC, as well as $(e, e'K^+)$ experiments at JLab.

As a starting typical example, in this article, we focus on a ${}^{20}_{\Lambda}$Ne hypernucleus whose nuclear core, ^{19}Ne, has an interesting level structure in low-lying states. (i) In a naive configuration picture, it is expected that $|(0d_{5/2})^2_{0^+}, 0d_{5/2}; J^\pi{=}\frac{5}{2}^+\rangle$ will be the ground state because the lowest sd-shell orbit is $0d_{5/2}$. However, actually, the ground state of ^{19}Ne has $J^\pi = \frac{1}{2}^+$ and is mainly described by $|(0d_{5/2})^2_{0^+}, 1s_{1/2}; J^\pi{=}\frac{1}{2}^+\rangle$ basis state, because an energy difference between $1s_{1/2}$ and $0d_{5/2}$ orbits is not so large and a mixing of $1s_{1/2}$ leads to strong attraction due to an NN interaction. (ii) Also, the negative-parity state with $J^\pi = \frac{1}{2}^-$ has a small excitation energy because of the strong and attractive four-body correlation which compensate the $1p$-$1h$ excitation energy creating a p-state hole in the ^{16}O closed core. These states with $J^\pi = \frac{1}{2}^+$, $\frac{5}{2}^+$ and $\frac{1}{2}^-$ are in a narrow energy region of few hundreds of keV, and then it is interesting to investigate effects of hyperon addition for these states in ${}^{20}_{\Lambda}$Ne. We calculate energy levels of ${}^{20}_{\Lambda}$Ne and investigate structures of low-lying states by using a multi-configuration shell model.

2. Shell-model framework

In present calculations, we take shell-model wave functions of ${}^{20}_{\Lambda}$Ne and ^{19}Ne as follows. The ^{16}O core consisting of sixteen nucleons is assumed to be inert and three valence nucleons are distributed in the sd-shell ($0d_{5/2}$, $1s_{1/2}$ and $0d_{3/2}$) orbits. In order to calculate negative-parity states, the $0p$-hole configuration with $1p$-$1h$ $1\hbar\omega$ excitation is employed. All the possible bases are taken into account in the diagonalization, although the Λ hyperon is confined tentatively to occupy the lowest $0s^{\Lambda}_{1/2}$ orbit. Basis states are schematically shown in Fig. 1, and the positive- and negative-parity wave functions are described, respectively, as ($A_c = 19$):

$$\Phi({}^A_\Lambda Z; J^+T) = \sum_k c_k \left[\{(0s)^4(0p)^{12}(0d_{5/2}1s_{1/2}0d_{3/2})^{A_c-16}\}_{J^k_c T^k_c} \otimes \phi(j^\Lambda)\right]_{JT}, \quad (1)$$

$$\Phi({}^A_\Lambda Z; J^-T) = \sum_k c_k \left[\{(0s)^4(0p)^{11}(0d_{5/2}1s_{1/2}0d_{3/2})^{A_c-15}\}_{J^k_c T^k_c} \otimes \phi(j^\Lambda)\right]_{JT}. \quad (2)$$

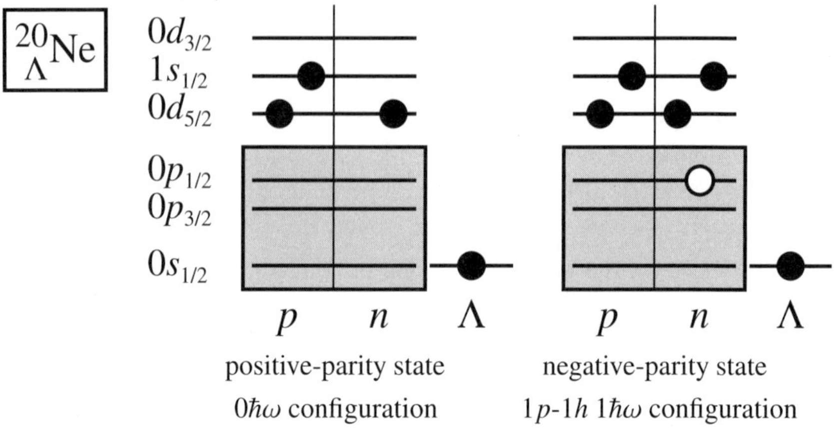

Fig. 1. Basis states of the $^{20}_{\Lambda}$Ne hypernucleus: (left) positive-parity states and (right) negative-parity states. In the course of calculations, the Λ particle occupies over $0s$, $0p$, and $1s0d$ orbits.

We adopt the Kuo G-matrix[7] for the sd-shell part of an NN effective interaction, and the Millener-Kurath interaction[8] for the p-sd cross-shell part. For the ΛN effective interactions, we try to adopt the Nijmegen Model-D potential (NHC-D)[9] and also a soft-core model NSC97f.[10] For single-particle energies with respect to the ^{16}O core, we use rather standard values of -21.86 ($0p_{3/2}$), -14.20 ($0p_{1/2}$), -4.14 ($0d_{5/2}$), -3.27 ($1s_{1/2}$) and $+0.94$ ($0d_{3/2}$) MeV, which are obtained from the experimental energy levels of ^{17}O,[11] except for the $0p_{1/2}$ orbit. For simplicity, we adjust the single-particle energy of $0p_{1/2}$ to reproduce the energy levels of the ^{19}Ne negative-parity states. Thus we take the satisfactory description with respect to the nuclear core excitations. This is a necessary option when we proceed next to couple the Λ particle to see what happens depending on the adopted ΛN interactions.

3. Numerical results of hypernuclear energy levels

Figure 2 shows the calculated energy levels of $^{20}_{\Lambda}$Ne, together with the experimental and calculated levels of ^{19}Ne. The calculation for ^{19}Ne well reproduces the experimental energy levels of three low-lying states with $J^{\pi} = \frac{1}{2}^{+}$, $\frac{5}{2}^{+}$ and $\frac{1}{2}^{-}$. The calculated excitation energies of $\frac{5}{2}^{+}$ and $\frac{1}{2}^{-}$ excited states are 0.36 and 0.43 MeV, respectively, which correspond to the experimental values, 0.24 and 0.28 MeV. However, the present shell model should be improved in order to describe the energies of the $\frac{3}{2}^{+}$ and $\frac{3}{2}^{-}$ states.

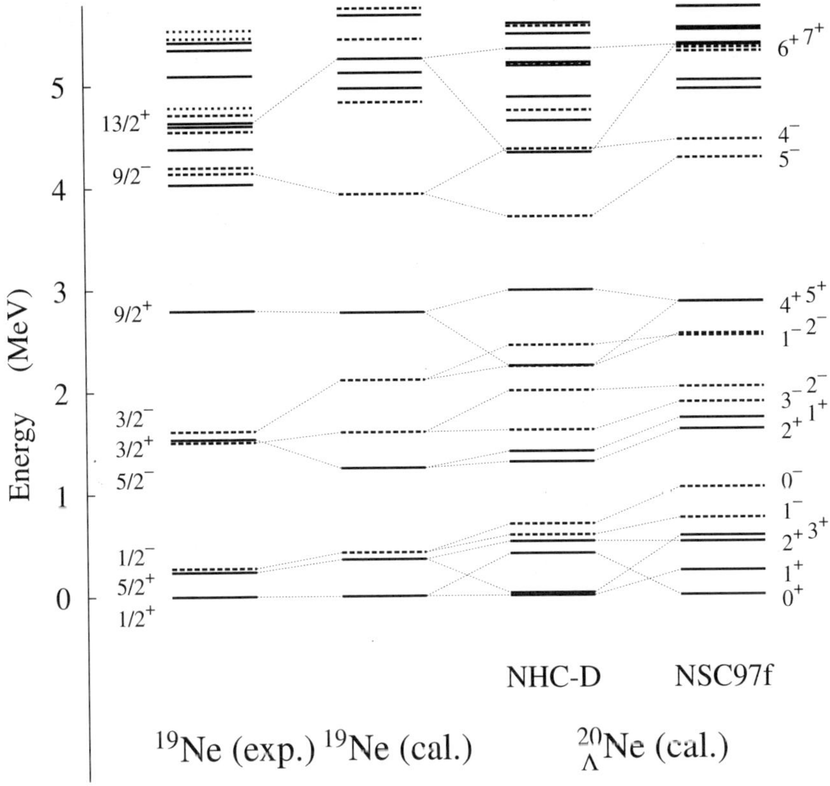

Fig. 2. Calculated energy levels of a $^{20}_{\Lambda}$Ne hypernucleus, together with experimental and calculated levels of ^{19}Ne. Solid and dashed lines denote positive- and negative-parity states, respectively.

In the $^{20}_{\Lambda}$Ne hypernucleus, we are presently interested in the energy splittings and order of the "spin-doublet" states ($J = J_c \pm \frac{1}{2}$) which usually attributed to the weak coupling of Λ to a nuclear core state (J_c). The calculation using the NHC-D interaction gives the 1^+ ground state, whose nuclear core is mainly described by the ^{19}Ne $\frac{1}{2}^+$ ground state. The spin-doublet partner (0^+) is obtained at the excitation energy of 0.41 MeV. On the other hand, the NSC97f ΛN interaction leads to the 0^+ ground state and the doublet energy splitting is small and the order is opposite in comparison with NHC-D.

In the p-shell cases such as $^{12}_{\Lambda}$C and $^{16}_{\Lambda}$O, it is known that the $J_<$ member comes lower in energy, so that spin-singlet interaction is more attractive than the triplet one, and the energy splitting is generally very small. Therefore, one believes that the NSC97f interaction gives a plausible prediction for the low-lying energy level sequence.

The Λ particle coupling to the $J_c = \frac{5}{2}^+$ core state ($E_x^{\rm cal} = 0.36$ MeV) leads to the 2^+ and 3^+ doublet in ${}^{20}_{\Lambda}$Ne. The energy of the 3^+ member is pushed down to 0.03 MeV with respect to the ground state, while the 2^+ member is pushed up probably due to the large spin-spin interaction of NHC-D. When the NSC97f interaction is used, then the order of the doublet states is reversed and the splitting energy becomes smaller. On the other hand, it is interesting to predict the negative-parity doublet of ${}^{20}_{\Lambda}$Ne, 1^- and 0^- states, at $E_x = 0.59$ MeV and 0.70 MeV, respectively. Both of these excitation energies are larger than that of the $\frac{1}{2}^-$ nuclear core excitation energy.

We analyze the matrix elements of the Λ-nuclear interaction for large configurations of the ${}^{20}_{\Lambda}$Ne states,

$$\Phi({}^{20}_{\Lambda}\mathrm{Ne}) = \sum_k c_k \left[\{ ({}^{16}\mathrm{O\ core}) \otimes \psi_N \} \otimes 0s^{\Lambda}_{1/2} \right], \tag{3}$$

where ψ_N consists of three valence nucleons for positive-parity states, and of four valence nucleons and a single-hole for negative-parity states. Table 1 shows amplitudes of the large configurations, where $(j)^n_J$ denotes a config-

Table 1. Wave functions calculated for the low-lying "spin-doublet" states of ${}^{20}_{\Lambda}$Ne. The amplitudes are expressed in terms of the nuclear-core configurations coupled with the $0s_{1/2}$-state Λ particle.

ψ_N	${}^{20}_{\Lambda}\mathrm{Ne}(J=0^+_1)$		${}^{20}_{\Lambda}\mathrm{Ne}(J=1^+_1)$	
	NHC-D	NSC97f	NHC-D	NSC97f
$[(0d^N_{5/2})^2_0(1s^N_{1/2})]_{1/2}$	$\sqrt{0.3371}$	$\sqrt{0.3744}$	$\sqrt{0.3691}$	$\sqrt{0.3558}$
$[(0d^N_{5/2})^2_1(1s^N_{1/2})]_{1/2}$	$-\sqrt{0.1554}$	$-\sqrt{0.1357}$	$-\sqrt{0.1394}$	$-\sqrt{0.1471}$
$(0d^N_{5/2})^3_{1/2}$	$-\sqrt{0.1437}$	$-\sqrt{0.1153}$	$-\sqrt{0.1350}$	$-\sqrt{0.1207}$
$[(0d^N_{5/2})\{(1s^N_{1/2})(0d^N_{3/2})\}_2]_{1/2}$	$\sqrt{0.1281}$	$\sqrt{0.1161}$	$\sqrt{0.1150}$	$\sqrt{0.1224}$
ψ_N	${}^{20}_{\Lambda}\mathrm{Ne}(J=2^+_1)$		${}^{20}_{\Lambda}\mathrm{Ne}(J=3^+_1)$	
	NHC-D	NSC97f	NHC-D	NSC97f
$(0d^N_{5/2})^3_{5/2}$	$\sqrt{0.4265}$	$-\sqrt{0.3663}$	$-\sqrt{0.4169}$	$-\sqrt{0.3735}$
ψ_N	${}^{20}_{\Lambda}\mathrm{Ne}(J=0^-_1)$		${}^{20}_{\Lambda}\mathrm{Ne}(J=1^-_1)$	
	NHC-D	NSC97f	NHC-D	NSC97f
$[(0p^N_{1/2})^{-1}(0d^N_{5/2})^4_0]_{1/2}$	$\sqrt{0.1595}$	$-\sqrt{0.1332}$	$\sqrt{0.1602}$	$-\sqrt{0.1332}$
$[(0p^N_{1/2})^{-1}(0d^N_{5/2})^2_0(1s^N_{1/2})^2_0]_{1/2}$	$-\sqrt{0.1400}$	$+\sqrt{0.1618}$	$-\sqrt{0.1394}$	$+\sqrt{0.1614}$

uration with the coupled angular momentum J that consists of n valence nucleons in a j orbit, and $(0p^N_{1/2})^{-1}$ denotes a single-hole configuration in the $0p^N_{1/2}$ orbit. The nuclear core in positive-parity states mainly consists of $(0d^N_{5/2})^3$ and $(0d^N_{5/2})^2(1s^N_{1/2})$ configurations. On the other hand, the nuclear core in negative-parity states is described as a product of a $(sd)^4$ valence-particle configuration and a $(0p^N_{1/2})^{-1}$ hole configuration. Table 2 shows values of matrix elements of the Λ-nuclear interaction for these configurations. We find that the NHC-D potential is much more attractive than the NSC97f one in the Λ-$(sd)^3$ and Λ-$(sd)^4$ interactions. In the negative-parity state, the Λ-$(sd)^4$ interaction is more attractive than the Λ-$(sd)^3$ interaction, while the Λ-$(0p^N_{1/2})^{-1}$ interaction is repulsive. As a result, the strength of the Λ-^{19}Ne$(\frac{1}{2}^-)$ interaction is slightly weaker than that of the Λ-^{19}Ne$(\frac{1}{2}^+)$ interaction.

For further analysis, we list in Table 3 the calculated wave functions of $^{20}_{\Lambda}$Ne as expressed simply in terms of the ^{19}Ne eigenfunctions coupled with the $0s^{\Lambda}_{1/2}$. In general, the NSC97f interaction gives rise to the purer wave functions than NHC-D interaction. This is due to the smaller spin-spin interaction with more attraction in the spin-singlet state.

Table 2. Matrix elements $\langle \psi_N, 0s^{\Lambda}_{1/2}|V|\psi_N, 0s^{\Lambda}_{1/2}\rangle$ of the Λ-nuclear interaction in unit of MeV. ψ_N denotes a nuclear configuration that is included in the low-lying $^{20}_{\Lambda}$Ne eigenstates.

ψ_N	$^{20}_{\Lambda}$Ne$(J=0^+_1)$		$^{20}_{\Lambda}$Ne$(J=1^+_1)$	
	NHC-D	NSC97f	NHC-D	NSC97f
$[(0d^N_{5/2})^2_0(1s^N_{1/2})]_{1/2}$	−3.65	−2.89	−4.45	−2.44
$[(0d^N_{5/2})^2_1(1s^N_{1/2})]_{1/2}$	−4.30	−2.45	−4.23	−2.58
$(0d^N_{5/2})^3_{1/2}$	−4.63	−2.51	−4.78	−2.50
$[(0d^N_{5/2})\{(1s^N_{1/2})(0d^N_{3/2})\}_2]_{1/2}$	−3.58	−1.79	−3.53	1.82
ψ_N	$^{20}_{\Lambda}$Ne$(J=2^+_1)$		$^{20}_{\Lambda}$Ne$(J=3^+_1)$	
	NHC-D	NSC97f	NHC-D	NSC97f
$(0d^N_{5/2})^3_{5/2}$	−4.49	−2.52	−4.93	−2.49
ψ_N	$^{20}_{\Lambda}$Ne$(J=0^-_1)$		$^{20}_{\Lambda}$Ne$(J=1^-_1)$	
	NHC-D	NSC97f	NHC-D	NSC97f
$(0d^N_{5/2})^4_0$	−6.33	−3.34	−6.33	−3.34
$[(0d^N_{5/2})^2_0(1s^N_{1/2})^2_0]_0$	−5.34	−3.43	−5.34	−3.43
$(0p^N_{1/2})^{-1}$	+1.45	+1.01	+1.33	+0.70

4. DWIA cross sections for light *sd*-shell hypernuclear production

The *sd*-shell nuclear targets used so far in the actual experiments are quite limited: The ^{28}Si nucleus is the favorite target often used in the (π^+, K^+) reactions[12] and in the recent $(e, e'K^+)$ reaction.[5] In fact, theoretical predictions for their reactions have been made for years.[6,13] Here, however, we focus our attention to the light *sd*-shell hypernuclei for which it seems easier to discuss hypernuclear structures in close connection with properties of ΛN interactions. Only one example exploited so far in light *sd*-shell region is the $^{18}\mathrm{O}\,(K^-, \pi^-)\,^{18}_{\Lambda}\mathrm{O}$ reaction experiment,[14] and its DWIA analysis was performed long ago.[15] Although it may be difficult to use ^{20}Ne as an actual target, we are now calculating the DWIA production cross sections on the basis of the similar theoretical prescriptions, so that we can have a concrete idea to be utilized in possible J-PARC projects for light *sd*-shell

Table 3. Wave functions calculated for the low-lying "spin-doublet" states of $^{20}_{\Lambda}$Ne. The amplitudes are reexpressed in terms of the ^{19}Ne calculated eigenstates coupled with the $0s_{1/2}$-state Λ particle. The asterisk indicates negligibly small number.

	$^{20}_{\Lambda}\mathrm{Ne}(J=0^+_1)$		$^{20}_{\Lambda}\mathrm{Ne}(J=1^+_1)$	
	NHC-D	NSC97f	NHC-D	NSC97f
$^{19}\mathrm{Ne}(\frac{1}{2}{}^+_1)\otimes 0s^{\Lambda}_{1/2}$	$-\sqrt{0.9955}$	$-\sqrt{0.9978}$	$-\sqrt{0.9928}$	$-\sqrt{0.9980}$
$^{19}\mathrm{Ne}(\frac{1}{2}{}^+_2)\otimes 0s^{\Lambda}_{1/2}$	$-\sqrt{0.0025}$	*	$-\sqrt{0.0022}$	*
$^{19}\mathrm{Ne}(\frac{1}{2}{}^+_3)\otimes 0s^{\Lambda}_{1/2}$	$\sqrt{0.0013}$	$-\sqrt{0.0005}$	*	*
$^{19}\mathrm{Ne}(\frac{3}{2}{}^+_1)\otimes 0s^{\Lambda}_{1/2}$			$-\sqrt{0.0034}$	$-\sqrt{0.0004}$
	$^{20}_{\Lambda}\mathrm{Ne}(J=2^+_1)$		$^{20}_{\Lambda}\mathrm{Ne}(J=3^+_1)$	
	NHC-D	NSC97f	NHC-D	NSC97f
$^{19}\mathrm{Ne}(\frac{5}{2}{}^+_1)\otimes 0s^{\Lambda}_{1/2}$	$\sqrt{0.9874}$	$-\sqrt{0.9892}$	$-\sqrt{0.9953}$	$-\sqrt{0.9978}$
$^{19}\mathrm{Ne}(\frac{5}{2}{}^+_2)\otimes 0s^{\Lambda}_{1/2}$	$\sqrt{0.0041}$	$\sqrt{0.0008}$	$-\sqrt{0.0023}$	$\sqrt{0.0005}$
$^{19}\mathrm{Ne}(\frac{5}{2}{}^+_3)\otimes 0s^{\Lambda}_{1/2}$	$-\sqrt{0.0003}$	*	$\sqrt{0.0008}$	*
$^{19}\mathrm{Ne}(\frac{3}{2}{}^+_1)\otimes 0s^{\Lambda}_{1/2}$	$\sqrt{0.0062}$	$\sqrt{0.0081}$		
	$^{20}_{\Lambda}\mathrm{Ne}(J=0^-_1)$		$^{20}_{\Lambda}\mathrm{Ne}(J=1^-_1)$	
	NHC-D	NSC97f	NHC-D	NSC97f
$^{19}\mathrm{Ne}(\frac{1}{2}{}^-_1)\otimes 0s^{\Lambda}_{1/2}$	$-\sqrt{0.9961}$	$\sqrt{0.9972}$	$-\sqrt{0.9948}$	$\sqrt{0.9948}$
$^{19}\mathrm{Ne}(\frac{1}{2}{}^-_2)\otimes 0s^{\Lambda}_{1/2}$	$-\sqrt{0.0021}$	$-\sqrt{0.0010}$	$-\sqrt{0.0022}$	$-\sqrt{0.0010}$
$^{19}\mathrm{Ne}(\frac{3}{2}{}^-_1)\otimes 0s^{\Lambda}_{1/2}$			$-\sqrt{0.0007}$	$-\sqrt{0.0018}$

hypernuclear production.

5. Summary and outlook

We have calculated the energy levels of $^{20}_{\Lambda}$Ne hypernucleus by using the multi-configuration shell model with the Nijmegen Model-D and NSC97f potentials for the ΛN effective interactions. The calculation using Model-D gives the 1^+ ground state, while that using NSC97f gives the 0^+ ground state. The difference of spin-dependence between these interactions leads to the order inversion of the 0^+-1^+ spin-doublet. The calculated wave functions are used for the estimates of cross sections of several hypernuclear production processes.

Acknowledgments

The authors are grateful to Y. Yamamoto for providing the YNG-type effective ΛN interactions derived from Nijmegen Model-D and NSC97f potentials. This work is supported by Grant-in-Aid for Scientific Research C (No. 21540284).

References

1. O. Hashimoto and H. Tamura, *Prog. Part. Nucl. Phys.* **57**, 564 (2006), and references therein.
2. H. Tamura, *Nucl. Phys.* **A827**, 153c (2009), and references therein.
3. D. J. Millener, *Nucl. Phys.* **A804**, 84 (2008), and references therein.
4. J. J. LeRose *et al.*, *Nucl. Phys.* **A804**, 116 (2008); F. Garibaldi *et al.*, in *Proc. of the Sendai International Symposium on Strangeness in Nuclear and Hadronic Systems* (World Scientific, 2010), p.195.
5. O. Hashimoto *et al.*, *Nucl. Phys.* **A804**, 125 (2008); Y. Fujii *et al.*, in *Proc. of the Sendai International Symposium on Strangeness in Nuclear and Hadronic Systems* (World Scientific, 2010), p.188.
6. T. Motoba, P. Bydžovský, M. Sotona, K. Itonaga, K. Ogawa, and O. Hashimoto, in *Proc. of the Sendai International Symposium on Strangeness in Nuclear and Hadronic Systems* (World Scientific, 2010), p.178.
7. T. T. S. Kuo, *Nucl. Phys.* **A103**, 71 (1967).
8. D. J. Millener and D. Kurath, *Nucl. Phys.* **A255**, 315 (1975).
9. M. M. Nagels, T. A. Rijken, and J. J. de Swart, *Phys. Rev. D* **12**, 744 (1975); **15**, 2547 (1977); **20**, 1633 (1979).
10. Th. A. Rijken, V. G. J. Stoks, and Y. Yamamoto, *Phys. Rev. C* **59**, 21 (1999).
11. *Table of Isotopes*, 8th ed., edited by R. B. Firestone and V. S. Shirley (John Wiley and Sons, Inc., New York, 1996).
12. T. Hasegawa *et al.*, *Phys. Rev. C* **53**, 1210 (1996); H. Hotchi *et al.*, *Phys. Rev. C* **64**, 044302 (2001).

13. T. Motoba, H. Bandō, R. Wünsch, and J. Žofka, *Phys. Rev. C* **38**, 1322 (1988).
14. M. May *et al.*, *Phys. Rev. Lett.* **47**, 1106 (1981).
15. T. Yamada, T. Motoba, K. Ikeda, and H. Bandō, *Prog. Theor. Phys. Suppl.* **81**, 104 (1985).

LIGHT Ξ HYPERNUCLEI IN FOUR-BODY CLUSTER MODELS

E. HIYAMA*

Nishina Center for Accelerator-Based Science, Institute for Physical and Chemical Research (RIKEN), Wako, Saitama 351-0198, Japan
**E-mail: hiyama@riken.jp*

Detailed structure calculations in ${}^{12}_{\Xi^-}$Be, ${}^{7}_{\Xi^-}$H and ${}^{10}_{\Xi^-}$Li are performed within the framework of the microscopic of four-body cluster models. We adopted effective ΞN interactions derived from the Nijmegen interaction models, which give rise to substantially attractive Ξ-nucleus potentials in accordance with the experimental indications. ${}^{7}_{\Xi^-}$H and ${}^{10}_{\Xi^-}$Li are predicted to have bound states. we propose to observe the bound states in future (K^-, K^+) experiments using ^{7}Li and ^{10}B targets in addition to the standard ^{12}C target. The experimental confirmation of these states will provide information on the spin- and isospin-averaged ΞN interaction.

1. Introduction

In studies of nuclear interactions, two-body scattering data are the primary input for characterizing interaction models. However, $S = -1$ hyperon (Y)-nucleon (N) scattering data are very limited because of experimental issues. For $S = -2$ interactions such as $\Lambda\Lambda$ and ΞN, there are currently no scattering data. Therefore, the existing YN and YY interaction models have a substantial degree of ambiguity. Some YN scattering experiments will be performed at the Japan Proton Accelerator Research Complex (J-PARC) in the near future. Even at this facility, however, the possibility of performing ΞN or $\Lambda\Lambda$ scattering experiments is very limited or practically impossible. Hence, in order to obtain useful information on $S = -2$ interactions, studies of many-body, hypernuclear structure are indispensable.

Our intention in this work is to investigate the possible existence of Ξ hypernuclei and to explore the properties of the underlying ΞN interac-

*The present work was done together with Y. Yamamoto, T. Motoba, Th. A. Rijken and M. Kamimura tions. Identification of Ξ hypernuclei in coming experiments at J-PARC will contribute significantly to understanding nuclear structure and interactions in $S=-2$ systems, which can lead to an entrance into the world of multi-strangeness. In order to encourage new experiments seeking Ξ hypernuclei, it is essential to make a detailed theoretical investigation of the possible existence of bound states, despite some uncertainty in contemporary ΞN interaction models.

We investigate here the binding energies and structure of Ξ hypernuclei produced by (K^-,K^+) reactions on light targets on the basis of microscopic cluster models. One of the primary issues is how to choose the ΞN interaction. Although there are no definitive data for any Ξ hypernucleus at present, a few experimental data indicate that Ξ-nucleus interactions are attractive. One example is the observed spectrum of the (K^-,K^+) reaction on a ^{12}C target, where the cross sections for Ξ^- production in the threshold region can be interpreted by assuming a Ξ-nucleus Wood-Saxon (WS) potential with a depth of $\sim$ 14 MeV.[1] Other indications of attractive Ξ-nucleus interactions are given by certain emulsion data, the events for twin-Λ hypernuclei, where the initial Ξ^- energies were determined by the identification of all fragments after the $\Xi^- p$-$\Lambda\Lambda$ conversion in nuclei. The inferred Ξ^- binding energies are substantially larger than those obtained using only the Coulomb interaction.[2] When these Ξ^- states are assumed to be $1p$ states, the WS potentials obtained from the binding energies are similar to the one above. These data suggest that the average ΞN interaction should be attractive, which we utilize to select the appropriate interaction models. In this work we adopt two types of ΞN interactions, the Nijmegen Hard-Core model D (ND)[3] and the Extended Soft-Core model (ESC04).[4,5]

The structure of light p-shell nuclei can be reasonably described in terms of cluster models composed of two- or three-body subunits. Here, we model the possible Ξ^- hypernuclei produced by (K^-,K^+) reactions on available light p-shell targets as four-body cluster structures: The possible targets ^{12}C, ^{11}B, ^{10}B, ^{9}Be and ^{7}Li naturally lead to such cluster configurations as $\alpha\alpha t\Xi^-(^{12}_{\Xi^-}\mathrm{Be})$, $\alpha\alpha 2n\Xi^-(^{11}_{\Xi^-}\mathrm{Li})$, $\alpha\alpha n\Xi^-(^{10}_{\Xi^-}\mathrm{Li})$, $\alpha t n\Xi^-(^{9}_{\Xi^-}\mathrm{He})$ and $\alpha nn\Xi^-(^{7}_{\Xi^-}\mathrm{H})$, respectively, by conversion of a proton into a Ξ^-. (In our model calculations, the $\alpha\Xi^-$ potential is generated from a G-matrix ΞN interaction via a folding procedure.) Here, among the above Ξ^- hypernuclei, $^{7}_{\Xi^-}\mathrm{H}(\alpha nn\Xi^-)$ is the lightest Ξ^- bound system, as shown in the following section. In the case of lighter targets, ^{6}Li, ^{4}He, ^{3}He and d, the Ξ^--hypernuclear states are composed of $\alpha n\Xi^-$, $pnn\Xi^-$ $(t\Xi^-)$, $pn\Xi^-$ and $n\Xi^-$ configurations, respectively. However, these systems are not expected to support bound states, considering the weakly attractive nature of the ΞN interactions suggested so far, except for Coulomb-bound (atomic) states. Thus, possible Ξ^- hypernuclear states to be investigated lie in the light p-shell region and may be considered to have basically a four-body cluster structure.

2. ΞN interaction

As for the ΞN interaction, we empoly two kinds of ΞN potentials, Namely, Nijmegen model D (ND) and new interaction model ESC04. In the past, the ND model has been popular for $S = -2$ interactions, because this model is compatible with the strong $\Lambda\Lambda$ attraction indicated by the older data on double Λ hypernuclei, and also it yields attractive Ξ-nucleus interactions. These aspects of ND are the result of its specific feature that the unitary-singlet scalar meson is included without any scalar-octet mesons. In this case, the strong ΞN attraction originates from this scalar-singlet meson which gives the same contributions in all YN and YY channels.

More recently, new interaction models ESC04 (a,b,c,d) have been introduced, models in which two-meson and meson-pair exchanges are taken into account, and in principle no *ad hoc* effective boson-exchange potentials are included.[4,5] The features of the ESC04 models differ significantly from those of the OBE models, especially in the $S = -2$ channels. Among the ESC04 models, ESC04d is distinguished, because the resulting Ξ-nucleus interaction gives attraction suggested by the above experimental situation. This is mainly due to the following mechanism: A remarkably strong attraction appears in the $T = 0$ triplet-even $(^{13}S_1)$ state, because the strongly repulsive contribution of vector mesons is cancelled by the attractive contributions from axial-vector mesons. In fact, the attraction in this state is so strong that peculiar Ξ bound states are produced in few-body systems,[6] though such considerations lie outside the scope of the work presented in this paper. In later calculations, the important points are the spin- and isospin-averaged even-state interactions, which are strongly attractive owing to the significant $^{13}S_1$-state attraction. Another important feature of ESC04d in the $S = -2$ channel is that the meson-pair exchange terms give rise to strong $\Lambda\Lambda$-ΞN-$\Sigma\Sigma$ and ΞN-$\Lambda\Sigma$-$\Sigma\Sigma$ coupling interactions. This feature of ESC04d makes the conversion widths of Ξ-hypernuclear states far larger than those for ND.

Our cluster models are composed of cluster units (α and t), n and Ξ^-, where the $\alpha(t)\Xi^-$ interactions are obtained by folding the ΞN G-matrix interactions into the density of $\alpha(t)$. According to the method described in Refs. 4,5, the ΞN G-matrix interactions are derived from ESC04d and ND in nuclear matter, where the imaginary parts arise from the energy-conserving transitions from ΞN to $\Lambda\Lambda$ channels in the nuclear medium. The resulting complex G-matrix interactions are represented as k_F-dependent local potentials

$$G^{(\pm)}_{TS}(r, k_F) = \sum_{i=1}^{3} (a_i + b_i k_F + c_i k_F^2) \exp\left(-r^2/\beta_i^2\right) , \qquad (1)$$

where k_F is the Fermi momentum of nuclear matter. The suffixes (+) and (−) specify even and odd, respectively. In our applications to finite Ξ systems, it is plausible to obtain the k_F values from the average density in the respective systems. In a similar G-matrix approach to Λ hypernuclei, for instance, the ΛN G-matrix interactions can be adopted to reproduce the observed Λ binding energy (B_Λ) by choosing appropriate k_F values. Such a procedure cannot be applied strictly in the case of Ξ hypernuclei, because there exist no definitive experimental data. In this work, we are obliged to choose the k_F values rather arbitrarily but within a reasonable range ($0.8 \sim 1.2$ fm^{-1} in light p-shell systems). Here, the experimental indication for the existence of ${}^{12}_{\Xi^-}$Be is used to adopt the ΞN G-matrix interactions, although it is not so definite because an experimental Ξ binding energy (B_Ξ) could not be extracted. As shown later, the adjustable parts included in ND and ESC04d are determined so that the Ξ s-state energy in our model of ${}^{12}_{\Xi^-}$Be has the value −2.2 MeV for an adequate value of $k_{\rm F}$, being obtained from the Ξ-nucleus WS potential with the depth −14 MeV[1] where the Coulomb interaction is switched off.

3. Results for ${}^{12}_{\Xi^-}$Be and the appropriate $k_{\rm F}$ parameter

As mentioned before, our ΞN interactions are adjusted so as to give the Ξ^- s-state energy −2.2 MeV in the ${}^{12}_{\Xi^-}$Be system. This value is consistent with the observed spectrum of the ^{12}C(K^-, K^+) reaction which suggests the WS potential depth of 14 MeV.[1] If we assume spin-nonflip dominance for the ^{12}C(K^-, K^+) reaction, the $[p^{-1}_{3/2}\, s^{\Xi^-}_{1/2}]_{J=1^-}$ state is naturally excited. Therefore, within the framework of the $\alpha\alpha t\Xi^-$ four-body model, the $k_{\rm F}$ parameters in the $\alpha\Xi^-$ and $t\Xi^-$ potentials, without Coulomb interaction, are tuned so that the 1^-_1 state energies agree with −2.2 MeV.

In the case of G_{ESC}, the 2^-_1 state is obtained at a lower energy than the 1^-_1 state, −3.18 MeV. On the other hand, the use of G_{ND} leads to the opposite order: the Ξ^- separation energy, B_{Ξ^-}, of 2^- is −1.56 MeV.

We found the appropriate k_F parameter values of the effective ΞN interactions to be $k_F = 1.055$ fm^{-1} (ESC) and $k_F = 1.025$ fm^{-1}(ND), which are consistent with the experimental indication in ${}^{12}_{\Xi^-}$Be. These interactions provide our basis to investigate the $A = 7$ and 10 Ξ^- hypernuclei.

4. Results for ${}^{7}_{\Xi^-}$H

Here we describe the results of the four-body calculations for ${}^{7}_{\Xi^-}$H($\alpha nn\Xi^-$) with $(T, J^\pi) = (3/2, 1/2^+)$. The basic question is whether this state is bound or not: The ^{6}He core is composed of an α and two weakly-bound ('halo') neutrons. Due to the weakness of the $\Xi^- n$ interaction, the binding between ^{6}He and Ξ^- is to a large extent determined by the $\alpha\Xi^-$ interaction.

The calculated energies in the $1/2^+$ ground state are demonstrated in Fig.1 as a function of k_F, for the two ΞN potential models without the imaginary part of the $\alpha\Xi^-$ interaction. These $1/2^+$ states are composed of the ground-state 0^+ configuration of ^{6}He coupled with the $0s$-state Ξ^- particle. The Coulomb interactions between α and Ξ^- are taken into account. In the figure, the dashed lines show the positions of threshold energies of $\alpha+n+n+\Xi^-$, ^{6}He$+\Xi^-$ and ${}^{5}_{\Xi^-}$H$(\alpha\Xi^-)_{\rm cal}+n+n$, respectively. One should be aware that the ${}^{5}_{\Xi^-}$H$(\alpha\Xi^-)_{\rm cal}+n+n$ threshold energy depends on the k_F value of the adopted ΞN interactions. This situation is unavoidable, because the calculated energies for ${}^{5}_{\Xi^-}$H have to be used instead of the unknown experimental value. We see that in the case (i) $k_F = 0.9$ fm^{-1} with ESC the lowest threshold is ${}^{5}_{\Xi^-}$H$(\alpha\Xi^-)_{\rm cal}+n+n$, and in the other cases the ^{6}He$+\Xi^-$ threshold is lower than the ${}^{5}_{\Xi^-}$H$(\alpha\Xi^-)_{\rm cal}+n+n$ threshold. On the other hand, in all k_F cases with ND the lowest threshold is ^{6}He$+\Xi^-$. The order of the ${}^{5}_{\Xi^-}$H$(\alpha\Xi^-)_{\rm cal}+n+n$ and ^{6}He$+\Xi^-$ threshold is determined by the competition between α-Ξ^- correlation and the α -(nn) correlation.

We see that we have bound state in both cases of ESC and ND at reasonable k_F values of around 0.9 fm^{-1}. If the experiment is carried out to observe the ${}^{7}_{\Xi^-}$H bound states, it is useful to extract information about the even-state spin- and isospin-averaged part of the ΞN interaction acting between the α and Ξ^-.

5. Results for ${}^{10}_{\Xi^-}$Li

The calculated results for ${}^{10}_{\Xi^-}$Li($\alpha\alpha n\Xi^-$) within the four-body model are displayed in Fig.2 for the lowest $T = 1$ doublet state energies ($J^\pi = 2^-, 1^-$). The $3/2^-$ ground state of the core nucleus ^{9}Be is bound by about 1.57 MeV with respect to the $\alpha+\alpha+n$ threshold. The ${}^{10}_{\Xi^-}$Li($\alpha\alpha n\Xi^-$) system is surely

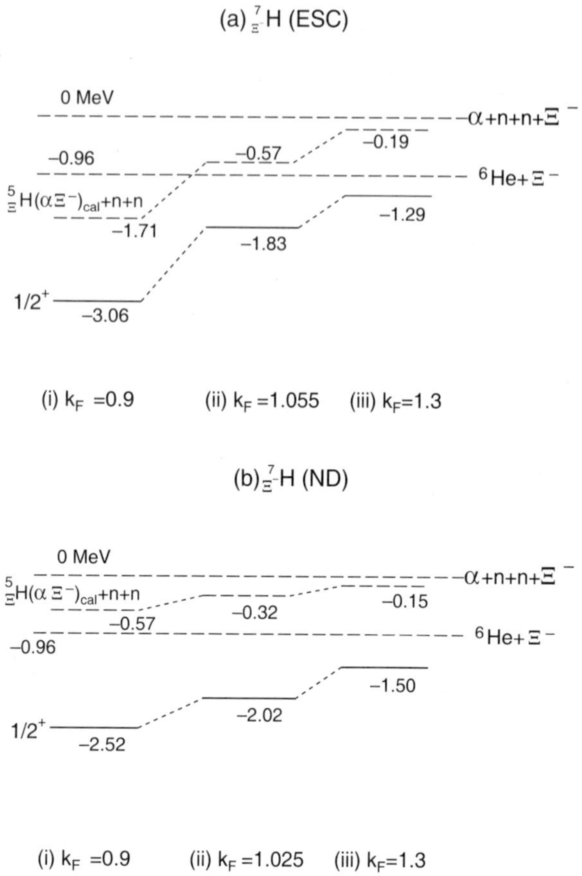

Fig. 1. (a)Calculated energy levels of $^{7}_{\Xi^-}$H for three $k_{\rm F}$ values using ESC. (b) Calculated energy levels of $^{7}_{\Xi^-}$H for for three $k_{\rm F}$ values using ND. The energies are shown when the imaginary part of the $\alpha\Xi^-$ interaction is switched off. The energies are measured from the $\alpha + n + n + \Xi^-$ breakup threshold. The dashed lines are threshold.

expected to be bound because the interaction between the Ξ^- and a p-orbit neutron is weakly attractive.

Although the binding energies of $^{10}_{\Xi^-}$Li ($J^\pi = 2^-, 1^-$) are found to be fairly sensitive to the choice of the $k_{\rm F}$ values, especially, in the case of ESC, we think the results with $k_{\rm F} \sim 1.0$ fm^{-1} are most acceptable. It is interesting to note that the $^{9}_{\Xi^-}$Li$(\alpha\alpha\Xi^-)_{\rm cal} + n$ threshold comes below the ^{9}Be+Ξ^- threshold in most cases. It is reasonable that the lowest breakup threshold is $^{9}_{\Xi^-}$Li$(\alpha\alpha\Xi^-)_{\rm cal} + n$, because the value of k_F in the $A = 10$

system has to be $\sim 1.0\ \mathrm{fm}^{-1}$, similar to the ${}^{12}_{\Xi^-}\mathrm{Be}$ system. It is notable that the binding energies of ${}^{10}_{\Xi^-}\mathrm{Li}$ measured from the ${}^{9}_{\Xi^-}\mathrm{Li}(\alpha\alpha\Xi^-)_{\mathrm{cal}} + n$ thresholds are similar to each other in both cases of ESC(2.6 MeV) and ND (2.7 MeV).

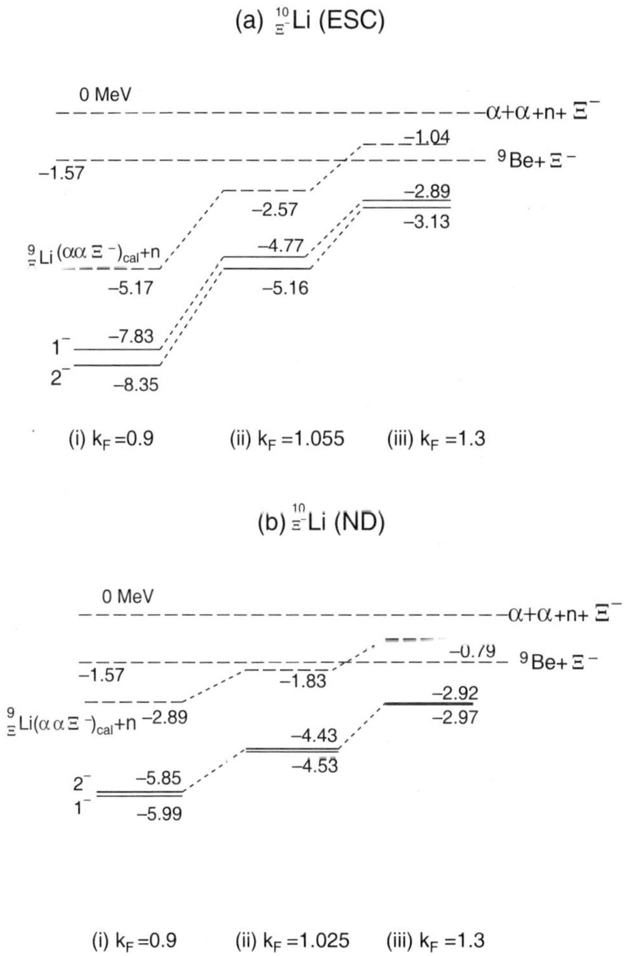

Fig. 2. (a)Calculated energy levels of ${}^{10}_{\Xi^-}\mathrm{Li}$ for three k_{F} values using ESC. (b)Calculated energy levels of ${}^{10}_{\Xi^-}\mathrm{Li}$ for three k_{F} values using ND. The energies are shown when the imaginary part of the $\alpha\Xi^-$ interaction is switched off. The energies are measured from the $\alpha + \alpha + n + \Xi^-$ threshold. The dashed lines indicate thresholds.

We expect such structure, a valence neutron coupled to ${}^{9}_{\Xi^-}\mathrm{Li}$ hypernucleus, since the lowest threshold is ${}^{9}_{\Xi^-}\mathrm{Li}(\alpha\alpha\Xi^-)_{\mathrm{cal}} + n$. While, if the k_{F} value becomes by chance much larger, then the Ξ^- particle is coupled to the ground state of ${}^{9}\mathrm{Be}$. Because the lowest threshold is ${}^{9}\mathrm{Be}+\Xi^-$ [See case (iii) in Fig.2 (a) and (b)].

The 2^- (1^-) state is dominated by the ${}^{33}S_1$ (${}^{31}S_0$) component of the two-body $n\Xi^-$ interaction. The ${}^{33}S_1$ interaction for ESC (ND) is more (less) attractive than the ${}^{31}S_0$ interaction. Therefore, the 2^- (1^-) state of ${}^{10}_{\Xi^-}\mathrm{Li}$ becomes the ground state in the case of ESC (ND), as shown in Fig.2 (a) and (b).

From these calculation, we can surely expect the existence of the bound state. To produce ${}^{10}_{\Xi^-}\mathrm{Li}$, we propose to perform the ${}^{10}\mathrm{B}(K^-, K^+)$ reaction experiment at J-PARC in addition to that with a ${}^{12}\mathrm{C}$ target.

We say that the $\alpha\alpha n\Xi^-$ (${}^{10}_{\Xi^-}\mathrm{Li}$) system produced by the (K^-, K^+) reaction on ${}^{10}\mathrm{B}$ is suitable to investigate $\alpha\Xi^-$ interactions, namely the spin-independent terms of even and odd-state ΞN interactions.

Acknowledgements

The numerical calculations were performed on the HITACHI SR11000 at KEK.

References

1. P. Khaustov *et al.*, Phys. Rev. **C61**, (2000) 054603.
2. S. Aoki et al., Prog. Theor. Phys. **89** (1993), 493.; S. Aoki et al., Phys. Lett. **B355** (1995), 45.; Y. Yamamoto, Genshikaku Kenkyu 39 (1996), 23.
3. M. M. Nagels, T. A. Rijken, and J. J. deSwart, Phys. Rev. **D15** (1977) 2547.
4. Th. A. Rijken and Y. Yamamoto, Phys. Rev. **C73**, 044008 (2006); [arXiv:nucl-th/0603042].
5. Th. A. Rijken and Y. Yamamoto, [arXiv:nucl-th/0608074].
6. Y. Yamamoto and Th. A Rijken, Nucl. Phys. A **804**, 139 (2008).

LARGE-AMPLITUDE QUADRUPOLE COLLECTIVE DYNAMICS IN Se AND Kr ISOTOPES

N. HINOHARA*

Theoretical Nuclear Physics Laboratory, RIKEN Nishina Center, Wako 351-0198, Japan
**E-mail: hinohara@riken.jp*

K. SATO

Department of Physics, Graduate School of Science, Kyoto University, Kyoto 606-8502, Japan

Theoretical Nuclear Physics Laboratory, RIKEN Nishina Center, Wako 351-0198, Japan

T. NAKATSUKASA

Theoretical Nuclear Physics Laboratory, RIKEN Nishina Center, Wako 351-0198, Japan

M. MATSUO

Department of Physics, Faculty of Science, Niigata University, Niigata 950-2181, Japan

Large-amplitude oblate-prolate shape coexistence dynamics in proton-rich Se and Kr isotopes is analyzed using the five-dimensional quadrupole collective Hamiltonian. The potential and inertial functions in the collective Hamiltonian are calculated using the constrained Hartree-Fock-Bogoliubov (CHFB) equation and the local quasiparticle random-phase approximation (LQRPA). The excitation spectra and the electric quadrupole transitions between the low-lying states in ^{68}Se are calculated by solving the collective Schrödinger equation, and the shape mixing dynamics in the low-lying collective states are discussed.

1. Introduction

Nuclei show various large-amplitude collective motions such as anharmonic vibrations, shape coexistence, and shape phase transitions in low-lying states. It is a challenge in the nuclear structure theory to construct a microscopic theory for such large-amplitude collective motions. The Bohr-

Mottelson collective Hamiltonian provides an adiabatic framework for large-amplitude quadrupole collective motion, which plays the most essential part in the low-lying collective dynamics, and the microscopic derivations of the collective potential and inertial functions in the Bohr-Mottelson collective Hamiltonian have been tried by Belyaev,[1] Kumar and Baranger,[2] and many other people over years. The collective potential can be calculated from the constrained Hartree-Fock-Bogoliubov (CHFB) theory, and the static properties of the large-amplitude collective phenomena have been widely discussed. As for the inertial functions, the Inglis-Belyaev cranking mass formula is widely used in the microscopic calculation. However, the contribution from the time-odd mean field induced by the residual two-body interactions is ignored in the cranking mass formula. It is well known that the Inglis-Belyaev inertial function is systematically too small to explain the excitation energies of vibrational and rotational collective states,[3] and it is necessary to include the contribution from time-odd mean field to the inertial functions for the description of the collective dynamics. The time-odd contribution can be taken into account with the aid of the quasi-particle random phase approximation (QRPA) in the small-amplitude limit. In the large-amplitude cases, however, the inertial functions should be calculated from the adiabatic time-dependent theory of large-amplitude collective motion. One of the successful time-dependent microscopic theory of large-amplitude collective motion is the adiabatic self-consistent collective coordinate (ASCC) method[4] developed as a microscopic theory capable of determining the collective degrees of freedom (collective submanifold) and the inertial functions in a self-consistent way. Since the ASCC method is formulated using the adiabatic expansion of the time-dependent Hartree-Fock-Bogoliubov (TDHFB) theory, the inertial functions derived from the ASCC method include the time-odd contribution of the moving mean field.[5]

In this paper, we propose the local QRPA (LQRPA) theory to calculate the inertial functions of the large-amplitude quadrupole vibrations which include the contribution of the time-odd mean-field on the basis of the ASCC method for a two-dimensional collective hypersurface. As an application of the theory, the five-dimensional collective Hamiltonian for the oblate-prolate shape coexistence/mixing dynamics in proton-rich ^{68}Se isotope is constructed, and the properties of the low-lying states are investigated. More detailed description of the method and the discussion of the numerical results will be reported in a forthcoming paper.

2. CHFB + LQRPA equations

We briefly introduce the CHFB + LQRPA theoretical framework to construct the five-dimensional classical quadrupole collective Hamiltonian in a microscopic way. The generalized quadrupole collective Hamiltonian of Bohr and Mottelson is written as

$$\mathcal{H}_{\text{coll}} = V(\beta,\gamma) + T_{\text{vib}} + T_{\text{rot}}, \tag{1}$$

$$T_{\text{vib}} = \frac{1}{2}D_{\beta\beta}(\beta,\gamma)\dot{\beta}^2 + D_{\beta\gamma}(\beta,\gamma)\dot{\beta}\dot{\gamma} + \frac{1}{2}D_{\gamma\gamma}(\beta,\gamma)\dot{\gamma}^2, \tag{2}$$

$$T_{\text{rot}} = \frac{1}{2}\sum_{k=1}^{3}\mathcal{J}_k(\beta,\gamma)\omega_k^2, \tag{3}$$

where $V(\beta,\gamma)$, $D(\beta,\gamma)$, and $\mathcal{J}_k(\beta,\gamma)$ are the collective potential, three vibrational collective masses, and three rotational moments of inertia as functions of two quadrupole deformation parameters β and γ, respectively.

The potential energy $V(\beta,\gamma)$ in Eq. (1) is determined by the CHFB equation

$$\delta\,\langle\phi(\beta,\gamma)|\,\hat{H}_{\text{CHFB}}\,|\phi(\beta,\gamma)\rangle = 0, \tag{4}$$

where the two constraints on quadrupole deformations (β,γ) and two constraints on the neutron and proton numbers are imposed in the variational calculation.

The vibrational inertial functions $D_{\beta\beta}(\beta,\gamma)$, $D_{\beta\gamma}(\beta,\gamma)$, and $D_{\gamma\gamma}(\beta,\gamma)$ in Eq. (2) are locally determined by the LQRPA equations

$$\delta\,\langle\phi(\beta,\gamma)|\,[\hat{H}_{\text{CHFB}},\hat{Q}(\beta,\gamma)] - \frac{1}{i}B(\beta,\gamma)\hat{P}(\beta,\gamma)\,|\phi(\beta,\gamma)\rangle = 0, \tag{5}$$

$$\delta\,\langle\phi(\beta,\gamma)|\,[\hat{H}_{\text{CHFB}},\frac{1}{i}\hat{P}(\beta,\gamma)] - C(\beta,\gamma)\hat{Q}(\beta,\gamma)\,|\phi(\beta,\gamma)\rangle = 0, \tag{6}$$

which are the QRPA equations on top of the CHFB states $|\phi(\beta,\gamma)\rangle$. Here $\hat{Q}$ and $\hat{P}$ are the LQRPA phonon operators, and B and C are the collective mass and the local stiffness with respect to the LQRPA mode. The quadrupole vibrational inertial functions are calculated from the transformation of two LQRPA modes to the (β,γ) quadrupole deformation degrees of freedom under the assumption that there is a one-to-one correspondence between them. The two LQRPA modes which give the smallest metric

$$W(\beta,\gamma) = D_{\beta\beta}(\beta,\gamma)D_{\gamma\gamma}(\beta,\gamma) - D^2_{\beta\gamma}(\beta,\gamma), \tag{7}$$

are chosen from all the possible pairs of low-lying modes.

The rotational moments of inertia $\mathcal{J}_k(\beta,\gamma)$ in Eq. (3) are evaluated from the "Thouless-Valatin" equations[6]

$$\delta \langle\phi(\beta,\gamma)| [\hat{H}_{\rm CHFB}, \Psi_k(\beta,\gamma)] - \frac{1}{i}\mathcal{J}_k^{-1}(\beta,\gamma)\hat{I}_k |\phi(\beta,\gamma)\rangle = 0, \quad (8)$$

$$\langle\phi(\beta,\gamma)| [\Psi_k(\beta,\gamma), \hat{I}_k] |\phi(\beta,\gamma)\rangle = i. \quad (9)$$

They are the QRPA equations for the collective rotation, and thus in the present approach, the time-odd mean-field contribution is taken into account in the vibrational and rotational inertial functions.

We use the pairing-plus-quadrupole (P+Q) Hamiltonian including the quadrupole-pairing interaction for the microscopic Hamiltonian, and adopt the P+Q interaction strengths given in Ref. 7. For the quadrupole-pairing interaction strengths, the self-consistent treatment proposed by Sakamoto and Kishimoto[8] is used.

3. Application to low-lying states in ^{68}Se

We have solved the CHFB and LQRPA equations numerically in the two-dimensional mesh points in the β and γ directions. Figure 1 shows the collective potential and the inertial functions for ^{68}Se calculated using the CHFB + LQRPA theory. The collective potential has two local minima: one is oblate and the other is prolate. It also possesses a triaxial potential valley. We note that the one-dimensional collective path extracted from the TDHFB phase space with the ASCC method in our previous work approximately follows this triaxial potential valley.[11] The vibrational inertial function shows a smooth behavior as a function of β and γ in the lower-half plane, indicating that the two quadrupole collective coordinates can be well separated from other degrees of freedom. However, it suddenly increases in the upper-half plane, and shows a complicated structure as a function of deformation. As for the rotational moment of inertia, $D_k(\beta,\gamma)$ which is given by

$$\mathcal{J}_k(\beta,\gamma) = 4\beta^2 D_k(\beta,\gamma)\sin^2(\gamma - 2\pi k/3), (k = 1,2,3), \quad (10)$$

is plotted in the figure instead of $\mathcal{J}_k(\beta,\gamma)$ itself. The $D_1(\beta,\gamma)$ becomes large in the small deformation region, where the moment of inertia $\mathcal{J}_1(\beta,\gamma)$ approaches zero.

In comparison with the Inglis-Belyaev inertial function, the LQRPA vibrational inertial function shows 20 $\sim$ 60% increase in the lower-half plane. It becomes approximately 10 times larger at the maximum value in the upper-half plane. The LQRPA rotational moment of inertia shows 5 $\sim$

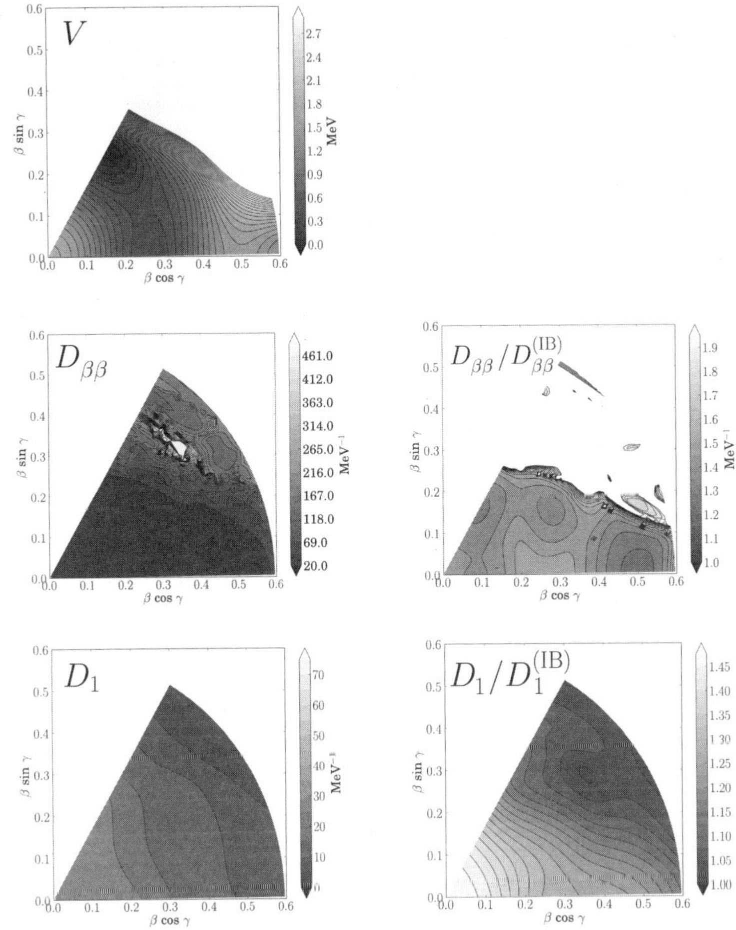

Fig. 1. Collective potential $V(\beta,\gamma)$ (upper left), the LQRPA inertial function $D_{\beta\beta}(\beta,\gamma)$ (middle left), its ratio to the Inglis-Belyaev value $D_{\beta\beta}/D_{\beta\beta}^{(\mathrm{IB})}$ (middle right), the moments of inertia about the intermediate axis in the region $0^\circ < \gamma < 60^\circ$ $D_1(\beta,\gamma)$ (lower left), and its ratio to the Inglis-Belyaev moment of inertia $D_1/D_1^{(\mathrm{IB})}$ (lower right).

50% increase. It is also worth mentioning that the deformation dependence of the Inglis-Belyaev inertial functions is different from that of the LQRPA inertial functions. In the case of the moment of inertia, the contribution from the time-odd mean field tends to become relatively large near the region where the moment of inertia vanishes. The sudden increase of the vibrational inertial function in the upper-half plane is not seen in the Inglis-

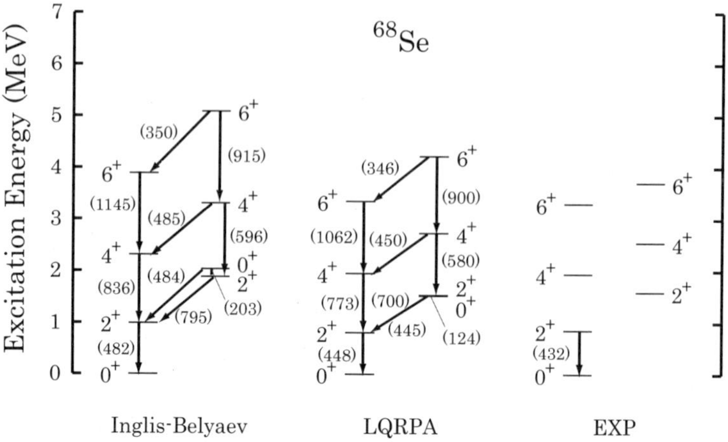

Fig. 2. Excitation energies and $B(E2)$ calculated for ^{68}Se by solving the collective Schrödinger equation. The results calculated with the Inglis-Belyaev and LQRPA inertial functions are compared with the experimental data.[9,10] The $B(E2)$ values are written in parentheses in units of e^2 fm^4.

Belyaev value. These results clearly suggest the importance of determining microscopically the contribution from the time-odd mean field.

The collective Schrödinger equation

$$\{V(\beta,\gamma)+\hat{T}_{\rm vib}+\hat{T}_{\rm rot}\}\Psi_{IMk}(\beta,\gamma,\Omega)=E_{Ik}\Psi_{IMk}(\beta,\gamma,\Omega), \tag{11}$$

$$\Psi_{IMk}(\beta,\gamma,\Omega)=\sum_{K\geq 0}\Phi_{IKk}(\beta,\gamma)\langle\Omega|IMK\rangle, \tag{12}$$

is solved for the collective Hamiltonian requantized according to Pauli's prescription. Here Ψ_{IMk} is the collective wave function in the laboratory frame, and Φ_{IKk} is its vibrational part in the body-fixed frame. The excitation energies and $E2$ transitions of the low-lying states in ^{68}Se are presented in Fig. 2. It shows a quantitatively good agreement between the experimental data and the theoretical values for the CHFB+LQRPA collective Hamiltonian. It should be noted that the adopted parameters of the P+Q model[7] are not fitted to these data. The Inglis-Belyaev inertial functions yield higher excitation energies than the LQRPA ones. In particular, the excitation energy of the 0_2^+ state is much affected by the time-odd component. The small value of $B(E2;2_2^+\to 0_2^+)$ also indicates a unique character of the 0_2^+ state. In Fig. 3, the collective wave functions squared are plotted. In the figure, they are multiplied by β^4 which carries the major β dependence of the volume element. The collective wave functions spread over the triaxial region as a consequence of the large-amplitude γ-vibrational motion. At the same time, we can see that the oblate (prolate) character dominates in the yrast (yrare) states except for the 0^+ states. Concerning the 0_2^+ state, there is a node in the β direction indicating a significant coupling between the large-amplitude γ vibration and the β-vibrational mode. These and other properties (not shown here) indicate a multi-facet character of low-lying states of ^{68}Se, exhibiting both the oblate-prolate shape coexistence and the γ-soft features.

4. Concluding remarks

We investigate the oblate-prolate shape coexistence dynamics in ^{68}Se using the Bohr-Mottelson collective Hamiltonian. All the potential and inertial functions in the collective Hamiltonian are microscopically determined from the CHFB + LQRPA theory. For the first time, the contribution of the time-odd mean field to the inertial functions is included for both rotational and large-amplitude vibrational motions with the full (β, γ) dependence. We show that the time-odd contribution of the mean field increases the vibrational and rotational inertial functions. Solving the collective Schrödinger equation, we have calculated the excitation spectra and $E2$ transition probabilities in the low-lying states. The effects of the time-odd component on the vibrational and rotational inertial function are shown that the time-odd mean field reduces the energies of the excited states. It is shown that the low-lying states in ^{68}Se possess a multi-facet character exhibiting both the oblate-prolate shape coexistence and the γ-soft features. A unique behavior of the 0_2^+ state is also pointed out.

Acknowledgments

One of the authors (N. H.) is supported by the Special Postdoctoral Researcher Program of RIKEN. This work is supported by the Grant-in-Aid for Scientific Research(B) (No. 21340073). The numerical calculations were carried out on RIKEN Cluster of Clusters (RICC) facility.

References

1. S. T. Belyaev, *Nucl. Phys.* **64**, 17 (1965).
2. K. Kumar and M. Baranger, *Nucl. Phys. A* **92**, 608 (1967).
3. J. Dudek, W. Dudek, E. Ruchowska and J. Skalski, *Z. Phys. A* **294**, 341 (1980).
4. M. Matsuo, T. Nakatsukasa and K. Matsuyanagi, *Prog. Theor. Phys.* **103**, 959 (2000).

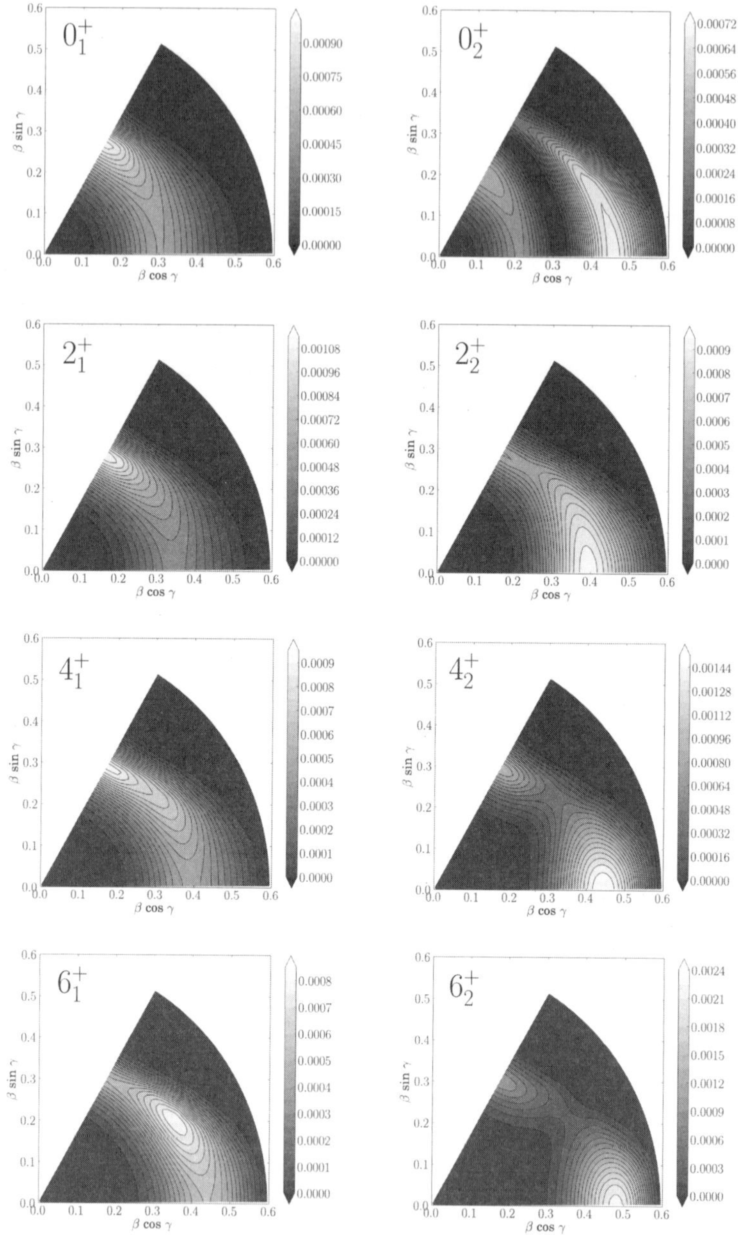

Fig. 3. The vibrational parts of the collective wave functions squared $\beta^4 \sum_K |\Phi_{IKk}(\beta,\gamma)|^2$, are plotted for the yrast and yrare states of ^{68}Se.

5. N. Hinohara, T. Nakatsukasa, M. Matsuo and K. Matsuyanagi, *Prog. Theor. Phys.* **115**, 567 (2006).
6. N. Hinohara, T. Nakatsukasa, M. Matsuo and K. Matsuyanagi, *Prog. Theor. Phys.* **119**, 59 (2008).
7. M. Kobayasi, T. Nakatsukasa, M. Matsuo and K. Matsuyanagi, *Prog. Theor. Phys.* **113**, 129 (2005).
8. H. Sakamoto and T. Kishimoto, *Phys. Lett. B* **245**, 321 (1990).
9. S. M. Fischer, C. J. Lister and D. P. Balamuth, *Phys. Rev. C* **67**, 064318 (2003).
10. A. Obertelli, *et al.*, *Phys. Rev. C* **80**, 031304 (2009).
11. N. Hinohara, T. Nakatsukasa, M. Matsuo and K. Matsuyanagi, *Phys. Rev. C* **80**, 014305 (2009).

ROLE OF PIONS ON FINITE NUCLEI WITH THE RELATIVISTIC CHIRAL MEAN FIELD MODEL

Y. OGAWA* and H. TOKI

Research Center for Nuclear Physics, Osaka University,
Ibaraki, 567-0047/Osaka, Japan
**E-mail: ogaway@rcnp.osaka-u.ac.jp*

We study the role of pion on finite nuclei using the relativistic chiral mean field (RCMF) model. We present how to treat the strong tensor correlation in a consistent way to the nuclear many-body problems and show calculated results in the RCMF model.

Keywords: finite nuclear many-body system; meson-exchange interaction; strong tensor-correlation; high-momentum component; 2p-2h variational space; chiral symmetry

1. Introduction

One of the fundamental subjects of nuclear physics is to understand a mutual relation between properties of nuclear many-body system and nuclear force. The most elementary meson, pion, was introduced by Yukawa as a mediator of the nuclear force.[1] The nature of pion is found as a Nambu-Goldstone boson of spontaneous chiral symmetry breaking.[2] The pion has the iso-vector and pseudo-scalar nature, which leads to strong tensor correlation and involves necessarily a large amount of momentum space for the construction of nuclei. Pion has the lightest meson mass, as a consequence, bringing significantly spatially large interaction range between pair nucleons, which is from 0.5 fm to 2.0 fm. Due to these complexities it is difficult to construct a framework for the description of the property of nuclear many-body system with this fundamental interaction, pion-exchange, in full strength and explicit way until recent years.

Recently, an *ab initio* variational calculation using Variational Monte-Carlo and Green's Function Monte-Carlo methods by Argonne-Illinois group for light nuclei shows an important role played by pion in nuclei.[3] The energy contribution of pion-exchange to the total two-body attractive force is about 80%. In addition to these light nuclei, the chiral perturbation theory for nuclear matter demonstrates that pion plays an important role on bulk property of nuclear system.[4] Especially, the Hartree term of the iterated one-pion exchange is very important to produce the binding energy.

For light nuclei we can calculate nuclear ground states precisely using realistic nucleon-nucleon interaction as much as we can without restriction of the model space as the case of the calculations of the Argonne-Illinois groups. However, if we wish to calculate finite nuclei heavier than ^{12}C, we need a tremendous computer power.

In this talk, we present a method to treat the strong tensor correlations due to pion-exchange interaction in a consistent way to the nuclear many-body problems. We use a self-consistent field theoretical method based on the single-particle picture or shell model, even though the correlations are strong and short-ranged. Furthermore, we can obtain wave functions that retains the high-momentum components. There is abundant evidence that the nuclear wave function retains large high-momentum components due to pionic degrees of freedom.[5–9] Since we have to calculate high-momentum components by using bare interaction, we have to take care of the short-range repulsion explicitly. We employ the unitary correlation operator method (UCOM)[10] for the central force.

2. Relativistic chiral mean field model

2.1. *Lagrangian*

We start from the linear σ model of Gell-Mann-Levy,[11] where the pion appears symmetrically with the scalar field as a chiral partner. We further introduced ω-nucleon coupling in order to produce necessary repulsive force to form nuclear system. The character of this Lagrangian is that all constituents obtain their masses by scalar meson condensation.[12] Adjustable parameters are only two, which are the mass of the scalar meson, m_σ, and the ω-nucleon coupling constants, g_ω. Others are decided by hadron physics in free space. We employ pseudo-vector type pion-nucleon coupling, so that the coupling between positive and negative energy states can be weakened and we do not need to treat seriously the negative energy states. We thus employ the non-linear realization of the chiral Lagrangian which is obtained by the Weinberg transformation of the linear σ model.[13] We take the lowest order term in the pion field.[14]

$$\mathcal{L} = \mathcal{L}_{\sigma,\omega} + \mathcal{L}_{\pi}, \tag{1}$$

where

$$\begin{aligned}\mathcal{L}_{\sigma,\omega} &= \bar{\psi}(i\gamma_\mu\partial^\mu - M - g_\sigma\sigma - g_\omega\gamma_\mu\omega^\mu)\psi \\ &+ \frac{1}{2}\partial_\mu\sigma\partial^\mu\sigma - \frac{1}{2}{m_\sigma}^2\sigma^2 - \lambda f_\pi\sigma^3 - \frac{\lambda}{4}\sigma^4 \\ &- \frac{1}{4}\omega_{\mu\nu}\omega^{\mu\nu} + \frac{1}{2}{m_\omega}^2\omega_\mu\omega^\mu \\ &+ \widetilde{g_\omega}^2 f_\pi\sigma\omega_\mu\omega^\mu + \frac{1}{2}\widetilde{g_\omega}^2\sigma^2\omega_\mu\omega^\mu,\end{aligned} \tag{2}$$

and

$$\mathcal{L}_{\pi} = -\frac{g_A}{2f_\pi}\bar{\psi}\gamma_5\gamma_\mu\partial^\mu\pi^a\tau^a\psi + \frac{1}{2}\partial_\mu\pi^a\partial^\mu\pi^a - \frac{1}{2}{m_\pi}^2{\pi^a}^2. \tag{3}$$

2.2. *Ground state wave function*

The total wave function of the whole nuclear system consists of nucleon and meson sectors.

$$\Psi = \Psi_N \otimes \Psi_M(\sigma,\omega). \tag{4}$$

For the nucleon wave function, Ψ_N, we take an extended variational model space than that of the first order Hartree-Fock model. Since the pion-exchange interaction accompanies spin and iso-spin flips, interacted pair-nucleons must jump into significantly higher momentum states than the Fermi level. Hence, we would like to write the total wave function for the nuclear part, Ψ_N, as 0p-0h state and the combination of 2p-2h states,

$$|\Psi_N\rangle = c_0|0p - 0h\rangle + \sum_\alpha c_\alpha|2p - 2h;\alpha, b_\alpha\rangle, \tag{5}$$

where α denotes various 2p-2h configurations. The possibility of a third particle interfering with the interacting pair is small. We sum up all the possible 2p-2h configurations. Due to this extension we secure the large amount of momentum space and configuration space. This part is the first key point of this model. For the sigma and omega meson sector, $\Psi_M(\sigma,\omega)$, we take the mean field approximation.

2.3. *Pionic energy minimization*

We take the total wave function, Ψ, as a trial wave function. The guiding principle is to make the expectation value of the total Hamiltonian minimized,

$$\delta \frac{\langle \Psi | \hat{H} | \Psi \rangle}{\langle \Psi | \Psi \rangle} = 0. \tag{6}$$

The variational parameters are

(1) the amplitude of 2p-2h configurations as a whole, c_α,
(2) the size of the particle-states, b_α,
(3) the strength of σ-, ω-meson mean fields,
(4) the single-particle states in the Fermi sea, namely, hole-states, $\{\psi_i, i = 1, ..., A\}$.

These conditions, (1) $\sim$ (4), are decided under the guiding principle. The second one is very important from a view point of the pseudo-scalar nature of the pion-exchange interaction. The finding of the charge- and parity-projected Hartree-Fock method[15] or charge- and parity-projected relativistic mean field method[16] is that the particle states of 2p-2h states have spatially compact distributions due to high-momentum components. In the second step of minimization, (2), we can achieve the full strength of pionic correlations by using small number of bases. We make "pionic optimization" in this size variation. Conditions, (1) and (2) give the particle-states wave function. This is the second key point of this model.

Since the nuclear many-body system is constructed self-consistently, the hole-states should be formed under the same guiding principle as shown in condition (4). The variation of total energy with respect to each single-particle orbit of hole-state under the orthonormalization condition, $\langle \psi_i | \psi_j \rangle = \delta_{ij}$, is written as

$$\frac{\partial}{\partial \psi_i^*} \Big\{ \langle \Psi | \hat{H} | \Psi \rangle - \sum_i \varepsilon_i \langle \psi_i | \psi_i \rangle \Big\} = 0. \tag{7}$$

This variation leads to the equations, which include the effect from the extended variational space of 2p-2h configurations. The hole states are influenced by the pionic correlation through the coupling term to the 2p-2h states.

We solve all these equations given by energy functional variations, (1) $\sim$(4), self-consistently. For finite nuclear system, it is very important that these variational conditions are solved iteratively until self-consistency is achieved. We note here that it is important not to use the condition (4) for particle states, but use pionic energy optimized Gaussian wave functions obtained by variational conditions of (1) and (2) for finite nuclear system. This is the third key point of this model.

3. Results and discussions

We apply the RCMF model to the ^{12}C and ^{16}O nuclei. For the sake of discussion of the pion contribution in Figures 1 and 2, we drop the coupling term of the hole states to the 2p-2h states in condition (4) until the discussion of the jj-close shell formation in Fig.3.

The convergence of energy constituents is realized with pion spin state $J^\pi = 10^-$ for ^{12}C as shown in Fig. 1 and 11^+ for ^{16}O. Those pion spin states correspond to $\Delta l_\mathrm{p} = 11$ for ^{12}C and $\Delta l_\mathrm{p} = 12$ for ^{16}O of the change of the single-particle orbital angular momenta, respectively.

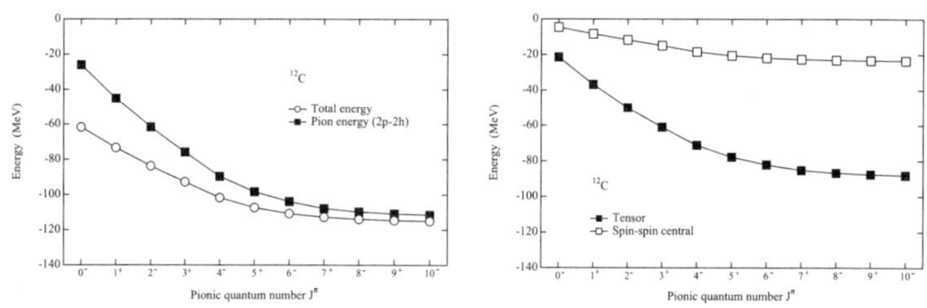

Fig. 1. Convergence of the total energy and pion energy with the pionic quantum number, J^π, for ^{12}C in left panel. The spin-spin central and tensor part of pion energy are shown in right one. The cutoff momentum, Λ, is taken to be 1000 MeV. The other free parameters, σ meson mass and the ω-nucleon coupling constant are 850 MeV and 6.81, respectively.

We show the momentum dependent part of pion-exchange interaction together with the form factor and the particle-hole transition density as a function of pion momentum, q in Fig. 2. In order to take into account the full strength of the pion-exchange interaction in the intermediate-range region we have to include these higher pion spin states.

We show the pionic energy per particle for ^{12}C and ^{16}O. The extra binding energy is obtained for ^{12}C in comparison with ^{16}O. This energy difference comes from the low pion spin states less than $J^\pi = 4$. This is the Pauli-blocking effect. Since the $p_{1/2}$ state of ^{16}O is already full occupied, jumping of pair nucleons from $0s_{1/2}$ into $0p_{1/2}$ is forbidden. The Pauli blocking effect dose not appear in high spin states any more. This energy

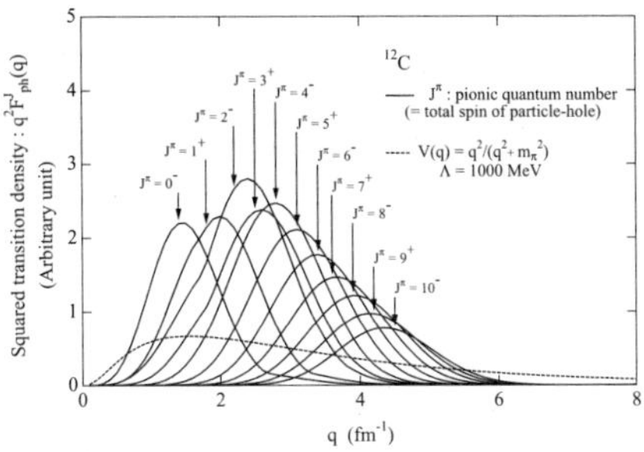

Fig. 2. The momentum dependent part of pion exchange interaction, $\frac{q^2}{m_\pi^2+q^2}$, together with the form factor, and the squared particle-hole transition density, $q^2F^J_{\rm ph}(q)$, as functions of q. The squared particle-hole transition densities are shown for various J^π and the pion exchange interaction is shown by solid curve for $\Lambda = 1000$ MeV. The parameters used for this presentation are σ meson mass, $m_\sigma = 850$ MeV and the ω-nucleon coupling constant, $g_\omega = 6.81$, respectively.

difference leads to formation of jj-magic structure. This is formed by the Pauli-blocking effect coming from the lower pion spin states.

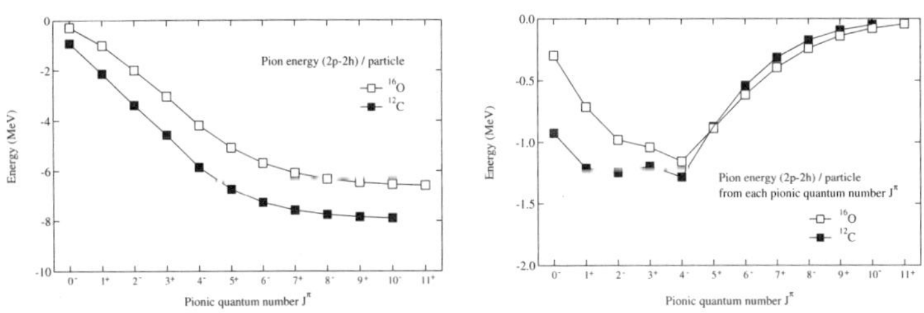

Fig. 3. Total energy and pionic energy are shown as functions of the pionic quantum number, J^π, for ^{12}C and ^{16}O. The cut-off momentum, Λ, are taken to be 1000 MeV. The other free parameters, σ meson mass and the ω-nucleon coupling constant are adjusted to reproduce simultaneously the binding energy and r.m.s. matter radius of the RMF calculation with the TM2 parameter set. We find $m_\sigma = 710$ MeV, $g_\omega = 8.344$ for ^{12}C, and 8.369 for ^{16}O.

In Figure 4 the effective masses for nucleon and ω meson are shown as function of the nuclear radius. The effective masses are given through the strength of the σ field for nucleon and ω meson as,

$$\begin{aligned} M^* &= M + g_\sigma \sigma = g_\sigma(f_\pi + \sigma), \\ m^*_\omega &= m_\omega + \widetilde{g}_\omega \sigma = \widetilde{g}_\omega(f_\pi + \sigma), \end{aligned} \tag{8}$$

respectively. Here, f_π is the expectation value of the scalar field in vacuum. The order parameter for the chiral symmetry in the linear σ model is the mean field value of the $(f_\pi + \sigma)$. The reduction of the masses of nucleon and ω meson in nuclear medium presents change of the scalar field. ter of the partial restoration of the chiral symmetry in nuclear medium in the linear σ model. The nucleon mass is reduced by 20% through the strength of scalar field in the interior of the nucleus.

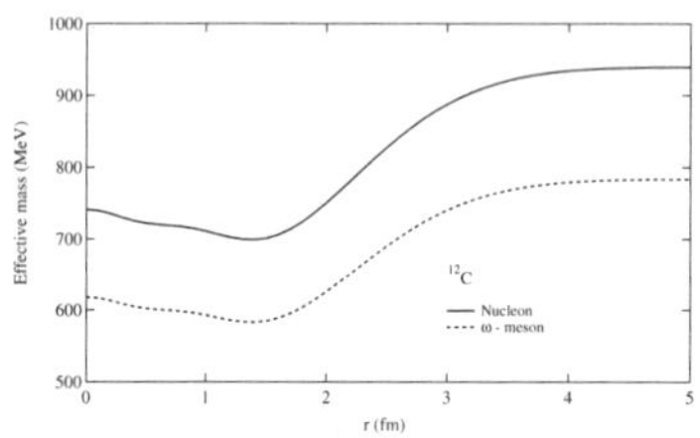

Fig. 4. The effective nucleon mass and effective ω meson mass are shown as a function of the radial distance, r, for ^{12}C. The parameters are adjusted to reproduce the binding energy and the matter r.m.s. radius simultaneously by the RMF calculation with the TM2 parameter set. We take $m_\sigma = 710$ MeV and $g_\omega = 8.344$. The pionic quantum number is taken up to $J^\pi_{\max} = 10^-$.

4. Summary

We study the role of pion on finite nuclei using the relativistic chiral mean field (RCMF) model. In the RCMF model, we start from the linear σ model Lagrangian, which has the chiral symmetry, and further introduce the omega meson coupling in order to produce necessary repulsion to form the nuclear system. We prepare first the mean field of σ and ω mesons and construct the meson fluctuation terms to include the dominant term of the pion-exchange interaction in the 2-particle 2-hole configurations to take into account the pseudo-scalar nature of the pion-exchange. The dominant term treated here is definitely the tensor part of pion-exchange interaction in second order or iterated one-pion in language of a perturbation theory. The effect of the short-range repulsion is included by unitary correlation operator method for central part of the pion-exchange interaction.

We applied the RCMF model to ^{12}C and ^{16}O nuclei. The extra binding energy is reproduced for ^{12}C as compared with ^{16}O by the Pauli-blocking effect coming from the nature of the pion-exchange interaction. This fact may lead to the jj-magic shell structure. The nucleon mass is reduced about 20% compared with free nucleon mass in the interior of nuclei through the strength of the scalar field.

References

1. H. Yukawa, *Proc. Phys.-Math. Soc. Jpn.* **17**, 48 (1935).
2. Y. Nambu and G. Jona-Lasinio, *Phys. Rev.* **122** 345 (1961); *ibid.* **124** 246 (1961).
3. R. B. Wiringa, S. C. Pieper, J. Carlson, and V. R. Pandharipande, *Phys. Rev. C* **62** 014001 (2000); S. C. Pieper and R. B. Wiringa, *Annu. Rev. Nucl. Part. Sci.* **51** 53 (2001).
4. N. Kaiser, S. Fritsch, W. Weise, *Nucl. Phys. A* **697** 255 (2002).
5. K. A. Brueckner, R. J. Eden, and N. C. Francis, *Phys. Rev.* **98** 1445 (1955).
6. T. Wakasa, H. Sakai, H. Okamura, H. Otsu, S. Fujita, S. Ishida, N. Sakamoto, T. Uesaka, Y. Satou, M. B. Greenfield, K. Hatanaka, *Phys. Rev. C* **55** 2909 (1997).
7. Y. Kalmykov, T. Adachi, G. P. A. Berg, H. Fujita, K. Fujita, Y. Fujita, K. Hatanaka, J. Kamiya, K. Nakanishi, P. von Neumann-Cosel, V. Yu. Ponomarev, A. Richter, N. Sakamoto, Y. Sakemi, A. Shevchenko, Y. Shimbara, Y. Shimizu, F. D. Smit, T. Wakasa, J. Wambach and M. Yosoi, *Phys. Rev. Lett.* **96** 012502 (2006).
8. V. R. Pandharipande, I. Sick and P.K.A. de Witt Huberts, *Rev. Mod. Phys.* **69** 981 (1997).
9. P.K.A. de Witt Huberts, *J. of Phys. G* **61** 507 (1990).
10. H. Feldmeier, T. Neff, R. Roth and J. Schnack, *Nucl. Phys. A* **632** 61 (1998); T. Neff, and H. Feldmeier, *Nucl. Phys. A* **713** 311 (2003).
11. M. Gell-Mann and M. Levy, *Nuovo Cimento* **16** 705 (1960).
12. J. Boguta, *Phys. Lett. B* **120** 34 (1983).
13. S. Weinberg, *Phys. Rev.* **166** 1568 (1968); *ibid.* **177** 2604 (1969).
14. Y. Ogawa, H. Toki, S. Tamenaga, H. Shen, A. Hosaka, S. Sugimoto, and K. Ikeda, *Prog. Theor. Phys.* **111** 75 (2004).
15. S. Sugimoto, K. Ikeda, H. Toki, *Nucl. Phys. A* **740** 77 (2004).
16. Y. Ogawa, H. Toki, S. Tamenaga, S. Sugimoto, and K. Ikeda, *Phys. Rev. C* **73** 034301 (2006).

PRESENT STATUS OF MICROSCOPIC THEORY FOR COMPLEX NUCLEUS-NUCLEUS INTERACTIONS

Y. SAKURAGI and T. FURUMOTO

Department of Physics, Osaka City University, Sumiyoshi-ku, Osaka 558-8585, Japan and RIKEN Nishina Center, RIKEN, Wako, Saitama 351-0198, Japan

Y. YAMAMOTO

Physics Section, Tsuru University, Tsuru, Yamanashi 402-8555, Japan

The present status of the microscopic theory for complex optical potentials of nucleus-nucleus systems is reviewed in the present talk, including the brief history and the latest achievement in the theoretical analysis of elastic scattering of light heavy-ion systems on the basis of double-folding model with the use of the CEG07 G-matrix interaction recently proposed by the present authors, with a special emphasis upon the role of the three-body force.

Keywords: Microscopic optical potential, Complex G-matrix, Three-body force

1. Introduction: A brief history of G-matrix folding model

Understanding the interactions between composite nuclei starting from the free-space nucleon-nucleon (NN) interaction is one of the fundamental subject in nuclear physics and also one of the key issues for explore nuclear structure through various nuclear reaction processes. Amongst the reaction processes, elastic scattering is the most important and fundamental process, which is the doorway to as well as the way out from all the possible reaction processes, and is usually described by a complex *optical potential* (OP) composed of real and imaginary parts.

When the elastic scattering data are available, one could find a complex *phenomenological optical potential* (POP) which is determined so as to reproduce the elastic scattering data by optimizing the potential parameters, such as the depth, radius and diffuseness parameters. However, for reactions with unstable nuclear beams, it is often the case that very poor or no elastic scattering data are available and, hence, there is little hope to obtain the POPs needed in the analyses of nuclear reaction data to extract a reliable information about unknown nuclear structure and reaction mechanism of unstable nuclei far from the stability line. Therefore, it is important to have a microscopic theory which can predict reliable OPs, particularly for nucleus-nucleus (AA) systems, starting from the basic NN interaction.

One of the promising theoretical approaches for predicting OPs for nucleon-nucleus (NA) and AA systems is the *G-matrix folding model* with the use of the G-matrix effective NN interaction on the basis of the lowest-order Brueckner theory. Here, the NN G-matrix is obtained by solving the Bethe-Goldstone equation for an NN pair in medium of infinite nuclear matter under scattering boundary condition. The obtained G-matrix interaction becomes a complex interaction composed of the real and imaginary parts, being dependent on the incident energy and the nucleon density in nuclar matter. The obtained complex, density-depnendent G-matrix is folded with the nucleon-density distribution of finite nuclei in the local-density approximation (LDA), leading to a complex potential for the NA or AA scattering systems.

The folding model with the complex G-matrices has so far been applied primarily to the NA systems. Various G-matrices starting from different kinds of free-space NN interactions were proposed and applied to the analyses of proton-nucleus elastic scattering with great success.[1–9] However, all of the G-matrices proposed so far were derived from just the two-body NN force and they were calculated for nucleon densities lower than the saturation density $\rho_0 \approx 0.17$ fm^{-3} of nuclear matter. This is because, in the case of NA scattering systems, the local density felt by the incident nucleon inside the target nucleus does not exceed ρ_0 even well inside the nucleus in usual cases.

As we go to higher-density region beyond ρ_0, it is well known that the effect of three-nucleon correlations (or *nuclear three-body force* (TBF)) become important. The empirical saturation point of symmetric nuclear matter (the binding energy per nucleon $E/A \approx 16$ MeV at the saturation density ρ_0) cannot be reproduced by any G-matrices obtained with only the two-body NN forces.[10] To obtain a reasonable saturation curve, it is indispensable to take into account the additional contributions of the TBF. The TBF consists of two components, the *three-body attraction* (TBA) and the *three-body repulsion* (TBR). It is important here that the saturation curve in the high-density region above ρ_0 is strongly pushed upward by the TBR contribution and, as the result, the nuclear-matter incompressibility becomes large and close to realistic values.[11–14] The role of TBF in complex nuclear systems is one of the key issues not only in nuclear physics but also in nuclear astrophysics relevant to high-density nuclear matter in neutron stars and supernova explosions.

In the case of the AA scattering system, the local density ($\rho_1 + \rho_2$) in the projectile-target overlap region may exceed the normal density of nuclear matter ρ_0 and could reach about twice this value, under the frozen-density approximation (FDA), as mentioned later. The TBR contributions are remarkably large in such high-density regions and, hence, one may expect a clear evidence of the TBR effect in *finite nuclear systems* through the calculated folding-model potential and the resultant elastic-scattering observables.

The importance of consistent description of nuclear saturation properties and elastic scattering of AA systems was first pointed out by Khoa *et al.*[21,22] on the basis of the folding-model analyses. They used density-dependent effective NN interactions such as DDM3Y, BDM3Y, and CDM3Y which were obtained from the density-independent effective interaction M3Y[23] by multiplying various kind of phenomenological density-dependent factors by hand, the parameters of which were chosen so as to represent various types of saturation curves in nuclear matter. The real part of the AA potential was calculated by the double-folding model (DFM) with those interactions, whereas the imaginary part was treated in a completely phenomenological way because M3Y was composed only of a real part. They showed the importance of using an effective interaction to be chosen to reproduce the realistic saturation curve in nuclear matter for the proper description of elastic scattering of AA systems. However, the density-dependence introduced in the effective interactions was purely artificial and had no explicit nor logical relation to the TBF in the nuclear medium. It is very recent that a direct and clear evidence of the important role of the repulsive TBF has been revealed in the collision of finite nuclear systems by the present authors[15,17] through the DFM analyses of elastic scattering of AA systems on the basis of the complex, density-dependent G-matrix interaction, called the CEG07, which includes the TBF effect.

In this paper, we present a major part of the recent achievement of the theoretical analyses of elastic scattering of AA systems on the basis of DFM with the CEG07 G-matrix interaction, with a special emphasis upon the role of TBF.

2. Formalism

2.1. *CEG07 G-matrices with the TBF effect*

The derivation of the G-matrix interaction was given in detail in our previous works[15–17] and here we only describe some essential points of our G-matrix interaction GEG07.

The G-matrix calculation is performed with the continuous choice for the intermediate nucleon spectra and the scattering boundary condition leads to the complex G-matrix. The present CEG07 G-matrix was derived from a free-space NN interaction model called the *extended soft-core* (ESC)[18,19] together with the TBF effect. The ESC model is designed so as to give a consistent description of interactions not only for NN system but also for nucleon-hyperon (NY) and hyperon-hyperon (YY) systems, where the TBR effect is taken into account effectively by changing vector-meson masses M_V in the nuclear matter according to $M_V(\rho) = M_V \exp(-\alpha_V \rho)$ with the parameter, α_V=0.18, which leads to an effective density-dependent two-body interaction. On the other hand, the TBA part is typically due to two-pion exchange with excitation of an intermediate Δ-resonance, that is the Fujita-Miyazawa diagram, which gives an important contribution at low densities. In Ref. 16, the TBA effect was included as an effective two-body interaction according to the formalism of Ref. 24 and added to the G-matrix obtained from the two-body interaction. Following Ref. 16, we call the G-matrix interaction derived from the two-body ESC part only as CEG07a and the version including further the effect of TBF (both the TBA and TBR components) as CEG07b. The inculusion of the TBF effect makes the saturation curve in symmetric nuclear matter realistic, especially in the high-density region ($\rho > \rho_0$), as shown in Fig. 1.

The obtained G-matrices are parameterized in a three-range Gaussian form as

$$v^{ST}(s, \rho, E/A) = \sum_{k=1}^{3} \sum_{i=1}^{4} v_{ik}^{ST}(E/A)\rho^{i-1} \exp\left(-\frac{s^2}{\lambda_k^2}\right), \tag{1}$$

with the fixed range parameters, λ_1=0.5 fm, λ_2=0.9 fm and λ_3=2.5 fm, and the strength v^{ST} for each spin (S=0 or 1) and isospin (T=0 or 1) component are given in the Appendix of Ref. 17 for the Fermi momentum k_F up to 1.8 fm^{-1} (which corresponds to $\rho \cong 0.39$ fm^{-3} $\approx 2\rho_0$) at energies of E/A = 70, 80, 100, 120, and 140 MeV.

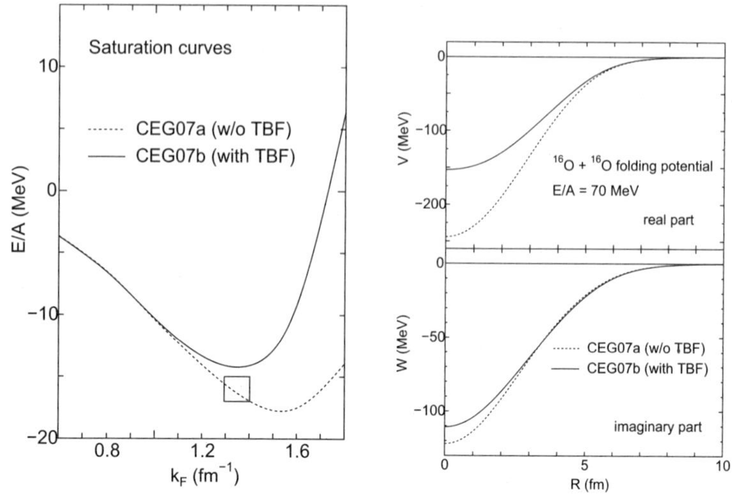

Fig. 1. Left: Effects of the three-body forces in the saturation caurves obtained by the G-matrix calculation. Right: The real and imaginary parts of the DFM potential for the ^{16}O + ^{16}O system at $E/A = 70$ MeV. The dotted and solid curves are the results with CEG07a (without the TBF) and CEG07b (with the TBF), respectively.

2.2. *Double-Folding model (DFM) potential*

The direct and exchange parts of the localized DFM potential, $U_{\rm D}(R)$ and $U_{\rm EX}(R)$, are written in the standard form of the DFM potential as

$$U_{\rm D}(R) = \int \rho_1(\boldsymbol{r}_1)\rho_2(\boldsymbol{r}_2)v_{\rm D}(s;\rho,E/A)d\boldsymbol{r}_1 d\boldsymbol{r}_2, \tag{2}$$

where $\boldsymbol{s} = \boldsymbol{r}_2 - \boldsymbol{r}_1 + \boldsymbol{R}$, and

$$U_{\rm EX}(R) = \int \rho_1(\boldsymbol{r}_1,\boldsymbol{r}_1+\boldsymbol{s})\rho_2(\boldsymbol{r}_2,\boldsymbol{r}_2-\boldsymbol{s})v_{\rm EX}(s;\rho,E/A) \times \exp\left[\frac{i\boldsymbol{k}(R)\cdot\boldsymbol{s}}{M}\right]d\boldsymbol{r}_1 d\boldsymbol{r}_2. \tag{3}$$

where $v_{\rm D}$ and $v_{\rm EX}$ are the direct and exchange parts of the complex G-matrix interaction and written as $v_{\rm D,EX} = (\pm v^{00} + 3v^{01} + 3v^{10} \pm +v^{11})/16$. Here, $\boldsymbol{k}(R)$ is the local momentum for nucleus-nucleus relative motion defined by $k^2(R) = (2mM/\hbar^2)[E_{\rm c.m.} - {\rm Re}U_{\rm F}(R) - V_{\rm coul}(R)]$ where $M = A_1A_2/(A_1+A_2)$, $E_{\rm c.m.}$ is the center-of-mass energy, E/A is the incident energy per nucleon, m is the nucleon mass and $V_{\rm coul}$ is the Coulomb potential. A_1 and A_2 are the mass numbers of the projectile and target, respectively. The exchange part is calculated self-consistently on the basis of the local energy approximation through the local momentum $k^2(R)$ defined above. The density matrix $\rho(\boldsymbol{r}, \boldsymbol{r}')$ is approximated[25] by

$$\rho(\boldsymbol{r}, \boldsymbol{r}') = \frac{3}{k_{\mathrm{F}}^{\mathrm{eff}} \cdot s} j_1(k_{\mathrm{F}}^{\mathrm{eff}} \cdot s) \rho\Big(\frac{\boldsymbol{r}+\boldsymbol{r}'}{2}\Big), \tag{4}$$

where $k_{\mathrm{F}}^{\mathrm{eff}}$ is the effective Fermi momentum[26] defined by

$$k_{\mathrm{F}}^{\mathrm{eff}} = \Big((3\pi^2\rho)^{2/3} + \frac{5C_{\mathrm{s}}[\nabla\rho^2]}{3\rho^2} + \frac{5\nabla^2\rho}{36\rho}\Big)^{1/2}, \tag{5}$$

where we adopt $C_{\mathrm{s}} = 1/4$ following Ref. 27. The detailed methods for calculating U_{D} and U_{EX} are the same as those given in Refs. 28 and 21, respectively.

In the present calculations, we employ the so-called frozen density approximation (FDA) for evaluating the local density. In the FDA, local density is defined as the sum of densities of colliding nuclei evaluated at the mid-point of the interacting nucleon pair, $\rho = \rho_1(\boldsymbol{r}_1 + \boldsymbol{s}/2) + \rho_2(\boldsymbol{r}_2 - \boldsymbol{s}/2)$. The FDA has been widely used also in the standard DFM calculations.[17,21,22,27,29,30] The prescriptions other than the FDA for evaluating the local density in AA systems have already been tested in Ref. 17 and the prescription of the FDA has been confirmed to be the best approximation in the use of CEG07 interactions.

3. Results

Let us apply the CEG07 G-matrix interactions to nucleus-nucleus (AA) systems through the DFM. Since the imaginary part of the optical potential for AA systems represents all excurrent flux escaping from elastic scattering channel through all the possible open reaction channels, it would be difficult to completely simulate those flux loss by the imaginary part of the G-matrix interaction originated from the pair-scattering correlations in the nuclear matter. So, we introduce a renormalization factor N_{W} for the imaginary part of the DFM potential and define the present microscopic optical potential with the CEG07 interaction as

$$U_{\mathrm{opt}}(R) = V(R) + iN_{\mathrm{W}}W(R). \tag{6}$$

Here, V and W denote the real and imaginary parts of the original DFM potential derived from the G-matrix interaction. We adjust the renormalization factor N_{W} so as to attain optimum fits to the experimental data for elastic-scattering cross sections.

First, we analyze elastic scattering of the ^{16}O + ^{16}O system in detail as a benchmark system for testing the interaction model, and then, we also analyze the ^{16}O scattering by other target nuclei as well as the ^{12}C + ^{12}C system. We adopt the nucleon density of ^{16}O calculated from the internal wave functions generated by the orthogonal condition model (OCM) by Okabe[31] based on the microscopic α + ^{12}C cluster picture. For other nuclei, we use the nucleon densities deduced from the charge densities[32] extracted from electron-scattering experiments by unfolding the charge form factor of a proton in the standard way.[33]

3.1. $^{16}O + {}^{16}O$ *elastic scattering*

In this paper, we make use of the CEG07 G-matrix interactions with and without the TBF effect, and analyze the elastic scattering of the ^{16}O + ^{16}O system to see how the TBF effect plays an important role in AA scattering systems. We choose elastic scattering of the ^{16}O + ^{16}O system for the following reasons: ^{16}O is one of the most stable double-magic nuclei and has no collective-excitation state strongly coupled with the ground state. This is important for a folding model based on the complex G-matrix, because the imaginary part of the G-matrix is expected to simulate the effect of single-particle-like excitations of finite nuclear systems through NN pair-scattering correlations and the effect of coherent, collective excitations of a finite nucleus may not be included in the imaginary part of the G-matrix. Therefore, the ^{16}O + ^{16}O system is one of the ideal benchmark systems to test the validity of interaction models; in fact, a number of interaction models, either purely phenomenological[34–36] or microscopic in various senses,[21,22,29,30] have been tested for decades on this system as a milestone to be cleared.

We now analyze the ^{16}O + ^{16}O elastic scattering at E/A = 70 MeV, paying a special attention to the roles of the TBF. The right panel of Fig. 1 shows the real (upper panel) and imaginary (lower panel) parts of the calculated DFM potential for the ^{16}O + ^{16}O elastic scattering at E/A = 70 MeV with the use of two types of CEG07 G-matrix interactions; CEG07a (without the TBF) and CEG07b (with the TBF). The effect of the TBF composed of the TBA and the TBR is clearly seen in the real part of the DFM potential over the whole range of internuclear distance, while the effect on the imaginary part is rather small except at short distances below 3 fm. The large difference in the real part is found to be mainly due to the effect of the repulsive part of TBF (TBR). While the attractive part of the TBF (TBA) is found to play only a minor role. The effects of the TBA and TBR were described in detail in Ref. 17.

We then calculate the ^{16}O + ^{16}O elastic-scattering cross section at E/A

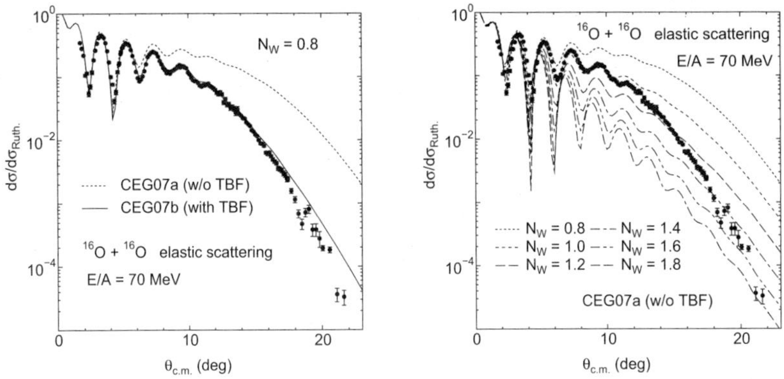

Fig. 2. Left: The cross sections for the ^{16}O + ^{16}O elastic scattering at $E/A = 70$ MeV, which are compared with the folding model calculations with (CEG07b) and without (CEG07a) the TBF effect. The $N_{\rm W}$ value is fixed to 0.8 for all of the calculations. The meaning of the curves is the same as in Fig. 2. The experimental data is taken from Ref. 37. Right: The N_W dependence of the cross section calculated with CEG07a. This figure is taken from Ref. 17.

= 70 MeV with the use of two types of DFM potentials shown in Fig. 1. It should be noted again that the only parameter in the present framework is the renormalization factor $N_{\rm W}$ for the imaginary part defined by Eq. (6). The results are shown in the left panel of Fig. 2. The solid and dotted curves are the results with the use of the DFM potentials obtained from CEG07b and CEG07a, respectively. Here, we take $N_{\rm W}$ to be 0.8 so that the solid curve (with the TBF effect) gives an optimum fit to the data. The solid curve with the TBF effect well reproduces the experimental data up to backward angles, while the dotted curve with CEG07a (without the TBF) overshoots the experimental data at middle and backward angles reflecting the too deep strength of the real part of the DFM potential. We found that no reasonable fit to the data was obtained by the DFM potential with CEG07a no matter how the imaginary part of the DFM potential is renormalized by changing the value of $N_{\rm W}$ as seen in the right panel of Fig. 2. Thus, the large difference between the solid and dotted curves in Fig. 2 (left) is a clear evidence of the decisive role of the TBF on elastic scattering of the ^{16}O + ^{16}O system. It was found that the large effect of the TBF seen in the figure was mainly due to the effect of repulsive part of the TBF (TBR) and that the effect of the attractive part (TBA) palyed only a minor role.[17]

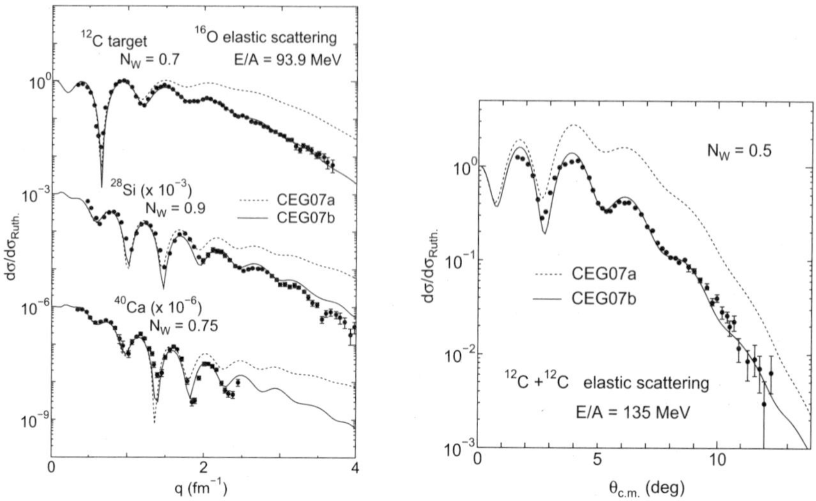

Fig. 3. Left: The cross sections for elastic scattering of ^{16}O by the ^{12}C, ^{28}Si and ^{40}Ca targets at $E/A = 93.9$ MeV, which are compared with the folding model calculations with the three types of complex G-matrix interactions. The dotted and solid curves are the results with CEG07a and CEG07b, respectively. The experimental data are taken from Ref. 38. The abscissa is the momentum transfer q defined as $q = 2k\sin(\frac{\theta}{2})$, where k being the asymptotic momentum. Right: The same as for the ^{12}C + ^{12}C scattering at $E/A = 135$ MeV. The experimental data are taken from Ref. 39.

3.2. *Elastic scattering of other systems*

Next, we analyze the elastic scattering of ^{16}O by the ^{12}C, ^{28}Si and ^{40}Ca target nuclei at $E/A = 93.9$ MeV with the use of CEG07a and CEG07b interactions. The left panel of Fig. 3 shows the resutls. The dotted and solid curves are the results with CEG07a and CEG07b, respectively. One can clear see the large effect of the TBF similar to the case of the ^{16}O target shown in Fig. 2. No reasonable fit to the data is obtained by the DFM calculation with CEG07a (without TBF effect), no matter how we change the value of $N_{\rm W}$, which is also the same as in the case of the ^{16}O target.

For the ^{28}Si target, the fit to the experimental data is not necessarily perfect at large angles. It may be related to the fact that the ^{28}Si nucleus presents a slightly stronger absorption ($N_{\rm W} = 0.9$) to the incident ^{16}O nucleus compared with other target nuclei ($N_{\rm W} \approx 0.75$). Since the ^{28}Si nucleus is known to be a very deformed nucleus that shows a typical rotational band in the excitation spectrum, it may be reasonable to expect that an additional absorption should be induced dynamically by collective excitation of the ^{28}Si nucleus in the collision with ^{16}O. This kind of dynamical effect may not be represented by the imaginary part of the G-matrix interaction evaluated in the nuclear matter. This would be one of the reason of a slightly larger value of the optimum N_W for the ^{28}Si target ($N_W = 0.9$) compared with that for other targets ($N_W \approx 0.75$). In fact, we have confirmed that a better fit up to the backward angles is obtained by a coupled-channel (CC) calculation based on the present DFM potential in the case of the ^{28}Si target with a smaller value of N_W. The CC analyses will be reported in forthcoming publications.

The right panel of Fig. 3 shows the ^{12}C + ^{12}C elastic scattering at $E/A = 135$ MeV. Again, the effect of the TBF is clearly seen in this case. One notices that a weaker absorption with $N_W = 0.5$ for this system compared with other systems discussed so far, although the origin of the weaker absorption for this system is not clear. It is worthwhile to mention that the similar trend of the weaker absorption for elastic scattering relevant to the ^{12}C nucleus was also pointed out previously.[40]

3.3. *Application to the higher energy cases*

When we further go to the higher energy region, the present folding model predicts that the real part of the DFM potential changes its sign from negative to positive, namely from *attractive* to *repulsive*. The left panel of Fig. 4 shows the real and imaginary parts of the DFM potentials for the ^{16}O + ^{16}O system at $E/A = 400$ MeV calculated with the three-types of CEG07 interactions (though the CEG07c vergion[17] is not discussed here). At this energy, the real part of the DFM potential become repulsive irrespective of the inclusion of the TBF effect, although the potential with CEG07b (with the TBF effect) is more repulsive than that with CEG07a (without the TBF effect), whereas the TBF effect is not prominent in the imaginary part which is the same as for lower energies. It is found that the real part changes its sign around $E/A = 200$ MeV in the case of CEG07b and around 300 MeV in the case of CEG07a. The different energy dependence of the DFM potentials with and without the TBF effect reflects upon the difference of the characteristic diffraction pattern of the elastic-scattering cross section, as shown in the right panel of Fig. 4. The large difference in the energy evolution of the characteristic diffraction pattern of the cross section may suggest the possibility to pin down, for the first time, an evidence of the repulsiveness of the heavy-ion optical potentials together with the role of the TBF at higher scattering energies.

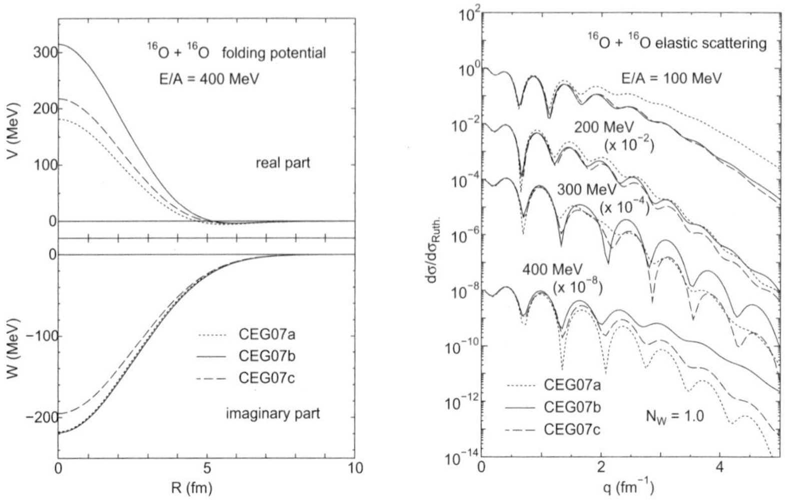

Fig. 4. Left: The real and imaginary parts of the DFM potentials for the $^{16}O + ^{16}O$ system at $E/A = 400$ MeV calculated with the three-types of CEG07 interactions. Right: The cross sections for the $^{16}O + ^{16}O$ elastic scattering at $E/A = 100$, 200, 300, and 400 MeV calculated with the three-types of CEG07 interactions, where the N_W value is fixed to 1.0 for all of the calculations. These figures are taken from Ref. 17.

4. Summary and remark

In this paper, we have presented the latest achievement of the theoretical study of optical potentials for the nucleus-nucleus (AA) systems on the basis of the double-folding model (DFM) with the use of the complex G-matrix obtained in nuclear-matter calculations starting from the free-space nucleon-nucleon (NN) interactions. The complex G-matrix CEG07 proposed recently by the present authors are shown to be quite successful to predict the complex optical potentials for AA systems at least in the energy range of $E/A = 70 \sim 135$ MeV and we have demonstrated the decisive role of the repulsive component of the three-body force (TBF) both in the real part of the DFM potential at high-density region and in the calculated elastic-scattering cross sections for the $^{16}O + ^{16}O$, $^{16}O + ^{12}C$, $^{16}O + ^{28}Si$, $^{16}O + ^{40}Ca$ and $^{12}C + ^{12}C$ systems at $E/A = 70{\sim}135$ MeV. The calculations with DFM potential only with the two-body force (TBF) contribution (with CEG07a) give no reasonable fit to the data no matter how the imaginary-strength parameter N_W is changed and the inclusion of the TBF effect is essential to reproduce the observed elastic-scattering cross sections. The real part of the calculated DFM potential is pushed upward remarkably by the repulsive contribution of TBF (TBR) and, as we goes to the higher energies above $E/A = 200{\sim}300$ MeV, it is predicted to change its sign and eventually becomes repulsive at higher energies.

We also have applied the CEG07 folding model to (1) the CDCC analysis[41,42] of the ^{6}Li elastic scattering with the $^6\text{Li} \to \alpha + d$ breakup process, (2) the elastic/quasielastic scattering of neutron-rich nuclei such as ^{9}Li and ^{11}Li (but the breakup processes being not included so far), and (3) the coulpled-channel (CC) calculation of inelastic scattering of the ^{16}O and ^{12}C projectiles by various targets, although they are not shown here due to the limited space and will be reported elsewhere, although they were presented in the talk.

References

1. F. A. Brieva and J. R. Rook, Nucl. Phys. A **291**, 299 (1977).
2. F. A. Brieva and J. R. Rook, Nucl. Phys. A **291**, 317 (1977).
3. F. A. Brieva and J. R. Rook, Nucl. Phys. A **297**, 206 (1978).
4. L. Rikus and V. V. Geramb, Nucl. Phys. A **426**, 496 (1984).
5. J. P. Jeukenne, A. Lejeune, and C. Mahaux, Phys. Rev. C **16**, 80 (1977).
6. E. Bauge, J. P. Delaroche, and M. Girod, Phys. Rev. C **58**, 1118 (1998).
7. N. Yamaguchi, S. Nagata, and T. Matsuda, Prog. Theor. Phys. **70**, 459 (1983).
8. N. Yamaguchi, S. Nagata, and J.Michiyama, Prog. Theor. Phys. **76**, 1289 (1986).
9. K. Amos, P. J. Dortmans, H. V. von Geramb, S. Karataglidis, and J. Raynal, Adv. Nucl. Phys. **25**, 275 (2000).
10. F. Coester, S. Cohen, B. D. Day and C. M. Vincent, Phys. Rev. C **1**, 769 (1970).
11. I. E. Lagaris and V. R. Pandharipande, Nucl. Phys. **A359**, 349 (1981).
12. R. B. Wiringa, V. Fiks and A. Fabrocini, Phys. Rev. C **38**, 1010 (1988).
13. M. Baldo, I. Bombaci and G. F. Burgio, Astron. Astrophys. **328**, 274 (1997).
14. A. Lejeune, U. Lombardo and W. Zuo, Phys. Lett. **B477**, 45 (2000)
15. T. Furumoto, Y. Sakuragi and Y. Yamamoto, Phys. Rev. C **79**, 011601(R) (2009).
16. T. Furumoto, Y. Sakuragi and Y. Yamamoto, Phys. Rev. C **78**, 044610 (2008).
17. T. Furumoto, Y. Sakuragi and Y. Yamamoto, Phys. Rev. C **80**, 044614 (2009).
18. T. A. Rijken, Phys. Rev. C **73**, 044007 (2006).
19. T. A. Rijken and Y. Yamamoto, Phys. Rev. C **73**, 044008 (2006).
20. J. Fujita and H. Miyazawa, Prog. Theor. Phys. **17**, 360 (1957).
21. D. T. Khoa, W. von Oertzen and H. G. Bohlen, Phys. Rev. C **49**, 1652 (1994).
22. D. T. Khoa, G. R. Satchler and W. von Oertzen, Phys. Rev. C **56**, 954 (1997).
23. G. Bertsch, J. Borysowicz, H. McManaus and W. G. Love, Nucl. Phys. **A284**,

399 (1977).
24. T. Kasahara, Y. Akaishi and H. Tanaka, Prog. Theor. Phys. Suppl. **56**, 96 (1974).
25. J. W. Negele and D. M. Vautherin, Phys. Rev. C **5**, 1472 (1972).
26. X. Campi and A. Bouyssy, Phys. Lett. **B73**, 263 (1978).
27. D. T. Khoa, Phys. Rev. C **63**, 034007 (2001).
28. S. Nagata, M. Kamimura and N. Yamaguchi, Prog. Theor. Phys. **73**, 512 (1985).
29. M. Katsuma, Y. Sakuragi, S. Okabe and Y. Kondo, Prog. Theor. Phys. **107**, 377 (2002).
30. G. R. Satchler and W. G. Love, Phys. Rep. **55**, 183 (1979).
31. S. Okabe, *Proc. of Tours Symposium on Nuclear Physics II* (eds. H. Utsunomiya et al., World Scientific, Singapore, 1995), p. 112.
32. H. DeVries, C. W. DeJager and C. DeVries, At. Data Nucl. Data Tables **36**, 495 (1987).
33. L. R. B. Elton, *Nuclear Size* (Oxford University, Oxford, 1961).
34. Y. Kondō, F. Michel, and G. Reidemeister, Phys. Lett. **B242**, 340 (1990).
35. Y. Kondō, Y. Sugiyama, Y. Tomita, Y. Yamanouchi, H. Ikezoe, K. Ideno, S. Hamada, T. Sugimitsu, M. Hijiya and H. Fujita, Phys. Lett. **B365**, 17 (1996).
36. M. E. Brandan and G. R. Satchler, Phys. Rep. **285**, 143 (1997), and references therein.
37. F. Nuoffer, *et al.*, Nuovo Cimento A **111**, 971 (1998).
38. P. Roussel-Chomaz, *et al.*, Nucl. Phys. **A477**, 345 (1988).
39. T. Ichimura, *et al.*, Nucl. Phys. **A569**, 287 (1994).
40. K. Schwarz, *et al.*, Eur. Phys. J. A **7**, 367 (2000).
41. Y. Sakuragi, M. Yahiro, M. Kamimura, Prog. Theor. Phys. Suppl. **89**, 136 (1986).
42. Y. Sakuragi, Phys. Rev. C **35**, 2161 (1987).

COMPLEX-SCALED CDCC METHOD FOR NUCLEAR BREAKUP REACTIONS

M. TAKASHINA

Graduate School of Medicine, Osaka University, Suita, Osaka 565-0871, Japan
Research Center for Nuclear Physics (RCNP),
Osaka University, Ibaraki, Osaka 567-0047, Japan
E-mail: takasina@sahs.med.osaka-u.ac.jp

T. MYO

General Education, Faculty of Engineering,
Osaka Institute of Technology, Osaka 535-8585, Japan
Research Center for Nuclear Physics (RCNP),
Osaka University, Ibaraki, Osaka 567-0047, Japan

Y. HIRABAYASHI

Information Initiative Center, Hokkaido University, Sapporo 060-0811, Japan

Y. KIKUCHI and K. KATO

Division of Physics, Graduate School of Science,
Hokkaido University, Sapporo 060-0810, Japan

In this talk, we propose applying the complex scaling method to CDCC, where the internal coordinate and momentum of projectile nucleus are complex-scaled. We show that the continuous S matrix element of breakup reaction is obtained without smoothing function.

Keywords: Complex scaling method, CDCC, breakup reaction

1. Introduction

Recently, various exotic phenomena of unstable nuclei have been revealed by nuclear reaction experiments. Nuclear breakup process is very important for light unstable nuclei induced reactions because of their weak binding nature. When the projectile nucleus breaks up into a two-body (three-body) system, one should solve the three-body (four-body) problem including the target nucleus. Although it is hard to solve the three- and four-body scattering problem exactly, many authors have attempted to describe the breakup reactions.

The continuum-discretized coupled-channel (CDCC) method is known to be one of the powerful methods to describe the nuclear breakup processes.[1] In CDCC, the breakup continuum states of the projectile (or target) nucleus are discretized in a proper way. Recently, the pseudo-state (PS) CDCC has been developed: one can obtain the continuous S matrix element by introducing *smoothing function* when the projectile breaks up into two-body system.[2] It also becomes possible to include the three-body breakup effects of the projectile such as ^{6}He only on the elastic channel.[3] More recently, several new methods of obtaining the *smoothing functions* have been proposed,[4,5] which are applicable to the three-body breakup.

In the present study, we propose another approach: we consider applying the complex-scaling method[6–8] to CDCC (complex-scaled CDCC, CS-CDCC), in which only the internal coordinate and momentum of the projectile are complex-scaled. The expected advantages of CS-CDCC are

(1) in spite of the discretization, the continuum level density can be obtained exactly,[9] and then, the continuous S matrix elements can also be obtained without introducing any *smoothing function*,
(2) in the complex scaling method, three-body scattering state can be solved properly,[8]
(3) the resonance state is strictly separated from the continuum states, and this fact is more advantageous for investigation of reaction mechanism than the ordinary CDCC method.[6,7]

2. Formalism

Here, we consider the deuteron (d) induced reaction on a target A including $d \to p + n$ breakup process. The coordinate system is shown in Fig.1.

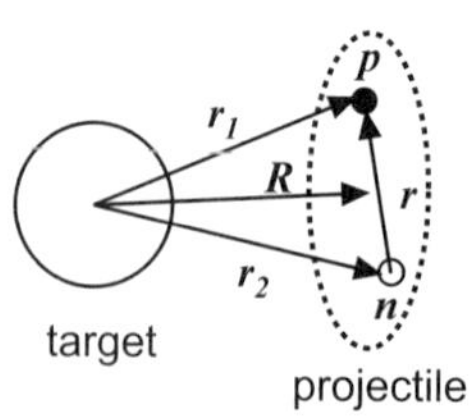

Fig. 1. Coordinate system used in the present calculation.

The scaling operator $U(\theta)$ transforms only the relative coordinate $\mathbf{r}$ of the projectile. Suppose that $h = T_{\mathbf{r}} + V(\mathbf{r})$ is the internal Hamiltonian of deuteron. The complex-scaled continuum states $\varphi_\alpha^\theta(\mathbf{r})$ of deuteron as well as its ground state are obtained by diagonalizing the complex-scaled Hamiltonian

$$h^\theta \;=\; U(\theta)\, h\, U^{-1}(\theta) \tag{1}$$

with the finite range basis functions. Using $\varphi_\alpha^\theta(\mathbf{r})$, the CS-CDCC equation for the total angular momentum J is written as

$$\left[\left(T - (E \;-\varepsilon_\alpha^\theta \;)\right]\; \chi_\alpha^{(J)\theta}(R) \;=\; -\sum_{\beta\neq\alpha} F_{\alpha,\beta}^{(J)\;\theta}(R)\; \chi_\beta^{(J)\theta}(R). \tag{2}$$

where α and β represent channels, $\varepsilon_\alpha^\theta$ represents the eigenenergy of discretized states, and

$$F_{\alpha,\beta}^{(J)\;\theta}(R) \;=\; \Big\langle \Big[\, \varphi_\alpha^\theta(\mathbf{r}) \otimes y_{L_\alpha}(\hat{\mathbf{R}}) \,\Big]_{JM} \;|\; U(\theta)\; [V_1(r_1) + V_2(r_2)\,] \\ \times\; U^{-1}(\theta) \;|\; \Big[\, \varphi_\beta^\theta(\mathbf{r}) \otimes y_{L_\beta}(\hat{\mathbf{R}}) \,\Big]_{JM} \Big\rangle . \tag{3}$$

$\chi_\alpha^{(J)\theta}(R)$ is the relative wave function between d and A, and is obtained by solving the coupled-equation Eq.(2). The boundary condition of the relative wave function is written as

$$\chi_\alpha^{(J)\theta}(R) \rightarrow \left[\; \delta_{\alpha 0}\; u_{L_0}^{(-)}(K_0 R) \;-\; \sqrt{\frac{v_0}{v_\alpha^\theta}}\; S_{\alpha 0}^{(J)\theta}\; u_{L_\alpha}^{(+)}(K_\alpha^\theta R) \right], \tag{4}$$

where $u_L^{(\pm)}(R)$ represents the Coulomb wave function, K_α^θ and v_α^θ are the relative momentum and velocity, respectively, between the projectile and target, and the subscript "0" indicates the elastic channel. By connecting the solved wave function $\chi_\alpha^{(J)\theta}(R)$ to the boundary condition Eq.(4), a set of *scaled* S matrix elements $S_{\alpha 0}^{(J)\theta}$ is obtained. As done in the calculation of strength function in Ref.,[8] the *scaled* S matrix elements are projected onto the real energy ε as

$$|S^{(J)}(\varepsilon)|^2 \;=\; \frac{1}{\pi}\, \mathcal{I}m \left[\sum_\alpha \frac{\tilde{S}_{\alpha 0}^{(J)\theta *}\; S_{\alpha 0}^{(J)\theta}}{\varepsilon_\alpha^\theta - \varepsilon} \right]. \tag{5}$$

Then, we obtain the continuous S matrix element as a function of ε.

3. Results

We apply CS-CDCC to the $d \rightarrow p + n$ breakup reaction on a ^{58}Ni target at the incident energy of E_d=80 MeV. The potential parameters are taken from Ref. 2. For simplicity, we neglect the coupling to the ℓ=2 states in deuteron, Coulomb potential between proton in deuteron and ^{58}Ni, and the imaginary potentials for the p-^{58}Ni and n-^{58}Ni systems.

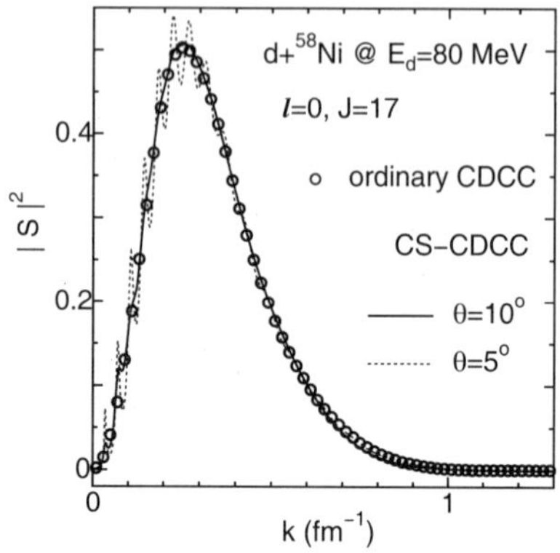

Fig. 2. Squared modulus of S matrix element as a function of k.

Figure 2 shows the calculated S matrix element for J=17 as a function of the relative momentum k between proton and neutron in deuteron. The open circles represent the result of the ordinary PS-CDCC calculation, and the solid and dotted curves represent the results of the present CS-CDCC calculations at the scaling angles θ=10° and 5°, respectively. It is found that the solid curve is almost equivalent to the result of ordinary PS-CDCC, although the dotted curve shows a oscillatory behavior as expected. This result indicates that CS-CDCC well works on describing the nuclear breakup reactions. It should be noted that continuous S matrix element is obtainable in CS-CDCC without any *smoothing function*.

The present CS-CDCC method is also applicable to the reactions where the projectile breaks up into three-body system, such as ^{6}He and ^{11}Li induced reactions.

References

1. M. Kamimura et al., *Prog. Theor. Phys. Suppl.* **89**, 1 (1986).
2. T. Matsumoto et al., *Phys. Rev. C* **68**, 064607 (2003).
3. T. Matsumoto et al., *Phys. Rev. C* **70**, 061601(R) (2004).
4. T. Egami et al., *Prog. Theor. Phys.* **121**, 789 (2009).
5. T. Matsumoto et al., *Prog. Theor. Phys.* **121**, 885 (2009).
6. J. Aguilar and J. M. Combes, *Commun. Math. Phys.* **22**, 269 (1971).

7. E. Balslev and J. M. Combes, *Commun. Math. Phys.* **22**, 280 (1971).
8. S. Aoyama et al., *Prog. Theor. Phys.* **116**, 1 (2001).
9. R. Suzuki et al., *Prog. Theor. Phys.* **113**, 1273 (2005).

^{10}Be+α CORRELATION IN ^{14}C

T. SUHARA
Department of Physics, Graduate School of Science, Kyoto University, Kyoto 606-8502, Japan
E-mail: suhara@ruby.scphys.kyoto-u.ac.jp

Y. KANADA-EN'YO
Yukawa Institute for Theoretical Physics, Kyoto University, Kyoto 606-8502, Japan
E-mail: yenyo@yukawa.kyoto-u.ac.jp

We investigated ^{14}C with β-γ constraint antisymmetrized molecular dynamics coupled with the generator coordinate method (β-γ constraint AMD+GCM). Above the ground band, three excited bands with the developed three-α-cluster structures were suggested. Characteristic structures such as a triaxial structure and a linear-chain structure appear in the excited bands. By estimating ^{10}Be+α components in the obtained ^{14}C states, it is suggested that the experimental state at 18.6 MeV is a candidate of the theoretical 4^+ linear-chainlike state.

1. Introduction

In C isotopes such as ^{14}C, one can expect that exotic structures emerge because of excess neutrons. In the stable carbon nucleus, ^{12}C, it is already well known that various structures appear in the ground and excited states. For example, the shell-model-like structure in the ground state, the α condensation of weakly interacting three α clusters in the 0_2^+ state, the equilateral-triangular structure of three α clusters in the 3_1^- state, and a linear-chainlike (or an obtuse-angle-triangular) structure of three α clusters in the 0_3^+ state were suggested.[1–7] It is natural to expect that rich phenomena may also appear in excited states of ^{14}C with two excess neutrons. Therefore, it is an challenging problem to investigate the structures of ^{14}C and to search for new types of structure.

For the excited states of ^{14}C, there are many experimental studies that suggest various cluster states.[8–12] Especially, some of highly excited states have been observed to decay via ^{10}Be+α channels. These states are candidates which have a new type of structure, the ^{10}Be+α cluster structure.

2. Framework of β-γ constraint AMD+GCM

To study various cluster and shell structures in ^{14}C, we adopted β-γ constraint AMD, which is efficient for preparing basis wave functions with various shell and cluster configurations.[13] In this method, we constraint on the quadrupole deformation parameters, β and γ. After the variation with the constraint, we obtain the optimized wave functions $|\Phi^{\pm}(\beta_0, \gamma_0)\rangle$ for each set of parameters, $(\beta, \gamma) = (\beta_0, \gamma_0)$. Here, $\pm$ means the optimized wave functions are the parity eigenstates.

In the calculations of energy levels, we superpose the total-angular-momentum projected wave functions $P^J_{MK}|\Phi^{\pm}(\beta, \gamma)\rangle$. Thus, the final wave function for the $J^{\pm}_n$ state is given by a linear combination of the basis wave functions as

$$|\Phi^{J\pm}_n\rangle = \sum_K \sum_i f_n(\beta_i, \gamma_i, K) P^J_{MK} |\Phi^{\pm}(\beta_i, \gamma_i)\rangle. \tag{1}$$

This corresponds to the GCM with the two-dimensional generator coordinates of the quadrupole deformation parameters, β and γ.

3. Results and discussion

We show the calculated energy levels of ^{14}C in Fig. 1 as well as the experimental levels.[9–12,14] Four columns on the left are the experimental data and five columns on the right are the theoretical results. States in the three columns from the left are the known positive-parity states. States which were discovered in the ^{10}Be+α break-up reactions are displayed in the fourth column from the left. The lower three states are those with a strong population of ^{10}Be(0^+_1) and upper three states are ^{10}Be(2^+_1). We classified the theoretical states into five groups by analyzing the intrinsic structures and $E2$ transition strengths. It is found that the ground band has a shell-model-like structure, while three α clusters develop well in most of the excited states. The second column is the $K^\pi = 0^+$ band, which has a triaxial structure, and the third column is the $K^\pi = 2^+$ side band of the band in the second column. In these states, three α clusters have the isosceles-triangle configuration and excess neutrons occupy the sd-like orbitals. The fourth column is the $K^\pi = 0^+$ band, in which developed three α clusters have a linear-chainlike structure. It is interesting that the excess neutrons distribute around two of the three α clusters. This indicates the ^{10}Be correlation in the linear-chain states. Other states are plotted in the fifth column.

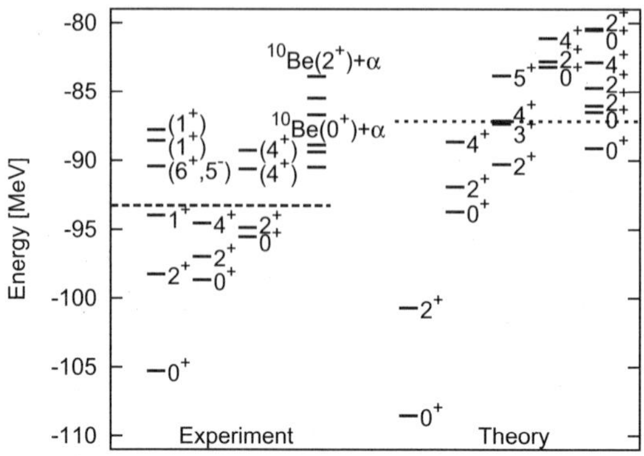

Fig. 1. Energy levels of ^{14}C. Four columns on the left are the experimental data and five columns on the right are the calculated results. The broken lines in the experimental-data and calculated-results regions show the experimental and theoretical ^{10}Be+α threshold energies, respectively. The details are described in the text.

Table 1. Overlaps between linear-chainlike states and artificial ^{10}Be+α states.

J^π of ^{14}C	(J', M') of ^{10}Be	Overlap
0^+	$(0,0)$	28.9%
	$(2,0)$	7.7%
2^+	$(0,0)$	28.2%
	$(2,2)$	0.3%
	$(2,1)$	6.7%
	$(2,0)$	10.1%
4^+	$(0,0)$	15.7%
	$(2,2)$	1.2%
	$(2,1)$	14.0%
	$(2,0)$	13.1%

To discuss ^{10}Be+α components, we calculated overlaps of theoretical states $|\Phi_n^{J\pm}\rangle$ with ^{10}Be+α cluster wave functions defined as

$$|^{10}\mathrm{Be}\left(J'^{\pm}, M'\right) + \alpha\rangle \equiv P^{J\pm}_{MM'} \left(P^{J'+}_{M'0} |^{10}\mathrm{Be}\rangle \otimes |\alpha\rangle \right), \tag{2}$$

which consists of a subprojected ^{10}Be and an α cluster. The distance between ^{10}Be and an α cluster is set to 5 fm. The calculated overlaps for linear-chainlike states are listed in Table 1. In other states, ^{10}Be+α components are small and less than a few percent. These results mean only linear-chainlike states have ^{10}Be+α correlation definitely. Particularly, it is interesting that 4^+ state has the larger ^{10}Be(2^+) component than the ^{10}Be(0^+) one. The simple sum of the ^{10}Be(2^+) components, 28.3%, is about two times of the ^{10}Be(0^+) component, 15.7%. In the experimental report of Ref. 12, the state at 18.6 MeV has larger branching ratio for ^{10}Be(2^+)+α than that for ^{10}Be(0^+)+α. This may suggest a significant ^{10}Be(2^+) component in the 18.6 MeV state. Therefore, it is suggested that the experimental state at 18.6 MeV is a candidate of the theoretical 4^+ linear-chainlike state.

4. Summary and outlook

We investigated ^{14}C with β-γ constraint AMD+GCM. Above the ground band, three excited bands with the developed three-α-cluster structures were suggested. Characteristic structures such as a triaxial structure and a linear-chain structure appear in the excited bands. By estimating ^{10}Be+α components in the theoretical ^{14}C states, it is suggested that the experimental states with a strong population of ^{10}Be(2^+)+α decays are candidates of linear-chainlike states.

As a future problem, we are estimating decay widths of ^{14}C into ^{10}Be+α. We will report the detailed results in a future paper.

Acknowledgments

The computational calculations of this work were performed by supercomputers in YITP and KEK. This work was supported by the YIPQS program in YITP. It was also supported by the Grant-in-Aid for the Global COE Program "The Next Generation of Physics, Spun from Universality and Emergence" from MEXT of Japan, and a Grant-in-Aid from JSPS.

References

1. E. Uegaki *et al.*, Prog. Theor. Phys. **57**, 1262 (1977).
2. Y. Fukushima and M. Kamimura, *in Proceedings of the International Conference on Nuclear Structure, Tokyo, 1977*, edited by T. Marumori [J. Phys. Sos. Jpn. **44**, 225 (1978)].
3. Y. Kanada-En'yo, Phys. Rev. Lett. **81**, 5291 (1998).
4. A. Tohsaki *et al.*, Phys. Rev. Lett. **87**, 192501 (2001).
5. Y. Funaki *et al.*, Phys. Rev. C **67**, 051306(R) (2003).
6. T. Neff and H. Feldmeier, Nuc. Phys. A **738**, 357 (2004).
7. Y. Kanada-En'yo, Prog. Theor. Phys. **117**, 655 (2007).
8. W. von Oertzen *et al.*, Eur. Phys. J. A **21**, 193 (2004).
9. M. Milin *et al.*, Nucl. Phys. A **730**, 285 (2004).
10. N. Soic *et al.*, Nucl. Phys. A **738**, 347 (2004).
11. D. L. Price *et al.*, Phys. Rev. C **75**, 014305 (2007).
12. P. J. Haigh *et al.*, Phys. Rev. C **78**, 014319 (2008).
13. T. Suhara and Y. Kanada-En'yo, Prog. Theor. Phys. **123** (2010), to be published.
14. F. Ajzenberg-Selove, Nucl. Phys. A **522**, 1 (1991).

BEYOND RELATIVISTIC HARTREE-FOCK MODEL FOR NUCLEAR MATTER

J. HU* AND H. TOKI

Research Center for Nuclear Physics (RCNP), Osaka University, Ibaraki, Osaka 567-0047, Japan
**E-mail: jinniuhu@rcnp.osaka-u.ac.jp*

W. WEN AND H. SHEN

Department of Physics, Nankai University, Tianjin 300071, China

Nuclear matter is studied by using beyond relativistic Hartree-Fock (RHF) model with the realistic nucleon-nucleon (NN) interaction. We use the unitary correlation operator method (UCOM) to treat the strong repulsive interaction in the short-range region between two nucleons. We find that the equation of state (EOS) of pure neutron matter is completely comparable with the results of Dirac-Brueckner-Hartree-Fock (DBHF) theory. However, for the symmetric nuclear matter, the EOS is largely different from the DBHF result particularly in the low density region.

Keywords: Nuclear matter; Short-range correlation; UCOM; Hartree-Fock

1. Introduction

Asymmetric nuclear matter is a very important research object which is the basic component of compact star in the universe such as neutron star. Especially, the properties of neutron rich-matter determine the whole evolutionary processes of neutron stars. Therefore, many phenomenological models as the relativistic mean-field (RMF) model[1] and microscopic theories as the DBHF model[2,3] predicted many quit different EOS's for asymmetric nuclear matter. The DBHF theory can obtain the reasonable saturation properties of symmetric nuclear matter by using the realistic Bonn interactions. It is very successful to deal with the short-rang and tensor correlations between two nucleons with Brueckner theory. In recent years, another very attractive method is presented by Feldmeier *et al.* in terms of the unitary correlation operator method (UCOM)[4] to treat the short-range correlation.

It is interesting to point our that the tensor interaction does not contribute much in neutron-rich matter and therefore in this work we will adopt the relativistic Hartree-Fock (RHF) model with the UCOM (RHFU) to investigate the properties of nuclear matter by using a realistic NN interaction, particulary for neutron-rich matter and pure neutron matter.

2. The relativistic Hartree-Fock model with UCOM

The RHF equation is written in the following form,[5,6]

$$T\psi_i(\mathbf{x}) + \sum_j \langle \cdot j|V|ij\rangle_A = \varepsilon_i \psi_i(\mathbf{x}), \tag{1}$$

where T is the Dirac kinetic energy and V is the NN interaction. The Bonn-A potential[2] is adopted to compare with the DBHF results. The interaction matrix element $\langle \cdot j|V|ij\rangle_A$ denotes the interaction matrix without the i state indicated by a dot in the bra-state with the anti-symmetrization indicated by the suffix A in the ket side. In the RHFU model, we replace the interaction V by $\widetilde{V}$ using the unitary correlation operator.

$$\widetilde{V} = U^\dagger V U + U^\dagger T U - T. \tag{2}$$

We assume $\widetilde{V}$ only by taking the unitary transformed operators up to the two-body operator part due to its extreme short-range nature of the repulsive core.[4] The original form of UCOM operator can be written as,

$$U = \exp\left\{ -\frac{i}{2}\left[\vec{q}\cdot\frac{\vec{r}}{r}\theta(r) + \theta(r)\frac{\vec{r}}{r}\cdot\vec{q}\right]\right\}, \tag{3}$$

where $\vec{q}$ is the relative momentum operator and $\theta(r)$ is a shift coordinate in the relative direction, which is made position dependent.[4]

The monopole form factor is chosen for each meson-nucleon vertex in the Bonn potential to consider the finite size effect of nucleon. The momentum dependent part of meson-exchange interaction with meson mass m can be calculated by taking the form factor term with the cut-off mass Λ,

$$\frac{1}{\vec{q}^2+m^2}\left(\frac{\Lambda^2-m^2}{\Lambda^2+\vec{q}^2}\right)^2 = \frac{1}{\vec{q}^2+m^2} - \frac{1}{\vec{q}^2+\Lambda^2} + (\Lambda^2-m^2)\frac{d}{d\Lambda^2}\frac{1}{\vec{q}^2+\Lambda^2}. \tag{4}$$

Once we include the form factor in the meson-exchange potential, the contact terms appearing in the meson-exchange interaction become momentum dependent. Hence, all the meson-exchange interactions in Bonn potential can be written in the form of Yukawa functions in coordinate space even after considering the form factor effect,

$$V(m,r) = \frac{e^{-mr}}{r}. \tag{5}$$

When the unitary correlation operator U is applied on the two-body NN interaction, we just take the following transformation in the potential $V(r)$ as,

$$U^{\dagger}V(m,r)U = V(m, R_{+}(r)), \tag{6}$$

where,

$$R_{+}(r) = r + \alpha \left(\frac{r}{\beta}\right)^{\eta} \exp(-\exp(r/\beta)). \tag{7}$$

Here, $R_{+}(r)$ is related with $\theta(r)$ in the UCOM operator and used to parameterize the form of the unitary operator.[4]

We write the kinetic energy part after applying the UCOM operator.

$$U^{\dagger}TU - T = \sum_{i<j} \left\{ (\vec{\alpha}_i - \vec{\alpha}_j) \cdot \frac{\vec{r}}{r} \frac{1}{\sqrt{R'_{+}(r)}} \frac{1}{r} q_r \left(\frac{r}{\sqrt{R'_{+}(r)}} \right) \right. \tag{8}$$
$$\left. + (\vec{\alpha}_i - \vec{\alpha}_j) \cdot \frac{\vec{r}}{r} \left(\frac{1}{R'_{+}(r)} - \frac{r}{R_{+}(r)} \right) q_r + \left(\frac{r}{R_{+}(r)} - 1 \right) (\vec{\alpha}_i - \vec{\alpha}_j) \cdot \vec{q} \right\}.$$

Here, $q_r = \vec{r} \cdot \vec{q}/r$ is defined as the radial momentum. The Dirac matrix α_i is an operator on the Dirac spinor of i-th nucleon and the prime denotes differentiation with respect to the relative coordinate r.

3. The result and summary

We show our results for the EOS's of symmetric nuclear matter and pure neutron matter within the RHFU model with Bonn-A potential in Fig. 1. We choose the UCOM parameters with $\alpha = 0.8$ fm, $\beta = 0.6$ fm and $\eta = 0.37$. We find that the EOS of pure neutron matter in DBHF is completely reproduced by our model. The short-range correlation largely decreases the repulsive contribution of realistic NN interaction comparing the case without the UCOM effect. However, for symmetric nuclear matter, we are not able to reproduce the EOS of the DBHF theory in particular in the low density region $\rho \leq 0.2$ fm^3. This behavior is caused by the missing tensor correlation. In the RHF framework, the effect of tensor interaction can not be included. However, it has a significant role on symmetric nuclear matter for the triplet S channel by providing a large attraction, which is large at low density and decreases with density due to the Pauli blocking effect.

In summary, we have successfully constructed a framework to treat neutron-rich neutron matter with a realistic NN interaction. For this purpose, it was essential to include the short-range correlation with UCOM in RHF model. We have completely reproduced the EOS of pure neutron

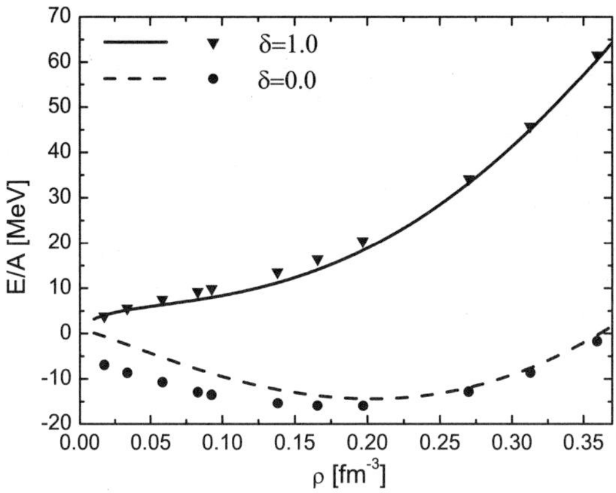

Fig. 1. The EOS's of symmetric nuclear matter and pure neutron matter calculated in the relativistic Hartree-Fock model with UCOM using the Bonn-A potential. The solid line is the pure neutron matter ($\delta = 1$), while dashed line is the symmetric nuclear matter ($\delta = 0$). The symbols are taken from the results of DBHF theory respectively.[3]

matter of the DBHF model with Bonn-A potential. For symmetric nuclear matter, the deviation becomes larger in particular in the low density part. This behavior is related with the structure of RHF model, which does not include the effect of tensor correlation.

Acknowledgments

The authors are grateful to the members of the theory group of RCNP for discussions on the role of pion in nuclear physics. J. Hu would like to thank Dr. E.N.E. van Dalen for providing DBHF results. This work is supported by JSPS research grand: 21540267. J. Hu is supported by the MEXT fellowship.

References

1. Y. Sugahara and H. Toki, Nucl. Phys. A **579**, 557 (1994).
2. R. Brockmann and R. Machleidt, Phys. Rev. C **42**, 1965 (1990).
3. E.N.E. van Dalen, C. Fuchs and A. Faessler, Eur. Phys. J. A **31**, 29 (2007).
4. H. Feldmeier, T. Neff, R. Roth and J. Schnack, Nucl. Phys. A **632**, 61 (1998).
5. P. Ring and P. Schuck, *The Nuclear Many-Body Problem* (Springer-Verlag, Berlin, 1980).
6. L. Savushkin and H. Toki, *The Atomic Nucleus As A Relativistic System* (Springer-Verlag, Berlin, 2004).

ELASTIC SCATTERING OF ^{8}B FROM ^{12}C WITH INTERNAL THREE CLUSTER STRUCTURE OF ^{8}B

K. HORII*, M. TAKASHINA and H. TOKI

Recerch Center for Nuclear Physics, Osaka Univercity,
Osaka, Japan
**E-mail: horii@rcnp.osaka-u.ac.jp*

T. FURUMOTO and Y. SAKURAGI

Department of Physics, Osaka City Univercity,
Osaka, Japan

We study theoretically the elastic scattering of ^{8}B from ^{12}C at $E_{lab} = 95$MeV, where the projectile has a three cluster structure α+^{3}He+p. We treat the halo proton in ^{8}B in the adiabatic recoil approximation[1] and take into account the excited states of ^{7}Be including resonance states by a coupled channel method with consideration of the cluster structure. We reproduce the differential cross section of elastic scattering of ^{8}B from ^{12}C. It turns out that the excitaion to the resonance state of ^{7}Be is important.

1. Introduction

With the development of radioactive nuclear beams we have now a powerful experimental tool to study the unstable nuclei. Particularly interesting is to use a halo nucleus as a projectile and to measure elastic and inelastic scattering cross sections. Now we treat the unstable nucleus ^{8}B which has a proton halo structure. The last proton is considered to make a halo structure with the core ^{7}Be nucleus because of the separation energy of the proton is 0.137 MeV. While the core ^{7}Be nucleus has also a cluster structure α+^{3}He. In this sense, it is interesting to study how all these weakly bound subsystems play their roles in the elastic scattering of ^{8}B with a target nucleus, where the break-up energy into the three body system α+^{3}He+p is 1.72 MeV. The purpose of this paper is to study theoretically the elastic scattering cross section of ^{8}B by ^{12}C at $E_{\rm lab}$=95 MeV[2]as an example of elastic scattering of halo nuclei with an internal cluster structure.

2. Method

We discuss here how to proceed for the description of the elastic scattering of ^{8}B by considering the specific featuere of the system. First we should take into account the halo property of the ^{8}B nucleus in the scattering process, because the binding energy of the proton with the rest is much smaller (E_B = 0.137MeV) than that of α+^{3}He in ^{7}Be (E_B = 1.583MeV) and the spatial extension of the halo proton is much larger than that of the α+^{3}He system. In this respect, there is an interesting study of th elastic scattering of the halo nucleus with a target nucleus ,which is suggested by Johnson *et al.*[1] Here, we call this method as an adiabatic recoil approximation. In the adiabatic recoil approximation, the halo nucleon is treated as a spectator and its effect is treated by the form factor. The elastic scattering cross section is written in terms of the elastic scattering cross section of the pointlike projectile nucleus with the target nucleus and the form factor of the halo nucleon in the projectile,as

$$\left(\frac{d\sigma}{d\Omega}\right)_{el} = |T_{el}|^2 = |F(\boldsymbol{Q})|^2 \left(\frac{d\sigma}{d\Omega}\right)_{point} , \tag{1}$$

$$F(\boldsymbol{Q}) \equiv \int d\boldsymbol{r}|\varphi(\boldsymbol{r})|^2 e^{i\boldsymbol{Q}\cdot\boldsymbol{r}} , \tag{2}$$

with $\boldsymbol{Q} = \frac{m_p}{m_p+m_{Be}}(\boldsymbol{K} \quad \boldsymbol{K}')$. The halo nucleon carries the fraction of the incident momentum and $\boldsymbol{r}$ is the relative coordinate proton and ^{7}Be. Since the halo proton is weakly coupled with ^{7}Be, we may be able to treat the halo proton as a spectator .

On the other hand, the cluster structure of the core nucleus ^{7}Be makes a dynamical contribution to the scattering process. In this process, we have to consider the fact that the internal cluster structure is bound with small binding energy and the clusters may be easily excited into continuum states. We describe this process in terms of the microscopic coupled channel method.[3,4] The ^{7}Be nucleus has the ground state with spin-parity $\frac{3}{2}^-$ and another bound state with $\frac{1}{2}^-$. There are two distinct resonance states $\frac{7}{2}^-$ and $\frac{5}{2}^-$. Here, we write ^{7}Be wave function with α+^{3}He cluster structure as

$$\Phi_{I\ell}(Be) = \phi(\alpha)[\phi_{\frac{1}{2}}(He) \times u_{I\ell}(\boldsymbol{\rho})] . \tag{3}$$

The wave functions are calculated by a simple potential model of α+^{3}He. The potential form is assumed to be the Woods-Saxon type and the parameters are determined so as to reproduce the binding energy, matter r.m.s. for the g.s. 2.40 fm, quadrupole moment for the g.s. Q=-5.36 fm^2, B(Eλ)

strength for the g.s. $\rightarrow \frac{1}{2}^-$ B(E2; $\uparrow$)=15.3 $e^2 \cdot fm^4$ and the resonance energies and α+^{3}He scattering phase shift as for the $\frac{7}{2}^-$ and $\frac{5}{2}^-$ resonance state . The coupled channel equation for the relative wave function $\chi_\beta(\boldsymbol{R})$ is written as

$$\left[-\frac{\hbar^2}{2\mu}\nabla^2 + U_{\beta\beta}(\boldsymbol{R}) - (E - \epsilon_\beta)\right]\chi_\beta(\boldsymbol{R}) = \sum_{\beta'\neq\beta} U_{\beta\beta'}(\boldsymbol{R})\chi_{\beta'}(\boldsymbol{R}) \quad (4)$$

where $U_{\beta\beta'}$ represents the diagonal ($\beta' = \beta$) or coupling ($\beta' \neq \beta$) potential. In the present study, $U_{\beta\beta'}$ is assumed to be

$$U_{\beta\beta'}(R) = (1 + iN_W)V_{\beta\beta'}(R) + V^{coul}_{\beta\beta'}(R)\ . \quad (5)$$

where $V_{\beta\beta'}(R)$ is the double-folded potential with the CDM3Y interaction[5] and $V^{coul}_{\beta\beta'}(R)$ is the double-folding Coulomb potential. Here, we assume that the shape of the imaginary potential is the same as that of the real part and the normalization factor N_W is determined phenomenologically.

3. Result and Conclusion

First we study the elastic scattering of ^{7}Be nucleus from the target nucleus by solving the coupled channel equation. The normalization factor N_W for the imaginary potential is set to 0.2 . The calculated results are shown in Fig.1. Since the experiment was performed with a mixed target of ^{12}C+^{14}N,[2] we calculate the elastic scattering cross sections for the ^{12}C and ^{14}N targets. As seen in Fig.1, the effect of the coupled channels is important to bring down the cross section at the backward angles, especially the excitation to the resonance state $\frac{7}{2}^-$ contributes most to the elastic scattering.

We finally present the calculated results for ^{8}B with ^{12}C at E = 95MeV by using the adiabatic recoil approximation. The calculated results are shown in Fig.2, where we have used the form factor corresponding to three cases; $\langle r^2\rangle^{\frac{1}{2}}$=4.51fm[6] with solid curve, 3.71fm[7] with dashed curve and 5.49fm[8] with dotted curve. We find the contribution of the proton halo to the elastic scattering with differernt radii is not so large and all the cases reproduce the experimental data except for the angular around 8°. Hence, it is essential to take into account the cluster structure of ^{7}Be and describe the inelastic excitation of the internal cluster states for the description of ^{8}B elastic scattering.

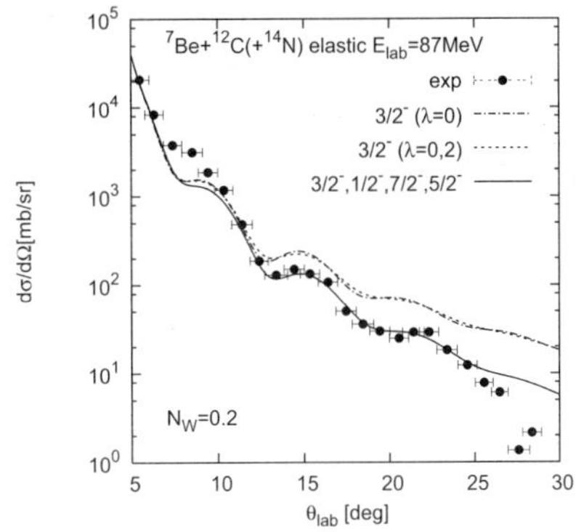

Fig. 1. ^{7}Be+^{12}C (^{14}N) elastic scattering cross section as a function of the laboratory angle. We see the effect of the quadrupole component and the other inelastic channels.

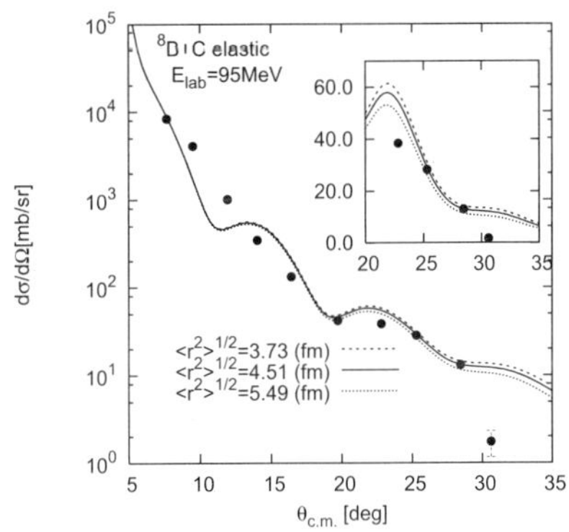

Fig. 2. The elastic scattering cross section of ^{8}B with ^{12}C calculated by using the adiabatic recoil approximation. We use the three form factors corresponding to $\langle r^2\rangle^{\frac{1}{2}}$=4.51fm[6] with solid curve, 3.71fm[7] with dashed curve and 5.49fm[8] with dotted curve.

References

1. R. C. Johnson, J. S. Al-Khalili and J. A. Tostevin,Phys. Rev. Lett. **79**, 2771 (1997).
2. G. Tabacaru, *et al.*, Phys. Rev. C **73**, 025808 (2006).
3. G. R. Satchler and W. G. Love, Phys. Rep. **55**, 183 (1979).
4. Dao. T. Khoa and G. R. Satchler, Nucl. Phys. A **668**, 3 (2000).
5. Dao.T.Khoa ,G.R.Satchler, and W.von Oertzen, Phys. Rev. C **56**, 954 (1997).
6. H. Esbensen , G.F. Bertsch *et al.*, Nucl. Phys. A **600**, 37 (1996).
7. J. S. Al-Khalili and J. A. Tostevin, Phys. Rev. Lett. **76**, 3903 (1996).
8. H. Esbensen and K. Hencken, Phys. Rev. C **61**, 054606 (2000).

NEW MULTICONDUCTOR TRANSMISSION-LINE THEORY AND THE MECHANISM OF NOISE GENERATION

H. TOKI* and K. SATO

Research Center for Nuclear Physics, Osaka University, Ibaraki, Osaka 567-0047, Japan
**E-mail: toki@rcnp.osaka-u.ac.jp*

We study the mechanism of electromagnetic noise generation in standard two-conductor transmission-line circuit due to its coupling to the circumstance by using the newly developed multiconductor transmission-line theory. We discuss then how to reduce the electromagnetic noise by introducing the third line and arrange all the electric loads symmetrically around the third line.

Keywords: multiconductor transmission-line theory, normal mode, common mode, electromagnetic noise

1. Introduction

We start with a statement that usual electric circuits have two transmission-lines and emit and absorb electromagnetic (EM) noise. Synchrotron accelerators use high electric power and they are sensitive to EM noise. In fact, the use of standard electric circuit provides large EM noise to magnets and it was not easy to make stable operation of synchrotron at J-PARC. The main ring accelerator introduced the Sato-Toki (ST) symmetrization and were able to reduce largely the noise level for successful synchrotron operation.[1]

In order to understand the generation of the electromagnetic noise, we have to consider the influence of the circumstance. The electric field produced by a charge placed in vacuum is totally changed when the charge is placed near the earth. This is true also to a current going through a transmission line. Hence, the performance of two-line electric circuit should get the influence from the circumstance. In order to simplify the problem and understand the influence of the circumstance, we consider a three-conductor transmission-line system, where two lines are the main lines and the third line is considered as the circumstance line. We further impose that the total current is zero so that the system is closed.[2] It is also interesting to study the case without this restriction of the total current being zero.

We formulate a new three-conductor transmission-line theory by constructing coupled differential equations, where the coefficients of differential equations are obtained explicitly. We utilize similar expressions of electric field and magnetic field due to charge (Coulomb's law) and current (Ampere's law), and determine the coefficients of potential and inductance. With the coupled differential equations, we can show that the common and normal modes decouple from each other when the first and second lines have the same geometrical shape and the same distance to the third line. On the other hand, when the symmetry of the first and second lines is broken, a coupling of the normal and common modes appears. The coupling of the common mode with the third line produces an electromagnetic field in the circumstance, if this third line represents the effect of the circumstance. On the other hand, if we consider this third line to be the middle line used in the HIMAC system, we can confine all the electromagnetic energy within the three transmission-lines.[3]

We shall discuss the construction of a new multiconductor transmission-line theory. We discuss then how can we understand the generation of the EM noise.

2. Multiconductor Transmission-Line System

We start writing the standard coupled differential equations for multiconductor transmission-line (MTL) system.[4,5]

$$\frac{\partial V_i(x,t)}{\partial x} = -\sum_j^N L_{ij} \frac{\partial I_j(x,t)}{\partial t} \,. \tag{1}$$

$$\frac{\partial I_i(x,t)}{\partial x} = -\sum_j^N C_{ij} \frac{\partial V_j(x,t)}{\partial t} \,. \tag{2}$$

These equations for currents $I_i(x,t)$ and potentials $V_i(x,t)$ for each transmission line $i = 1, 2, ..N$ at position and time x, t are derived from the Maxwell equation for the TEM mode. The coefficients of capacity C_{ij} and the coefficients of inductance L_{ij} are defined as

$$Q_i(x,t) = \sum_{j=1}^{N} C_{ij} V_j(x,t) \tag{3}$$

$$A_i(x,t) = \sum_{j=1}^{N} L_{ij} I_j(x,t)$$

Here, the charge $Q_i(x,t)$ is related with the scalar potential $V_i(x,t)$ through the coefficients of capacity. The vector potential $A_i(x,t)$ is related with the current through the coefficients of inductance. These two coefficients are written by the geometries of the transmission lines. The first important step for a new MTL theory is to reverse the relation for the charge and the scalar potential as

$$V_i(x,t) = \sum_{j=1}^{N} P_{ij} Q_j(x,t) \tag{4}$$

Here, P_{ij} are called the coefficients of potential.

To be specific, we consider the case of three parallel conductor transmission-lines (parallel cables) and explicitly work out the behavior of this coupled system. We write electric potentials, V_1, V_2 and V_3, and the corresponding currents, I_1, I_2 and I_3, for these three lines. To be concrete in our discussion, we impose the condition that the sum of the three currents is zero, $I_1(x,t) + I_2(x,t) + I_3(x,t) = 0$. This relation for the currents ensures that the three transmission-lines form a closed system.

We have introduced the coefficients of potential P_{ij} as the inverse of the C-matrix, $P = C^{-1}$. We can then write the above equation in terms of the coefficients of potential as

$$\frac{\partial V_i(x,t)}{\partial t} = -\sum_{j}^{N} P_{ij} \frac{\partial I_j(x,t)}{\partial x} , \tag{5}$$

with the condition that the sum of the three currents zero. The next important step is to introduce the concept of the normal and common modes.

$$\begin{aligned} V_n &= V_1 - V_2 \\ V_c &= \frac{1}{2}(V_1 + V_2) - V_3 \\ I_n &= \frac{1}{2}(I_1 - I_2) \\ I_c &= I_1 + I_2 \end{aligned} \tag{6}$$

Here, those quantities with the suffix n are for the normal-mode and the other with the suffix c are for the common-mode. We then obtain the relation $I_3 = -I_c$ owing to the imposed condition. We note here that normal-mode quantities are usually used for electric circuit. However, the common-mode quantities have not been discussed clearly in the literature.

We obtain then the coupled differential equations for the normal and common modes as

$$\begin{aligned} \frac{\partial V_n}{\partial t} &= -P_n \frac{\partial I_n}{\partial x} - P_{nc} \frac{\partial I_c}{\partial x} \\ \frac{\partial V_c}{\partial t} &= -P_{cn} \frac{\partial I_n}{\partial x} - P_c \frac{\partial I_c}{\partial x} \,. \end{aligned} \tag{7}$$

The coefficients P_n, P_c and P_{nc} are written as

$$\begin{aligned} P_n &= P_{11} + P_{22} - 2P_{12} \\ P_c &= \frac{1}{4}(P_{11} + P_{22} + 2P_{12} - 4P_{13} - 4P_{23} + 4P_{33}) \\ P_{nc} &= P_{cn} = \frac{1}{2}(P_{11} - P_{22}) - P_{13} + P_{23} \,. \end{aligned} \tag{8}$$

We can work out the other differential equations for the normal mode and the common mode.

$$\begin{aligned} \frac{\partial V_n}{\partial x} &= -L_n \frac{\partial I_n}{\partial t} - L_{nc} \frac{\partial I_c}{\partial t} \\ \frac{\partial V_c}{\partial x} &= -L_{cn} \frac{\partial I_n}{\partial t} - L_c \frac{\partial I_c}{\partial t} \end{aligned} \tag{9}$$

We have expressions for the L coefficients similar to the P coefficients.

$$\begin{aligned} L_n &= L_{11} + L_{22} - 2L_{12} \\ L_c &= \frac{1}{4}(L_{11} + L_{22} + 2L_{12} - 4L_{13} - 4L_{23} + 4L_{33}) \\ L_{nc} &= L_{cn} = \frac{1}{2}(L_{11} - L_{22}) - L_{13} + L_{23} \end{aligned} \tag{10}$$

In general, therefore, the normal and common modes are coupled.

We want to understand first the most symmetric case in which the coupling terms are zero: $P_{nc} = 0$ and $L_{nc} = 0$. This condition is fulfilled when the sizes and shapes of lines 1 and 2 are the same ($P_{11} = P_{22} : L_{11} = L_{22}$), and the relations of lines 1 and 2 to 3 are the same ($P_{13} = P_{23} : L_{13} = L_{23}$). In this case, the normal and common modes are completely decoupled. We have then differential equations for the normal mode as

$$\begin{aligned} \frac{\partial V_n}{\partial t} &= -P_n \frac{\partial I_n}{\partial x} \\ \frac{\partial V_n}{\partial x} &= -L_n \frac{\partial I_n}{\partial t} \,. \end{aligned} \tag{11}$$

This expression with the coefficients P_n and L_n corresponds to the one written in textbook. We can eliminate I_n and obtain the conventional wave equation for V_n as

$$\frac{\partial^2 V_n}{\partial x^2} = \frac{L_n}{P_n}\frac{\partial^2 V_n}{\partial t^2} = L_n C_n \frac{\partial^2 V_n}{\partial t^2} \ , \tag{12}$$

where we have defined the capacity for the normal mode as $C_n = \frac{1}{P_n}$. Hence, the capacity for the normal mode is written as the inverse of the special sum of the coefficients of potential. We can work out these coefficients and the results are $\frac{L_n}{P_n} = L_n C_n = \frac{1}{c^2}$, where c is the velocity of light. This means that the normal-mode propagates through the transmission-lines at the speed of light.

We can work out the coefficients of potential and inductance in the similar manner in the quasi-stationary approximation using the Neumann's formula.[2] We write only the results.

$$\begin{aligned} L_{ij} &= \frac{\mu}{2\pi}(\ln 2l - 1 - \ln \tilde{a}_{ij}) = \frac{\mu}{2\pi}(\ln \frac{2l}{\tilde{a}_{ij}} - 1) \ , \\ P_{ij} &= \frac{1}{2\pi\varepsilon}(\ln 2l - 1 - \ln \tilde{a}_{ij}) = \frac{1}{2\pi\varepsilon}(\ln \frac{2l}{\tilde{a}_{ij}} - 1) \ . \end{aligned} \tag{13}$$

We use the geometrical mean distance (GMD) as $\tilde{a}_{ij}$.

$$\ln \tilde{a}_{ij} = \frac{1}{S_i S_j}\int\int \ln |\vec{s}_i - \vec{s}_j| ds_i ds_j \tag{14}$$

Here, l is related with the length of the transmission-line. The large S_i is the cross section of the transmission line i and the vector $\vec{s}_i$ is the vector in the cross section area.

We calculate now all the physical coefficients of inductance for the three conductor lines.

$$\begin{aligned} L_n &= L_{11} + L_{22} - 2L_{12} = \frac{\mu}{2\pi}\ln\frac{\tilde{b}^2}{\tilde{a}_1\tilde{a}_2} \\ L_c &= \frac{1}{4}(L_{11} + L_{22} + 2L_{21} - 4L_{31} - 4L_{32} + 4L_{33}) = \frac{\mu}{8\pi}\ln\frac{\tilde{d}_1^4\tilde{d}_2^4}{\tilde{a}_1\tilde{a}_2\tilde{b}^2\tilde{a}_3^4} \\ L_{nc} &= L_{cn} = \frac{1}{2}(L_{11} - L_{22}) - L_{13} + L_{23} = \frac{\mu}{4\pi}\ln\frac{\tilde{a}_2\tilde{d}_1^2}{\tilde{a}_1\tilde{d}_2^2} \end{aligned} \tag{15}$$

We see that all the coefficients of inductance exclude the length of the conductor transmission-lines. We can get similar expressions for $P's$, where the factors in front of the geometrical expressions are written in terms of $1/\varepsilon$ instead of μ.

We mention here that the coefficients of potential and inductance can be calculated using the same equations (the Neumann's formula) and they all have the proportional relation

$$P_{ij}/L_{ij} = c^2 \,. \tag{16}$$

We obtain the coefficients of characteristic impedance through

$$Z_{ij} = \sqrt{P_{ij}L_{ij}} \,. \tag{17}$$

We have the same relation for the common and normal mode coefficients.

We introduce characteristic impedance as $Z_i = cL_i = \frac{P_i}{c}$. We can then write the above differential equations as

$$\begin{aligned}
\frac{\partial V_n}{\partial t} &= -cZ_n \frac{\partial I_n}{\partial x} - cZ_{nc}\frac{\partial I_c}{\partial x} \\
\frac{\partial V_n}{\partial x} &= -\frac{Z_n}{c}\frac{\partial I_n}{\partial t} - \frac{Z_{nc}}{c}\frac{\partial I_c}{\partial t} \\
\frac{\partial V_c}{\partial t} &= -cZ_c \frac{\partial I_c}{\partial x} - cZ_{cn}\frac{\partial I_n}{\partial x} \\
\frac{\partial V_c}{\partial x} &= -\frac{Z_c}{c}\frac{\partial I_c}{\partial t} - \frac{Z_{cn}}{c}\frac{\partial I_n}{\partial t} \,.
\end{aligned} \tag{18}$$

We will try to find out a general solution to the coupled differential equation by introducing the plane wave solution

$$V_n(x,t) = V_n \exp(j\omega t - jkx) \tag{19}$$

for the normal-mode potential. We introduce the same relations for the common-mode potential and also for both normal- and common-mode currents.

We now write this equation in matrix form;

$$\begin{pmatrix} \omega & -ckZ_n & 0 & -ckZ_{nc} \\ ck & -\omega Z_n & 0 & -\omega Z_{nc} \\ 0 & -ckZ_{cn} & \omega & -ckZ_c \\ 0 & -\omega Z_{cn} & ck & -\omega Z_c \end{pmatrix} \begin{pmatrix} V_n \\ I_n \\ V_c \\ I_c \end{pmatrix} = 0 \,. \tag{20}$$

We can get the determinant of the 4×4 matrix as

$$det|A| = (Z_n Z_c - Z_{cn} Z_{nc})(c^2k^2 - \omega^2)^2 \,. \tag{21}$$

This equation for $det|A| = 0$ gives the solution for k as

$$k = \pm \frac{\omega}{c} \,. \tag{22}$$

This is a very interesting result, where the dispersion relations of the coupled system are not different from those of the decoupled system.

We can now work out the power of the system. The total power is position-independent. This is a natural consequence of making the total current zero so that the system is closed. The common-mode power is zero when there is no coupling term, that is, $Z_{nc} = 0$. Hence, when the transmission line system is constructed symmetrically, no common mode appears. On the other hand, when Z_{nc} is finite, the power of the common mode appears as a function of the position x. This means that when the transmission-line system is constructed asymmetrically, the power in the normal mode is transferred to the common mode and also that in the common mode is transferred to the normal mode. This finding is exactly the same as that in the case of electric circuit in the framework of a lumped circuit. When the electric circuit system is constructed asymmetrically, the coupling of the normal and common modes appears.[3]

3. Discussion on the Role of the Third Line and the Sato-Toki Symmetrization

The common mode in the three-conductor transmission-line system provides a current going through the third line, $I_3 = -I_c$. Hence, an electromagnetic field develops between the main two lines and the third line. If this third line represents the mirror line of a perfect conductor plane (e.g. earth surface), an electromagnetic field is produced on the earth surface. Any electric objects or devices placed on the earth surface near the transmission-lines are influenced by the electromagnetic field. This electromagnetic field should be interpreted as a source of electromagnetic noise.

We consider now the case in which a new conductor line (line 4) is inserted very close to the two main conductor lines in the space made between the two-main-line system and the third line. In this case, owing to the very strong electromagnetic field produced by the common-mode current of the two main conductor lines in the inserted line, the common mode interacts more strongly with the newly inserted line, but weakly with the third line owing to the large distance. In this case, we may consider that the role of the third line can be neglected owing to the fact that the two main lines couple more strongly now with the newly inserted line. Hence, when the inserted line is placed very close to the two main lines, the electromagnetic field is approximately confined within the small space made by the two main lines with the inserted line. In this case, we may neglect the effect of the circumstance represented by the third line. If we further consider that the inserted line is the middle line of the electric circuit of the symmetrically arranged circuit as the HIMAC system, we can control electromagnetic field and hence can suppress electromagnetic noise.[3] The concept of the three-line electric circuit with the symmetrization around the middle line is shown in Fig.1.

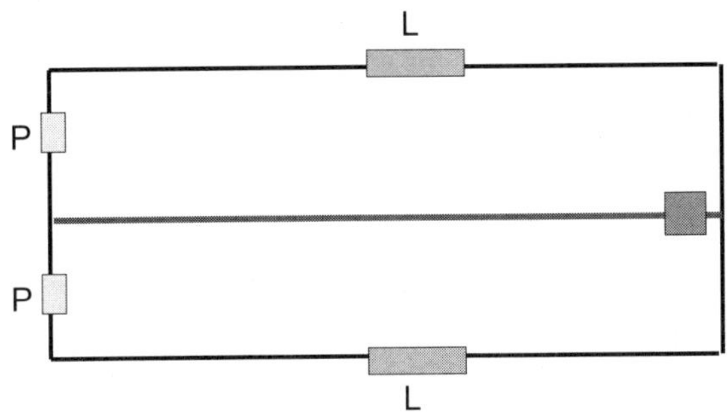

Fig. 1. Three-line electric circuit with the arrangement of powers and electric loads arranged symmetrically around the middle line.

We have discussed the performance of symmetrically arranged circuit with the third line used as the central line for the symmetrization.[3] The MR synchrotron of J-PARC introduced the ST symmetrization for their accelerator magnet circuit. Shown in Fig.2 is the configuration of the J-PARC MR before and after the introduction of the ST symmetrization. The symmetrization was performed in the accelerator magnet part and the accelerator magnets use the normal mode magnet.[3] It is advisable to introduce further the ST symmetrization in the power supply circuit also where the coupling of the normal mode to the common mode still exists.

4. Conclusions

We have studied a new MTL theory. An important step to develop the new MTL theory is the use of the coefficients of potential instead of the coefficients of capacity. There is a symmetry between the coefficients of potential and the coefficients of inductance. At the same time, we can express both the coefficients in terms of the geometrical quantities of the conductor lines. We should introduce the concept of the normal and common modes for the description of the EM noise, since the common mode is related with the role

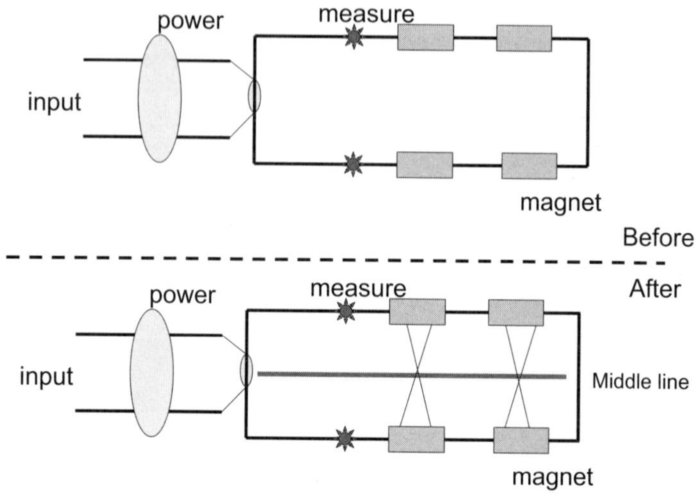

Fig. 2. Accelerator magnet circuit of J-PARC MR. The upper arrangement is the configuration of the accelerator magnet circuit before the introduction of the Sato-Toki symmetrization. The lower arrangement is the configuration after the symmetrization. The star places are the positions where the measurement of the current is performed. The power circuit part is not yet symmetrised where the normal and common mode coupling still exists.

of the circumstance. In the usual two-line electric circuit, the normal mode couples with the common mode. It is important to introduce a third line to a usual electric circuit and introduce all the electric loads symmetrically around the third line. In this way, we are able to make a noise free circuit. It would be important to introduce the third line and the symmetrization in the power supply side.[6] All the details of the performance of J-PARC MR synchrotron and the electromagnetic noise are provided in the recent publications written in Japanese.[7,8] We are working out the general case, where the sum of the currents is non-zero and the radiation and absorption take place in the electric circuit.

References

1. H. Kobayashi: Proc. of Particle Accelerator Conf. (2009) PAC09-WE1GRI02.
2. H. Toki and K. Sato, J. Phys. Soc. Jap. **78** (2009) 094201.
3. K. Sato and H. Toki: Nuclear Instrument and Method (NIM) A **565** (2006) 351-357.
4. J. C. Maxwell: A Treatise on Electricity & Magnetism (Dover Publication Inc.,

New York, 1876) Vols. 1 and 2.
5. C. R. Paul: Analysis of Multiconductor Transmission Lines, (Wiley-Internscience (IEEE), New Jersey, 2008) 1.
6. K. Sato: Accelerator **4** (2007) 271-287 (in Japanese).
7. S. Nakamura, F. Akimura, H. Matsumoto and H. Kobayashi: Accelerator **6** (2009) 292-301 (in Japanese).
8. H. Toki and K. Sato: Accelerator **6** (2009) 302-311 (in Japanese).

NUCLEAR AND PARTICLE PHYSICS WITH ULTRA-INTENSE LASERS

H. TAKABE

Institute of Laser Energetics, Osaka University
**E-mail: takabe@ile.osaka-u.ac.jp*

With the maturity of high-power laser technology, we can realize an extremely high intensity of photons in a tiny space. About 10 kJ of 1eV photons (10^{22}photons) can be squeezed in a space of several micro-meter during 10 fs (10^{-14}s). This means the density of photons of $n_{photon}=10^{34}/cm^3$. Photon is boson and can be squeezed to a tiny space. We expect that we can design unique experiments of the collective effects of elementary particles and such lasers can also be used for a new type of experiments. The physics will be more wider with the combination of the conventional accelerators. In the present paper, I briefly describe what we have carried out as academic research related to so-called laser nuclear physics.

1. Introduction

It was early 90s that Ultra-intense lasers became familiar in our society after the invention of CPA technique. With such lasers the electron oscillation by focused laser electric and magnetic field entered into the relativistic regime and the physics shown in Fig. 1 can be expected in extremely higher intensity regimes. At that time, I thought of two things. One is the possibility of laboratory high-energy astrophysics experiment with use of the electron-positron plasma produced by laser irradiation on high-Z material [1]. The other is the application of the ultra-intense laser to the nuclear transmutation of radio-active waste from nuclear reactors. This is motivated by comparison to the same application of the neutron source by proton accelerator such as J-PARC[2] Regarding the electron-positron plasma, we had started the analysis of the experimental data obtained by Tom Cowan et al in LLNL with two-beam Peta-Watt (10^{15} W) laser of NOVA system[3]. It was around the year of 2000. We have carried out a computational simulation of the positron production based on the Fokker-Planck type kinetic equation [4]. In such a situation, most of the laser energy is converted to

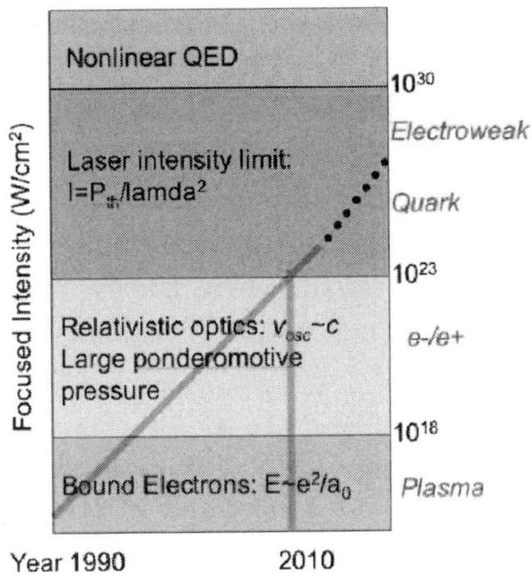

Fig. 1. The peak intensity of short pulse lasers reached in the past 2 decades. Selected physics topics are highlighted for the related laser intensity Region (Courtesy by Hui Chen and Scott Wilks)

the energy of relativistic electrons, while it was found that the direct production of the positrons via the Trident process is relatively in-efficient. The most of positrons are found to be produced, however, via Bethe-Hitler process. In this process, the electron energy is once converted to that of γ-ray via bremmstralung process and γ-ray efficiently produces the pairs via interaction with nuclei[4]. In the later work, we found that the plasma effect through the am-bipolar electric field is found to be essential in comparing with the Cowan's experimental data.

As for the nuclear transmutation, I have organized a symposium on "Laser Nuclear Physics" at the JPS annual meeting in 2001. Many of my friends in abroad are also interested in what we have discussed in this symposium, I asked all invited speakers to write their papers in English to publish from a Journal [5]. My decision was correct because Ledingham et al recently published a review in Rev. Mod. Phys. [6] and our papers are cited in the abstract as a pioneering work.

In the symposium at JPS, I wished to compare the efficiency of nuclear transmutation between with J-PARC and with lasers, while the speaker from JAEA in charge of this project could not show the number and I knew that it is too early to compare at that time. In our scheme, we aimed at the excitation of Gamov-Teller giant resonance [7] by the collision of about 10 MeV relativistic electrons quivering by laser irradiation [8]. What we found after theoretical analysis is that the electrons excite the giant resonance mode via the virtual photon, and its cross-section is about 100 times smaller than that via real photon. As the result, we need many collisions of an electron and nucleus, and it is concluded that for the 100% transmutation, we need NIF-class huge laser system [9]. It is still open question if this is in-efficient or not in comparison to the case of the neutron source like J-PARC. In the countries like Japan where nuclear waste increases rapidly, such discussion on the efficiency would be seriously discussed in the near future.

So-called Active Galactic Nuclei (AGN) are one of the most important subjects to be studied in the high-energy astrophysics[10]. The AGN is the central part of a very active galaxy and it is said that there is a very massive Black Hole at the center and extremely high energy continues to be released to produce the electron-positron bi-polar jets propagating over longer distance than the size of the galaxy[10]. We have studied the propagation of such highly relativistic (Lorentz factor of about 10-20) bi-polar jets [11,12]. At the same time, the experimental data were published by Tom et al [13]. This data was a good target for checking our theoretical work and we started the theoretical analysis of the experimental data.

We were lucky that Mike Campbell, Vice-President of General Atomics, asked me almost at the same time to begin the collaboration on the pair creation with intense laser, because Tom also moved to GA and need a help of theoretician. Then, I showed a paper by K. Nakashima [4] and proposed to send him to GA. During his stay, I also had a chance to stay in GA during the summer of 2002. During my stay, I also asked Dr. Y. Sentoku to carry out the particle simulation (PIC) in order to compare the Tom's experimental spectrum of positron to 2D simulation. The simulation clearly showed that the positron gains about 10 MeV by the am-bipolar field produced by the expansion of the relativistic electrons at the target rear surface. This is a clear evidence of Plasma Effect in pair creation by high-energy-density photon field. During my stay, Tom and we have completed a paper to be submitted to Phys. Rev. Lett. But, it is not submitted yet[14].

I have almost forgotten this paper because of daily busy time, but I remember when I met Dr. Sentoku at EPS Plasma Physics Conference held at Create, Greek in June, 2008. With phone number of Tom in Germany (he moved to Germany soon after the completion of the paper), I called him from Paris where I stayed for joint research after EPS. It has passed six years already, but the paper contains still very new results and I pushed him to submit the paper. Tom also agreed that it is very much important to publish the paper and he promised me he will submit in a week after minor polishing up. But, the paper is not published yet.

During IFSA conference in September, 2009[15], Scott Wilks of LLNL came to me and told me that they found how important the plasma effect is in anti-matter production with PIC code. I of course replied that that is exactly what we found eight years ago, but the paper is not published; it is my great regret. Scott was happy to know that we also have carried out such simulation and he kindly invited me to give an invited talk at Workshop on Antimatter Creation Using Intense Lasers - Investigating laser-based particle creation for accelerator, astrophysical, and antimatter science to be held for April 27-28, 2010 at Berkeley, CA[16].

In order to promote Laser Nuclear Physics at our institute, I asked my friends in RCNP to introduce a theoretician of nuclear physics who can stay in our institute as guest professor. With recommendation of Prof. Fujiwara, I invited Dr. Alexander Titov for 4 months in 2009. As the collaboration we have proceeded our research from the pair of electron to the pair of muon. In this scheme, we used the laser accelerated electron experimentally demonstrated by Leemans [17]. The electrons of about 1GeV are now available with laser acceleration and we proposed di-muon production with such high-energy-density electrons [18].

The description so far is only related to my activities and there are of course another topics related to nuclear physics with intense lasers. One big example is described in a paper [19] and a book [20]. With use of NIF [9], Livermore scientists will demonstrate fusion ignition and burn in laboratory. They plan to do a model experiments to clarify r-process and s-process of nucleon-synthesis in supernova explosion. In addition, they also plan to study the density effects on nuclear reaction rate.

2. Nuclear Physics with Ultra-Intense Lasers

There are three ways to excite nuclei or trigger nuclear reactions with intense lasers.

(1) Direct laser nuclear reactions: At extremely high field, laser can directly interact with nucleus, and the excitation and the following nuclear decay would be observed. It requires, however, the laser intensity of 5×10^{28} W/cm^2; too high intensity compared to the value expected near future; namely, 10^{23} - 10^{26} W/cm^2. Around such extremely high intensity, we can observe electron pair creation in the vacuum. So-called "vacuum-breakdown" which is, for example, thought to be very active near the event horizon of Black Hole[21] can be demonstrated in laboratory. The vacuum-

breakdown requires that the potential difference generated by laser field is about $2mc^2$ over the electron Compton wavelength λ_C (=h/mc). This gives us the threshold intensity of $5x10^{28}$ W/cm^2[4]. It is interesting topics itself, but we do not discuss this topic in this paper.

(2) Nuclear processes via interaction with bound electrons of atoms: Several processes including NEET are enumerated, while all are characterized by the process that the electric shells of atoms work as mediators between laser and nuclei. In the present paper, we do not deal with such processes. The details is reviewed in Ref. [6].

(3) Nuclear processes triggered by relativistic electrons generated by ultra-intense lasers irradiated on matter: In most of the recent laser nuclear experiments, these processes have been reported. In addition, most of nuclear reactions are triggered by γ-rays generated by the Bremmstrahlung process in a target[4,14,22,23]. Such nuclear reactions are schematically shown in Fig. 2. In the present paper, we focus on such nuclear reaction processes.

As shown in Fig. 2, there are three important reactions, when an ultra-intense laser is irradiated on a gold foil attached by uranium. The first one is the photo-nuclear reaction of gold;

$$^{197}\mathrm{Au}(\gamma, \mathrm{xn})^{197-\mathrm{x}}\mathrm{Au} \tag{1}$$

where x is integer (x= $1, 2\cdots$). In the reaction (1), x neutrons are kicked out by the collision of γ-photon. Such reaction happens if the energy of γ-photon is more than 8.5 MeV. In the case of x=1, for example, the reaction product ^{196}Au β-decays to ^{196}Pt with the line x-ray emission of 356 keV, which has been observed experimentally [13].

The second reaction is photo-induced nuclear fission. By the collision of energetic γ-ray, the fission of ^{238}U is induced as schematically shown in Fig. 2. In the experiment [13], the fission events of $1.8x10^7$ were observed and the distribution of fission fragments is confirmed to well coincide with the theory. This reaction is written as ^{258}U(γ, f). The threshold energy of γ-photon inducing the fission is relatively low, 5 MeV. In addition, the nuclear fission triggered by photo-nuclear reaction was also observed in the same experiment. This fission is driven by the neutrons generated, for example, by the photo-nuclear reaction ^{197}Au(γ,n) ^{196}Au.

The third reaction is the electron-pair creation. The pair creation by impinging relativistic electrons in high-Z matter is due to the following two processes.

$$\text{(Trident process) } \mathrm{Z} + \mathrm{e}^- \rightarrow \mathrm{Z} + 2\mathrm{e}^- + \mathrm{e}+ \tag{2}$$

$$\text{(Bethe-Heitler process) } \mathrm{Z} + \gamma \rightarrow \mathrm{Z} + \mathrm{e}^- + \mathrm{e}^+ + \gamma' \tag{3}$$

The trident process is of one step, while Bethe-Heitler (BH) process is of two steps, namely electron energy must be once converted to -photons by the Bremmstrahlung process. Although BH process is the two-step process, the cross-section of pair creation is about two orders of magnitude higher than that of Trident process. Which process is dominant, therefore, depends upon the thickness of the gold foil. According to Ref. 4, BH process is dominant for the gold foil thicker than about 20 μm.

The positron creation by lasers will open a new field in astrophysics, We call it High Energy Laser Astrophysics (HELA), where baryon-free electron positron plasmas produced by laser are used to study the relativistic phenomena in relation to the cosmological jets, fireballs of γ-ray bursts, and so on. The details are described in the authors review paper on laser astrophysics[1]. Energetic and collimated proton beam with energy of MeV are

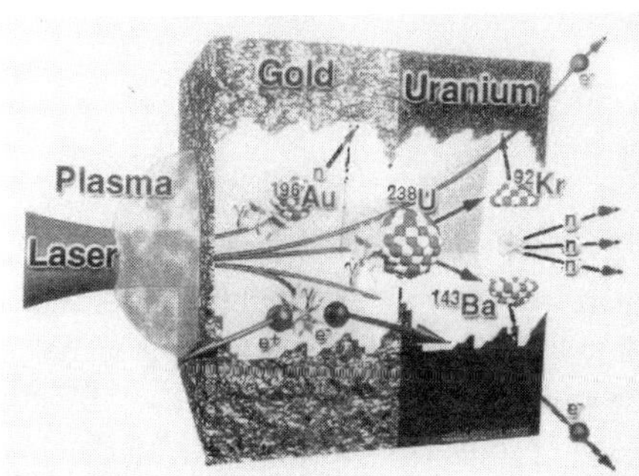

Fig. 2. Nuclear processes driven by relativistic electrons produced by laser and high-Z matter interaction

generated from the rear side of a target foil[24]. The angle of the ejected protons is normal to the rear surface, because the protons are accelerated by the am-bipolar electric field due to the relativistic electrons so that the electric field is formed normal to the target surface. G. Mourou, the founder of CPA, insisted a future dream to be realized by the increase of the energy of the proton beam over GeV range. He dreams using the proton beam for creation of neutrinos. Compact neutrino source is getting to be of near future non-fiction.

In order to increase the energy of generated high-energy electrons for given laser intensity, a variety of ideas have been proposed. In the framework of the laser acceleration activity, the controlled pre-formed plasma is useful to increase the energy. In Fig. 3, three-dimensional simulation results are shown for the case with the plasma whose density profile is exponential [25]. Such trial will enrich the nuclear physics induced by lasers. In Fig.

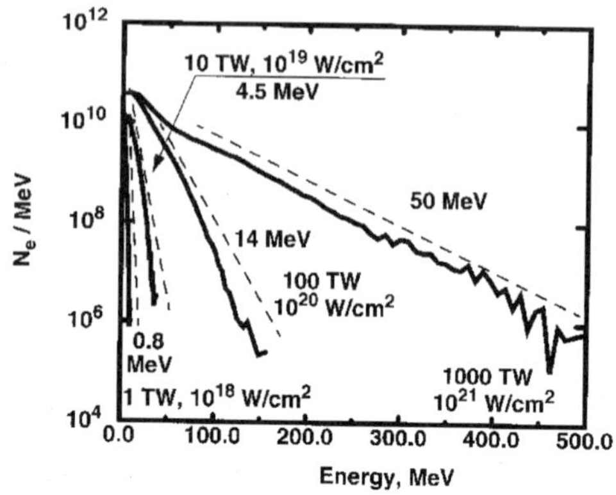

Fig. 3. Simulation results of electron distribution functions as a function of laser intensity

4, a typical electron spectrum is shown for the case when an ultra-intense laser is focused on a gold target[13]. The laser intensity is 10^{21} W/cm^2 and very high energy electrons more than 10 MeV are observed. In this section, let us consider about the physical mechanism of the high-energy electron production.

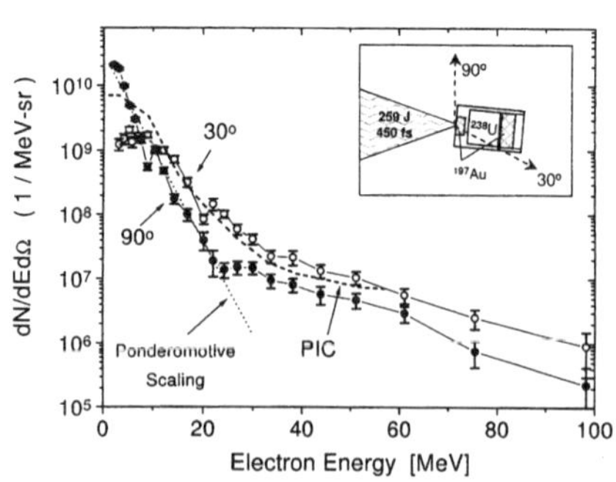

Fig. 4. Experimental data of electron spectrum [13]

3. Electron-Positron Production and Plasma Effect Forming Relativistic Jets

Figure 5 shows the electron and positron spectra for shot with 622 J incident on a 125 μm Au foil, at an intensity of 3x10^{20} W/cm^2. The data points with error bars show the positron yield based on track counting in the emulsion, and the blue points show the electron spectrum deduced from the optical density recorded in a film placed behind the electron emulsion strip. The solid blue lines show the observed and predicted positron spectrum using the relativistic FP code [6], assuming the electrons pass once through the Au foil from the laser-focus. The FP code includes all transport effects: bremsstrahlung, BH and Trident pair production, and energy-loss of all charged particles. The measured positron distribution appears to have an excess of events both at high energies above 10 MeV, and at low energies, below 4 MeV. A larger discrepancy between measured and predicted positron yields was found for thinner targets In Fig. 5, the green dotted line is the data obtained by Monte-Carlo type calculation like EGS code widely used in High-Energy Physics. It is also the same as FP result and cannot explain the experimental data. We considered a variety of possible

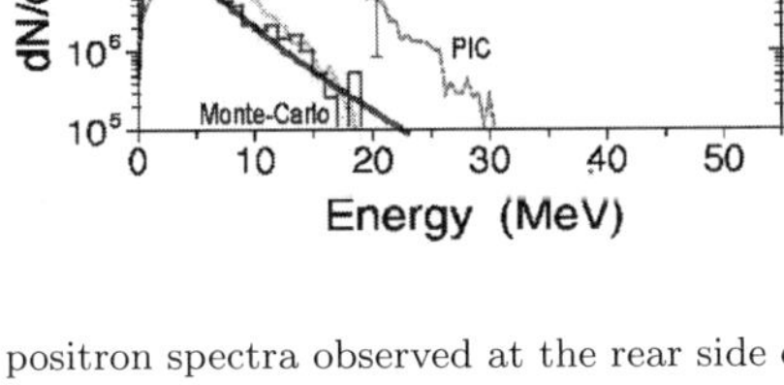

Fig. 5. Electron and positron spectra observed at the rear side of a gold foil irradiated with ultra-intense laser. The blue and red points are experimental data of electron and positron. The solid blue lines are electron and positron spectra obtained with FP calculation. The red histogram is from EGS code. The green solid lines are the result from 2-D PIC simulation.

origins for the enhanced positron yield and average energy for thin foils. We could rule out that neither a pronounced energy-angle correlation in the differential cross-section of BH or Trident positrons, nor the ponderomotive acceleration of positrons produced in, or scattered into, the laser focus, could explain the data. By performing Particle-in-Cell (PIC) simulations of the evolution of the cloud of positrons and electrons as they leave the target, we found that acceleration in the transiently produced ambipolar electric field is consistent with the observed increase in positron energy. Moreover the PIC simulations indicate a strong reflection of the high-energy electrons by the space charge field created as they begin the leave the target foil. As described below, the resulting recirculation of electrons is sufficient to explain the enhanced total positron yield.

The PIC result is shown in Fig. 6. We took as input an initial source diameter of 80 μm (fwhm), consistent with an initial electron and positron angular spread. We estimate an initial average high-energy electron density in the target volume of 10^{20} cm^{-3}, and a positron density of 10^{16} cm^{-3} (corresponding to the measured total yields), with the initial positron spectrum taken from the FP simulation (shown in Fig.5). Figure 6(a) shows the longitudinal momentum distributions of positrons and electrons versus distance from the target. The positron momentums normal to the foil surface are observed to be boosted as seen in Fig. 6(a) and Jet-like injection is seen as shown in Fig. 6(c). The outward electron flow is inhibited by the space charge electric field. The electric field magnitude is 10^{12} V/m, consistent with many recent experiments on ion acceleration in the same am-bipolar sheath [26, 27]. The affect on the escaping positron energy distribution is to produce a shift of positrons to higher energy, in well agreement with the measured data as shown in Fig. 5 with green solid line. The magnitude of the positron yield calculated from electrons passing only once through the foil, is insufficient to explain the observations in the lower energy range ($>$ 5MeV). Since the PIC simulation points toward "refluxing" and recirculation of the electrons, we included this effect in the FP code by including partially reflecting boundary conditions at the front and rear surface of the foil. The distribution function of positrons created differs from the single-pass result in that the relevant parent electron energies include the lower, slowing-down component, and the positrons lose their weak directionality and become isotropically directed within the target. We expect that acceleration of positrons generated from refluxing electrons is smaller in magnitude both because the positrons are created over longer times, past the peak of the accelerating field (note $\sim$100 fs as peak of acceleration time in Fig. 6-(a)), and because the refluxing process naturally leads to a radial diffusion of the electrons and daughter positrons in the target foil, into regions of reduced hot electron density and hence reduced ambipolar field strength.

This analysis explains for the first time all the salient features of positron

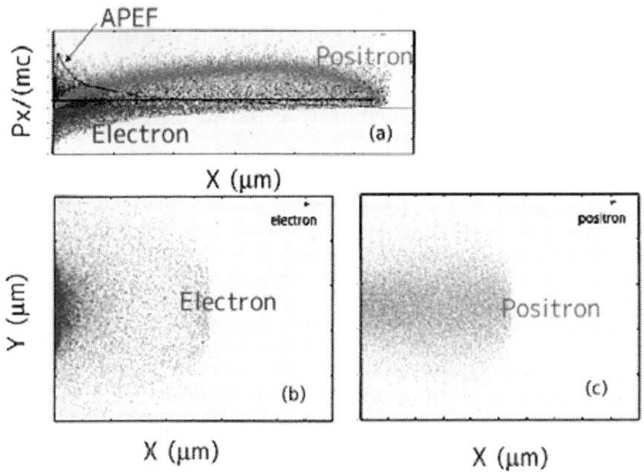

Fig. 6. (a) The longitudinal momentum distributions of positron (red and electrons green) at back of the target. The target rear is at X=0. The solid line indicates the sheath field normalized by 10^{12} V/m. (b) and (c) shows snapshots of the electrons(b) and positrons(c) escaping to the rear side vacuum.

production in laser-produced plasma measured to date. In particular, both the energy distribution and the absolute yield of positrons is consistent with our model of Bethe-Heitler and Trident production, with the necessary inclusion of the plasma-specific effects of the ambipolar acceleration of positrons away from the target foil and the re-circulation of hot electrons refluxing through the target foil. Because of the dynamical nature of these plasma effects, the positron production is considerably more complex than from a conventional electromagnetic shower produced by the passage of an electron beam through a solid target. We predict, for example, that the escaping positrons should be accompanied by a hot electron cloud, and that the most strongly accelerated positrons could be directed somewhat more normal to the target plane for thin foils. A careful measurement of the angular distribution of positrons would provide important confirmation and additional information in a next generation experiment.

It is shown in Ref. 34 that the positron energy spectra have peaks in lower energy side (see Fig. 7). This can be easily understood that the pair creation cross-section as the function of γ-ray energy has the threshold near 1 MeV, but the γ-ray spectrum produced by the high-energy X-rays stemming from the bremsstrahlung process shows rapid decrease as the increase of the photon-energy as shown in Fig. 8. The positron production is proportional to the product of these two values and as the result it has a peak near several MeV as schematically shown in Fig. 8. It should be noted that the data in Fig.7 is of course affected by the acceleration by the ambiplar electric field and its original peak is lower than the experimental one. For possible future studies of pair plasma physics, it is important to

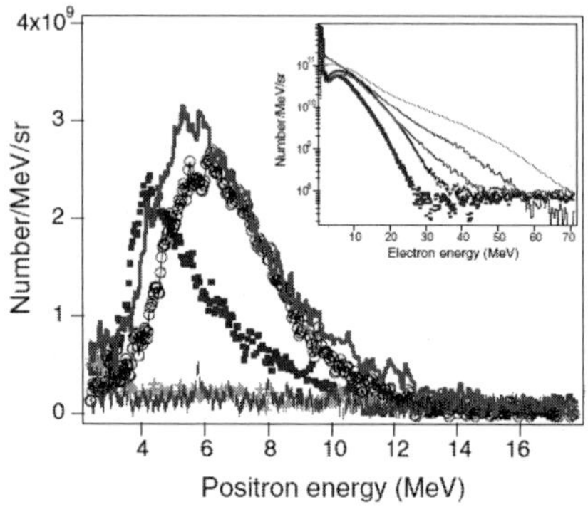

Fig. 7. Positron spectra from experiments with targets of Au (three upper traces), Sn and Cu (two lower traces). No positrons were detected for Cu and Sn targets. The inset shows the corresponding electron spectra for each shot. For the complete color figure, see Ref. [23]. The peak energy is determined as in Fig. 8 and the acceleration by ambipolar field.

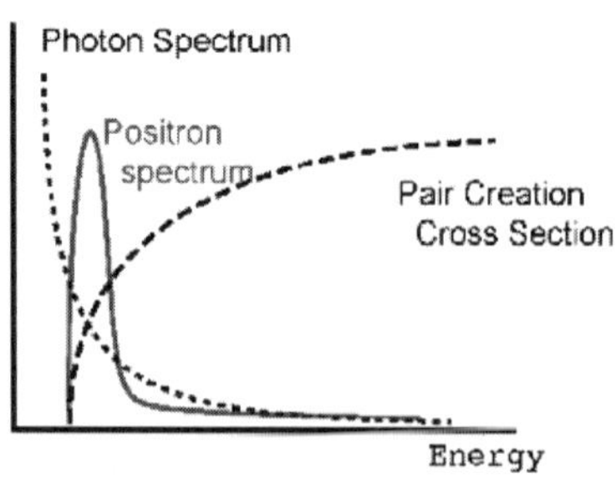

Fig. 8. Positron spectrum is derived from the product of the pair creation cross section and photon spectrum.

quantify our progress towards producing relativistic positron plasma, by comparing the volume of the positron-electron cloud with the calculated

Debye length of the positrons. As noted above, the average positron density appears to be approximately 10^{16} cm^{-3} distributed in a volume of ~100 μm dia. by ~200 μm length. Taking 2 MeV as an effective mean temperature (note that the somewhat higher observed energy is due to a quasi-1D boost normal to the target foil), the hot positron Debye length is $\lambda_{\mathrm{D,e^+}} \sim (\mathrm{kT_{e^+}}/4\pi e^2 \mathrm{n_{e^+}})^{1/2} \sim 110\mu$m, only slightly larger than the production volume. This provides some optimism that with more sophisticated target design we may in future experiments approach the conditions of the truly relativistic high-density positron-electron pair plasma in the laboratory and we can design model experiments of a variety of high-energy astrophysical phenomena in laboratory.

4. Giant Dipole Resonance Excitation by High Energy Electrons

It may be possible to do nuclear transmutation with the high-energy electrons produced by ultra-intense lasers [5]. We can use the gamma-rays produced by Bremmstrahlung of the high energy electrons [5]. Thinking of efficiency and trying to compete another way by use of neutrons generated by target spallation (fragmentation) with irradiation of high-intensity proton beam [28], it is cleverer to confine electrons in a standing wave by two laser beams and maintain the collisions of relativistic electrons with nuclei many times. In a relativistic laser filed of standing wave, an electron has an 8-figure orbit. The laser electric field causes the up-down oscillation, and the Lorentz force gives the forward-backward motion.

If the electron is well confined in the 8-figure orbit, the electron has a chance to excite the nucleus through a virtual photon and giant dipole resonance (GDR) may be excited. In general, the cross-section of GDR of heavy element is very broad over 10 MeV. The cross-section of ^{238}U is, for example, shown in Fig. 9. The absolute value of GDR is very small, but if the electron wave function is squeezed initially, we can expect high reaction rate via electron impact. We skip the detail and do rough estimation on the possibility of laser nuclear transmutation[8]. The fractional event F of GDR excitation is given to be

$$\mathrm{F} = \frac{\sigma_E}{\pi r^2}\eta_{\mathrm{vp}} \tag{4}$$

Where σ_E is the GDR excitation cross-section, $\mathrm{r_e}$ is the effective radius of the electron wave function, and η_{vp} is the virtual photon efficiency. Although it is too much optimistic to assume that the electron is in the same

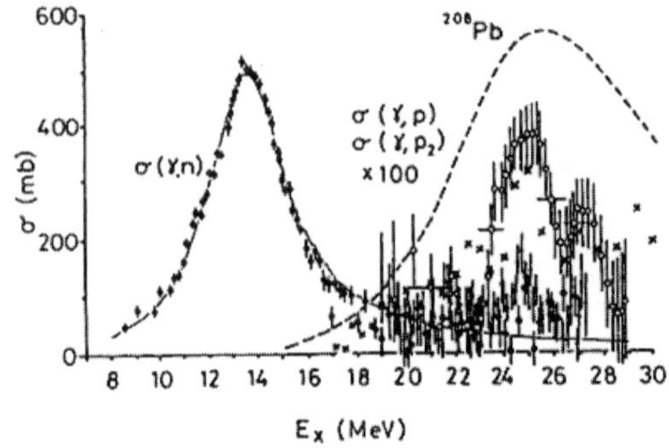

Fig. 9. The giant resonance excitation spectrum of ^{238}U

orbit over 10^3 times oscillations. The number of the excitation events per laser pulse is calculated to be

$$N = 2f\tau_p/\tau_L \tag{5}$$

where $\tau_L = 2\pi/\omega$is the laser one period and equal to 3 fs for the wavelength 1 μm. In Eq. (5), on the other hand, τ_p is the laser pulse duration.

Let us set parameters.

$$\begin{aligned} \sigma_E &= 500[\text{mb}] = 5\text{x}10^{-25}\text{cm}^2 \\ r_e &= r_B/Z, \ Z = 50 \\ \tau_p &= 1 \text{ ps} \end{aligned} \tag{6}$$

where r_B is the Bohr radius and we assumed hydrogen-like atom with Z = 50 initially. The value of σ_E was taken roughly from Fig. 9. Then, the number of event N is calculated.

$$N = 0.01\eta_{vp} \tag{7}$$

If we take 0.01 as a reasonable value of η_{vp}, we need to keep the 8-figure orbit for 10 ns in order to get N = 1. This is unrealistic and if this estimation is reasonable, we need NIF-class huge ultra-intense laser system.

We do not go any forward, but it should be noted that if such orbit-keeping scheme would not work, we think of higher density case, where one electron interacts with many nuclei. If one quivering electron interacts with

104 nuclei per one cycle, the excitation number N approaches unity even for the case of 1 ps laser pulse. Since the size of 8-figure obit is about 1 m, the required density of nuclei is roughly $10^4(10^{-4})^{-3} = 10^{16}$ cm^{-3}. Not so bad. But, reality is not so simple.

The electrons are confined by laser fields and have average oscillation energies of a few tens MeV. Roughly speaking, the traveling distance of an electron $\mathrm{L} = c\tau_\mathrm{p}$ and this should be compared with the GDR excitation mean free path l_E. When the density of nuclei is n, then

$$l_\mathrm{E} = 1/(\mathrm{n}\sigma_E) \tag{8}$$

In this case, the number of excitation event is;

$$\mathrm{N} = \mathrm{L}/l_\mathrm{E} = c\tau_\mathrm{p}\mathrm{n}\sigma_\mathrm{E} \tag{9}$$

Inserting the parameters in Eq. (6), the required density for N=1 is

$$\mathrm{n} = 10^{28}\mathrm{cm}^{-3} \tag{10}$$

This is also too hard to realize. We need another idea.

5. Summary

With recent dramatic progress of laser technology, the laser intensity has reached to 10^21 W/cm^2 and is approaching to 10^{23} 10^{26}W/cm^2 in several years. The laser electric filed is 10 TV/cm and oscillates the electrons to relativistic energy. Such ultra-intense lasers can be used to produce relativistic electrons in the range of MeV efficiently.

The most interesting topics in the present paper are the relativistic electron-positron plasma production in the form of a jet. This suggest us the future laboratory experiment of Gamma-Ray-Bust (GRB), Active Galactic Nuclei (AGN) jet, and Binary Black Hole jet in laboratory. We have to design the target so that generated high-energy electron energy is efficiently converted to the energy of γ-rays. Then, the γ-ray should interact efficiently with high-Z nucleus. In order to increase the density of the pair plasma jetting from the rear side, we can use the geometrical focusing scheme as used in MeV ion beam production. It is very exciting to see what physics is measured through the interaction of the jet and the ambient gas or plasmas.

In the present paper, we have proposed to use such electrons to excite the giant dipole resonance for the purpose of nuclear transmutation of radioactive waste. Rough estimation has been done, while the efficiency is not as high as expected from small cross-section of DGR excitation. We believe, however, that there should be the another ideas and ways to do nuclear physics and its applications with such ultra-intense lasers and hope the readers will suggest many good schemes to open a new scientific field called Laser Nuclear Physics.

References

1. H. Takabe, Progress of Theoretical Physics, Supplement No. 143, PP. 202-265 (2001).
2. J-PARC URL: http://j-parc.jp/index-e.html
3. T. E. Cowan et al., Laser Part. Beams 17, 773 (1999).
4. K. Nakashima and H. Takabe, Phys. Plasmas 9, 1505-1512 (2002).
5. H. Takabe (editor), Laser Nuclear Physics, J. Plasma Fusion Research 77, 1018-1026 (2001).
6. K. W. D. Ledingham and W. Calster, Laser Driven Nuclear and Particle Physics, to be published in Rev. Mod. Phys. (2010).
7. Y. Sumi, Introduction to Nuclear Physics, Chap.6 (1997, Shokabo, Tokyo) in Japanese.
8. H. Takabe, Laser Nuclear Physics, AAPPS Bulletin, Vol. 13, No. 1, 18-25 (2003); http://www.aapps.org/
9. National Ignition Facility URL: https://lasers.llnl.gov/
10. S. Rosswog and M. Buggen, Introduction to High-Energy Astrophysics (Cambridge Univ. Pres, 2007), Chap.8.
11. A. Mizuta, S. Yamada, and H. Takabe, Astrophysical J. 567, 535-642 (2002).
12. A. Mizuta, S. Yamada, and H. Takabe, Astrophysical Journal, 606, 804-818 (2004)
13. T. E. Cowan et al., Phys. Rev. Lett. 84, 903 (2000).
14. T. E. Cowan et al., Plasma-induced enhancement of positron creation in relativistic laser-plasmas, Phys. Rev. Lett. (2002), not submitted yet.
15. IFSA conference URL: https://ifsa09.llnl.gov/
16. https://ilsa.llnl.gov/html/conferences/future/antimatter/antimatter.html
17. W. P. Leemans et al., Nature Phys. 2, 696 (2006); W. Leemans and E. Esarey, Phys. Today 62_3_33_2009_.
18. A.I. Titov, B. Keampfer, and H. Takabe, Physical Review Special Topics - Accelerators and Beams, Accepted Wed Oct 7, 2009
19. E. I. Moses et al., Physics Plasmas 16, 041006 (2009).
20. R. N. Boyd, L. Bernstein, and C. Brune, Physics Today, August (2009) p. 60.
21. S. L. Shapiro and S. A. Teukolsky, Black Holes, White Dwarfs, and Neutron Stars, (Wiley, 1983) Chapter 12.
22. H. Chen, S. C. Wilks, et al, Phys. Rev. Lett. 102, 105001 (2009)
23. H. Chen, S. C. Wilks, et al, Phys. Plasmas 16, 122702 (2009)
24. M. Roth et al., in M. Lontano et al. (edited), Superstrong Fields in Plasmas (AIP Conf. Series 611, AIP, 2002).
25. A. Pukov and J. Meyer-ter-Vehn, Laser Particle Beams 17, 571 (1999).

26. M. Hegelich et al., Phys. Rev. Lett. 89, 085002 (2002).
27. A. MacKinnon et al., Phys. Rev. Lett. 88, 215006 (2002).
28. Nuclear Transmutation at J-PARC: http://j-parc.jp/Transmutation/en/ads.html

LIST OF PARTICIPANTS

NAME	*INSTITUTION*
AHN, Jung Keun	Pusan National University
ANDO, Shung-ichi	The University of Manchester
BITO, Jun	RCNP, Osaka University
CHEN, Hua-Xing	Department of Physics, Peking University
DMITRASINOVIC, Veljko	Vinca Institute of Nuclear Studies
DOTE, Akinobu	IPNS/KEK
FUJIOKA, Hiroyuki	Department of Physics, Kyoto University
FUJIWARA, Mamoru	RCNP, Osaka University
FUNAKI, Yasuro	University of Tsukuba
FURUTACHI, Naoya	Hokkaido University
HINOHARA, Nobuo	RIKEN Nishina Center
HIRENZAKI, Satoru	Nara Women's Univ.
HIYAMA, Emiko	RIKEN
HORIUCHI, Hisashi	RCNP, Osaka University
HOSAKA, Atsushi	RCNP, Osaka University
HOTTA, Tomoaki	RCNP, Osaka University
HU, Jinniu	RCNP, Osaka University
HYODO, Tetsuo	Tokyo Institute of Technology
IKENO, Natsumi	Nara Women's Univ.
IMAI, Shotaro	RCNP, Osaka University
ITO, Makoto	Kansai University
KANEKO, Hiromi	RCNP, Osaka University
KANETA, Masashi	Tohoku University
KAORI, Horii	RCNP, Osaka University
KIM, Hyun-Chul	Inha University
KIM, Kyung-il	Yonsei University
KISHIMOTO, Tadafumi	RCNP, Osaka University
KOHRI, Hideki	RCNP, Osaka University
LIU, Yanrui	Tokyo Institute of Technology
MYO, Takayuki	Osaka Institute of Technology
MENG, Jie	Pekin University
NAGAHIRO, Hideko	Nara Women's University
NAKANO, Takashi	RCNP, Osaka University
NAM, Seung-il	CYCU

NAWA, Kanabu	RCNP, Osaka University
NIIYAMA, Masayuki	RIKEN
OGAWA, Yoko	RCNP, Osaka University
OKA, Makoto	Tokyo Institute of Technology
OHKODA, Syunsuke	RCNP, Osaka University
OHTA, Takeshi	RCNP, Osaka University
ONO Satoshi	RCNP, Osaka University
OUYANG, Zhen	IMP, Chinese Academy of Sciences
OZAKI, Sho	RCNP, Osaka university
REN, Zhongzhou	Nanjing University
RIJKEN, Thomas	IMAPP, University of Nijmegen
RYU, Huiyoung	RCNP, Osaka University
RYU, Sun Young	RCNP, Osaka University
SAITO, Takuya	Kochi Univ.
SAKURAGI, Hiroyuki	Osaka City University
SATO, Toru	Dep. of Phys. Osaka Univ.
SHEN, Xiaoyan	Institute of High Energy Physics
SHIN, Jaewon	Sungkyunkwan University
SUHARA, Tadahiro	Kyoto University
TAKABE, Hideaki	ILE, Osaka University
TAKASHINA, Masaaki	Department of Medicine, Osaka Univ.
TANIHATA, Isao	RCNP, Osaka University
TOGASHI, Tomoaki	Hokkaido University
TOHSAKI, Akihiro	RCNP, Osaka University
TANIHATA, Isao	RCNP, Osaka University
TOKI, Hiroshi	RCNP, Osaka University
TSUBAKIHARA, Kohsuke	Hokkaido University
UECHI, Shun	RCNP, Osaka University
UEDA, Keisuke	RCNP, Osaka University
UMEYA, Atsushi	RIKEN
VALVERDE, Manuel	RCNP, Osaka University
YAKHSHIEV, Ulugbek	Inha University
YAMADA, Taiichi	Kanto Gakuin University
YAMAGUCHI Yasuhiro	RCNP, Osaka University
YAMANAKA, Nodoka	Department of physics, Osaka University
YASUI, Shigehiro	KEK
YOSHIDA, Tooru	CNS, University of Tokyo
YOSOI, Masaru	RCNP, Osaka University
ZHANG, Ai-Lin	Shanghai University
ZHANG, Pengming	IMP, Chinese Academy of Sciences
ZHU, Lizi	RCNP, Osaka University